BARRON'S

SAT® SUBJECT TEST CHEMISTRY

9TH EDITION

Joseph A. Mascetta, M.A., C.A.S.
Former Principal
Chemistry Teacher and Coordinator, Science Department
Mount Lebanon High School, Pittsburgh, Pennsylvania
Science Educational Consultant

BARRON'S

About the Author:

Joe Mascetta has taught high school chemistry for twenty years. He was the science department coordinator and principal of Mt. Lebanon High School in Pittsburgh, Pennsylvania. He also served as a science consultant to the area schools and is a past-president of the Western Pennsylvania Association of Supervision and Curriculum Development (ASCD) and the State Advisory Committee of ASCD. He holds degrees from the University of Pittsburgh, the University of Pennsylvania, and Harvard University, and was a participant in Harvard Project Physics, a General Electric Science Fellowship to Union College in Schenectady, New York, the Chemical Bond Approach Curriculum Study at Kenyon College, Ohio, and the Engineering Concepts Curriculum Project and Science Curriculum Supervision at the University of Colorado.

All inquiries should be addressed to:
Barron's Educational Series, Inc.
250 Wireless Boulevard
Hauppauge, New York 11788
http://www.barronseduc.com

ISBN-13: 978-0-7641-3881-2
ISBN-10: 0-7641-3881-2
ISBN-13 (with CD-ROM): 978-0-7641-9419-1
ISBN-10 (with CD-ROM): 0-7641-9419-4

ISSN: 1559-3487

PRINTED IN THE UNITED STATES OF AMERICA
16 15 14 13 12 11 10 09

Contents

Introduction: About the Test

The College Board, which is a national nonprofit membership organization, sponsors the Admissions Testing Program (ATP). Educational Testing Service (ETS) develops and administers the tests for the College Board and also prepares a booklet about the testing program, of which the SAT Subject Tests are a part. Copies of this booklet, which describes each of the SAT Subject Tests and gives sample questions, are automatically sent at the beginning of each academic year to secondary schools for free distribution to students who plan to register for the SAT Subject Tests. The name of this publication is *SAT Subject Tests Preparation Booklet*, and requests from schools or individuals about how to obtain free copies should be addressed to College Board SAT Program, P.O. 025505, Miami, FL 32102. The phone number (Monday–Friday) is (866) 756-7346, 8:30 a.m. to 9:30 p.m. Eastern Time; or you can order online at **www.collegeboard.com**.

All of the SAT Subject Tests are contained in the same test booklet. Each takes 1 hour of testing time, and you may choose any one, two, or three tests to take at one sitting. They all consist of multiple-choice questions.

Many colleges require or recommend one or more Subject Tests for admission or placement. The scores are used in conjunction with your high school record, results on the SAT, teacher recommendations, and other background information to provide a reliable measure of your academic achievements and a good predication of your future performance.

In addition to obtaining a standardized assessment of your achievement from your scores, some colleges use the test results for placement into their particular programs in the freshman year. At others, advisers use the results to guide freshmen in the selection of courses.

Is the SAT Subject Test in Chemistry Required?

The best information on whether SAT Subject Tests are required and, if so, which ones is found in the individual college catalogs or a directory of colleges. Some colleges specify which tests you must take, while others allow you to choose. Obviously, if you have a choice and you have done well in chemistry, you should pick the SAT Subject Test in Chemistry as one of your tests. Even if the test is not required by the colleges to which you are applying, you can add the result to your record to support your achievement level. The College Board publishes *The College Handbook*, which is a good source of information about colleges' requirements with regard to taking SAT Subject Tests. A copy of this book should be in your school library. You can also order it by using your registration form or by visiting the College Board's online store at **www.collegeboard.com**.

When Should You Take the Test?

You will undoubtedly do best if you take the test after completing the high school chemistry course or courses that you plan to take. At this time, the material will be fresh in your mind. Forgetting begins very quickly after you are past a topic or have finished the course. You should plan a review program for at least the last 6 weeks before the test date. (A plan is provided later in this book for such a review.) Careful review definitely helps—cramming just will not do if you want to get the best score of which you are capable!

Colleges that use SAT Subject Test results as part of the admissions process usually require that you take the test no later than December or January of your senior year. For early-decision programs, the test time is June of your junior year. Since chemistry is often a junior year course, June of that year is the optimum time to take the test.

When Is the Test Offered?

The chemistry test is available every time the SAT Subject Tests are given, that is, on the first Saturday of October, November, December, May, and June. They are also given on the last Saturday of January. Be sure that the testing site for which you plan to register offers the SAT Subject Tests on each of these six times. Remember that you may choose to take one or two additional tests besides chemistry on any one test day. You do have to specify in advance which tests you plan to take on the test date you select; however, except for the Language Test with Listening, you may change your mind on the day of the test.

How Do You Register?

The *SAT Registration Bulletin* contains all the information you need to register and to have your scores sent to the college(s) of your choice. Copies of this publication should be available in every high school guidance office. If you have a problem getting one, write to College Board Admissions Testing Program (see the address on page vii) or go online to **www.collegeboard.com**

The deadline for registration is approximately one month before the test date.

How Should You Prepare for the Test?

Barron's SAT Subject Test in Chemistry will be especially helpful. The more you know about the test, the more likely you are to get the best score possible for you. This book provides you with a diagnostic test, scoring information, four practice tests and the equivalent of one more test incorporated with the chapter review tests that allow you to become familiar with the question types and the wording of directions, and to gain a feel for the degree of emphasis on particular topics and the ways in which information may be tested. Each of these aspects should be consciously pursued as you use this book.

What Topics Appear on the Test, and to What Extent?

The following charts show the content of the test and the levels of thinking skills tested:

Topics		Percent of Test (Approx.)	Number of Questions (Approx.)
Structure of Matter		25	21
	I. Atomic theory and structure, energy levels, quantum numbers, orbitals, electron configurations, periodic trends, nucleonics		
	II. Molecular structures, shapes, Lewis structures, polarity		
	III. Bonding (ionic, covalent, metallic), relationships to properties and structures, intermolecular forces, hydrogen bonding, London dispersion forces, dipole-dipole forces		
States of Matter		16	14
	IV. Gases, kinetic molecular theory, gas law relationships, molar volume, density and related problems		
	V. Liquids, solids, forces in these, types of solids, phase diagrams, phase changes		
	VI. Solutions; molarity; molality; percent by mass; solubility factors for solids, liquids, and gases; colligative properties		

	Topics	Percent of Test (Approx.)	Number of Questions (Approx.)
Reaction Types		14	12
	VII. Acids and bases, theories, strong and weak forms, pH, titration problems, indicators		
	VIII. Oxidation-reduction, combustion, using oxidation numbers, activity series		
Stoichiometry		14	12
	IX. Mole concept, molar mass, Avogadro's number, empirical and molecular formulas, chemical equations, balancing equations, solving related problems, determining yield, limiting factors		
Equilibrium and Reaction Rates		5	4
	X. Equilibrium systems, factors affecting, Le Châtelier's principle in gaseous and aqueous, constants, expressions, rates of reactions, factors affecting rates, activation energies, reaction diagrams		
Thermochemistry		6	5
	XI. Conservation of energy, calorimetry, specific heat, thermal curves, enthalpy changes, entropy changes		

	Topics	Percent of Test (Approx.)	Number of Questions (Approx.)
Descriptive Chemistry		12	10
	XII. Physical and chemical properties, nomenclature of elements, compounds, ions, properties related to the periodic table, examples of basic organic compounds, environmental concerns		
Laboratory		8	7
	XIII. Laboratory safety rules, nomenclature, use of equipment, observations, data to analyze, interpreting graphical data, drawing conclusions		
			Total Questions (85)

Note: Each test contains approximately five questions on equation balancing and/or predicting products of chemical reactions. These are distributed among the various content categories.

Thinking Skills Tested	Percent of Test (approx.)
Recalling fundamental concepts, specific pieces of information, and basic terminology (low-level skill)	20
Showing a *comprehension of the basics* and the *ability to apply this information* in a rather straightforward manner to questions, situations, and the solution of qualitative or quantitative problem-oriented questions (medium-level skill)	45
Using the ability to *analyze* quantitative and/or qualitative data and to *synthesize* the knowledge learned to *evaluate* how and what ideas or relationships should be used to draw conclusions or to solve problems (high-level skill)	35

The first chart gives you a general overview of the content of the test. Your knowledge of the topics and your skills in recalling, applying, and synthesizing this knowledge are evaluated through 85 multiple-choice questions. This material is that generally covered in an introductory course in chemistry at a level suitable for college preparation. While every test covers the topics listed, different aspects of each topic are stressed from year to year. Add to this the differences that exist in high school courses with respect to the percentage of time devoted to each major topic and to the specific subtopics covered, and you may find that there are questions on topics with which you have little or no familiarity.

Each of the sample tests in this book is constructed to match closely the distribution of topics shown in the preceding chart so that you will gain a feel for the makeup of the actual test. After each test, a chart will show you which questions relate to each topic. This will be very helpful to you in planning your review because you can identify the areas on which you need to concentrate in your studies. Another chart enables you to see which chapters correspond to the various topic areas.

What General Information Should You Have About the Test?

1. A periodic chart is provided in this test as a resource and as the source of atomic numbers and atomic masses of the elements.
2. You will not be allowed to use an electronic calculator during the test.
3. Numerical calculations are limited to simple arithmetic.
4. You should be familiar with the concepts of ratios and of direct and inverse proportions, scientific notation, and exponential functions.
5. Metric system units are used in this test.
6. The test is composed of three types of questions as explained in the next section.

What Types of Questions Appear on the Test?

There are three general types of questions on the SAT Subject Test in Chemistry—matching questions, true/false and relationship analysis questions, and general five-choice questions. This section will discuss each type and give specific examples of how to answer these questions. You should learn the directions for each type so that you will be familiar with them on the test day. The directions in this section are similar to those on the test.

TYPE 1. MATCHING QUESTIONS IN PART A. In each of these questions, you are given five lettered choices that you will use to answer all the questions in that set. The choices may be in the form of statements, pictures, graphs, experimental findings, equations, or specific situations. Answering a question may be as simple as recalling information or as difficult as analyzing the information given to establish what you need to do qualitatively or quantitatively to synthesize your answer. The directions for this type of question specifically state that a choice may be used once, more than once, or not at all in each set.

Part A

Directions: Every set of the given choices below refers to the numbered statements or formulas immediately following it. Choose the one lettered choice that best fits each statement or formula and then fill in the corresponding oval on the answer sheet. Each choice may be used once, more than once, or not at all in each set.

EXAMPLE

Questions 1–3 refer to the following graphs:

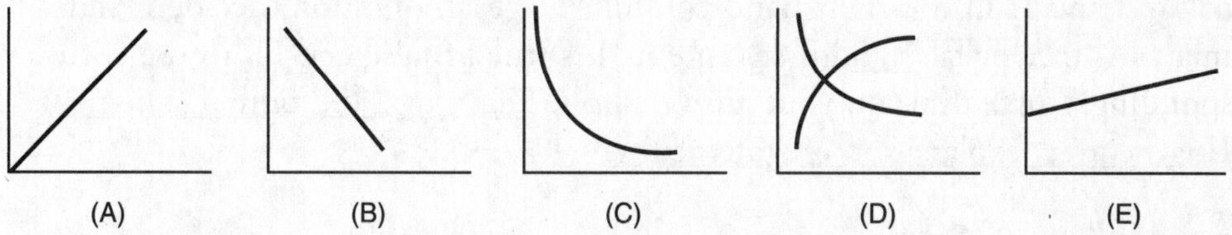

1. The graph that best shows the relationship of volume to temperature for an ideal gas while the pressure is held constant
2. The graph that best shows the relationship of volume to pressure for an ideal gas while the temperature is held constant
3. The graph that best shows the relationship of the number of grams of solute that is soluble in 100 grams of water at varying temperatures if the solubility begins as a small quantity and increases slowly as the temperature is increased

These three questions require you to recall the basic gas laws and the graphic depiction of the relationship expressed in each law, as well as how solubility can be shown graphically.

To answer question 1, you must recognize that the relationship of gas volume to changes in temperature is a direct relationship that is depicted by graphing Charles's Law: $V_1/T_1 = V_2/T_2$. The only graph that shows that type of direct relationship with the appropriate slope is (A).

To answer question 2, you need to understand that Boyle's Law states that the pressure of a gas is inversely proportional to the volume at constant temperature. Mathematically, this means that pressure (P) times volume (V) is a constant, or $P_1V_1 = P_2V_2$. This inversely proportional relationship is accurately depicted in (C). Although (B) shows the values on the x-axis increasing as the y-axis values decrease, it does not fit the graph for an inverse proportion.

Question 3 requires that you have knowledge about solubility curves and can apply the solubility relationship given in words to graph (E).

TYPE 2. TRUE/FALSE AND RELATIONSHIP QUESTIONS IN PART B. On the actual SAT Subject Test in Chemistry, this type of question must be answered in a special section of your answer sheet labeled "chemistry." Type 2 questions are numbered beginning with 101. Each question consists of a statement or assertion in column I and, on the other side of the word BECAUSE, another statement or assertion in column II. Your first task is to determine whether each of the statements is true or false and to record your answer for each in the answer blocks for column I and column II in the answer grid by darkening either the Ⓣ or the Ⓕ oval. Here you must use your reasoning skills and your understanding of the topic to determine whether there is a cause-and-effect relationship between the two statements.

Here are the directions and two examples of a relationship analysis question.

Part B

Directions: Every question below contains two statements, I in the left-hand column and II in the right-hand column. For each question, decide if statement I is true or false and if statement II is true or false and fill in the corresponding T or F ovals on your answer sheet. Fill in oval CE only if statement II is a correct explanation of statement I.

Sample Answer Grid:

CHEMISTRY * Fill in oval CE only if II is a correct explanation of I.

	I	II	CE*
101.	Ⓣ Ⓕ	Ⓣ Ⓕ	◯

EXAMPLE 1

101. When 2 liters of oxygen gas react completely with 2 liters of hydrogen gas, the limiting factor is the volume of the oxygen	BECAUSE	the coefficients in the balanced equation of a gaseous reaction give the volume relationship of the reacting gases.

The reaction that takes place is

$$2H_2 + O_2 \rightarrow 2H_2O$$

The coefficients of this gaseous reaction show that 2 L of hydrogen react with 1 L of oxygen, leaving 1 L of unreacted oxygen. The limiting factor is the quantity of hydrogen.

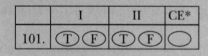

The ability to solve this quantitative relationship shows that statement I is not true. However, statement II does give a true statement of the relationship of coefficients in a balanced equation of gaseous chemical reaction. Therefore, the answer blocks would be completed like this:

	I	II	CE*
101.	(T) (F●)	(T●) (F)	( )

EXAMPLE 2

102. Water is a good solvent of ionic and polar compounds BECAUSE the water molecule has polar properties due to the factors involved in the bonding of the hydrogen and oxygen atoms.

Statement I is true because water is such a good solvent that, as you have probably learned, it is sometimes referred to as the universal solvent. This property is attributed mostly to its polar structure. The polar covalent bond between the oxygen and hydrogen atoms and the angular orientation of the hydrogens at 105 degrees between them contribute to the establishment of a permanent dipole moment in the water molecule. This also gives rise to a high degree of hydrogen bonding. These properties combine to make water a powerful solvent for both polar and ionic compounds. Because of your familiarity with these concepts and the processes by which substances go into solution, you know that statement II not only is true but also is the reason that statement I is true. There is a cause-and-effect relationship between the two statements. Therefore, the answer blocks would be marked like this:

	I	II	CE*
102.	(T●) (F)	(T●) (F)	(●)

TYPE 3: GENERAL FIVE-CHOICE QUESTIONS IN PART C. The five-choice items in Part C are written usually as questions but sometimes as incomplete statements. You are given five suggested answers or completions. You must select the one that is best in each case and record your choice in the appropriate oval. In some questions you are asked to select the one inappropriate answer. Such questions contain a word in capital letters, such as NOT, LEAST, or EXCEPT.

In some of these questions, you may be asked to make an association between a graphic, pictorial, or mathematical representation and a stated explanation or problem. The solution may involve solving a scientific problem by correctly interpreting the representation. In some cases the same representation may be used for a series of two or more questions. In no case, however, is the correct answer to one question necessary for answering a subsequent question correctly. Each question in the set is independent of the others.

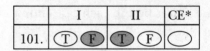

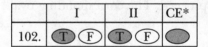

Part C

Directions: Every question or incomplete statement below is followed by five suggested answers or completions. Choose the one that is best in each case and then fill in the corresponding oval on the answer sheet.

EXAMPLE 1

40. In this graphic representation of a chemical reaction, which arrow depicts the activation energy?

(A) A
(B) B
(C) C
(D) D
(E) E

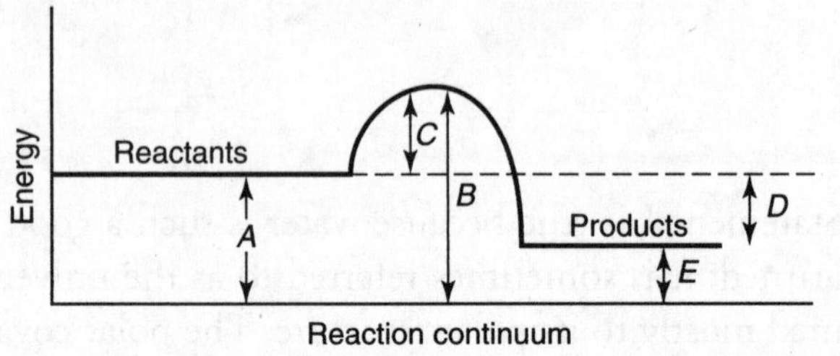

To answer this question, you need to know how to interpret the energy levels in this graphic representation of energy-level changes along the time continuum of the reaction. The activation energy is the minimum energy required for a chemical reaction to take place. The reactant molecules come together, and chemical bonds are stretched, broken, and formed in producing the products. During this process the energy of the system increases to a maximum, then decreases to the energy of the products. The activation energy is the difference between the maximum energy and the energy of the reactants. Choice (C) in the graphic depiction shows this energy barrier that has to be overcome for the reaction to proceed. The corresponding oval on the answer sheet should be darkened.

EXAMPLE 2

41. If the molar mass of NH_3 is 17 g/mol, what is the density of this compound at STP?

(A) 0.25 g/L
(B) 0.76 g/L
(C) 1.25 g/L
(D) 3.04 g/L
(E) 9.11 g/L

The solution of this quantitative problem depends on the application of several principles. One principle is that the molar mass of a gas expressed in grams/mole will occupy 22.4 L at standard temperature and pressure (STP). The other is that the density of a gas at STP is the mass of 1 L of the gas. Therefore, 17 g of ammonia (NH_3) will occupy 22.4 L, and 1 L is equal to 17 g/22.4 L or 0.76 g/L. The correct answer is (B).

EXAMPLE 3

Some questions in this part are followed by three or four bits of information labeled by Roman numerals I through III or IV. One or more of these statements may correctly answer the question. You must select from the five lettered choices the one that best answers the question.

42. Which bond(s) is (are) ionic?

I. H–Cl (g)
II. S–Cl (g)
III. Cs–F (g)

(A) I only
(B) III only
(C) I and II only
(D) II and III only
(E) I, II, and III

To determine the type of bonding that exists in these three substances, you must use your knowledge of ionic bonds and the way they are formed. You must also use your knowledge of the relationship of the electronegativity of an element and the position of that element in the periodic chart. Compounds I and II are formed from elements that do not have enough difference in their respective electronegativities to cause the formation of an ionic bond. This can be inferred by checking the positions of the elements (H, Cl, and S) in the Periodic Table and noting how electronegativity varies with an element's position in the table. Compound III, cesium fluoride, consists of elements that appear in the lower right corner and the upper left corner, respectively, of the Periodic Table; therefore, the difference in their electronegativity values is sufficient so that an ionic bond can be predicted between them. Of the choices given, only (B) is a correct answer.

How Can You Use This Book to Prepare for the Test?

The best way to use this book is a two-stage approach, and the next sections are arranged accordingly. First, you should take the diagnostic test. This will give you a preliminary exposure to the type of test you are planning to take, as well as a measure of how well you achieve on each of the three parts. You will also become aware of the types of questions that the test includes. Use the test-scoring information following the diagnostic test to determine your raw score and your strengths and weaknesses in the specific areas of the test.

Having taken the diagnostic test, you should then follow a study program. A study plan covering the 6 weeks before the test has been developed for you and is given in detail on page 35. It requires a minimum of 1 or 2 hours per night on weekdays but leaves your weekends free.

How Can You Improve Your Problem-Solving Skills?*

Chemistry is a subject that deals with many problem situations that you, the student, must be able to solve. Solving problems may seem to be a natural process when the degree of difficulty is not very great, and you may not need a structured method to attack these problems. However, for complex problems an orderly process is required.

The following is such a problem-solving process. Each step is vital to the next step and to the final solution of the problem.

Step 1. Clarify the problem: to separate the problem into the facts, the conditions, and the questions that need to be answered, and to establish the goal.

Step 2. Explore: to examine the sufficiency of the data, to organize the data, and to apply previously acquired knowledge, skills, and understanding.

Step 3. Select a strategy: to choose an appropriate method to solve the problem.

Step 4. Solve: to apply the skills needed to carry out the strategy chosen.

Step 5. Review: to examine the reasonableness of the solution and to evaluate the effectiveness of the process.

The steps of the problem-solving process listed above should be followed in sequence. The subskills listed below for each step, however, are not in sequence. The order in which subskill patterns are used will differ with the nature of the problem and/or with the ways in which the individual problem solver thinks. Also, not every subskill need be employed in solving every problem.

CLARIFY THE PROBLEM

a. Identify the facts. What is known about the problem?

b. Identify the conditions. What is the current situation?

c. Identify the questions. What needs to be answered before the problem can be solved?

d. Visualize the problem.

1. Make mental images of the problem.
2. If desirable or necessary, draw a sketch or diagram, make an outline, write down symbols or equations that correspond to the mental images.

e. Establish the goal. The goal defines the specific result to be accomplished through the problem-solving process. It defines the purpose or function the solution is expected to achieve and serves as the basis for evaluating the solution.

EXPLORE

a. Review previously acquired knowledge, skills, and understanding. Determine whether the current problem is similar to a previously seen type.

**Adapted with permission from* Thinking Skills Resource Guide, *a noncopyrighted publication of Mount Lebanon School District, Pittsburgh, PA.*

b. Estimate the sufficiency of the data. Does there seem to be enough information to solve the problem?

c. Organize the data. There are many ways in which data can be organized. Some examples are outline, written symbols and equations, chart, table, graph, map, diagram, and drawing. Determine whether the data organized in the way(s) you have chosen will enable you to partially or completely solve the problem.

d. Determine what new data, if any, need to be collected. What additional information may be needed to solve the problem? Can the existing data be reorganized to generate new information? Do other resources need to be consulted? This step may suggest possible strategies to be used to solve the problem.

SELECT A STRATEGY

A strategy is a goal-directed sequence of mental operations. Selecting a strategy is the most important and also the most difficult step in the problem-solving process. Although there may be several strategies that will lead to the solution of a problem, the skilled problem solver uses the most efficient strategy. The choice of the most efficient strategy is based on knowledge and experience as well as a careful application of the clarify and explore steps of the problem-solving method. Some problems may require the use of a combination of strategies.

The following search methods may help you to select a strategy. They do not represent all of the possible ways in which this can be done. Other methods of strategy selection are related to specific content areas.

a. Trial-and-error search: Such a search either doesn't have or doesn't use information that indicates that one path is more likely to lead to the goal than any other path.

Trial-and-error search comes in two forms, blind and systematic. In *blind search,* the searchers pick paths to explore blindly, without considering whether they have already explored these paths. A preferable method is *systematic search,* in which the searchers keep track of the paths they have already explored and do not duplicate them. Because this method avoids multiple searches, systematic search is usually twice as efficient as blind search.

b. Reduction method: This involves breaking the problem into a sequence of smaller parts by setting up subgoals. Subgoals make problem solving easier becaue they reduce the amount of search required to find the solution.

You can set up subgoals by working part way into a problem and then analyzing the partial goal to be achieved. In doing this, you can drop the problem restrictions that do not apply to the subgoal. By adding up all the subgoals, you can solve the "abstracted" problem.

c. Working backward: When you have trouble solving a problem head-on, it is often useful to try to work backward. Working backward involves a simple change in representation or point of view. Your new starting point is the original goal. Working backward can be helpful because problems are often easier to solve in one direction than another.

d. Knowledge-based method: This strategy uses information stored in the problem solver's memory, or newly acquired information, to guide the search for the solution. The problem solver may have solved a similar problem and can use this knowledge in a new situation. In other cases, problem solvers may have to acquire needed knowledge. For example, they may solve an auxiliary problem to learn how to solve the one they are having difficulty with.

Searching for analogous (similar) problems is a very powerful problem-solving technique. When you are having difficulty with a problem, try to pose a related, easier one and hope thereby to learn something that will help you solve the harder problem.

SOLVE

Use the strategy chosen to actually solve the problem. Executing the solution provides you with a very valuable check on the adequacy of your plan. Sometimes students will look at a problem and decide that, since they know how to solve it, they need not bother with the drudgery of actually executing the solution. Sometimes the students are right, but at other times they miss an excellent opportunity to discover that they were wrong.

REVIEW

a. Evaluation. The critical question in evaluation is this: "Does the answer I propose meet all of the goals and conditions set by the problem?" Thus, after the effort of finding a solution, you must turn back to the problem statement and check carefully to be sure your solution satisfies it.

With easy problems there is a strong temptation to skip evaluation because the probability of error seems small. In some cases, however, this can be costly. Evaluation may prove that errors were present.

b. Verification of the reasonableness of the answer. It is easy to become so involved with the process and mathematics of a problem that an answer is recorded that is totally illogical. To avoid this mistake, you should simplify the numbers involved and solve for an answer. Having done this, compare your estimated result with your answer to ensure that your answer is feasible.

For example, a problem requires the following operations:

$$5.12 \times 10^5 \times 3.98 \times 10^6 \text{ divided by } 910$$

And doing all the math, you get an answer of

$$0.02239 \times 10^{11} \text{ or } 2.24 \times 10^9$$

To estimate the answer, first simplify the numbers to one significant figure (significant figures are discussed in Chapter 1). This gives

$$5 \times 10^5 \times 4 \times 10^6 \text{ divided by } 9 \times 10^2$$

which is

$$20 \times 10^{11} \text{ divided by } 9 \times 10^2 = 2.2 \times 10^9$$

This is the estimated answer, which validates the answer above.

When you are dealing with test items that provide multiple-choice answers, you can often use estimation to arrive at the answer without doing the more complicated mathematics.

c. Consolidation. Here the basic question to be answered is: "What can I learn from the experience of solving this problem?" The following more specific questions may help you to answer this general one:

1. Why was this problem difficult?
2. Was it difficult to follow a plan?
3. Was it difficult to decide on a plan? If so, why?
4. Did I take the long way to the answer?
5. Can I use this plan again in similar problems?

The important thing is to reflect on the process that you used in order to make future problem solving easier.

CHEMISTRY TEST

Material in the following table may be useful in answering the questions in this examination

Periodic Table of the Elements

1 **H** 1.0078																	2 **He** 4.0026
3 **Li** 6.941	4 **Be** 9.012											5 **B** 10.811	6 **C** 12.011	7 **N** 14.007	8 **O** 16.00	9 **F** 19.00	10 **Ne** 20.179
11 **Na** 22.99	12 **Mg** 24.30											13 **Al** 26.98	14 **Si** 28.09	15 **P** 30.974	16 **S** 32.06	17 **Cl** 35.453	18 **Ar** 39.948
19 **K** 39.10	20 **Ca** 40.08	21 **Sc** 44.96	22 **Ti** 47.90	23 **V** 50.94	24 **Cr** 52.00	25 **Mn** 54.938	26 **Fe** 55.85	27 **Co** 58.93	28 **Ni** 58.69	29 **Cu** 63.55	30 **Zn** 65.39	31 **Ga** 69.72	32 **Ge** 72.59	33 **As** 74.92	34 **Se** 78.96	35 **Br** 79.90	36 **Kr** 83.80
37 **Rb** 85.47	38 **Sr** 87.62	39 **Y** 88.91	40 **Zr** 91.22	41 **Nb** 92.91	42 **Mo** 95.94	43 **Tc** (98)	44 **Ru** 101.1	45 **Rh** 102.91	46 **Pd** 106.42	47 **Ag** 107.87	48 **Cd** 112.41	49 **In** 114.82	50 **Sn** 118.71	51 **Sb** 121.75	52 **Te** 127.60	53 **I** 126.91	54 **Xe** 131.29
55 **Cs** 132.91	56 **Ba** 137.33	57 **La** 138.91	72 **Hf** 178.49	73 **Ta** 180.95	74 **W** 183.85	75 **Re** 186.21	76 **Os** 190.2	77 **Ir** 192.2	78 **Pt** 195.08	79 **Au** 196.97	80 **Hg** 200.59	81 **Tl** 204.38	82 **Pb** 207.2	83 **Bi** 208.98	84 **Po** (209)	85 **At** (210)	86 **Rn** (222)
87 **Fr** (223)	88 **Ra** 226.02	89 **†Ac** 227.03	104 **Rf** (261)	105 **Db** (262)	106 **Sg** (263)	107 **Bh** (262)	108 **Hs** (265)	109 **Mt** (266)	110 **§** (269)	111 **§** (272)	112 **§** (277)	§ Not yet named					

*Lathanides Series	58 **Ce** 140.12	59 **Pr** 140.91	60 **Nd** 144.24	61 **Pm** (145)	62 **Sm** 150.4	63 **Eu** 151.97	64 **Gd** 157.25	65 **Tb** 158.93	66 **Dy** 162.50	67 **Ho** 164.93	68 **Er** 167.26	69 **Tm** 168.93	70 **Yb** 173.04	71 **Lu** 174.97
†Actinides Series	90 **Th** 232.04	91 **Pa** 231.04	92 **U** 238.03	93 **Np** 237.05	94 **Pu** (244)	95 **Am** (243)	96 **Cm** (247)	97 **Bk** (247)	98 **Cf** (251)	99 **Es** (252)	100 **Fm** (257)	101 **Md** (258)	102 **No** (259)	103 **Lr** (260)

PART 1

A DIAGNOSTIC TEST

A Diagnostic Test

The following test of 85 questions is a sample of the actual test you will take to measure your chemistry achievement. It has basically the same distribution of topics, directions, and number and types of questions. Before taking this test, read the advice given in the section entitled "Final Preparation—The Day Before the Test" (p. 35). Use the answer sheet provided, and limit the test time to 1 hour.

A Periodic Table of the Elements has been included for your use on problems requiring this source of information. Use this table also with the practice tests at the end of the book.

Turn now to the test.

Answer Sheet

DIAGNOSTIC TEST

Determine the correct answer for each question. Then, using a No. 2 pencil, blacken completely the oval containing the letter of your choice.

1	(A)	(B)	(C)	(D)	(E)
2	(A)	(B)	(C)	(D)	(E)
3	(A)	(B)	(C)	(D)	(E)
4	(A)	(B)	(C)	(D)	(E)
5	(A)	(B)	(C)	(D)	(E)
6	(A)	(B)	(C)	(D)	(E)
7	(A)	(B)	(C)	(D)	(E)
8	(A)	(B)	(C)	(D)	(E)
9	(A)	(B)	(C)	(D)	(E)
10	(A)	(B)	(C)	(D)	(E)
11	(A)	(B)	(C)	(D)	(E)
12	(A)	(B)	(C)	(D)	(E)
13	(A)	(B)	(C)	(D)	(E)
14	(A)	(B)	(C)	(D)	(E)
15	(A)	(B)	(C)	(D)	(E)
16	(A)	(B)	(C)	(D)	(E)
17	(A)	(B)	(C)	(D)	(E)
18	(A)	(B)	(C)	(D)	(E)
19	(A)	(B)	(C)	(D)	(E)
20	(A)	(B)	(C)	(D)	(E)
21	(A)	(B)	(C)	(D)	(E)
22	(A)	(B)	(C)	(D)	(E)
23	(A)	(B)	(C)	(D)	(E)
24	(A)	(B)	(C)	(D)	(E)
25	(A)	(B)	(C)	(D)	(E)

ON THE ACTUAL CHEMISTRY TEST, THE FOLLOWING TYPE OF QUESTION MUST BE ANSWERED ON A SPECIAL SECTION (LABELED "CHEMISTRY") AT THE LOWER LEFT-HAND CORNER OF PAGE 2 OF YOUR ANSWER SHEET. THESE QUESTIONS WILL BE NUMBERED BEGINNING WITH 101 AND MUST BE ANSWERED ACCORDING TO THE DIRECTIONS.

CHEMISTRY* Fill in oval CE only if II is a correct explanation of I.

	I		II		CE*
101	(T)	(F)	(T)	(F)	○
102	(T)	(F)	(T)	(F)	○
103	(T)	(F)	(T)	(F)	○
104	(T)	(F)	(T)	(F)	○
105	(T)	(F)	(T)	(F)	○
106	(T)	(F)	(T)	(F)	○
107	(T)	(F)	(T)	(F)	○
108	(T)	(F)	(T)	(F)	○
109	(T)	(F)	(T)	(F)	○
110	(T)	(F)	(T)	(F)	○
111	(T)	(F)	(T)	(F)	○
112	(T)	(F)	(T)	(F)	○
113	(T)	(F)	(T)	(F)	○
114	(T)	(F)	(T)	(F)	○

ON THE ACTUAL CHEMISTRY TEST, THE REMAINING QUESTIONS MUST BE ANSWERED BY RETURNING TO THE SECTION OF YOUR ANSWER SHEET YOU STARTED FOR CHEMISTRY.

26	(A)	(B)	(C)	(D)	(E)
27	(A)	(B)	(C)	(D)	(E)
28	(A)	(B)	(C)	(D)	(E)
29	(A)	(B)	(C)	(D)	(E)
30	(A)	(B)	(C)	(D)	(E)
31	(A)	(B)	(C)	(D)	(E)
32	(A)	(B)	(C)	(D)	(E)
33	(A)	(B)	(C)	(D)	(E)
34	(A)	(B)	(C)	(D)	(E)
35	(A)	(B)	(C)	(D)	(E)
36	(A)	(B)	(C)	(D)	(E)
37	(A)	(B)	(C)	(D)	(E)
38	(A)	(B)	(C)	(D)	(E)
39	(A)	(B)	(C)	(D)	(E)
40	(A)	(B)	(C)	(D)	(E)
41	(A)	(B)	(C)	(D)	(E)
42	(A)	(B)	(C)	(D)	(E)
43	(A)	(B)	(C)	(D)	(E)
44	(A)	(B)	(C)	(D)	(E)
45	(A)	(B)	(C)	(D)	(E)
46	(A)	(B)	(C)	(D)	(E)
47	(A)	(B)	(C)	(D)	(E)
48	(A)	(B)	(C)	(D)	(E)
49	(A)	(B)	(C)	(D)	(E)
50	(A)	(B)	(C)	(D)	(E)
51	(A)	(B)	(C)	(D)	(E)
52	(A)	(B)	(C)	(D)	(E)
53	(A)	(B)	(C)	(D)	(E)
54	(A)	(B)	(C)	(D)	(E)
55	(A)	(B)	(C)	(D)	(E)
56	(A)	(B)	(C)	(D)	(E)
57	(A)	(B)	(C)	(D)	(E)
58	(A)	(B)	(C)	(D)	(E)
59	(A)	(B)	(C)	(D)	(E)
60	(A)	(B)	(C)	(D)	(E)
61	(A)	(B)	(C)	(D)	(E)
62	(A)	(B)	(C)	(D)	(E)
63	(A)	(B)	(C)	(D)	(E)
64	(A)	(B)	(C)	(D)	(E)
65	(A)	(B)	(C)	(D)	(E)
66	(A)	(B)	(C)	(D)	(E)
67	(A)	(B)	(C)	(D)	(E)
68	(A)	(B)	(C)	(D)	(E)
69	(A)	(B)	(C)	(D)	(E)
70	(A)	(B)	(C)	(D)	(E)
71	(A)	(B)	(C)	(D)	(E)

THE DIAGNOSTIC TEST

Note: For all questions involving solutions, you should assume that the solvent is water unless otherwise noted. **Reminder: You may not use a calculator on this test!**

The following symbols have the meanings listed unless otherwise noted.

H = enthalpy	atm = atmosphere
M = molar	g = gram(s)
n = number of moles	J = joules(s)
P = pressure	kJ = kilojoules
R = molar gas constant	L = liter(s)
S = entropy	mL = milliliter(s)
T = temperature	mm = millimeter(s)
V = volume	V = volt(s)

Part A

Directions: Every set of the given lettered choices below refers to the numbered statements or formulas that immediately follow it. Choose the one lettered choice that best fits each statement or formula; then fill in the corresponding oval on the answer sheet. Each choice may be used once, more than once, or not at all in each set.

<u>Questions 1–4</u> refer to the periodic group shown as these elements are aligned in the Periodic Table. Letter choices are substituted for the respective elements.

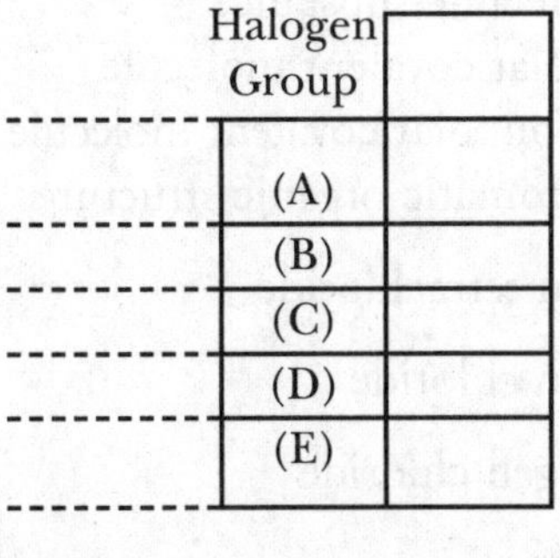

1. The element that is most active chemically
2. The element with the smallest ionic radius
3. The element with the lowest first ionization potential
4. The element that first shows some visible metallic properties at room temperature

<u>Questions 5–7</u>

(A) X^{+}
(B) X^{2+}
(C) X^{3+}
(D) XO_3^{2-}
(E) XO_4^{2-}

5. A type of ion found in sodium acetate
6. A type of ion found in aluminum oxide
7. A type of ion found in potassium phosphate

GO ON TO THE NEXT PAGE.

Questions 8–12

(A) Avogadro's number
(B) $P_1V_1 = P_2V_2$
(C) $V_1T_2 = V_2T_1$
(D) Dalton's Theory
(E) $P_1T_2 = P_2T_1$

8. Proposes basic postulates concerning elements and atoms

9. Proposes a relationship between the combining volumes of gases with respect to the reactants and gaseous products

10. Proposes a temperature-volume relationship of gases

11. Proposes a concept regarding the number of particles in a mole

12. Proposes a volume-pressure relationship of gases

Questions 13–16

(A) R–OH
(B) R–O–R*
(C) R–C(=O)–H
(D) R–C(–H)–OH
(E) R–C(=O)–O–R*

(*Alkyl group that is not necessarily the same as R)

13. The organic structure designation that includes the functional group of an aldehyde

14. The organic structure designation that includes the functional group of an acid

15. The organic structure designation that includes the functional group of an ester

16. The organic structure designation that includes the functional group of an ether

Questions 17–21

(A) $H_2(g)$
(B) $CO_2(g)$
(C) $2N_2O(g)$
(D) $2NaCl$ (aq)
(E) H_2SO_4(dilute aq)

17. The expression that can be used to designate a linear nonpolar molecule

18. The expression that can be used to designate 2 moles of atoms

19. The expression that can be used to designate 3 moles of atoms

20. The expression that can be used to designate a maximum of 3 moles of ions

21. The expression that can be used to designate 6 moles of atoms

Questions 22–25

(A) Ionic substance
(B) Metallic substance
(C) Polar covalent molecule
(D) Nonpolar covalent molecule
(E) Aromatic organic structure

22. Carbon tetrachloride

23. Cesium chloride

24. Hydrogen chloride

25. Benzene

Part B

ON THE ACTUAL SAT SUBJECT TEST IN CHEMISTRY, THE FOLLOWING TYPE OF QUESTION MUST BE ANSWERED ON A SPECIAL SECTION (LABELED "CHEMISTRY") AT THE LOWER LEFT-HAND CORNER OF PAGE 2 OF YOUR ANSWER SHEET. THESE QUESTIONS ARE NUMBERED BEGINNING WITH 101 AND MUST BE ANSWERED ACCORDING TO THE FOLLOWING DIRECTIONS.

Directions: Every question below contains two statements, I in the left-hand column and II in the right-hand column. For each question, decide if statement I is true or false and whether statement II is true or false, and fill in the corresponding T or F ovals on your answer sheet. *Fill in oval CE only if statement II is a correct explanation of statement I.

Sample Answer Grid:

CHEMISTRY *Fill in oval CE only if II is a correct explanation of I.

	I		II		CE*
101.	(T)	(F)	(T)	(F)	( )

	I		II
101.	A catalyst can accelerate a chemical reaction	BECAUSE	a catalyst can decrease the activation energy required for the reaction to occur.
102.	Molten sodium chloride is a good electrical conductor	BECAUSE	sodium chloride in the molten state allows ions to move freely.
103.	Ice is less dense than liquid water	BECAUSE	water molecules are nonpolar.
104.	Two isotopes of the same element have the same mass number	BECAUSE	isotopes have the same number of protons.
105.	A saturated solution can be classified as dilute	BECAUSE	a solute can have a very low solubility in a solvent.
106.	Two liters of CO_2 can be produced by 1 gram of carbon burning completely	BECAUSE	the amount of gas evolved in a chemical reaction can be determined by using the mole relationship of the coefficients in the balanced equation.

GO ON TO THE NEXT PAGE.

I		II
107. A reaction is at equilibrium when it reaches completion	BECAUSE	the concentrations of the reactants in a state of equilibrium equal the concentrations of the products.
108. The anions in an electrolytic cell migrate to the cathode	BECAUSE	positively charged ions are attracted to the negatively charged cathode in an electrolytic cell.
109. A solution with pH = 5 has a higher concentration of hydronium ions than a solution with a pH = 3	BECAUSE	pH is defined as $-\log [H^+]$.
110. An endothermic reaction can be spontaneous	BECAUSE	both the enthalpy and the entropy changes affect the Gibbs free-energy change of the reaction.
111. Weak acids have small values for the equilibrium constant, K_a	BECAUSE	the concentration of the hydronium ion is in the numerator of the K_a expression.
112. One mole of NaCl contains 2 moles of ions	BECAUSE	NaCl is a stable salt at room temperature.
113. A pi bond is formed between the lobes of adjacent *p* orbitals in the same plane of two atoms that contain only one electron each	BECAUSE	each of the two lobes of a single *p* orbital can hold two electrons of opposite spin.
114. H_2S and H_2O have a significant difference in their boiling points	BECAUSE	hydrogen sulfide has a higher degree of hydrogen bonding than water.

Part C

Directions: Every question or incomplete statement below is followed by five suggested answers or completions. Choose the one that is best and then fill in the corresponding oval on the answer sheet.

26. Two immiscible liquids, when shaken together vigorously, may form
    (A) a solution
    (B) a tincture
    (C) a sediment
    (D) a hydrated solution
    (E) a colloidal dispersion

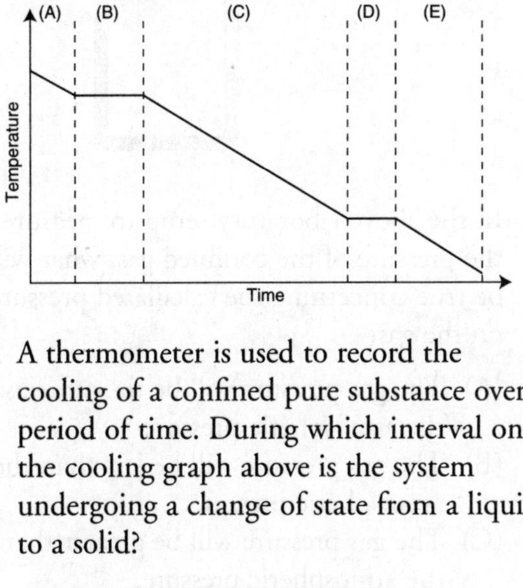

27. A thermometer is used to record the cooling of a confined pure substance over a period of time. During which interval on the cooling graph above is the system undergoing a change of state from a liquid to a solid?

28. If a principal energy level of an atom in the ground state contains 18 electrons, they will be arranged in orbitals according to the pattern
    (A) $s^6p^6d^6$
    (B) $s^2p^6d^{10}$
    (C) $s^2d^6f^{10}$
    (D) $s^2p^6f^{10}$
    (E) $s^2p^2f^{14}$

29. For a particular organic compound, which of the following pairs can represent the empirical and the molecular formula, respectively?
    (A) CH and CH_4
    (B) CH and C_6H_6
    (C) CH_2 and C_2H_2
    (D) CH_2 and C_2H_3
    (E) CH_3 and C_3H_6

30. A liter of hydrogen is at 5°C temperature and under 640 torr pressure. If the temperature were raised to 60°C and the pressure decreased to 320 torr, how would the liter volume be modified?
    (A) $1\text{L} \times \frac{5}{60} \times \frac{640}{32}$
    (B) $1\text{L} \times \frac{60}{5} \times \frac{320}{640}$
    (C) $1\text{L} \times \frac{278}{333} \times \frac{640}{320}$
    (D) $1\text{L} \times \frac{333}{278} \times \frac{640}{320}$
    (E) $1\text{L} \times \frac{333}{278} \times \frac{320}{640}$

GO ON TO THE NEXT PAGE.

31. Of the following statements about the number of subatomic particles in an ion of ${}^{32}_{16}S^{2-}$, which is (are) true?

I. 16 protons
II. 14 neutrons
III. 18 electrons

(A) II only
(B) III only
(C) I and II only
(D) I and III only
(E) I, II, and III

32. The most active metallic elements are found in

(A) the upper right corner of the periodic chart
(B) the lower right corner of the periodic chart
(C) the upper left corner of the periodic chart
(D) the lower left corner of the periodic chart
(E) the middle of the periodic chart, just beyond the transition elements

33. If 1 mole of each of the following substances was dissolved in 1,000 grams of water, which solution would have the highest boiling point?

(A) NaCl
(B) KCl
(C) $CaCl_2$
(D) $C_6H_{10}O_5$
(E) $C_{12}H_{22}O_{11}$

34. A tetrahedral molecule, XY_4, would be formed if X were using the orbital hybridization

(A) p^2
(B) s^2
(C) sp
(D) sp^2
(E) sp^3

35. In the following reaction, how many liters of SO_2 at STP will result from the complete burning of pure sulfur in 8 liters of oxygen?

$$S(s) + O_2(g) \rightarrow SO_2(g)$$

(A) 1
(B) 4
(C) 8
(D) 16
(E) 32

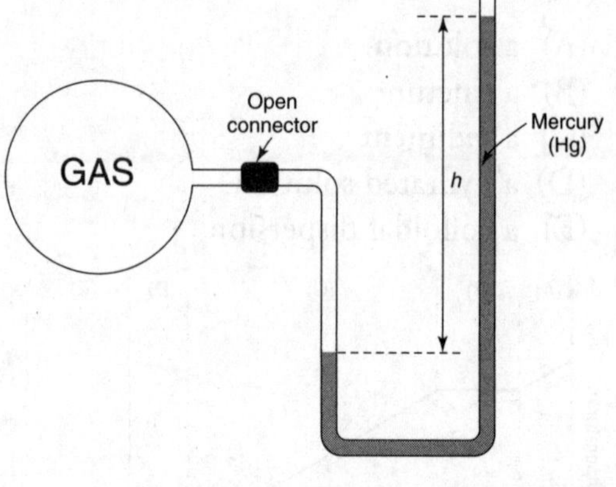

36. In the above laboratory setup to measure the pressure of the confined gas, what will be true concerning the calculated pressure on the gas?

(A) The gas pressure will be the same as the atmospheric pressure.
(B) The gas pressure will be less than the atmospheric pressure.
(C) The gas pressure will be greater than the atmospheric pressure.
(D) The difference in the height (*h*) of mercury levels is equal to the pressure of the gas.
(E) The height (*h*) of mercury has no effect on the pressure calculation since the column of mercury is only used to enclose the gas volume.

37. Which of the following changes in the experiment shown in question 36 would cause the pressure in the glass container to vary from that shown?

(A) Use a U-tube of a greater diameter and maintain the height of mercury.
(B) Increase the temperature of gas in the tube.
(C) Increase the length of the upper portion of the right side of tubing.
(D) Use a U-tube of a smaller diameter and maintain the height of mercury.
(E) Replace the flask with one that has the same volume but has a flat bottom.

38. Which of the following can be classified as amphoteric?

(A) Na_3PO_4
(B) HCl
(C) NaOH
(D) HSO_4^-
(E) $C_2O_4^{2-}$

39. Standard conditions (STP) are

(A) 0°C and 2 atm
(B) 32°F and 76 torr
(C) 273 K and 760 mm Hg
(D) 4°C and 7.6 cm Hg
(E) 0 K and 760 mm Hg

40. Laboratory results showed the composition of a compound to be 58.81% barium, 13.73% sulfur, and 27.46% oxygen. What is the empirical formula of the compound?

(A) $BaSO_4$
(B) BaS_2O
(C) Ba_2SO_3
(D) BaS_2O_4
(E) Ba_2SO_4

41. What is the percentage composition of calcium in calcium hydroxide, $Ca(OH)_2$? (1 mol = 74 g)

(A) 40%
(B) 43%
(C) 54%
(D) 69%
(E) 74%

42. How many grams of hydrogen gas can be produced from the following reaction if 65 grams of zinc and 65 grams of HCl are present in the reaction?

$$Zn(s) + 2HCl(aq) \rightarrow ZnCl_2(aq) + H_2(g)$$

(A) 1
(B) 1.8
(C) 3.6
(D) 7
(E) 58

43. The following statements were recorded while preparing carbon dioxide gas in the laboratory. Which one involves an interpretation of the data rather than an observation?

(A) No liquid was transfered from the reaction bottle to the beaker.
(B) The quantity of solid minerals decreased.
(C) The cloudiness in the last bottle of limewater was caused by the product of the reaction of the colorless gas and the limewater.
(D) The bubbles of gas rising from the mineral remained colorless throughout the experiment.
(E) There was a 4°C rise in temperature in the reaction vessel during the experiment.

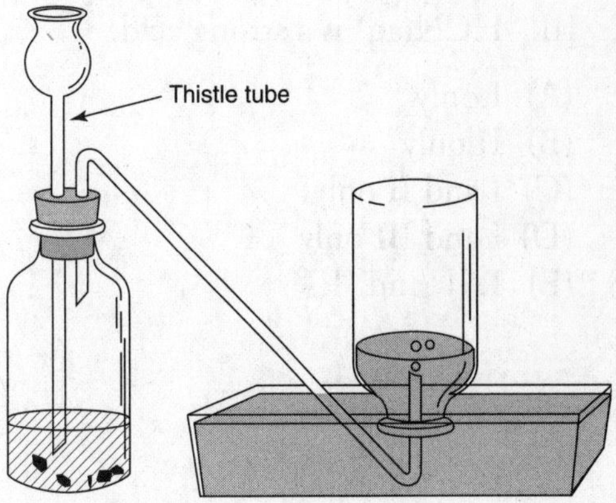

GO ON TO THE NEXT PAGE.

44. The above laboratory setup can be used to prepare which of the following?

I. $CO_2(g)$
II. $H_2(g)$
III. NO(g)

(A) I only
(B) III only
(C) I and III only
(D) II and III only
(E) I, II, and III

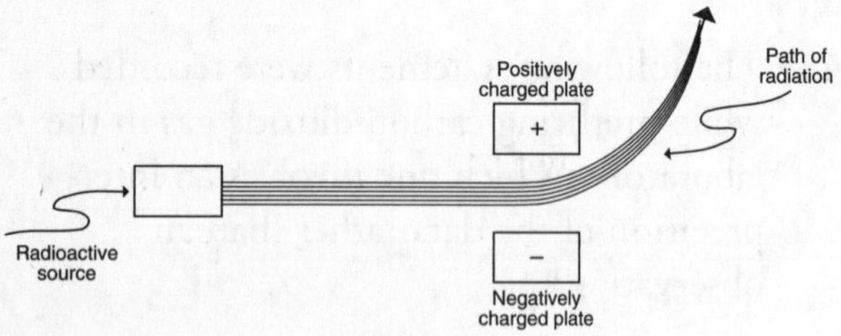

45. In the experiment diagrammed above, the radiation emitted by the source is probably

(A) $^{0}_{-1}e$
(B) $^{4}_{2}He$
(C) $^{1}_{2}H$
(D) a gamma ray
(E) $^{0}_{1}e$

46. Which of the following is (are) true regarding the aqueous dissociation of HCN, $K_a = 4.9 \times 10^{-10}$, at 25°C?

I. At equilibrium, $[H^+] = [CN^-]$.
II. At equilibrium, $[H^+] > [CN^-]$.
III. HCN(aq) is a strong acid.

(A) I only
(B) II only
(C) I and II only
(D) I and III only
(E) I, II, and III

47. This question pertains to the reaction represented by the following equation:

$$2NO(g) + O_2(g) \rightleftharpoons 2NO_2(g) + 150 \text{ kJ}$$

Suppose that 0.8 mole of NO is converted to NO_2 in the above reaction. What amount of heat will be evolved?

(A) 30 kJ
(B) 60 kJ
(C) 80 kJ
(D) 130 kJ
(E) 150 kJ

48. How does a Brönsted-Lowry acid differ from its conjugate base?

(A) The acid has one more proton.
(B) The acid has one less proton.
(C) The acid has one more electron.
(D) The acid has one less electron.
(E) The acid has more than one additional proton.

49. Two containers having 1 mole of hydrogen gas and 1 mole of oxygen gas, respectively, are opened. What will be the ratio of the rate of effusion of the hydrogen to that of the oxygen?

(A) $\sqrt{2} : 1$
(B) 4 : 1
(C) 8 : 1
(D) 16 : 1
(E) $\sqrt{32} : 1$

50. A molecule in which the electron configuration is a resonance hybrid is

(A) SO_2
(B) C_2H_6
(C) Cl_2
(D) HBr
(E) NaCl

51. What is the pH of a solution in which the $[OH^-]$ is 1.0×10^{-4}?

(A) −4
(B) +4
(C) +7
(D) −10
(E) +10

52. If 0.365 gram of hydrogen chloride is dissolved to make 1 liter of solution (Cl = 35.5 and H = 1), the pH concentration of the solution is

(A) 0.001
(B) 0.01
(C) 1
(D) 2
(E) 12

53. In the laboratory, a sample of hydrated salt was heated at 110°C for 30 minutes until all the water was driven off. The data were as follows:

Mass of the hydrate before heating = 250 grams

Mass of the hydrate after heating = 160 grams

From these data, what was the percent of water by mass in the original sample?

(A) 26.5
(B) 36
(C) 47
(D) 56
(E) 90

54. Which of the following oxides dissolves in water to form an acidic solution?

(A) Na_2O
(B) CaO
(C) Al_2O_3
(D) ZnO
(E) SO_3

55. In the laboratory, 20.0 milliliters of an aqueous solution of calcium hydroxide, $Ca(OH)_2$, was used in a titration. A drop of phenolphthalein was added to it to indicate the end point. The solution turned colorless after 20.0 milliliters of a standard solution of 0.050 M HCl solution was added. What was the molarity of the $Ca(OH)_2$?

(A) 0.01 M
(B) 0.025 M
(C) 0.50 M
(D) 0.75 M
(E) 1.00 M

56. Which of the following reactions will NOT spontaneously go to completion?

(A) $Zn(s) + 2HCl(aq) \rightarrow ZnCl_2(aq) + H_2(g)$
(B) $CaCO_3(s) + 2HCl(aq) \rightarrow CaCl_2(aq) + H_2O(aq) + CO_2(g)$
(C) $Ag^+(aq) + HCl(aq) \rightarrow AgCl(s) + H^+(aq)$
(D) $Cu(s) + 2H^+(aq) \rightarrow Cu^{2+}(aq) + H_2(g)$; $E^0 = -0.34$ V
(E) $H_2SO_4(aq) + 2NaOH(aq) \rightarrow 2Na_2SO_4(aq) + 2H_2O(aq)$

57. For a laboratory experiment, a student placed sodium hydroxide crystals on a watch glass, assembled the titration equipment, and prepared a solution of 0.10 M sulfuric acid. Then he weighed 4 grams of sodium hydroxide and added it to enough water to make 1 liter of solution. What might be a source of error in the results of the titration?

(A) Some sulfuric acid evaporated.
(B) The sulfuric acid became more concentrated.
(C) The NaOH solution gained weight, thus increasing its molarity.
(D) The NaOH crystals gained H_2O weight, thus making the solution less than 0.1 M.
(E) The evaporation of sulfuric acid solution countered the absorption of H_2O by the NaOH solution.

GO ON TO THE NEXT PAGE.

58. If 60 grams of NO is reacted with sufficient O_2 to form NO_2 that is removed during the reaction, how many grams of NO_2 can be produced? (Molar masses: NO = 30 g/mol, NO_2 = 46 g/mol)

(A) 46
(B) 60
(C) 92
(D) 120
(E) 180

59. Based on the information shown, each of the following equations represents a reaction in which the entropy, ΔS, is positive EXCEPT

(A) $CaCO_3(s) \rightarrow CaO(s) + CO_2(g)$
(B) $Zn(s) + 2H^+(aq) \rightarrow H_2(g) + Zn^{2+}(aq)$
(C) $2C_2H_6(g) + 7O_2(g) \rightarrow 4CO_2(g) + 6H_2O(g)$
(D) $NaCl(s) \rightarrow Na^+(aq) + Cl^-(aq)$
(E) $N_2(g) + 3H_2(g) \rightarrow 2NH_3(g)$

60. $Cl_2(g) + 2Br^-$ (excess) $\rightarrow$?

When 1 mole of chlorine gas reacts completely with excess KBr solution, as shown above, the products obtained are

(A) 1 mol of Cl^- ions and 1 mol of Br^-
(B) 1 mol of Cl^- ions and 2 mol of Br^-
(C) 1 mol of Cl^- ions and 1 mol of Br_2
(D) 2 mol of Cl^- ions and 1 mol of Br_2
(E) 2 mol of Cl^- ions and 2 mol of Br_2

61. Two water solutions are made in the laboratory, one of glucose (molar mass = 180 g/mol), the other of sucrose (molar mass = 342 g/mol). If the glucose solution had 180 grams in 1,000 grams of water and the sucrose had 342 grams in 1,000 grams of water, which statement about the freezing points of the solutions is the most accurate?

(A) The glucose solution would have the lower freezing point.
(B) The sucrose solution would have the lower freezing point.
(C) The freezing point of the sucrose solution would be lowered twice as much as that of the glucose solution.
(D) Both solutions would have the same freezing point.
(E) The freezing points of the solutions would not be affected, because both solutes are nonpolar.

62. Which K_a value indicates the strongest acid?

(A) 1.3×10^{-2}
(B) 6.7×10^{-5}
(C) 5.7×10^{-10}
(D) 4.4×10^{-7}
(E) 1.8×10^{-16}

63. What mass of $CaCO_3$ is needed to produce 11.2 liters of CO_2 when the calcium carbonate is reacted with an excess amount of hydrochloric acid? (Molar masses: $CaCO_3$ = 100 g/mol, HCl = 36.5 g/mol, CO_2 = 44 g/mol)

(A) 25 g
(B) 44 g
(C) 50 g
(D) 100 g
(E) None of the above

64. By experimentation it is found that a saturated solution of $BaSO_4$ at 25°C contains 3.9×10^{-5} mole/liter of Ba^{2+} ions. What is the K_{sp} of the $BaSO_4$?

(A) 1.5×10^{-4}
(B) 1.5×10^{-9}
(C) 1.5×10^{-10}
(D) 3.9×10^{-10}
(E) 39×10^{-9}

65. What is the ΔH^0 value for the decomposition of sodium chlorate, given the following information?

$$NaClO_3(s) \rightarrow NaCl(s) + \frac{3}{2}O_2(g)$$

(ΔH^0_f values: $NaClO_3(s) = -85.7$ kcal/mol, $NaCl(s) = -98.2$ kcal/mol, $O_2(g) =$ 0 kcal/mol)

(A) 12.5 kcal
(B) −12.5 kcal
(C) 173.9 kcal
(D) −173.9 kcal
(E) $\frac{3}{2}$(173.9 kcal)

66. To the equilibrium reaction shown below:

$$AgCl(s) \rightleftharpoons Ag^+ + Cl^-$$

a beaker of concentrated HCl (12 M) is slowly added. Which is the best description of what will occur?

(A) More salt will go into solution, and the K_{sp} will remain the same.
(B) More salt will go into solution, and the K_{sp} will increase.
(C) Salt will come out of the solution, and the K_{sp} will remain the same.
(D) Salt will come out of the solution, and the K_{sp} will decrease.
(E) No change in concentration will occur, and the K_{sp} will increase.

67. When the following redox equation is balanced and all coefficients are reduced to lowest whole-number terms, what is the coefficient of H_2O?

$$HCl + KMnO_4 \rightarrow H_2O + KCl + MnCl_2 + Cl_2$$

(A) 1
(B) 2
(C) 5
(D) 8
(E) 16

68. Each of the following systems is at equilibrium in a closed container. A decrease in the total volume of each container will increase the number of moles of product(s) for which system?

(A) $2NH_3(g) \rightleftharpoons N_2(g) + 3H_2(g)$
(B) $H_2(g) + Cl_2(g) \rightleftharpoons 2HCl(g)$
(C) $2NO(g) + O_2(g) \rightleftharpoons 2NO_2(g)$
(D) $CO(g) + H_2O(g) \rightleftharpoons CO_2(g) + H_2(g)$
(E) $Fe_3O_4(s) + 4H_2(g) \rightleftharpoons 3Fe(s) + 4H_2O(g)$

69. Which element(s), when forming an ionic bond, has (have) the following electron configuration?

$$1s^2\ 2s^2\ 2p^6\ 3s^2\ 3p^6$$

I. Potassium
II. Sulfur
III. Chlorine

(A) I only
(B) III only
(C) I and III only
(D) II and III only
(E) I, II, and III

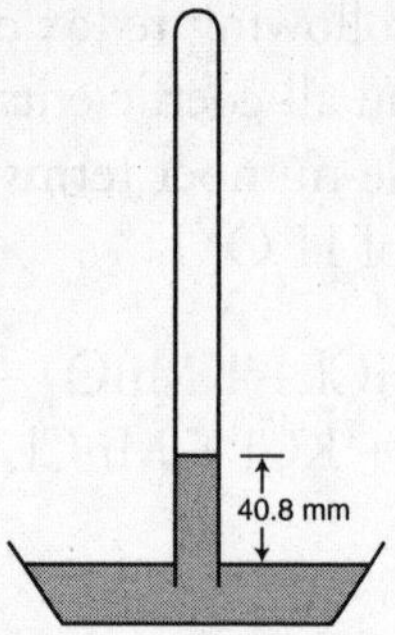

70. Hydrogen gas is collected in a eudiometer tube over water as shown above. The water level inside the tube is 40.8 millimeters higher than that outside. The barometric pressure is 730 millimeters Hg. The water vapor pressure at the room temperature of 29°C is found in a handbook to be 30.0 millimeters Hg. What is the pressure of the dry hydrogen?

(A) 659.2 mm Hg
(B) 689.2 mm Hg
(C) 697.0 mm Hg
(D) 740.8 mm Hg
(E) 800.8 mm Hg

71. How many moles of electrons are required to reduce 2.93 grams of nickel ions from melted $NiCl_2$? (Molar mass of Ni = 58.7 g/mol)

(A) 0.05
(B) 0.10
(C) 1.0
(D) 1.5
(E) 2.0

STOP

If you finish before one hour is up, you may go back to check your work or complete unanswered questions.

Answers and Explanations

1. **(A)** In the halogen family, the most active nonmetal would be the top element, fluorine, because it has the highest electronegativity.

2. **(A)** As you proceed down a group, the ionic radius increases as additional energy levels are filled farther from the nucleus. Therefore fluorine, the top element, has the smallest ionic radius.

3. **(E)** Since astatine has the largest atomic radius and its outer electrons are shielded from the protons by a large number of interior electrons, it has the lowest first ionization potential.

4. **(D)** Some physical characteristics of metal are found in iodine, the fourth halogen down in the group.

5. **(A)** Sodium acetate has Na^+ and $C_2H_3O_2^-$ ions.

6. **(C)** Aluminum oxide has Al^{3+} and O^{2-} ions.

7. **(A)** Potassium phosphate has a K^+ and a PO_4^{3-} ion.

8. **(D)** John Dalton is credited with the basic postulates of atomic theory.

9. **(E)** Gay-Lussac is credited with the statement that, when gases combine, they do so in ratios of small whole numbers that are in relationship to the volumes of the reactants and the volumes of the products under the same conditions.

10. **(C)** Charles is credited with this temperature-volume relationship of gases: $\frac{V_1}{V_2} = \frac{T_1}{T_2}$ or $V_1T_2 = V_2T_1$ as shown in the question.

11. **(A)** Avogadro is credited with the concept regarding the number of particles in a mole, 6.02×10^{23}, and this number bears his name.

12. **(B)** Boyle is credited with the $P_1V_1 = P_2V_2$ relationship of gases.

13. **(C)** This includes the functional group of an aldehyde.

14. **(D)** This includes the functional group of an organic acid.

15. **(E)** An ester is the equivalent of an organic salt since it is usually formed from an organic alcohol,

R–OH, plus an organic acid, $R^*-\overset{\overset{O}{\|}}{C}-OH$.

The bonding is $R-O\boxed{H\,HO}-\overset{\overset{O}{\|}}{C}-R^*$, which gives

$R-O-\overset{\overset{O}{\|}}{C}-R^* + H_2O$. ($R^*$ indicates that this hydrogen branch need not be the same as R.)

16. **(B)** This includes the functional group of an ether. It can be formed by the dehydration of two alcohol molecules. The reaction is

$$R-O\boxed{H + HO}-R^* \rightarrow R-O-R^* + H_2O.$$

17. **(B)** The molecular structure of carbon dioxide is O=C=O, where the oxygens are 180° apart and, although the bonding is polar to the carbon, counteract each other to constitute a nonpolar molecule.
18. **(A)** A hydrogen gas molecule is diatomic; it has 2 mol of atoms in each mole of molecules, represented by H_2.
19. **(B)** Each CO_2 has three atoms per molecule; hence the expression can represent 3 mol of atoms in 1 mol of molecules.
20. **(E)** With complete ionization $H_2SO_4 \rightarrow 2H^+ + SO_4^{2-}$, or 3 mol of ions per mole of H_2SO_4.
21. **(C)** The expression $2N_2O$ represents two triatomic molecules or 6 mol of atoms.
22. **(D)** Carbon tetrachloride has four hydrogens, each bonded to an sp^3 hybridized orbital. The four sp^3 orbitals of carbon are in a tetrahedral orientation so that the resulting configuration balances the polarity of each bond, forming a nonpolar covalent molecule.
23. **(A)** Because cesium is a very active metal, the chloride formed has an ionic bond and is an ionic substance.
24. **(C)** Chlorine's stronger electronegativity draws the hydrogen electron closer to itself more often, so that the charge distribution in the molecule is uneven, causing a negative and positive polarity. Therefore, hydrogen chloride is a polar covalent molecule.
25. **(E)** Benzene is an aromatic (ring) structure classified as an organic substance.

101. **(T, T, CE)** A catalyst can accelerate a chemical reaction by lowering the activation energy required for the reaction to occur.

102. **(T, T, CE)** Sodium chloride is an ionic substance and when molten is a good electrical conductor, and the reason is that in the molten state the ions are free to migrate to the anode and cathode.

103. **(T, F)** Ice is less dense than liquid water. Water molecules, however, are polar, not nonpolar, and water expands as these molecules arrange themselves into a crystal lattice.

104. **(F, T)** The statement that isotopes of the same element have the same mass number is false. Isotopes of the same element have the same number of protons but vary in the number of neutrons in the nucleus. Therefore, these isotopes have the same atomic number but different mass numbers.

105. **(T, T, CE)** A solute that has a very low solubility in a particular solvent can reach the saturated point in a solution with that solvent, but the solution can still be classified as dilute. The term *dilute* merely refers to a small amount of solute in a solvent.

106. **(T, T, CE)** The reaction is:

$$C + O_2 \rightarrow CO_2$$

and shows 1 mol or 12 g of carbon produces 1 mol or 22.4 L of CO_2. Then

$$2\ \cancel{L\,CO_2} \times \frac{12\text{ g C}}{22.4\ \cancel{L\,CO_2}} = 1.07\text{ g C.}$$

So the statement is correct, and the assertion explains it correctly.

107. **(F, F)** A reaction at equilibrium has reached a point where the forward and reverse reactions are occurring at equal rates. The concentrations of the reactants and products, however, are not necessasrily equal and are described by the K_{eq} value at the temperature of the reaction.

108. **(F, T)** In an electrolyte reaction, anions migrate to the anode. It is true that positively charged ions (cations) are attracted to the negatively charged cathode.

109. **(F, T)** The value pH = 5 can be expressed as

$$\left[H_3O^+\right] = 1 \times 10^{-5} \frac{\text{mol}}{\text{L}}$$

and pH = 3 as

$$\left[H_3O^+\right] = 1 \times 10^{-3} \frac{\text{mol}}{\text{L}}.$$

Thus pH = 3 represents a larger concentration of hydronium ions, H_3O^+.

110. **(T, T, CE)** The first statement is true. The change in Gibbs free energy, ΔG, depends on the enthalpy change, ΔH, and the entropy change, ΔS, from the equation $\Delta G = \Delta H - T\Delta S$. Thus, statement II is also true and explains the first statement.

111. **(T, T, CE)** In the expression for the equilibrium constant of an acid, $[H_3O^+]$ is in the numerator:

$$K_a = \frac{[H_3O^+][Y^-]}{[HY]}.$$

As $[H_3O^+]$ decreases in the weak acids, the numerator becomes smaller as the denominator gets larger, therefore giving smaller K_a values.

112. **(T, T)** Both statements are true, but they are not related.

113. **(T, F)** A pi bond is formed between *p* lobes of adjacent atoms and in the same plane:

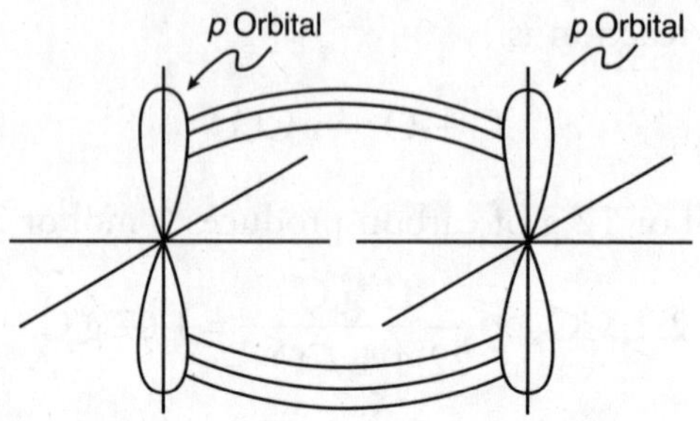

However, each *p* orbital consisting of two lobes can hold a total of two electrons, so the reason is false.

114. **(T, F)** The assertion that the boiling points of H_2S and H_2O are significantly different is true, but the reason is false. Water has the higher degree of hydrogen bonding.

26. **(E)** Two immiscible liquids, when shaken together vigorously, may each form small, colloidal-size particles, dispersed in the other liquid.

27. **(D)** In the graph, the first plateau must represent the condensation from gas to liquid because there is a second, lower plateau, which would represent the second change of state, from liquid to solid.

28. **(B)** If the electrons have the same principal energy level, they will fill the s^2, p^6, then the d^{10} level. This progression is from the lowest energy sublevel to the highest, to accommodate 18 electrons.

29. **(B)** The empirical formula identifies the atoms present and the lowest-whole-number ratio of their occurrence as CH. The molecular formula, which gives the actual composition of the compound, must then be a multiple of the empirical formula: C_6H_6.

30. **(D)** Considering the concepts behind Charles's Law and Boyle's Law, you can arrive at the fraction to be used in kelvins and torrs. The volume must increase with an increase in temperature and also increase with a decrease in pressure. Therefore the fractions would be $\frac{333\text{ K}}{278\text{ K}}$ and $\frac{640\text{ torr}}{320\text{ torr}}$.

Using the combined gas law equation,

$$\frac{V_1P_1}{T_1} = \frac{V_2P_2}{T_2}$$

and solving for V_2, you get

$$V_2 = V_1 \times \frac{T_2}{T_1} \times \frac{P_1}{P_2}$$

which is the same answer.

31. **(D)** The mass number, 32, is the total number of neutrons and protons. Since the atomic number, 16, gives the number of protons, 32 – 16 = 16, or the number of neutrons, statement II is false. Since this ion has a charge of 2^-, it has two more electrons than protons, or 18 electrons. Statements I and III are true.

32. **(D)** The most active metallic elements are found in the lower left corner of the Periodic Table.

33. **(C)** The rise in the boiling point depends on the number of particles in solution. One mole of $CaCl_2$ gives 3 mol of ions, more than any other substance listed:

$$CaCl_2 \rightarrow Ca^{2+} + 2Cl^-.$$

The number of moles of ions given by the other substances are as follows:

(A) = 2, (B) = 2, (D) = 1, (E) = 1.

34. **(E)** The sp^3 hybrid has the tetrahedron configuration. The sp^2 (D) is trigonal planar. The sp (C) is linear. The s (B) and p (A) are the usual orbital structures.

35. **(C)** The reaction is

$$\begin{array}{ccc} & 8\text{L} & x\text{L} \\ S(s) + & O_2(g) \rightarrow & SO_2(g) \end{array}$$

The given (8 L) and the unknown (x L) are shown above. Since the equation, according to Gay-Lussac's Law, shows that 1 volume of oxygen yields 1 volume of sulfur dioxide, then

$$8\ \cancel{\text{L}\ O_2} \times \frac{1\text{ L } SO_2}{1\ \cancel{\text{L}\ O_2}} = 8\text{ L } SO_2$$

36. **(C)** The fact that the mercury level in the U-tube is higher in the right side of the tube indicates that the pressure in the flask is higher than the atmospheric pressure exerted on the open end of the tube on the right side. If the pressure inside the flask were the same as the atmospheric pressure, the height of the mercury would be the same in both sides of the U-tube.

37. **(B)** The only change listed that would change the pressure of the gas inside the flask is to increase the temperature of the gas. This would cause the pressure to rise.

38. **(D)** An amphoteric substance must be able to be a proton, (H^+), donor, and a proton receiver. The bisulfate ion, HSO_4^-, is the only choice that can either accept a proton and become H_2SO_4, or lose a proton and become the sulfate ion, SO_4^{2-}.

39. **(C)** Standard conditions are 273 K and 760 mm Hg.

40. **(A)** Dividing the percentage of each element by the atomic mass of that element gives the basic ratio of atoms, but not necessarily in whole numbers. Thus, (Ba) 58.8 ÷ 137 = 0.43, (S) 13.7 ÷ 32 = 0.43, (O) 27.5 ÷ 16 = 1.72.

    Because atoms occur in whole numbers, you now must manipulate these numbers mathematically to get whole numbers. Usually dividing each number by the smallest one helps to accomplish this: (Ba) 0.43 ÷ 0.43 = 1, (S) 0.43 ÷ 0.43 = 1, (O) 1.72 ÷ 0.43 = 4. The empirical formula is $BaSO_4$.

41. **(C)** The percentage composition can be found by dividing each total atomic mass in the formula by the molar mass of the compound.

CA = 40	40 ÷ 74 × 100% = 54% Ca
2O = 32	32 ÷ 74 × 100% = 43% O
2H = <u>2</u>	2 ÷ 74 × 100% = 2.7% H
74	

42. **(B)** The solution setup is:

$$\overset{65\text{ g}}{\underset{65\text{ g}}{Zn}} + \overset{65\text{ g}}{\underset{73\text{ g}}{2HCl}} \rightarrow ZnCl_2 + \overset{x\text{ g}}{\underset{2\text{ g}}{H_2(g)}}$$

Note that the equation mass is calculated under the substances that have mass units above them. According to the calculated equation masses, 73 g of HCl would be needed to react with 65 g of zinc. Since there is only 65 g of HCl, we use this and disregard the 65 g of Zn.

$$65\text{ g }\cancel{HCl} \times \frac{2\text{ g }H_2}{73\text{ g }\cancel{HCl}}$$

$= 1.78$ or 1.8 g of H_2

43. **(C)** All the other statements represent observations because they merely record what was seen.

44. **(E)** The setup is appropriate for the collection of a basically nonsoluble gas by the displacement of water. All three gases fit this description.

45. **(A)** Because the paths of the particles from the radioactive source bend toward the positive plate, the source must be negatively charged. The only radioactive particle that fits this criterion is the beta particle, ${}_{-1}^{0}e$.

46. **(A)** The small K_a indicates that this is a weak acid, so statement III is false. When HCN ionizes, it can be shown that $HCN \rightleftharpoons H^+ + CN^-$. This is a molar ratio of 1 : 1, so statement I must be true.

47. **(B)** Because the equation shows that 2 mol of NO react to release 150 kJ, the solution is

$$0.8\ \cancel{\text{mol NO}} \times \frac{150\ \text{kJ}}{2\ \cancel{\text{mol NO}}} = 60\ \text{kJ}$$

This problem would be solved in the same manner if the heat had been expresssed in kilocalories. To convert one unit to the other, use $4.18 \times 10^3\ J = 1$ kcal.

48. **(A)** In the Brönsted-Lowry acid-base theory:

HCl	$+ NH_3$	$\rightarrow$	NH_4^+	$+$	Cl^-
acid	base		conjugate acid		conjugate base

49. **(B)** According to Graham's Law, the rates of effusion of two gases are inversely proportional to the square roots of their molar masses. Since 1 mol $O_2 = 32$ g and 1 mol $H_2 = 2$ g

$$\frac{H_2 \text{ rate}}{O_2 \text{ rate}} = \frac{\sqrt{32}}{\sqrt{2}} = \sqrt{\frac{16}{1}} = \frac{4}{1}$$

Therefore, the effusion rate of hydrogen is four times faster than that of oxygen.

50. **(A)** The structure of SO_2 is a resonance hybrid, shown as follows:

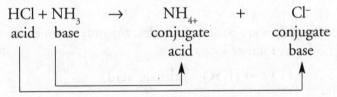

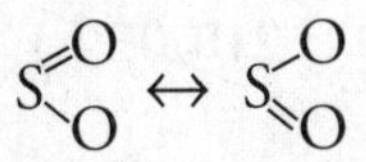

51. **(E)** The K_w of water is

$$K_w = [H^+][OH^-] = 10^{-14}$$

$$\text{If } [OH^-] = 1.0 \times 10^{-4}, \text{ then}$$

$$[H^+] = \frac{10^{-14}}{10^{-4}} = 10^{-10}$$

and

$$pH = -\log[H^+] = 10$$

52. **(D)** $0.365 \text{ g } \cancel{HCl} \times \frac{1 \text{ mol HCl}}{36.5 \text{ g } \cancel{HCl}}$

= 0.01 mol HCl

0.01 mol HCl → 0.01 mol H^+

+ 0.01 mol Cl^-

If $[H^+] = 0.01 = 1 \times 10^{-2}$ mol/L, then pH = 2.

53. **(B)** Hydrate mass before heating = 250 g

− Hydrate mass after heating = 160 g

Water loss = 90 g

$$\frac{90 \text{ g mass loss}}{250 \text{ g original mass}} \times 100\%$$

= 36% by mass

54. **(E)** Because metallic oxides are basic anhydrides, the only nonmetallic oxide is SO_3. The reaction is as follows

$$SO_3 + H_2O \rightarrow H_2SO_4 \text{ (sulfuric acid)}$$

55. **(B)** In the titration, the reaction is:

$$2HCl + Ca(OH)_2 \rightarrow CaCl_2 + 2H_2O$$

The acid to base ratio is 2 : 1, or (moles acid used) = 2(moles base used), so $M_aV_a = 2M_bV_b$, where M is the molarity and V is the volume expressed in liters, then

$$M_b = \frac{M_aV_a}{2V_b}$$

$$M_b = \frac{0.05 \text{ M} \times \cancel{0.02 \text{ L}}}{2(\cancel{0.02 \text{ L}})} = 0.025 \text{ M}$$

56. **(D)** All the reactions will go to completion except (D), which will not occur spontaneously at all. If E^0 had been positive, however, this redox reaction would occur.

57. **(D)** Sodium hydroxide is hygroscopic and will attract water to its surface area. This water will influence its mass; consequently there will be less sodium hydroxide in the mass used.

58. **(C)** $\overset{60\,g}{\underset{60\,g}{2NO}} + O_2 \rightarrow \overset{x\,g}{\underset{92\,g}{2NO_2}}$

This is a mass stoichiometry problem. The equation masses are placed beneath the substances that have quantities above them.

Using the equation relationship gives

$$60\ g\ \cancel{NO} \times \frac{92\ g\ NO_2}{60\ g\ \cancel{NO}} = 92\ g\ NO_2$$

Using the mole method gives

$$60\ g\ \cancel{NO} \times \frac{1\ mol\ NO}{30\ g\ \cancel{NO}} = 2\ mol\ NO$$

Using the equation coefficients gives

$$2\ \cancel{mol\ NO} \times \frac{2\ mol\ NO_2}{2\ \cancel{mol\ NO}} = 2\ mol\ NO_2$$

then

$$2\ \cancel{mol\ NO_2} \times \frac{46\ g\ NO_2}{1\ \cancel{mol\ NO_2}} = 92\ g\ NO_2$$

59. **(E)** Since in this reaction 4 volumes of gases are forming 2 volumes of product, the randomness of the system is decreasing. Therefore, entropy is decreasing and ΔS is negative. In all the other reactions randomness is increasing.

60. **(D)** The reaction is

$$Cl_2(g) + 2Br^-(excess) \rightarrow 2Cl^- + Br^2(g)$$

Here 2 mol of chloride ions and 1 mol of Br_2 molecules are produced.

61. **(D)** Since both solutions were 1 molal and neither compound ionized into more particles, the freezing points would be the same.

62. **(A)** $HY \rightleftharpoons H^+ + Y^-$

Since $[H^+]$ is in the numerator of K_a:

$$K_a = \frac{[H^+][Y^-]}{[HY]}$$

The stronger the acid, the greater are $[H^+]$ and the K_a value. Of the choices given, 1.3×10^{-2} is the largest.

63. **(C)** $\underset{100\text{ g}}{\overset{x\text{ g}}{CaCO_3}} + 2HCl \rightarrow CaCl_2 + H_2O + \underset{22.4\text{ L}}{\overset{11.2\text{ L}}{CO_2}}$

This is a mass-volume problem with the mass and volume indicated below the equation.

Using the factor-label method gives

$$11.2\ \text{L}\ CO_2 \times \frac{100\ \text{g}\ CaCO_3}{22.4\ \text{L}\ CO_2}$$

$= 50\ \text{g}\ CaCO_3$

Using the mole method gives

$$11.2\ \text{L}\ CO_2 \times \frac{1\ \text{mol}\ CO_2}{22.4\ \text{L}\ CO_2}$$

$= 0.5\ \text{mol}\ CO_2$

$$0.5\ \text{mol}\ CO_2 \times \frac{1\ \text{mol}\ CaCO_3}{1\ \text{mol}\ CO_2}$$

$= 0.5\ \text{mol}\ CaO_3$

$$0.5\ \text{mol}\ CaCO_3 \times \frac{100\ \text{g}\ CaCO_3}{1\ \text{mol}\ CaCO_3}$$

$= 50\ \text{g}\ CaCO_3$

64. **(B)** $K_{sp} = [Ba^{2+}][SO_4^{\ 2-}]$

If $[Ba^{2+}] = 3.9 \times 10^{-5}$, then $[SO_4^{\ 2-}]$ must also equal the same amount, so

$$K_{sp} = [3.9 \times 10^{-5}][3.9 \times 10^{-5}]$$
$$= 15.2 \times 10^{-10} \text{ or } 1.5 \times 10^{-9}.$$

65. **(B)** $\Delta H^0_{\text{reaction}} = \Delta H^0_f(\text{products}) - \Delta H^0_f(\text{reactants})$

$$\Delta H^0_{\text{reaction}} = -98.2\ \text{kcal} - (-85.7\ \text{kcal})$$

$$\Delta H^0_{\text{reaction}} = -12.5\ \text{kcal}$$

66. **(C)** Because the HCl solution will add to the chloride ion concentration, according to Le Châtelier's Principle the equilibrium will shift in the direction to reduce this disturbance, so the K_{sp} will remain the same but salt will come out of solution. This process:

$$AgCl \rightleftharpoons Ag^+ + Cl^-$$

will continue until the K_{sp} is reestablished. This phenomenon is called the "common ion effect."

67. **(D)** $HCl + KMnO_4 \rightarrow H_2O + KCl + MnCl_2 + Cl_2$
oxidation: $2Cl^- \rightarrow Cl_2 + 2e^-$
reduction: $8H^+ + MnO_4^- + 5e^- \rightarrow Mn^{2+} + 4H_2O$
5 (oxidation reaction) and 2 (reduction reaction) will balance the e^- gain and loss, giving
$10Cl^- \rightarrow 5Cl_2 + \cancel{10}e^-$
$16H^+ + 2MnO_4^- + \cancel{10}e^- \rightarrow 2Mn^{2+} + 8H_2O$
Therefore
$16HCl + 2KMnO_4 \rightarrow 8H_2O + 2KCl + 2MnCl_2 + 5Cl_2$

68. **(C)** A decrease in volume will cause the equilibrium to shift in the direction that has less volume(s) of gas(es). In every case except (C) this is the reverse reaction, which decreases the product. The coefficients give the volume relationships.

69. **(E)** Each of the elements when becoming ions has this configuration of 18 electrons.
K with 19 electrons – $e^- \rightarrow K^+$ with 18 electrons
S with 16 electrons + $2e^- \rightarrow S^{2-}$ with 18 electrons
Cl with 17 electrons + $e^- \rightarrow Cl^-$ with 18 electrons

70. **(C)** Change the water height to the equivalent Hg height:

$$40.8\ \cancel{mm\ H_2O} \times \frac{1\ mm\ Hg}{13.6\ \cancel{mm\ H_2O}}$$

= 3 mm Hg

Adjust for the difference in height to get the gas pressure. Pressure on the gas is

730 mm Hg – 3 mm Hg
= 727 mm Hg.

Vapor pressure of H_2O at 29°C accounts for 30 mm Hg pressure. Therefore

727 mm Hg – 30 mm Hg
= 697 mm Hg

71. **(B)** The reaction is

$$Ni^{2+} + 2e^- \rightarrow Ni\ (s)$$

Since 2 mol of electrons are required to form 1 mol of nickel, and

$$2.9\ \cancel{g\,Ni} \times \frac{1\ \text{mol Ni}}{58.7\ \cancel{g\,Ni}} = 0.05\ \text{mol Ni}$$

then

$$0.05\ \cancel{\text{mol Ni}} \times \frac{2\ \text{mol electrons}}{1\ \cancel{\text{mol Ni}}}$$

$= 0.1$ mol of electrons

CALCULATING YOUR SCORE

Your score on the diagnostic test can now be computed manually. The actual test will be scored by machine, but the same method is used to arrive at the raw score. You get one point for each correct answer. For each wrong answer, you lose one-fourth of a point. Questions that you omit or for which you have indicated more than one answer are not counted. On your answer sheet, mark all correct answers with a "C" and all incorrect answers with an "X."

Determining Your Raw Test Score

Total the number of correct answers you have recorded on your answer sheet. It should be the same as the total of all the numbers you place in the block in the lower left corner of each area of the Subject Area summary in the next section.

A. Enter the total number of correct answers here: ________
   Now count the number of wrong answers you recorded on your answer sheet.
B. Enter the total number of wrong answers here: ________
   Multiply the number of wrong answers in B by 0.25.
C. Enter that product here: ________
   Subtract the result in C from the total number of right answers in A.
D. Enter the result of your subtraction here: ________
E. Round the result in D to the nearest whole number: ________.
   This is your raw test score.

Conversion of Raw Scores to Scaled Scores

Your raw score is converted by the College Board into a scaled score. The College Board scores range from 200 to 800. This conversion is done to ensure that a score earned on any edition of a particular SAT Subject Test in Chemistry is comparable to the same scaled score earned on any other edition of the same test. Because some editions of the tests may be slightly easier or more difficult than others, scaled scores are adjusted so that they indicate the same level of performance regardless of the edition of the test taken and the ability of the group that takes it. Consequently, a specific raw score on one edition of a particular test will not necessarily translate to the same scaled score on another edition of the same test.

Since the practice tests in this book have no large population of scores with which they can be scaled, scaled scores cannot be determined.

Results from previous SAT Chemistry tests appear to indicate that the conversion of raw scores to scaled scores GENERALLY follows this pattern:

Raw Score	Scaled Score	Raw Score	Scaled Score
85–75	800–800	30–25	560–540
75–70	800–780	25–20	540–520
70–65	780–750	20–15	520–490
65–60	750–720	15–10	490–460
60–55	720–700	10–5	460–430
55–50	700–680	5–0	430–410
50–45	680–650	0 to –5	410–380
45–40	650–620	–5 to –10	380–360
40–35	620–590	–10 to –15	360–330
35–30	590–560		

Note that this scale provides only a general idea of what a raw score may translate into on a scaled score range of 800–200. Scaling on every test is usually slightly different. Some students who have taken the SAT Subject Test in Chemistry after using this book have reported that they have scored slightly higher on the SAT test than on the practice tests in this book. They all reported that preparing well for the test paid off in a better score!

DIAGNOSING YOUR NEEDS

This section will help you to diagnose your need to review the various categories tested by the SAT Subject Test in Chemistry.

After taking the diagnostic test, check your answers against the correct ones. Then fill in the chart below. In the space under each question number, place a check (✓) if you answered that question correctly.

Next, total the checks for each section and insert the number in the designated block. Now do the arithmetic indicated, and insert your percentage for each area.

Subject Area*	(✓) Questions Answered Correctly
I. Atomic Theory and Structure, including periodic relationships ☐ No. of checks ÷ 8 × 100 = _____%	1, 2, 3, 5, 6, 7, 8, 28
II. Nucleonics ☐ No. of checks ÷ 3 × 100 = _____%	104, 31, 45
III. Chemical Bonding and Molecular Structure ☐ No. of checks ÷ 9 × 100 = _____%	17, 20, 22, 23, 24, 113, 114, 34, 69
IV. States of Matter and Kinetic Molecular Theory of Gases ☐ No. of checks ÷ 8 × 100 = _____%	9, 10, 12, 27, 30, 35, 39, 49
V. Solutions, including concentration units, solubility, and colligative properties ☐ No. of checks ÷ 5 × 100 = _____%	105, 26, 33, 55, 61
VI. Acids and Bases ☐ No. of checks ÷ 8 × 100 = _____%	109, 111, 38, 48, 51, 52, 54, 62
VII. Oxidation-Reduction and Electrochemistry ☐ No. of checks ÷ 6 × 100 = _____%	102, 108, 56, 60, 67, 71
VIII. Stoichiometry ☐ No. of checks ÷ 11 × 100 = _____%	11, 18, 19, 106, 112, 29, 40, 41, 42, 58, 63
IX. Reaction Rates ☐ No. of checks ÷ 2 × 100 = _____%	101, 107

**The subject areas have been expanded to identify specific areas in the text.*

Subject Area*	(✓) Questions Answered Correctly
X. Equilibrium ☐ No. of checks ÷ 4 × 100 = _____%	46, 64, 66, 68
XI. Thermodynamics: energy changes in chemical reactions, randomness, and criteria for spontaneity ☐ No. of checks ÷ 4 × 100 = _____%	110, 47, 59, 65
XII. Descriptive Chemistry: physical and chemical properties of elements and their familiar compounds; organic chemistry; periodic properties ☐ No. of checks ÷ 11 × 100 = _____%	4, 13, 14, 15, 16, 21, 25, 103, 32, 37, 50
XIII. Laboratory: equipment, procedures, observations, safety, calculations, and interpretation of results ☐ No. of checks ÷ 6 × 100 = _____%	36, 43, 44, 53, 57, 70

**The subject areas have been expanded to identify specific areas in the text.*

PLANNING YOUR STUDY

The percentages give you an idea of how you have done on the various major areas of the test. Because of the limited number of questions on some parts, these percentages may not be as reliable as the percentages for parts with larger numbers of questions. However, you should now have at least a rough idea of the areas in which you have done well and those in which you need more study. (There are four more practice tests in the back of this book, which may be used in a diagnostic manner as well.)

Start your study with the areas in which you are the weakest. The corresponding chapters are indicated on the next page.

Subject Area	Chapters to Review
I. Atomic Theory and Structure, including periodic relationships	2, 15
II. Nuclear Reactions	15
III. Chemical Bonding and Molecular Structure	3, 4
IV. States of Matter and Kinetic Molecular Theory of Gases	1
V. Solutions, including concentration units, solubility, and colligative properties	7
VI. Acids and Bases	11
VII. Oxidation-Reduction and Electrochemistry	12
VIII. Stoichiometry	5, 6
IX. Reaction Rates	9
X. Equilibrium	10
XI. Thermodynamics: energy changes in chemical reactions, randomness, and criteria for spontaneity	8
XII. Descriptive Chemistry: physical and chemical properties of elements and their familiar compounds; organic chemistry; periodic properties	1, 2, 13, 14
XIII. Laboratory: equipment, procedures, observations, safety, calculations, and interpretation of results.	All lab diagrams, 16

After you have spent some time reviewing your weaker areas, plan a schedule of work that spans the 6 weeks before the test. Unless you set up a regular study pattern and goals, you probably will not prepare sufficiently.

The following schedule provides such a plan. Note that weekends are left free, and the time spans are held to 1- or 2-hour blocks. This will be time well spent!

	Monday	Tuesday	Wednesday	Thursday	Friday
First Week:	Ch. 1: 1 hr	Ch. 2: 2 hr	Ch 2: 1 hr.	Ch. 3: 2 hr	Ch. 3: 1 hr
Second Week:	Ch. 4: 1 hr	Ch. 4: 1 hr	Ch. 5: 2 hr	Ch. 5: 1 hr	Ch. 6: 1 hr
Third Week:	Ch. 6: 2 hr	Ch. 7: 2 hr	Ch. 8: 1 hr	Ch. 9: 2 hr	Ch. 10: 1 hr
Fourth Week:	Ch. 10: 1 hr	Ch. 11: 2 hr	Ch. 12: 2 hr	Ch. 13: 1 hr	Ch. 14: 2 hr
Fifth Week:	Chs. 15, 16: 1 hr	Take Practice Test 1: 1 hr Study Ans.: 1 hr	Review weakest areas: 1 hr	Take Practice Test 2: 1 hr Study Ans.: 1 hr	Review weakest areas: 1 hr
Sixth Week:	Take Practice Test 3: 1 hr Study Ans.: 1 hr	Review weakest areas: 1 hr	Take Practice Test 4: 1 hr Study Ans.: 1 hr	Review weakest areas: 1 hr	Review a practice test already taken: 1 hr *Go to bed early.*

FINAL PREPARATION—THE DAY BEFORE THE TEST

The day before the test, review one of the practice tests you have already taken. Study again the directions for each type of question. Long hours of study at this point will probably only heighten your anxiety, so just look over the answer section of the practice test and refer to any chapter in the book if you need more information. This type of limited, relaxed review will probably make you feel more comfortable and better prepared.

Get together the materials that you will need. They are:

- Your admission ticket. Check the time your admission ticket specifies for arrival.
- Your identification. (You will not be admitted without some type of positive identification such as a student I.D. card with picture or a driver's license.)
- Two No. 2 pencils with erasers.
- Know the route to the test center and the entrance indicated.

Note that calculator use is ***not*** allowed during the SAT Subject Test in Chemistry.
You should also go over this checklist:

A. Plan your activities so that you will have time for a good night's sleep.

B. Lay out comfortable clothes for the next day. You may want to bring a snack.

C. Review the following helpful tips about taking the test:

1. Read the directions carefully.

2. In each group of questions, answer first those that you know. Temporarily skip difficult questions, but mark them in the margin so you can go back if you have time. Keep in mind that an easy question answered correctly counts as much as a difficult one.
3. Avoid haphazard guessing since this will probably lower your score. Instead, guess smart! If you can eliminate one or more of the choices to a question, it will generally be to your advantage to guess which of the remaining answers is correct. Your score will be based on the number right minus a fraction of the number answered incorrectly.
4. Omit questions when you have no idea of how to answer them. You neither gain nor lose credit for questions you do not answer.
5. Keep in mind that you have 1 hour to complete the test, and pace yourself accordingly. If you finish early, go back to questions you skipped.
6. Mark the answer grid clearly and correctly. Be sure each answer is placed in the proper space and within the oval. *Erase all stray marks completely.*
7. Write as much as you like in the test booklet. Use it as a scratch pad. Only the answers on the answer sheet are scored for credit.

D. Set your alarm clock so as to allow plenty of time to dress, eat your usual (or even a better) breakfast, and reach the test center without haste or anxiety.

AFTER THE TEST

Your scores will be mailed to you, your high school, and the college and scholarship programs you entered on your registration form (unless you selected "Score Choice") about 3 weeks after the test date. Your score report includes your scores, percentiles, and interpretive information. You can get your scores for a fee by phone several weeks after each test. Call customer service at 866-756-7346 or go to **www.collegeboard.com.** Check the Registration Bulletin for a complete description of these services and dates when scores are available.

PART 2

REVIEW OF MAJOR TOPICS

Introduction to Chemistry

CHAPTER 1

- Distinguish types of matter: i.e., elements, mixtures, compounds, and distinctive substances.
- Identify chemical and physical properties and changes.
- Explain how energy is involved in these changes.
- Identify and use the SI units of measurements.
- Do mathematical calculations by using scientific notation, the factor-label method, and proper significant figures.

Anyone who studies chemistry must understand basic concepts and terms. When a chemist discusses matter, what does he mean? When another chemist writes a scientific paper about energy, what does she mean? What method do these two use in their research? How do they take measurements and make calculations? All of these topics must be understood when studying chemistry.

MATTER

Matter is defined as anything that occupies space and has mass. **Mass** is the quantity of matter which a substance possesses and, depending on the gravitational force acting on it, has a unit of weight assigned to it. Its formula is $w = mg$. Although the **weight** then can vary, the mass of the body is a constant and can be measured by its resistance to a change of position or motion. This property of mass to resist a change of position or motion is called **inertia.** Since matter does occupy space, we can compare the masses of various substances that occupy a particular unit volume. This relationship of mass to a unit volume is called the **density** of the substance. It can be shown in a mathematical formula as $D = \frac{m}{V}$. The basic unit of mass(m) in chemistry is the gram (g), and of volume (V) is the cubic centimeter (cm^3) or milliliter (mL).

TIP

Matter occupies space and has mass.

An example of how density varies can be shown by the difference in the volumes occupied by 1 gram of a metal, such as gold, and 1 gram of Styrofoam. Both have the same mass, 1 gram, but the volume occupied by the Styrofoam is much larger. Therefore, the density of the metal will be much larger than that of the Styrofoam. In chemistry, the standard units for density of gases are grams/liter at a standard temperature and pressure. This aspect of the density of gases is discussed in Chapter 6. Basically then, density can be defined as the mass per unit volume.

$$\text{Density} = \frac{\text{Mass}}{\text{Volume}}$$

States of Matter

Matter occurs in three states: solid, liquid, and gas. A **solid** has both a definite size and a definite shape. A **liquid** has a definite volume but takes the shape of the container, and a **gas** has neither a definite shape nor a definite volume. These states of matter can often be changed by the addition or subtraction of heat energy. An example is ice changing to liquid water and finally steam.

Composition of Matter

Matter can be subdivided into two general categories: distinct substances and mixtures. A **distinct substance** can be subdivided into the smallest particle that still has the properties of that substance. At that point, if the substance is made up of only one kind of atom, it is called an **element.** Atoms are considered to be the basic building blocks of matter that cannot be easily created nor destroyed. The word **atom** comes from the Greeks and means the smallest possible piece of something. Today, scientists recognize approximately 109 different kinds of atoms, each with its own unique composition. These atoms then are the building blocks of elements when only one kind of atom makes up the substance. If, however, two or more kinds of atoms join together in a definite grouping, this distinct substance is called a **compound.** Compounds are made by combining atoms of two or more elements in a definite proportion (or ratio) by mass, according to the **Law of Definite Composition (or Proportions)**. The smallest naturally occurring unit of a compound is called a **molecule** of that compound. A molecule of a compound has a definite shape that is determined by how the atoms are bonded to or combined with each other, as described in Chapter 3. An example is the compound water: it always occurs in a two hydrogen atoms to one oxygen atom relationship. **Mixtures**, however, can vary in their composition.

In general, then:

TIP

Know how to separate mixtures by using their properties.

Mixtures	Distinct Substances
1. Composition is indefinite (generally **heterogeneous**).* (Example: marble) 2. Properties of the constituents are retained. 3. Parts of the mixture react differently to changed conditions. *Solutions are mixtures, such as sugar in water, but since the substance, like sugar, is distributed evenly throughout the water, the mixture can be said to be homogeneous.	ELEMENTS 1. Composition is made up of one kind of atom. (Examples: nitrogen, gold, neon) 2. All parts are the same throughout **(homogeneous)**. COMPOUNDS 1. Composition is definite (homogeneous). (Examples: water, carbon dioxide) 2. All parts react the same. 3. Properties of the compound are distinct and different from the properties of the individual elements that are combined in its make-up.

The following chart shows a classification scheme for matter.

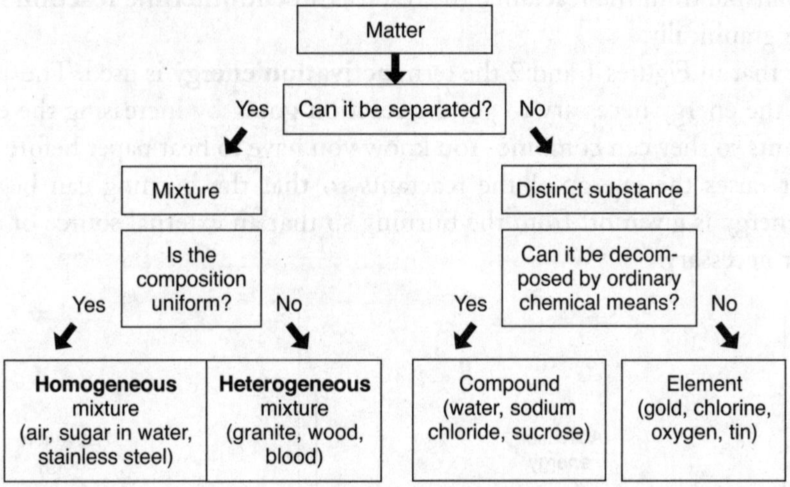

Chemical and Physical Properties

Physical properties of matter are those properties that can usually be observed with our senses. They include everything about a substance that can be noted when no change is occurring in the type of structure that makes up its smallest component. Some common examples are physical state, color, odor, solubility in water, density, melting point, taste, boiling point, and hardness.

Chemical properties are those properties that can be observed in regard to whether or not a substance reacts with other substances. Some common examples are: iron rusts in moist air, nitrogen does not burn, gold does not rust, sodium reacts with water, silver does not react with water, and water can be decomposed by an electric current.

Chemical and Physical Changes

TIP

Physical change does not alter the identity of the substance. Chemical change does.

The changes matter undergoes are classified as either physical or chemical. In general, a **physical change** alters some aspect of the physical properties of matter, but the composition remains constant. The most often altered properties are form and state. Some examples of physical changes are breaking glass, cutting wood, melting ice, and magnetizing a piece of metal. In some cases, the process that caused the change can be easily reversed and the substance regains its original form. Water changing its state is a good example of physical changes. In the solid state, ice, water has a definite size and shape. As heat is added, it changes to the liquid state, where it has a definite volume but takes the shape of the container. When water is heated above its boiling point, it changes to steam. Steam, a gas, has neither a definite size, because it fills the containing space, nor shape, because it takes the shape of the container.

Chemical changes are changes in the composition and structure of a substance. They are always accompanied by energy changes. If the energy released in the formation of a new structure exceeds the chemical energy in the original substances, energy will be given off, usually in the form of heat or light or both. This is called an **exothermic reaction**. If, however, the new structure needs to absorb more energy than is available from the reactants, the result is an **endothermic reaction**. This can be shown graphically.

Notice that in Figures 1 and 2 the term **activation energy** is used. The activation energy is the energy necessary to get the reaction going by increasing the energy of the reactants so they can combine. You know you have to heat paper before it burns. This heat raises the energy of the reactants so that the burning can begin; then enough energy is given off from the burning so that an external source of energy is no longer necessary.

TIP

Know how to read these graphic representations.

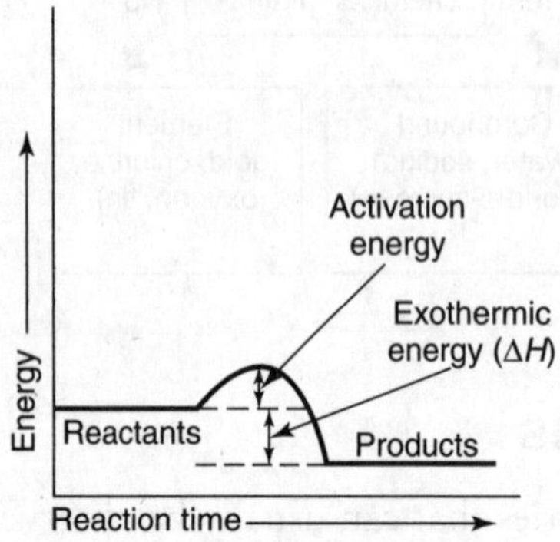

Figure 1. An Exothermic Reaction

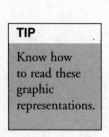

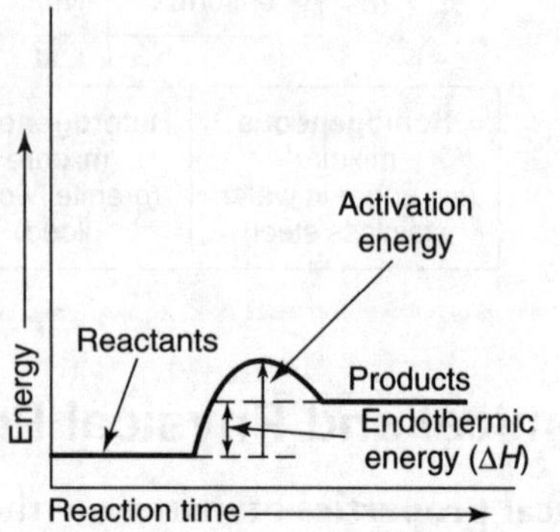

Figure 2. An Endothermic Reaction

Conservation of Mass

When ordinary chemical changes occur, the mass of the reactants equals the mass of the products. This can be stated another way: In a chemical change, matter can neither be created nor destroyed, but only changed from one form to another. This is referred to as the **Law of Conservation of Matter** (Lavoisier—1785). This law is extended by the Einstein mass-energy relationship, which states that matter and energy are interchangeable (see page 44).

ENERGY

Definition of Energy

The concept of energy plays an important role in all of the sciences. In chemistry, all physical and chemical changes have energy considerations associated with them. To understand how and why these changes happen, an understanding of energy is required.

Energy is defined as the capacity to do work. Work is done whenever a force is applied over a distance. Therefore, anything that can force matter to move, to change speed, or to change direction has energy. The following example will help you understand this definition of energy. When you charge a battery with electricity, you are storing energy in the form of chemical energy. The charged battery has a capacity to do work. If you use the battery to operate a toy car, the stored energy is transformed into mechanical energy that exerts a force on the mechanism that turns the wheels and makes the car move. This process continues until the **charge**

or stored energy is completely used. In its uncharged condition, the battery no longer has the capacity to do work.

Work itself is measured in **joules**, and so is energy. In some problems, however, energy may be expressed in **kilocalories**. The relationship between these two units is that 4.18×10^8 joules (J) equals 1 kilocalorie (kcal).

Forms of Energy

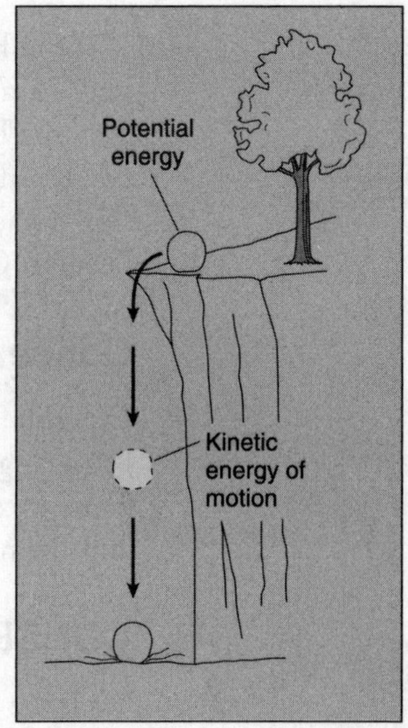

Energy may appear in a variety of forms. Most commonly, energy in reactions is evolved as **heat**. Some other forms of energy are **light**, **sound**, **mechanical energy**, **electrical energy**, and **chemical energy**. Energy can be converted from one form to another, as when the heat from burning fuel is used to vaporize water to steam. The energy of the steam is used to turn the wheels of a turbine to produce mechanical energy. The turbine turns the generator armature to produce electricity, which is then available in homes for use as light or heat, or in the operation of many modern appliances.

Two general classifications of energy are **potential energy** and **kinetic energy**. Potential energy is due to position; kinetic energy is energy of motion. The difference can be illustrated by a boulder sitting on the side of a mountain. It has a high potential energy due to its position above the valley floor. If it falls, however, its potential energy is converted to kinetic energy. This illustration is very similar to the situation of electrons cascading to lower energy levels in the atomic model described in Chapter 2.

This concept can also be applied at the molecular level. For example, the molecule of N_2 in the diagram below shows three variations of kinetic energy. The N_2 molecule also possesses potential energy because of the chemical bond within the molecule. To break this bond would require energy. The potential energy in each isolated atom of nitrogen would then be greater. This would be similar to raising the boulder to a higher position on the mountainside. Both systems have higher potential energy (than their prior state) and have a tendency to fall back to the lower state, which is more stable.

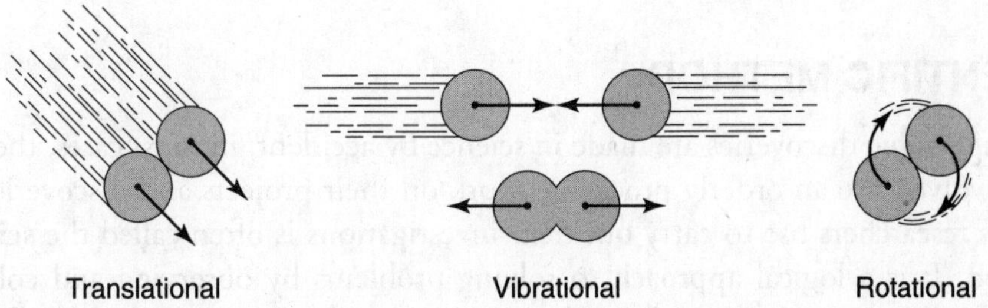

Types of Reactions (Exothermic Versus Endothermic)

When physical or chemical changes occur, energy changes are involved. Change of heat content can be designated as ΔH. The heat content (H) is sometimes referred to as the **enthalpy**. Every system has a certain amount of heat. This changes during the course of a physical or chemical change. The change in heat content, ΔH, is the difference between the heat content of the products and that of the reactants. The equation is:

$$\Delta H = H_{\text{products}} - H_{\text{reactants}}$$

If the heat content of the products is greater than the heat content of the reactants, ΔH is a positive quantity ($\Delta H > 0$) and the reaction is **endothermic**. If, however, the heat content of the products is less than the heat content of the reactants, ΔH is a negative quantity ($\Delta H < 0$) and the reaction is **exothermic**. This relationship is shown graphically in Figures 1 and 2 on page 42. This topic is developed in detail in Chapter 8.

Conservation of Energy

Experiments have shown that energy is neither gained nor lost in physical or chemical changes. This principle is known as the **Law of Conservation of Energy** and is often stated as follows: Energy is neither created nor destroyed in ordinary physical and chemical changes.

CONSERVATION OF MASS AND ENERGY

With the introduction of atomic theory and a more complete understanding of the nature of both mass and energy, it was found that a relationship exists between these two concepts. Einstein formulated the **Law of Conservation of Mass and Energy**. This states that mass and energy are interchangeable under special conditions. The conditions have been created in nuclear reactors and accelerators, and the law has been verified. This relationship can be expressed by Einstein's famous equation:

$$E = mc^2$$
$$\text{Energy} = \text{Mass} \times (\text{Velocity of light})^2$$

SCIENTIFIC METHOD

Although some discoveries are made in science by accident, in most cases, the scientists involved use an orderly process to work on their projects and discoveries. The process researchers use to carry out their investigations is often called the **scientific method**. It is a logical approach to solving problems by observing and collecting data, formulating a hypothesis, and constructing theories supported by the data. The formulating of a hypothesis consists of carefully studying the data collected and organized to see if a testable statement can be made with regard to the data. The hypothesis takes the form of an "if . . . then" statement. If certain data are true, then a prediction can be made concerning the outcome. The next step is to test the prediction to see if it withstands the experimentation. The hypothesis can go through several revisions as the process continues. If the data from experimentation show that the predictions of the hypothesis are successful, scientists usually try to explain the phenomena by constructing a **model**. The model can be a visual, verbal, or mathematical means of explaining how the data is related to the phenomena.

The stages of this process can be illustrated by the diagram shown below:

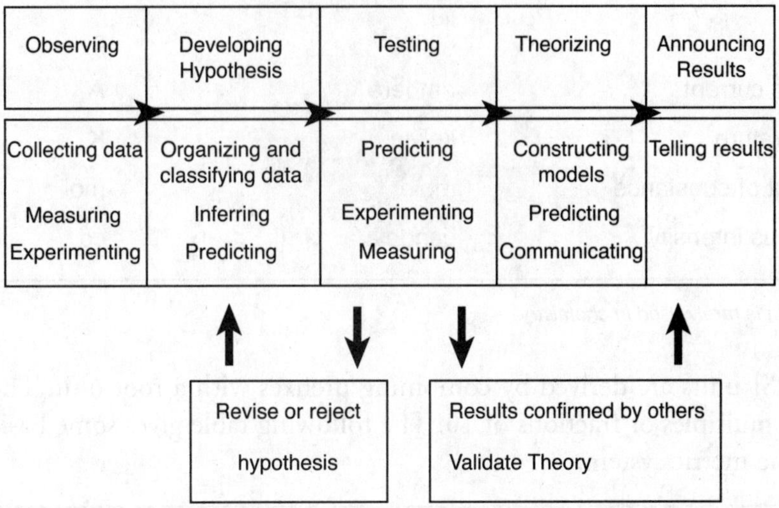

The student of chemistry must be able to use the correct measurement terms accurately and to solve problems that require mathematical skill, as well as proper terminology, for their correct solution. The following sections review these topics.

MEASUREMENTS AND CALCULATIONS

The student of chemistry must be able to use the correct measurement terms accurately and to solve problems that require mathematical skill, as well as proper terminology, for their correct solution. The following sections review these topics.

Metric System

It is important that scientists around the world use the same units when communicating information. For this reason, scientists use the modernized metric system, designated in 1960 by the General Conference on Weights and Measures as the International System of Units. This is commonly known as **SI**, an abbreviation for the French name Le Système International d'Unités. It is now the most common system of measurement in the world.

TIP

Only metric units are used on the SAT test.

The reason SI is so widely accepted is twofold. First, it uses the decimal system as its base. Second, many units for various quantities are defined in terms of units for simpler quantities.

There are seven basic units that can be used to express the fundamental properties of measurement. These are called the SI base units and are shown in the table that follows.

SI Base Units

Property	Unit	Abbreviation
mass	kilogram	kg
length	meter	m
time	second	s
electric current	ampere	A
temperature	kelvin	K
amount of substance	mole	mol
luminous intensity	candela*	cd

The candela is rarely used in chemistry.

Other SI units are derived by combining prefixes with a root unit. The prefixes represent multiples or fractions of 10. The following table gives some basic prefixes used in the metric system.

Prefixes Used with SI Units

Prefix	Symbol	Meaning	Scientific Notation
exa-	E	1,000,000,000,000,000,000	10^{18}
peta-	P	1,000,000,000,000,000	10^{15}
tera-	T	1,000,000,000,000	10^{12}
giga-	G	1,000,000,000	10^{9}
mega-	M	1,000,000	10^{6}
kilo-	k	1,000	10^{3}
hecto-	h	100	10^{2}
deka-	da	10	10^{1}
—	—	1	10^{0}
deci-	d	0.1	10^{-1}
centi-	c	0.01	10^{-2}
milli-	m	0.001	10^{-3}
micro-	µ	0.000 001	10^{-6}
nano-	n	0.000 000 001	10^{-9}
pico-	p	0.000 000 000 001	10^{-12}
femto-	f	0.000 000 000 000 001	10^{-15}
atto-	a	0.000 000 000 000 000 001	10^{-18}

For an example of how a prefix works in conjunction with the root word, consider the term *kilometer.* The prefix *kilo-* means "multiply the root word by 1,000," so a kilometer is 1,000 meters. By the same reasoning, a millimeter is 1/1,000 meter.

Because of the prefix system, all units and quantities can be easily related by some factor of 10. Here is a brief table of some metric unit equivalents.

Length
10 millimeters (mm) = 1 centimeter (cm)
100 cm = 1 meter (m)
1,000 m = 1 kilometer (km)
Volume
1,000 milliliters (mL) = 1 liter (L)
1,000 cubic centimeters (cm^3) = 1 liter
1 mL = 1 cm^3
Mass
1,000 milligrams (mg) = 1 gram (g)
1,000 g = 1 kilogram (kg)

A unit of length, used especially in expressing the length of light waves, is the nanometer, abbreviated as nm and equal to 10^{-9} meter.

Because in the United States measurements are occasionally reported in units of the English system, it is important to be aware of some metric to English system equivalents. Some common conversion factors are shown in the following table.

2.54 centimeters = 1 inch
1 meter = 39.37 inches (10% longer than 1 yard)
28.35 grams = 1 ounce
454 grams = 1 pound
1 kilogram = 2.2 pounds
0.946 liter = 1 quart
1 liter (5% larger than 1 quart) = 1.06 quarts

TIP

For your information only; metric is used on the test.

The metric system standards were chosen as natural standards. The meter was first described as the distance marked off on a platinum-iridium bar but now is defined as the length of the path traveled by light in a vacuum during a time interval of $1/2.99792458 \times 10^8$ second.

There are some interesting relationships between volume and mass units in the metric system. Because water is most dense at 4°C, the gram was intended to be 1 cubic centimeter of water at this temperature. This means, then, that:

1,000 cm^3 = 1 L of water @ 4°C

1,000 cm^3 of water weighs 1,000 g @ 4°C.

Therefore

1 L of water @ 4°C weighs 1 kg,

and

1 mL of water @ 4°C weighs 1 g.

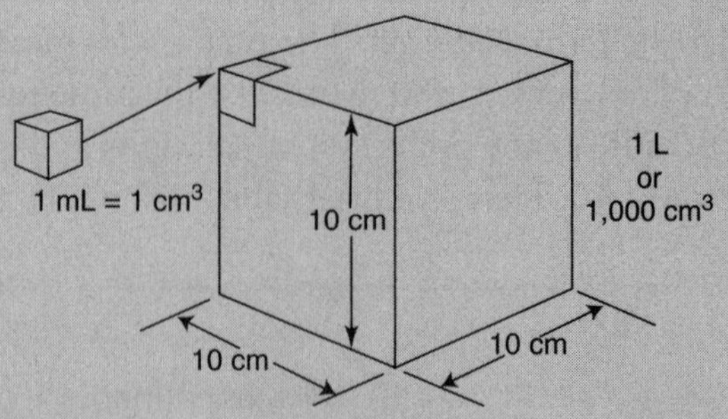

When 1 L is filled with water @ 4°C, it has a mass of 1 kg.

Temperature Measurements

The most commonly used temperature scale in scientific work is the **Celsius** scale. It gets its name from the Swedish astronomer Anders Celsius and dates back to 1742. For a long time it was called the centigrade scale because it is based on the concept of dividing the distance on a thermometer between the freezing point of water and its boiling point into 100 equal markings or degrees.

There is another scale that is based on the lowest theoretical temperature (called absolute zero). This temperature has never actually been reached, but scientists in laboratories have reached temperatures within about a thousandth of a degree above absolute zero. Lord William Kelvin proposed this scale, on which a degree is the same size as a Celsius degree and which is referred to as the **Kelvin** scale. Through experiments and calculations, it has been determined that absolute zero is 273.15 degrees below zero on the Celsius scale. This figure is usually rounded off to −273°C.

The diagram and conversion formulas that follow give the graphic and algebraic relationships among three temperature scales: the Celsius and Kelvin, commonly used in chemistry, and the Fahrenheit.

TIP

Kelvin units are used on the SAT test.

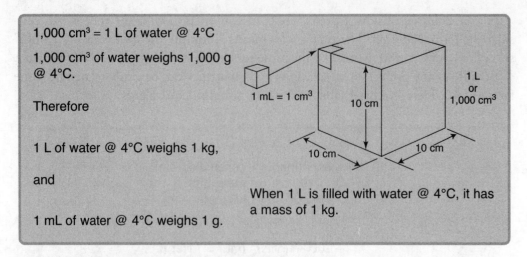

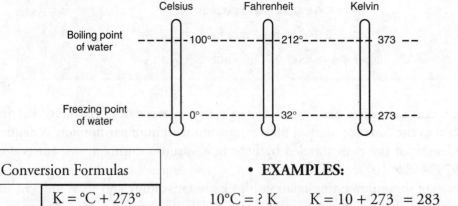

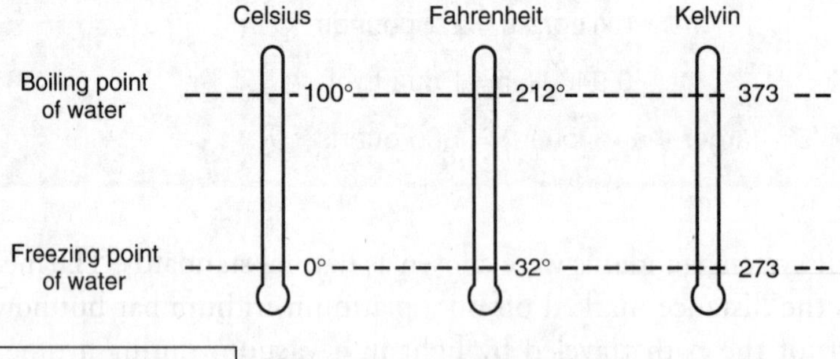

Conversion Formulas

K = °C + 273°
°C = K − 273°

• **EXAMPLES:**

10°C = ? K   K = 10 + 273 = 283
200 K = ?°C   °C = 200 − 273° = −73°

Note: In Kelvin notation, the degree sign is omitted: 283 K. The unit is the kelvin, abbreviated as K.

Heat Measurements

The scales above are used to measure the degree of heat. A pail and a thimble can both be filled with water at 100° Celsius. The water in both measures the same degree of heat. However, the pail of water has a greater quantity of heat. This could be easily demonstrated by the amount of ice that could be melted by the water in these two containers. Obviously, the pail of water at 100° Celsius will melt more ice than will a thimble full of water at the same temperature. Therefore, the pail of water contains a greater number of **calories** of heat. The calorie unit is used to measure the quantity of heat. It is defined as the amount of heat needed to raise the temperature of 1 gram of water by 1 degree on the Celsius scale. This is a rather small unit to measure the quantities of heat involved in most chemical reactions. Therefore, the **kilocalorie** is more often used. The kilocalorie equals 1,000 calories. It is the quantity of heat that will increase the temperature of 1 kilogram of water by 1 degree on the Celsius scale. Although the calorie is commonly used in everyday usage with regard to food, the SI unit for heat energy is the **joule**. It is abbreviated as J and, because it is a rather small unit, it is commonly given in kilojoules (kJ). The relationship between the calorie and the joule is that 1 calorie equals 4.18 joules.

TIP

Because the joule is rather small, kJ is used most often.

Problems involving heat transfers in water are called water calorimetry problems and are explained on page 186.

Scientific Notation

When students must do mathematical operations with numerical figures, the **scientific notation system** is very useful. Basically this system uses the exponential means of expressing figures. With large numbers, such as 3,630,000., move the decimal point to the left until only one digit remains to the left (3.630000) and then indicate the number of moves of the decimal point as the exponent of 10 (3.63×10^6). With a very small number such as 0.000000123, move the decimal point to the right until only one digit is to the left (0000001.23) and then express the number of moves as the negative exponent of 10 (1.23×10^{-7}).

TIP

Scientific notation is based on exponents of 10.

With numbers expressed in this exponential form, you can now use your knowledge of exponents in mathematical operations. An important fact to remember is that in multiplication you add the exponents of 10, and in division you subtract the exponents. Addition and subtraction of two numbers expressed in scientific notation can be performed only if the numbers have the same exponent.

EXAMPLES:

Multiplications:
$(2.3 \times 10^5)(5.0 \times 10^{-12})$. Multiplying the first numbers, you get 11.5, and addition of the exponents gives 10^{-7}. Now, changing to a number with only one digit to the left of the decimal point gives you 1.15×10^{-6} for the answer.

Try these:
$(5.1 \times 10^{-6})(2 \times 10^{-3}) = 10.2 \times 10^{-9} = 1.02 \times 10^{-8}$
$(3 \times 10^5)(6 \times 10^3) = 18 \times 10^8 = 1.8 \times 10^9$

Divisions:
$(1.5 \times 10^3) \div (5.0 \times 10^{-2}) = 0.3 \times 10^5 = 3 \times 10^4$
$(2.1 \times 10^{-2}) \div (7.0 \times 10^{-3}) = 0.3 \times 10^1 = 3$
(Notice that in division the exponents of 10 are subtracted.)

Addition and subtraction:
$(4.2 \times 10^4 \text{ kg}) + (7.9 \times 10^3 \text{ kg}) =$
$(4.2 \times 10^4 \text{ kg}) + (0.79 \times 10^4 \text{ kg})$ (note that the exponents of 10 are now the same) $= 4.99 \times 10^4$ kg
This can be rounded to 5.0×10^4 kg.

$(6.02 \times 10^{-3}) - (2.41 \times 10^{-4}) = (6.02 \times 10^{-3}) - (.241 \times 10^{-3})$ (note that the exponents of 10 are now the same) $= 5.779 \times 10^{-3}$ or 5.8×10^{-3} when rounded to two significant figures.

TIP

Cancel out all units except the one for the answer.

Factor-Label Method of Conversion (Dimensional Analysis)

When you are working problems that involve numbers with units of measurement, it is convenient to use this method so that you do not become confused in the operations of multiplication or division. For example, if you are changing 0.001 kilogram to milligrams, you set up each conversion as a fraction so that all the units will factor out except the one you want in the answer.

$$1 \times 10^{-3} \cancel{\text{kg}} \times \frac{1 \times 10^3 \cancel{\text{g}}}{1 \cancel{\text{kg}}} \times \frac{1 \times 10^3 \text{ mg}}{1 \cancel{\text{g}}} = 1 \times 10^3 \text{ mg}$$

Notice that the kilogram is made the denominator in the first fraction to be factored with the original kilogram unit. The numerator is equal to the denominator except that the numerator is expressed in smaller units. The second fraction has the gram unit in the denominator to be factored with the gram unit in the preceding fraction. The answer is in milligrams because this is the only unit remaining and it assures you that the correct operations have been performed in the conversion.

ANOTHER EXAMPLE:

1 ft = ? cm

$$1 \cancel{\text{ft}} \times \frac{12 \cancel{\text{in.}}}{1 \cancel{\text{ft}}} \times \frac{2.54 \text{ cm}}{1 \cancel{\text{in.}}} = 30.48 \text{ cm}$$

The factor-label method is used in examples throughout this book.

Precision, Accuracy, and Uncertainty

Two other factors to consider in measurement are **precision** and **accuracy**. Precision indicates the reliability or reproducibility of a measurement. Accuracy indicates how close a measurement is to its known or accepted value.

> **TIP**
>
> Accuracy is how close you have come to the true value.

> **TIP**
>
> Precision is how repeatable the results are.

For example, suppose you were taking a reading of the boiling point of pure water at sea level. Using the same thermometer in three trials, you record 96.8, 96.9, and 97.0 degrees Celsius. Since these figures show a high reproducibility, you can say that they are precise. However, the values are considerably off from the accepted value of 100 degrees Celsius, so we say they are not accurate. In this example we probably would suspect that the inaccuracy was the fault of the thermometer.

Regardless of precision and accuracy, all measurements have a degree of **uncertainty**. This is usually dependent on one or both of two factors—the limitation of the measuring instrument and the skill of the person making the measurement. Uncertainty can best be shown by example.

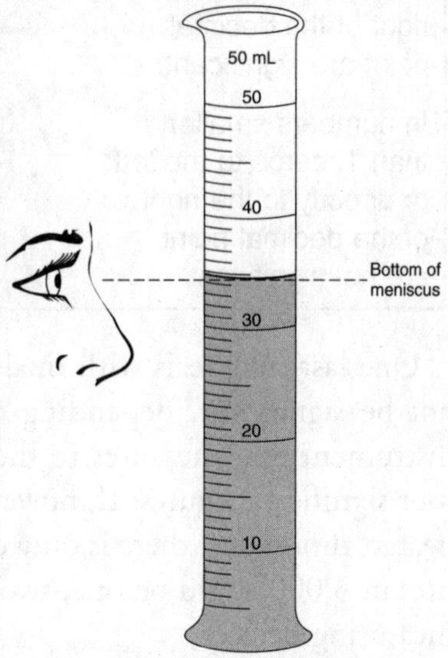

Graduated Cylinder Reading 34.3 mL

The graduated cylinder in the illustration contains a quantity of water to be measured. It is obvious that the quantity is between 30 and 40 milliliters because the meniscus lies between these two marked quantities. Now, checking to see where the bottom of the **meniscus** lies with reference to the ten intervening subdivisions, we see that it is between the fourth and fifth. This means that the volume lies between 34 and 35 milliliters. The next step introduces the uncertainty. We have to guess how far the reading is between these two markings. We can make an approximate guess, or estimate, that the level is more than 0.2 but less than 0.4 of the distance. We therefore report the volume as 34.3 milliliters. The last digit in any measurement is an estimate of this kind and is uncertain.

Significant Figures

Any time a measurement is recorded, it includes all the digits that are certain plus one uncertain digit. These certain digits plus the one uncertain digit are referred to as **significant figures**. The more digits you are able to record in a measurement, the less relative uncertainty there is in the measurement. The following table summarizes the rules of significant figures.

Rule	Example	Number of Significant Figures
All digits other than zeros are significant.	25 g 5.471 g	2 4
Zeros between nonzero digits are significant.	309 g 40.06 g	3 4
Final zeros to the right of the decimal point are significant.	6.00 mL 2.350 mL	3 4
In numbers smaller than 1, zeros to the left or directly to the right of the decimal point are not significant.	0.05 cm 0.060 cm	1 The zeros merely mark the position of the decimal point. 2 The first two zeros mark the position of the decimal point. The final zero is significant.

One last rule deals with final zeros in a whole number. These zeros may or may not be significant, depending on the measuring instrument. For instance, if an instrument that measures to the nearest mile is used, the number 3,000 miles has four significant figures. If, however, the instrument in question records miles to the nearest thousands, there is only one significant figure. The number of significant figures in 3,000 could be one, two, three, or four, depending on the limitation of the measuring device.

This problem can be avoided by using the system of scientific notation. For this example, the following notations would indicate the numbers of significant figures:

3×10^3	one significant figure
3.0×10^3	two significant figures
3.00×10^3	three significant figures
3.000×10^3	four significant figures

Calculations with Significant Figures

RULE:

Your answer cannot have more significant figures than the quantity having the fewest number of significant figures.

When you do calculations involving numbers that do not have the same number of significant figures in each, keep the following two rules in mind.

First, in multiplication and division, the number of significant figures in a product or a quotient of measured quantities is the same as the number of significant figures in the quantity having the smaller number of significant figures.

EXAMPLE 1

Problem	Unrounded answer	Answer rounded to the correct number of significant figures
4.29 cm × 3.24 cm =	13.8996 cm^2 =	13.9 cm^2

Explanation: Both measured quantities have three significant figures. Therefore, the answer should be rounded to three significant figures.

EXAMPLE 2

Problem	Unrounded answer	Answer rounded to the correct number of significant figures
4.29 cm × 3.2 cm =	13.728 cm^2 =	14 cm^2

Explanation: One of the measured quantities has only two significant figures. Therefore, the answer should be rounded to two significant figures.

EXAMPLE 3

Problem	Unrounded answer	Answer rounded to the correct number of significant figures
8.47 cm^2/4.26 cm =	1.9882629 cm =	1.99 cm

Explanation: Both measured quantities have three significant figures. Therefore, the answer should be rounded to three significant figures.

Second, when adding or subtracting measured quantities, the sum or difference should be rounded to the same number of decimal places as the quantity having the least number of decimal places.

EXAMPLE 1

Problem	Unrounded answer	Answer rounded to the correct number of significant figures
3.56 cm 2.6 cm + 6.12 cm Total =	12.28 cm =	12.3 cm

Explanation: One of the quantities added has only one decimal place. Therefore, the answer should be rounded to only one decimal place.

EXAMPLE 2

Problem	Unrounded answer	Answer rounded to the correct number of significant figures
3.514 cm −2.13 cm Difference =	1.384 cm =	1.38 cm

Explanation: One of the quantities has only two decimal places. Therefore, the answer should be rounded to only two decimal places.

Chapter Summary

The following terms summarize all the concepts and ideas that were introduced in this chapter. You should be able to explain their meaning and how you would use them in chemistry. They appear in boldface type in this chapter to draw your attention to them. The boldface type also makes it easier for you to look them up if you need to. You could also use the search or "Google" action on your computer to get a quick and expanded explanation of these terms, laws, and formulas.

accuracy
activation energy
calorie
Celsius
chemical change
chemical property
compound
density
element
endothermic
exothermic
gas
heterogeneous
homogeneous
inertia
joule
Kelvin
kilocalorie
kilojoule
kinetic energy
liquid
mass
meniscus
matter
mixture
physical change
physical property
potential energy
precision
scientific notation
significant figures
SI units
solid
uncertainty

Law of Conservation of Energy
Law of Conservation of Matter
Law of Conservation of Mass and Energy
Law of Definite Composition or Proportion

Practice Exercises

1. 1.2 mg = __________ g
2. 6.3 cm = __________ mm
3. 5.12 m = __________ cm
4. 32°C = __________ K
5. 6.111 mL = __________ L
6. 1 km = __________ mm
7. 1.03 kg = __________ g
8. 0.003 g = __________ kg
9. 22.4 L = __________ mL
10. 10,013 cm= __________ km
11. The density of CCl4 (carbon tetrachloride) is 1.58 grams/milliliter. What will 100. milliliters of CCl4 weigh?
12. A piece of sulfur weighs 227 grams. When it was submerged in a graduated cylinder containing 50. milliliters of H_2O, the level rose to 150. milliliters. What is the density (g/mL) of the sulfur?
13. (a) A box 20.0 centimeters × 20.0 centimeters × 5.08 inches has what volume in cubic centimeters?

    (b) What weight, in grams, of H_2O @ 4°C will the box hold?

14. Set up the following using the *factor-label method:*

$$\frac{5 \text{ cm}}{\text{s}} = \frac{\text{km}}{\text{h}}$$

15. How many significant figures are in each of the following?

(a) 1.01
(b) 200.0
(c) 0.0021
(d) 0.0230

16. A baking powder can carries the statement, "Ingredients: corn starch, sodium bicarbonate, calcium acid phosphate, and sodium aluminum sulfate." Therefore, this baking powder is

(A) a compound
(B) a mixture
(C) a molecule
(D) a mixture of elements

17. Which of the following is a physical property of sugar?

(A) It decomposes readily.
(B) Its composition is carbon, hydrogen, and oxygen.
(C) It turns black with concentrated H_2SO_4.
(D) It can be decomposed with heat.
(E) It is a white crystalline solid.

18. A substance that can be further simplified using ordinary means may be either

(A) an element or a compound
(B) an element or a mixture
(C) a mixture or a compound
(D) a mixture or an atom

19. Chemical action may involve all of the following EXCEPT

(A) combining of atoms of elements to form a molecule
(B) separation of the molecules in a mixture
(C) breaking down compounds into elements
(D) reacting a compound and an element to form a new compound and a new element

20. The energy of a system can be

(A) easily changed to mass
(B) transformed into a different form
(C) measured only as potential energy
(D) measured only as kinetic energy

21. If the ΔH of a reaction is a negative quantity, the reaction is definitely

(A) endothermic
(B) unstable
(C) exothermic
(D) reversible

22. Write E for each element, C for each compound, and M for each mixture in the following list:

Water	Aluminum oxide	Hydrochloric acid
Wine	Hydrogen	Nitrogen
Soil	Carbon dioxide	Tin
Silver	Air	Potassium chloride

The following questions are in the format that is used on the **SAT Subject Test in Chemistry**. If you are not familiar with these types of questions, study pages xii–xv before doing the remainder of the review questions.

23. If the graphic representation of the energy levels of the reactants and products in a chemical reaction looks like this:

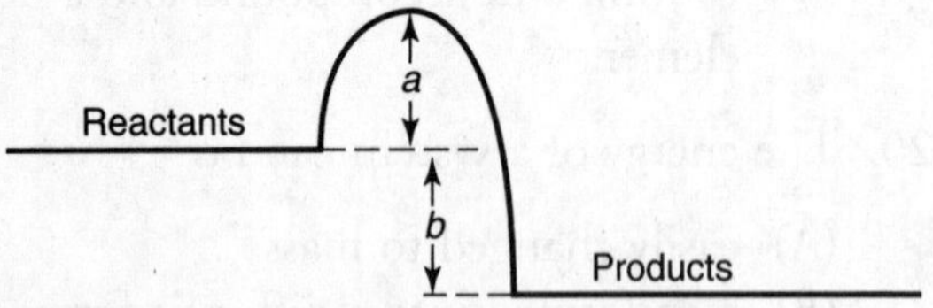

Which of the following statements are true?

I. The activation energy for the forward reaction is represented by the "a" portion.
II. The activation energy for the forward reaction is represented by the "b" portion of the graph.
III. The "a" portion is the energy given off in the forward reaction.

(A) I only
(B) II only
(C) I and III only
(D) II and III only
(E) I, II, and III

24. A substance that can be further simplified by ordinary chemical means may be which of the following?

I. An element or a compound
II. A mixture or a compound
III. An element of a mixture

(A) I only
(B) II only
(C) III only
(D) I and II only
(E) I, II, and III

<u>Questions 25–29</u> refer to the following terms.

(A) Density
(B) A solid
(C) Volume
(D) Weight
(E) Matter

25. Gives the mass per unit volume
26. Has mass and a definite size and shape
27. Gives the space occupied
28. Has mass and occupies space
29. Defined as a measure of the mass times the gravitational force

Directions:

(See the explanation for this type of question and the examples on pages xiv–xv before attempting questions 30–32.)

Every question below contains two statements, I in the left-hand column and II in the right-hand column. For each question, decide if statement I is true or false and whether statement II is true or false, and fill in the corresponding T or F ovals in the answer spaces. *Fill in oval CE only if statement II is a correct explanation of statement I.

	I		II
30.	A substance composed of two or more elements chemically combined is called a mixture	BECAUSE	the properties of the constituents of a mixture are retained.
31.	A chemical change involves change in the composition and molecular structure of the reactants	BECAUSE	in a chemical reaction bonds are broken and new substances and new bonds are formed.
32.	The burning of paper is a physical change	BECAUSE	when a chemical change occurs energy is either gained or lost by the reactants.

*Fill in oval CE only if II is a correct explanation of I.

	I	II	CE*
30.	(T) (F)	(T) (F)	( )
31.	(T) (F)	(T) (F)	( )
32.	(T) (F)	(T) (F)	( )

ANSWERS AND EXPLANATIONS

1. 0.0012 g

$$1.2\ \cancel{\text{mg}} \times \frac{7\ \text{g}}{1{,}000\ \cancel{\text{mg}}} = 0.0012\ \text{g}$$

2. 63 mm

$$6.3\ \cancel{\text{cm}} \times \frac{10\ \text{mm}}{1\ \cancel{\text{cm}}} = 63\ \text{mm}$$

3. 512 cm

$$5.12\ \cancel{m} \times \frac{100\text{ cm}}{1\ \cancel{m}} = 512\text{ cm}$$

4. 305 K
$32°C + 273 = 305$ K

5. 0.006111 L

$$6.111\ \cancel{mL} \times \frac{1\text{ L}}{1{,}000\ \cancel{mL}} = .006111\text{ L}$$

6. 1,000,000 or 1×10^6 mm

$$1\ \cancel{km} \times \frac{1{,}000\ \cancel{m}}{1\ \cancel{km}} \times \frac{1{,}000\text{ mm}}{1\ \cancel{m}} = 1{,}000{,}000 \text{ or } 1 \times 10^6\text{ mm}$$

7. 1.03×10^3 kg

$$1.03\ \cancel{kg} \times \frac{1{,}000\text{ g}}{\cancel{kg}} = 1030\text{ g or } 1.03 \times 10^3\text{ g}$$

8. .000003 kg or 3×10^{-6} kg

$$.003\ \cancel{g} \times \frac{1\text{ kg}}{1{,}000\ \cancel{g}} = .000003\text{ kg or } 3 \times 10^{-6}\text{ kg}$$

9. 22,400 or 2.24×10^4 mL

$$22.4\ \cancel{L} \times \frac{1{,}000\text{ mL}}{1\cancel{L}} = 22{,}400 \text{ or } 2.24 \times 10^4\text{ mL}$$

10. 1.0013 km

$$10{,}013\ \cancel{cm} \times \frac{1\ \cancel{m}}{100\ \cancel{cm}} \times \frac{1\text{ km}}{100\ \cancel{m}} = 1.0013\text{ km}$$

11. 158 g Because the density is 1.58 g/mL and you want the weight of 100 mL, you use the formula that density × volume = weight.

Inserting the values, gives

$$1.58\text{ g}/\cancel{mL} \times 100\ \cancel{mL} = 158\text{g}$$

12. 2.27 g/mL

To find the volume of 227 g of sulfur, subtract the volume of water before from the volume after immersion.

$$150.\ \text{mL} - 50.\ \text{mL} = 100.\ \text{mL}$$

Then

$$\frac{227\ \text{g}}{100.\ \text{mL}} = \underline{\underline{2.27 \text{g/mol}}}$$

13. (a) 5,160 cm^3 (b) 5,160 g

(a) Convert 5.08 in. to centimeters:

$$5.08\ \cancel{\text{in.}} \times \frac{2.54\ \text{cm}}{1\ \cancel{\text{in.}}} = 12.9\ \text{cm}$$

Then 20.0 cm × 20.0 cm × 12.9 cm = <u>5,160</u> cm^3

(b) Since 1 cm^3 of water <u>at 4°C</u> has a mass of 1 g, then 5,160 cm^3 = <u>5,160 g</u>

14. $$\frac{5\ \cancel{\text{cm}}}{\cancel{\text{s}}} \times \frac{1\ \cancel{\text{m}}}{100\ \cancel{\text{cm}}} \times \frac{1\ \text{km}}{1{,}000\ \cancel{\text{m}}} \times \frac{60\ \cancel{\text{s}}}{1\ \cancel{\text{min}}} \times \frac{60\ \cancel{\text{min}}}{1\ \text{h}} =$$

15. (a) 3 (b) 4 (c) 2 (d) 3

16. **(B)** Because all the substances are compounds.

17. **(E)** Because all the other statements refer to chemical poperties of sugar.

18. **(C)** Neither an element nor an atom can be simplified into anything else by ordinary means.

19. **(B)** Separating molecules in a mixture (e.g., evaporating water from a sugar solution) does not involve a chemical change action.

20. **(B)** Because the other choices are not true. Transforming energy into different forms can be demonstrated by batteries in a flashlight, changing chemical energy into electrical energy and then into light energy.

21. **(C)** Anytime the ΔH of a reaction is negative, the reaction gives off energy and, it is classified as exothermic.

22.

C	C	C
M	E	E
M	C	E
E	M	C

23. **(A)** In the diagram part of question 23

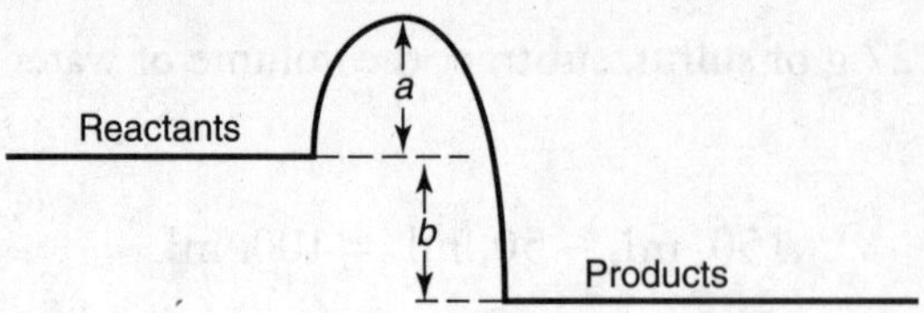

"a" is the activation energy for the forward reaction. The sum of "a" and "b" is the activation energy for the reverse reaction, and "b" alone is the energy given off by the forward reaction.

24. **(B)** Because an element is a basic building block in the periodic table, it cannot be further simplified by ordinary chemical means. A compound and a mixture can be simplified by chemical and physical means, respectively.

25. **(A)** The mass per unit volume is the definition of density.

26. **(B)** Matter that has mass and a definite size and shape is called a solid.

27. **(C)** The space occupied by a substance is called its volume.

28. **(E)** That a substance occupies space and mass is the basic definition of all matter.

29. **(D)** Weight is the product of mass times a gravitational force. In outer space where there is no gravitational force, we say a mass is weightless.

30. **(F, T)** The column I statement is false, because a mixture is elements and/or compounds in physical contact (without bonds) with one another.

31. **(T, T, C E)** Chemical reactions involve the changing of molecular structures by the breaking of bonds and the formation of new bonds.

32. **(F, T)** Burning paper is a chemical reaction not a physical one. The statement in column II is true, because a chemical change always involves an energy change.

Atomic Structure and the Periodic Table of the Elements

CHAPTER 2

- Describe the history of the development of atomic theory.
- Explain the structure of atoms, their main energy levels, sublevels, orbital configuration, and the rules that govern how they are filled.
- Place atoms in groups and periods based on their atomic structure.
- Explain how chemical and physical properties are related to positions in the Periodic Table.

The idea of small, invisible particles being the building blocks of matter can be traced back more than 2,000 years to the Greek philosophers Democritus and Leucippus. These particles, considered to be so small and indestructible that they could not be divided into smaller particles, were called **atoms**, the Greek word for indivisible. The English word **atom** comes from this Greek word. This early concept of atoms was not based upon experimental evidence but was simply a result of thinking and reasoning on the part of the philosophers. It was not until the eighteenth century that experimental evidence in favor of the atomic hypothesis began to accumulate. Finally, around 1805, John Dalton proposed some basic assumptions about atoms based on what was known through scientific experimentation and observation at that time. These assumptions are very closely related to what scientists presently know about atoms. For this reason, Dalton is often referred to as the father of modern atomic theory. Some of these basic ideas were:

1. All matter is made up of very small, discrete particles called atoms.
2. All atoms of an element are alike in weight, and this weight is different from that of any other kind of atom.
3. Atoms cannot be subdivided, created, or destroyed.
4. Atoms of different elements combine in simple whole-number ratios to form chemical compounds.
5. In chemical reactions, atoms are combined, separated, or rearranged.

TIP

Know Dalton's five basic ideas about atoms.

> **TIP**
> Know Niels Bohr's model based on a planetary model as opposed to quantum theory based on a probability model.

By the second half of the 1800s, many scientists believed that all the major discoveries related to the elements had been made. The only thing left for young scientists to do was to refine what was already known. This came to a suprising halt when J. J. Thomson discovered the electron beam in a cathode ray tube in 1897. Soon afterward, Henri Becquerel announced his work with radioactivity, and Marie Curie and her husband, Pierre, set about trying to isolate the source of radioactivity in their laboratory in France.

During the late nineteenth and early twentieth centuries, more and more physicists turned their attention to the structure of the atom. In 1913 the Danish physicist Niels Bohr published a theory explaining the line spectrum of hydrogen. He proposed a planetary model that quantized the energy of electrons to specific orbits. The work of Louis de Broglia and others in the 1920s and 1930s showed that quantum theory had a more probabilistic model of where the electrons could be found that resulted in the theory of orbitals. Enough was learned about nuclear structure to make practical use of atomic nuclei, as in nuclear power generators that use fission reactions and in fission and fusion bombs in the 1940s and 1950s. Today the search still goes on to study the particle physics of the atom and to attempt to control nuclear fusion reactions as a source of energy.

ELECTRIC NATURE OF ATOMS

From around the beginning of the twentieth century, scientists have been gathering evidence about the structure of atoms and fitting the information into a model of the atomic structure.

Basic Electric Charges

> **TIP**
> The electron was discovered by J. J. Thomson.

The discovery of the electron as the first subatomic particle is credited to **J. J. Thomson** (England, 1897). He used an evacuated tube connected to a spark coil as shown in Figure 3. As the voltage across the tube was increased, a beam of light became visible. This was referred to as a cathode ray. Thomson found that the beam was deflected by both electrical and magnetic fields. Therefore, he concluded that cathode rays are made up of very small, negatively charged particles, which he named **electrons**.

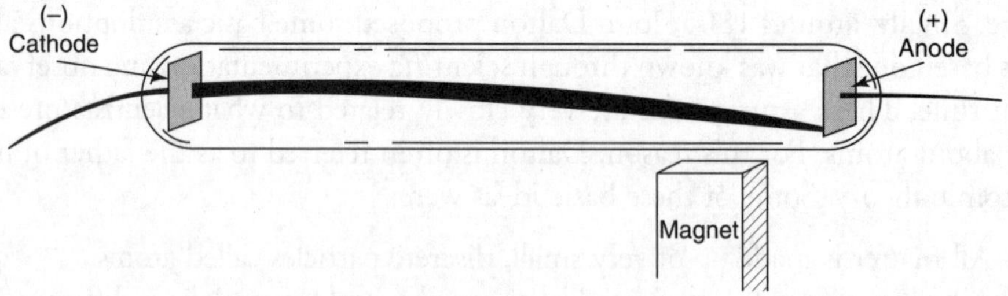

Figure 3. Electron Equipment

Further experimentation led Thomson to find the ratio of the electrical charge of the electron to its mass. This was a major step toward understanding the nature of the particle. He was awarded a Nobel Prize in 1906 for his accomplishment.

It was an American scientist, **Robert Millikan**, who in 1909 was able to measure the charge on an electron using the apparatus pictured in Figure 4.

> **TIP**
> Millikan's experiment determined the mass of an electron.

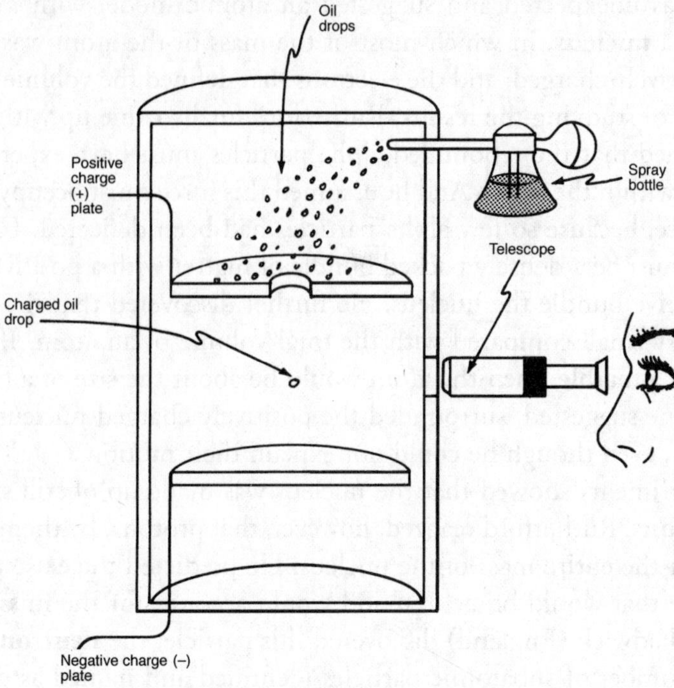

FIGURE 4. Millikan Oil Drop Experiment

Oil droplets were sprayed into the chamber and, in the process, became randomly charged by gaining or losing electrons. The electric field was adjusted so that a negatively charged drop would move slowly upward in front of the grid in the telescope. Knowing the rate at which the drop was rising, the strength of the field, and the weight of the drop, Millikan was able to calculate the charge on the drop. Combining the information with the results of Thomson, he could calculate a value for the mass of a single electron. Eventually, this number was found to be 9.11×10^{-28} gram.

Ernest Rutherford (England, 1911) performed a gold foil experiment (Figure 5) that had tremendous implications for atomic structure.

> **TIP**
> Rutherford's experiment using alpha particles confirmed that there was mostly empty space between the nucleus and electrons.

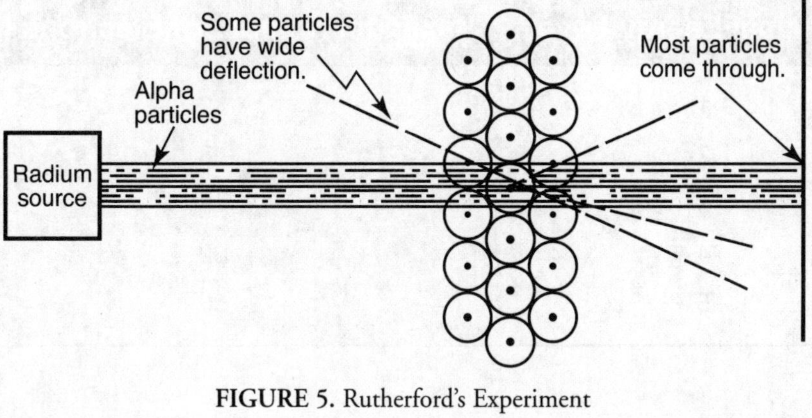

FIGURE 5. Rutherford's Experiment

Alpha particles (helium nuclei) passed through the foil with few deflections. However, some deflections (1 per 8,000) were almost directly back toward the source. This was unexpected and suggested an atomic model with mostly empty space between a **nucleus,** in which most of the mass of the atom was located and which was positively charged, and the electrons that defined the volume of the atom. After two years of studying the results, Rutherford finally came up with an explanation. He reasoned that the rebounded alpha particles must have experienced some powerful force within the atom. And he assumed this force must occupy a very small amount of space, because so few alpha particles had been deflected. He concluded that the force must be a densely packed bundle of matter with a positive charge. He called this positive bundle the nucleus. He further discovered that the volume of a nucleus was very small compared with the total volume of an atom. If the nucleus were the size of a marble, then the atom would be about the size of a football field. The electrons, he suggested, surrounded the positively charged nucleus like planets around the sun, even though he could not explain their motion.

Further experiments showed that the nucleus was made up of still smaller particles called **protons**. Rutherford realized, however, that protons, by themselves, could not account for the entire mass of the nucleus. He predicted the existence of a new nuclear particle that would be neutral and would account for the missing mass. In 1932, James Chadwick (England) discovered this particle, the **neutron**.

Today the number of subatomic particles identified and named as discrete units has risen to well over 90.

Bohr Model of the Atom

In 1913, **Niels Bohr** (Denmark) proposed his model of the atom. This pictured the atom as having a dense, positively charged nucleus and negatively charged electrons in specific spherical orbits, also called energy levels or shells, around this nucleus. These energy levels are arranged concentrically around the nucleus, and each level is designated by a number: 1, 2, 3, . . . The closer to the nucleus, the less energy an electron needs in one of these levels, but it has to gain energy to go from one level to another that is farther away from the nucleus.

Because of its simplicity and general ability to explain chemical change, the Bohr model still has some usefulness today.

TIP

Bohr's electron distribution to principal energy levels has the formula $2n^2$.

Principal Energy Level	Maximum Number of Electrons ($2n^2$)
1	2
2	8
3	18
4	32
5	50

Components of Atomic Structure

The chart below lists the basic particles of the atom. For the latest information about subatomic particles, see "New Subatomic Particles" in Chapter 15.

Particle	Charge	Symbol	Actual Mass	Relative Mass Compared to Proton	Discovery
Electron	– (e^-)	${}^{0}_{-1}e$	9.109×10^{-28} g	1/1,837	J. J. Thomson–1897
Proton	+ (p^+)	${}^{1}_{1}H$	1.673×10^{-24} g	1	——— early 1900s
Neutron	0 (n^0)	${}^{1}_{0}n$	1.675×10^{-24} g	1	J. C. Chadwick–1932

(There are now some 30 or more named particles or units of atomic structure, but the above are the most commonly used.)

When these components are used in the model, the protons and neutrons are shown in the nucleus. These particles are known as **nucleons.** The electrons are shown outside the nucleus.

The number of protons in the nucleus of an atom determines the **atomic number.** All atoms of the same element have the same number of protons and therefore the same atomic number; atoms of different elements have different atomic numbers. Thus, the atomic number identifies the element. An English scientist, **Henry Moseley**, first determined the atomic numbers of the elements through the use of x rays.

Since the actual masses of subatomic particles and atoms themselves are very small numbers when expressed in grams, scientists use atomic mass units (amu) instead. An **atomic mass unit** is defined as 1/12 the mass of a carbon-12 atom. Thus, the mass of any atom is expressed relative to the mass of one atom of carbon-12, which is sometimes called the **unified mass unit** (u) or the **dalton**.

The sum of the number of protons and the number of neutrons in the nucleus is called the **mass number.**

Table 1 (see page 66) summarizes the relationships just discussed. Notice that the outermost energy level can contain no more than eight electrons. The explanation of this is given in the next section.

In some cases, different types of atoms of the same element have different masses. For example, three types of hydrogen atoms are known. The most common type of hydrogen, sometimes called protium, accounts for 99.985% of the hydrogen atoms found on Earth. The nucleus of a protium atom contains one proton only, and it has one electron moving about it. The second form of hydrogen, known as deuterium, accounts for 0.015% of Earth's hydrogen atoms. Each deuterium atom has a nucleus containing one proton and one neutron. The third form of hydrogen, tritium, is radioactive. It exists in very small amounts in nature, but it can be prepared artificially. Each tritium atom contains one proton, two neutrons, and one electron.

Protium, deuterium, and tritium are isotopes of hydrogen. **Isotopes** are atoms of the same element that have different masses. The isotopes of a particular element all have the same number of protons and electrons but different numbers of neutrons.

TIP

Isotopes have the same atomic number but a different atomic mass. This means they differ in the number of neutrons, not protons.

In all three isotopes of hydrogen, the positive charge of the single proton is balanced by the negative charge of the electron. Most elements consist of mixtures of isotopes. Tin, for example, has ten stable isotopes, the most of any element.

The percentage of each isotope in the naturally occurring element on Earth is nearly always the same, no matter where the element is found. The percentage at which each of an element's isotopes occurs in nature is taken into account when calculating the element's average atomic mass. **Average atomic mass** is the weighted average of the atomic masses of the naturally occurring isotopes of an element.

Table 1. Table of the First 21 Elements*

Element	Atomic No.	Mass No.	Number of Protons	Number of Neutrons	Number of Electrons	Electrons in PELs** 1	2	3	4
Hydrogen	1	1	1	0	1	1			
Helium	2	4	2	2	2	2			
Lithium	3	7	3	4	3	2	1		
Beryllium	4	9	4	5	4	2	2		
Boron	5	11	5	6	5	2	3		
Carbon	6	12	6	6	6	2	4		
Nitrogen	7	14	7	7	7	2	5		
Oxygen	8	16	8	8	8	2	6		
Fluorine	9	19	9	10	9	2	7		
Neon	10	20	10	10	10	2	8		
Sodium	11	23	11	12	11	2	8	1	
Magnesium	12	24	12	12	12	2	8	2	
Aluminum	13	27	13	14	13	2	8	3	
Silicon	14	28	14	14	14	2	8	4	
Phosphorus	15	31	15	16	15	2	8	5	
Sulfur	16	32	16	16	16	2	8	6	
Chlorine	17	35	17	18	17	2	8	7	
Argon	18	40	18	22	18	2	8	8	
Potassium	19	39	19	20	19	2	8	8	1
Calcium	20	40	20	20	20	2	8	8	2
Scandium	21	45	21	24	21	2	8	9	2

**A complete list of the names and symbols of the known elements can be found in the Tables for Reference section.*
*** PEL is used to represent the principal energy levels.*

Calculating Average Atomic Mass

The average atomic mass of an element depends on both the mass and the relative abundance of each of the element's isotopes. For example, naturally occurring copper consists of 69.17% copper-63, which has an atomic mass of 62.919 598 amu, and 30.83% copper-65, which has an atomic mass of 64.927 793 amu. The average atomic mass of copper can be calculated by multiplying the atomic mass of each isotope by its relative abundance (expressed in decimal form) and adding the results.

$$0.6917 \times 62.919\,598 \text{ amu} + 0.3083 \times 64.927\,793 \text{ amu} = 63.55 \text{ amu}$$

Therefore, the calculated average atomic mass of naturally occurring copper is 63.55 amu. Average atomic masses of the elements listed in the Periodic Table, rounded to one decimal place for use in calculations and also in full to four decimal places, are given in the Chemical Elements table in the Tables for Reference section at the back of the book.

Oxidation Number and Valence

Each atom attempts to have its outer energy level complete and accomplishes this by borrowing, lending, or sharing its electrons. The electrons found in the outermost energy level are called **valence electrons.** The absolute number of electrons gained, lost, or borrowed is referred to as the valence of the atom. When valence electrons are lost or partially lost by sharing, the valence number of the atom is assigned a + sign and is called the **oxidation number**. If valence electrons are gained or partially gained by an atom, its valence is assigned a – sign for the oxidation number. These + and – signs are placed to the left of the number.

EXAMPLE: $_{17}Cl$ = ● (nucleus))2)8)7← valence electrons

This picture can be simplified to $\cdot\ddot{\underset{\cdot\cdot}{Cl}}:$, showing only the valence electrons as dots in an electron dot notation. This is called the **Lewis structure** of the atom. To complete its outer orbit to eight electrons, chlorine must borrow an electron from another atom. Its valence number then is 1. As stated above, when electrons are gained, we assign a – sign to this number, so the oxidation number of chlorine is –1. (More on pages 78 and 79.)

TIP

A Lewis structure shows the atomic symbol to represent the nucleus and inner shell electrons. It shows dots to represent the valence electrons.

ANOTHER EXAMPLE: $_{11}Na$ = ● (nucleus))2)8)1← valence electrons

Na • (Lewis dot structure)

Since sodium tends to lose this electron, its oxidation number is +1.

Some Basic Rules for Oxidation Numbers

1. Atoms of free elements have an oxidation number of zero.
2. Hydrogen has an oxidation number of +1 except in metallic hydrides, where it is –1.
3. Oxygen has an oxidation number of –2 except in peroxides, where it is –1. In combination with fluorine, it is +2.

A table of oxidation numbers is given on page 115. More information about oxidation numbers in chemical combinations is provided on page 116.

Metallic, Nonmetallic, and Noble Gas Structures

On the basis of atomic structure, atoms are classified as **metals** if they tend to lend electrons; as **nonmetals** if they tend to borrow or gain electrons; and as noble gases if they tend neither to borrow nor to lend electrons and have a complete outer energy level.

Reactivity

The fewer electrons an atom tends to borrow, or lend, or share to fill its outer energy level, the more reactive it tends to be in chemical reactions. This is elaborated on in the Periodic Table section.

ATOMIC SPECTRA

The Bohr model was based on a simple postulate. Bohr applied to the hydrogen atom the concept that the electron can exist only in certain energy levels without an energy change but that, when the electron changes its state, it must absorb or emit the exact amount of energy that will bring it from the initial state to the final state. The **ground state** is the lowest energy state available to the electron. The **excited state** is any level higher than the ground state. The formula for a change in energy (ΔE) is:

$$\Delta E_{\text{electron}} = E_{\text{final}} - E_{\text{initial}}$$

When an electron moves from the ground state to an excited state, it must absorb energy. When it moves from an excited state to the ground state, it emits energy. This release of energy is the basis for **atomic spectra**. (See Figure 6.)

The energy values shown were calculated from Bohr's equation.

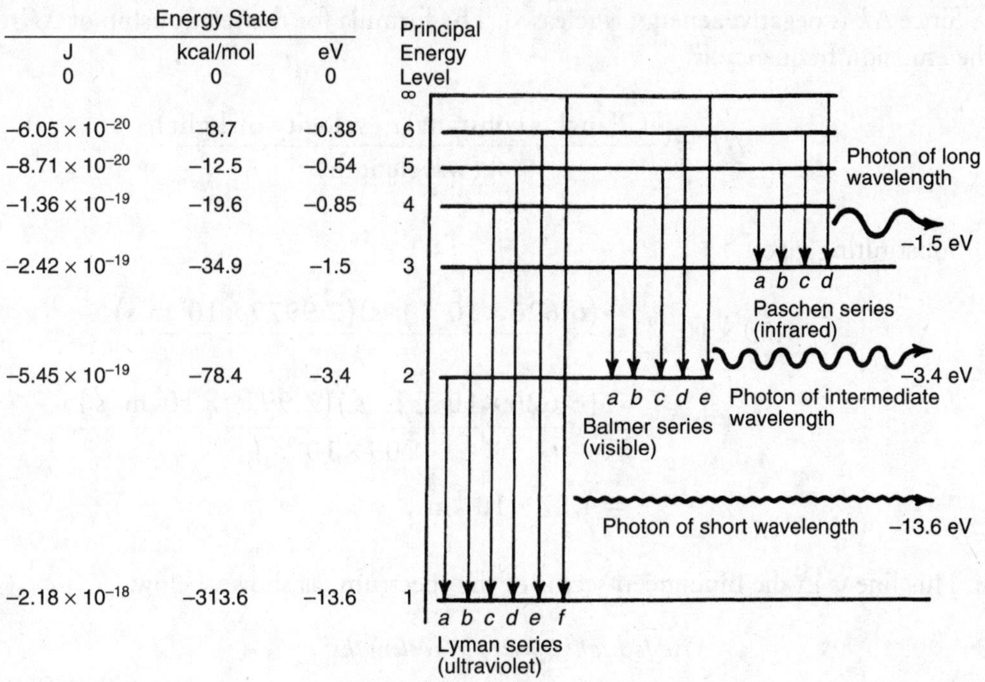

Figure 6. Atomic Spectra Chart

TIP

When electrons drop to the lowest available energy level, they release energy.

When energy is released in the "allowed" values, it is released in the form of discrete radiant energy called **photons**. Each of the first three levels has a particular name associated with the emissions that occur when an electron reaches its ground state on that level. The emissions, consisting of ultraviolet radiation, that occur when an electron cascades from a level higher than the first level down to $n = 1$ are known as the **Lyman series**. Note in Figure 6 that the next two higher levels have the names **Balmer** (for $n = 2$) and **Paschen** ($n = 3$) **series**, respectively.

Spectroscopy

When the light emitted by energized atoms is examined with an instrument called a **spectroscope**, the prism or diffraction grating in the spectroscope disperses the light to allow an examination of the **spectra** or distinct colored lines. Since only particular energy jumps are available in each type of atom, each element has its own unique emission spectra made up of only the lines of specific wavelength that correspond to its atomic structure.

EXAMPLE: Hydrogen can have an electron drop from the $n = 4$ to $n = 2$ level. What visible spectral line in the Balmer series will result from this emission of energy?

$$\Delta E_{\text{evolved}} = E_{n=2} - E_{n=4}$$

From Figure 6, $E_2 = -5.45 \times 10^{-19}$ J and $E_4 = -1.36 \times 10^{-19}$ J. Then

$$E_{\text{evolved}} = -5.45 \times 10^{-19}\,\text{J} - (-1.36 \times 10^{-19}\,\text{J}) = -4.09 \times 10^{-19}\,\text{J}$$

Since ΔE is negative, energy is released. The formula for the relationship of ΔE to the emission frequency is

$$\Delta E = \frac{h(\text{Planck's constant})\, c\,(\text{velocity of light})}{\lambda\,(\text{wavelength})}$$

Substituting gives

$$-4.09\times10^{-19}\,\text{J} = \frac{(6.626\times10^{-34}\,\text{J}\cdot\text{s})(2.9979\times10^{8}\,\text{m/s})}{\lambda}$$

$$\lambda = \frac{(6.626\times10^{-34}\,\cancel{\text{J}}\cdot\cancel{\text{s}})(2.9979\times10^{8}\,\text{m}/\cancel{\text{s}})}{4.09\times10^{-19}\,\cancel{\text{J}}}$$

$$= 4.87\times10^{-7}\,\text{m}$$

This line is in the blue-green sector of the spectrum, as shown below.

Visible Light Spectra Wavelengths

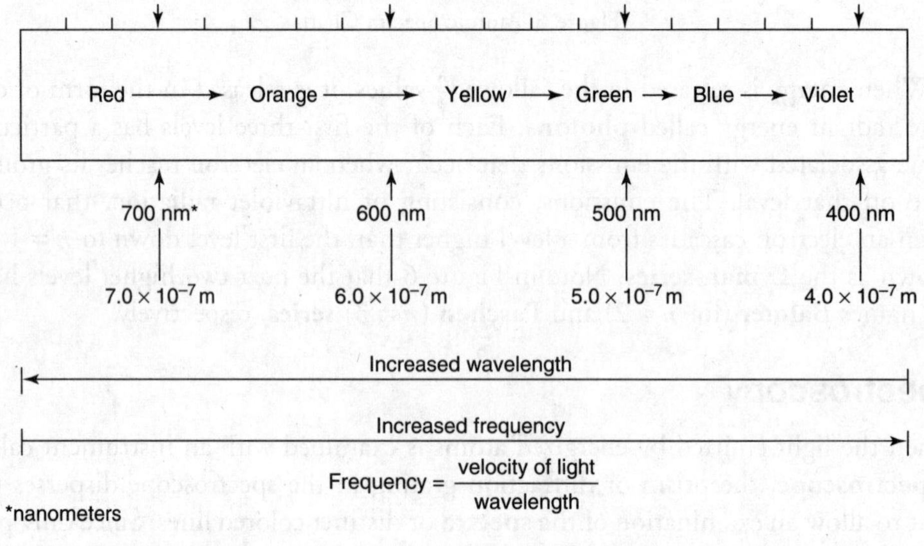

A partial atomic spectrum for hydrogen would look like this:

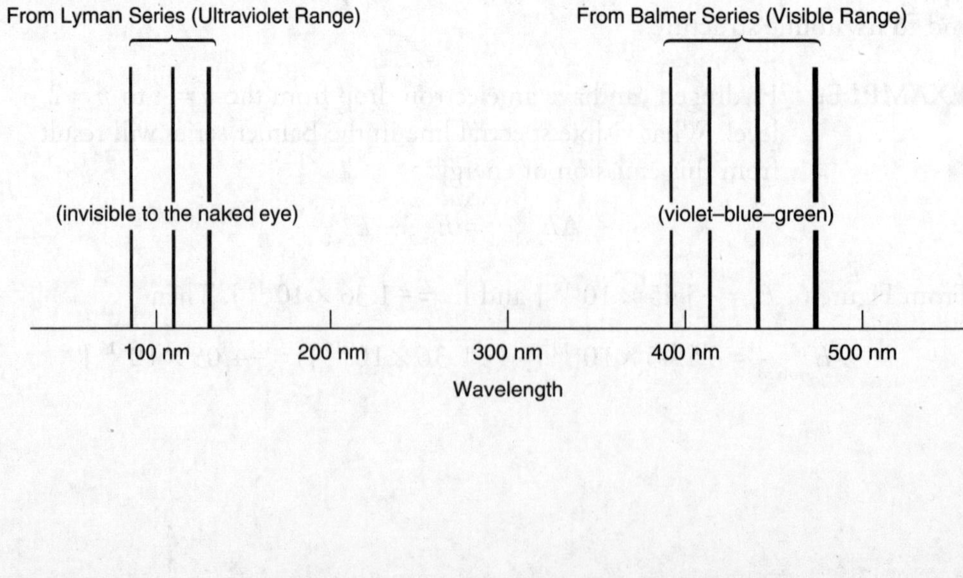

The right-hand group is in the visible range and is part of the Balmer series. The left-hand group is in the ultraviolet region and belongs to the Lyman series.

Spectral lines like these can be used in the identification of unknown specimens.

Mass Spectroscopy

Another tool used to identify specific atomic structures is mass spectroscopy, which is based on the concept that differences in mass cause differences in the degree of bending that occurs in a beam of ions passing through a magnetic field. This is shown in Figure 7.

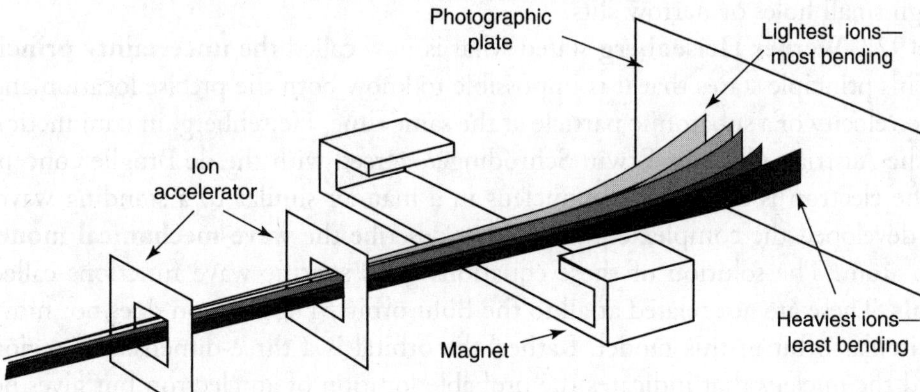

Figure 7. Mass Spectroscope

TIP

A mass spectroscope separates isotopes of the same element based on differences in their mass.

The intensity on the photographic plate indicates the amount of each particular **isotope**. Other collectors may be used in place of the photographic plate to collect and interpret these data.

THE WAVE-MECHANICAL MODEL

In the early 1920s, some difficulties with the Bohr model of the atom were becoming apparent. Although Bohr used classical mechanics (which is the branch of physics that deals with the motion of bodies under the influence of forces) to calculate the orbits of the hydrogen atom, this discipline did not serve to explain the ability of electrons to stay in only certain energy levels without the loss of energy. Nor could it explain why a change of energy occurred only when an electron "jumped" from one energy level to another and why the electron could not exist in the atom at any energy level between these levels. According to Newton's laws, the kinetic energy of a body always changes smoothly and continuously, not in sudden jumps. The idea of only certain **quantized** energy levels being available in the Bohr atom was a very important one. The energy levels explained the existence of atomic spectra, described in the preceding sections.

Another difficulty with the Bohr model was that it worked well only for the hydrogen atom with its single electron. It did not work with atoms that had more electrons. A new approach to the laws governing the behavior of electrons inside the atom was needed, and such an approach was developed in the 1920s by the combined work of many scientists. Their work dealt with a more mathematical model usually referred to as **quantum mechanics** or **wave mechanics**. By this time, Albert Einstein had already proposed a relativity mechanics model to deal with the relative nature of mass as its speed approaches the speed of light. In the same manner, a quantum/wave mechanics model was now needed to fit the data of the atomic model. **Max Planck** suggested in his quantum theory of light that light has both particlelike properties and wavelike characteristics. In 1924, **Louis de Broglie**, a young French physicist, suggested that, if light can have both wavelike and particle-like characteristics as Planck had suggested, then perhaps particles can also have wavelike characteristics. In 1927, de Broglie's ideas were verified experimentally when investigators showed that electrons could produce diffraction patterns, a property associated with waves. Diffraction patterns are produced by waves as they pass through small holes or narrow slits.

In 1927, **Werner Heisenberg** stated what is now called the **uncertainty principle**. This principle states that it is impossible to know both the precise location and precise velocity of a subatomic particle at the same time. Heisenberg, in conjunction with the Austrian physicist Erwin Schrödinger, agreed with the de Broglie concept that the electron is bound to the nucleus in a manner similar to a standing wave. They developed the complex equations that describe the **wave-mechanical model** of the atom. The solution of these equations gives specific wave functions called **orbitals**. These are not related at all to the Bohr orbits. The electron does not move in a circular orbit in this model. Rather, the orbital is a three-dimensional region around the nucleus that indicates the probable location of an electron but gives no information about its pathway. The drawings in Figures 8*a* and 8*b* are only probability distribution representations of where electrons in these orbitals might be found.

Quantum Numbers

Each electron orbital of an atom may be described by a set of four quantum numbers in the wave-mechanical model. These numbers give the position with respect to the nucleus, the shape of the orbital, its spatial orientation, and the spin of the electron in the orbital.

TIP

The principal quantum number refers to the principal energy level: 1, 2, 3, and so on. The angular momentum quantum number refers to shape.

Principal quantum number (*n*) 1, 2, 3, 4, 5, etc.	This number refers to average distance of the orbital from the nucleus. 1 is closest to the nucleus and has the least energy. The numbers correspond to the orbits in the Bohr model. They are called energy levels.
Angular momentum (ℓ) *quantum number* *s*, *p*, *d*, *f* (in order of increasing energy)	This number refers to the shape of the orbital. The number of possible shapes is limited by the principal quantum number. The first energy level has only one possible shape, the *s* orbital. The second has two possible shapes, the *s* and *p*. See Figures 8*a* and 8*b* for representations of these shapes.

Magnetic quantum number (m_ℓ)
$s = 1$ space-oriented orbital
$p = 3$ space-oriented orbitals
$d = 5$ space-oriented orbitals
$f = 7$ space-oriented orbitals

Spin quantum number (m_s)
+ spin − spin

The drawings in Figure 8a show the *s*-orbital shape, which is a sphere, and the *p* orbitals, which have dumbbell shapes with three possible orientations on the axis shown. The number of spatial orientations of orbitals is referred to as the magnetic quantum number. The possible orientations are listed. Figure 8b represents the *d* orbitals. Electrons are assigned one more quantum number called the spin quantum number. This describes the spin in either of two possible directions. Each orbital can be filled by only two electrons with opposite spins. The main significance of electron spin is explained by the postulate of Wolfgang Pauli. It states that in a given atom no two electrons can have the same set of four quantum numbers (n, ℓ, m_ℓ, and m_s). This is referred to as the **Pauli Exclusion Principle.** Therefore each orbital in Figures 8a and 8b can hold only two electrons.

Pauli Exclusion Principle:

No two electrons can have the same four quantum numbers.

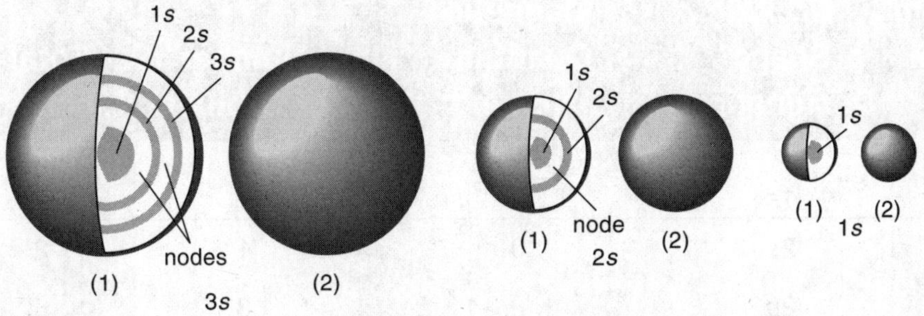

Two representations of the hydrogen 1*s*, 2*s*, and 3*s* orbitals. (1) The electron probability distribution; the nodes indicate regions of zero probability. (2) The surface that contains 90% of the total electron probability (the size of the orbital, by definition).

TIP

s orbitals are spherical and can hold 2 electrons.

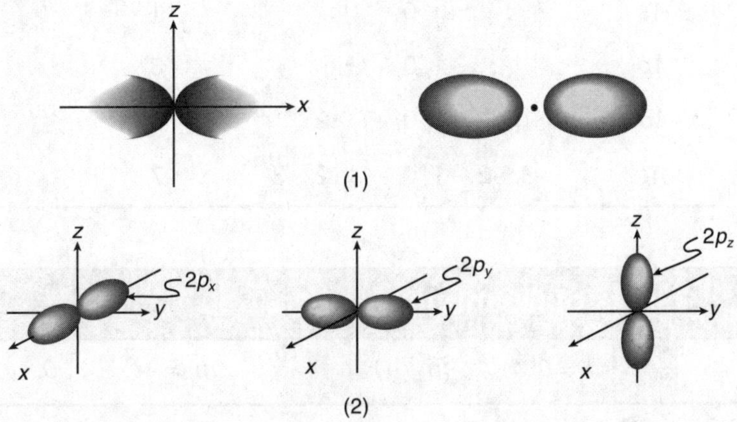

Representation of the 2 *p* orbitals. (1) The electron probability distribution and (2) the boundary surface representations of all three orbitals.

FIGURE 8a. Representations of *s* and *p* Orbitals

TIP

p orbitals have a dumbell shape, oriented on the *x*, *y*, and *z* axes, and can hold a total of 2 electrons each, making a total of 6.

TIP

d orbitals have 5 orientations and can hold 2 electrons each, making a total of 10.

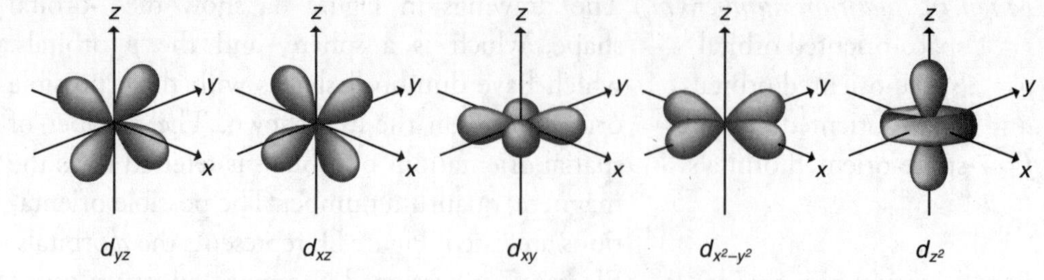

(Indicates orbitals in *y* and *z* planes)
Representations of the 3*d* orbitals in terms of their boundary surfaces.
The subscripts of the first four orbitals indicate the planes in which the four lobes are centered.

FIGURE 8b. Representations of *d* Orbitals

Quantum numbers are summarized in the table below.

Summary of Quantum Numbers for the First Four Levels of Orbitals in the Hydrogen Atom

Principal Quantum No., n	Angular Momentum Quantum No., ℓ	Orbital Shape Designation	Magnetic Quantum No., m_ℓ	Number of Orbitals	Total Electrons
1	0	1*s*	0	1	2
2	0	2*s*	0	1	2
	1	2*p*	−1, 0, +1	3	6
3	0	3*s*	0	1	2
	1	3*p*	−1, 0, +1	3	6
	2	3*d*	−2, −1, 0, +1, +2	5	10
4	0	4*s*	0	1	2
	1	4*p*	−1, 0, +1	3	6
	2	4*d*	−2, −1, 0, +1, +2	5	10
	3	4*f*	−3, −2, −1, 0, +1, +2, +3	7	14

Limits of Quantum Numbers

$n = 1, 2, 3, \ldots$	$\ell = 0, 1, \ldots (n - 1)$	$m_\ell = -\ell, \ldots, 0, \ldots, +\ell$

Hund's Rule of Maximum Multiplicity

It is important to remember that, when there is more than one orbital at a particular energy level, such as three *p* orbitals or five *d* orbitals, only one electron will fill each orbital until each has one electron. This principle, that an electron occupies the lowest energy orbital that can receive it, is called the **Aufbau Principle.** After this, pairing will occur with the addition of one more electron to each orbital. This principle, called **Hund's Rule of Maximum Multiplicity**, is shown in Table 2, where each arrowhead indicates an electron ([↑]).

TIP

Know the Aufbau Principle and Hund's Rule of Maximum Multiplicity.

Table 2. Orbital Notations

Chemical Symbol	Atomic No.	Orbital Notation 1s	2s	2p	Electron Orbital Notation
H	1	[↑]	[ ]	[ ][ ][ ]	$1s^1$
He	2	[↑↓]	[ ]	[ ][ ][ ]	$1s^2$
Li	3	[↑↓]	[↑]	[ ][ ][ ]	$1s^2\ 2s^1$
Be	4	[↑↓]	[↑↓]	[ ][ ][ ]	$1s^2\ 2s^2$
B	5	[↑↓]	[↑↓]	[↑][ ][ ]	$1s^2\ 2s^2\ 2p^1$
C	6	[↑↓]	[↑↓]	[↑][↑][ ]	$1s^2\ 2s^2\ 2p^2$
N	7	[↑↓]	[↑↓]	[↑][↑][↑]	$1s^2\ 2s^2\ 2p^3$
O	8	[↑↓]	[↑↓]	[↑↓][↑][↑]	$1s^2\ 2s^2\ 2p^4$
F	9	[↑↓]	[↑↓]	[↑↓][↑↓][↑]	$1s^2\ 2s^2\ 2p^5$
Ne	10	[↑↓]	[↑↓]	[↑↓][↑↓][↑↓]	$1s^2\ 2s^2\ 2p^6$

Maximum electrons in orbitals at a particular sublevel:

s = 2 (one orbital)
p = 6 (three orbitals)
d = 10 (five orbitals)
f = 14 (seven orbitals)

If each orbital is indicated in an energy diagram as a square (□), we can show relative energies in a chart such as Figure 9. If this drawing represented a ravine with the energy levels as ledges onto which stones could come to rest only in numbers equal to the squares for orbitals, then pushing stones into the ravine would cause the stones to lose their kinetic energy as they dropped to the lowest level available to them. Much the same is true for electrons.

SUBLEVELS AND ELECTRON CONFIGURATION

Order of Filling and Notation

The sublevels do not fill up in numerical order, and the pattern of filling is shown on the right side of the approximate relative energy levels chart (Figure 9). In the first instance of failure to follow numerical order, the 4*s* fills before the 3*d*. (Study Figure 9 carefully before going on.)

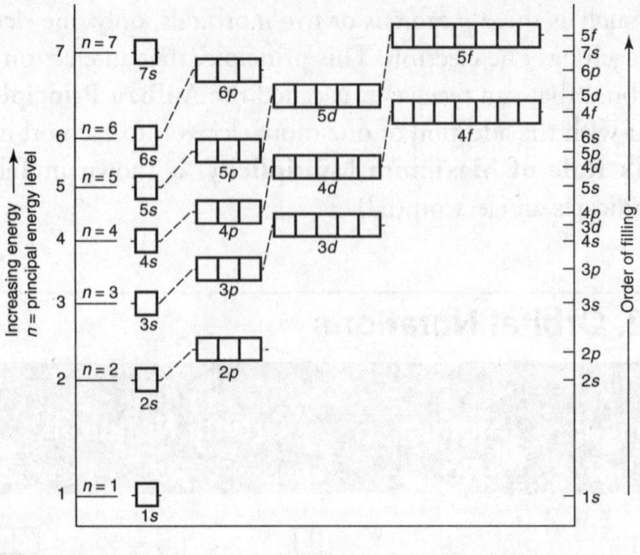

Figure 9. Approximate Relative Energy Level of Subshells

$_{19}$K $1s^2\ 2s^2\ 2p^6\ 3s^2\ 3p^6\ 4s^1$
$_{20}$Ca $1s^2\ 2s^2\ 2p^6\ 3s^2\ 3p^6\ 4s^2$
$_{21}$Sc $1s^2\ 2s^2\ 2p^6\ 3s^2\ 3p^6\ 4s^2\ 3d^1$ (note 4*s* filled before 3*d*)

There is a more stable configuration to a half-filled or filled sublevel, so at atomic number 24 the 3*d* sublevel becomes half-filled by taking a 4*s* electron;

$_{24}$Cr $1s^2\ 2s^2\ 2p^6\ 3s^2\ 3p^6\ 3d^5\ 4s^1$

and at atomic number 29 the 3*d* becomes filled by taking a 4*s* electron:

$_{29}$Cu $1s^2\ 2s^2\ 2p^6\ 3s^2\ 3p^6\ 3d^{10}\ 4s^1$

Table 3 shows the electron configurations of the elements. A triangular mark ◤ indicates an outer-level electron dropping back to a lower unfilled orbital. These phenomena are exceptions to the Aufbau Principle. By following the atomic numbers throughout this chart, you will get the same order of filling as shown in Figure 9.

Table 3. Electron Configuration of the Elements

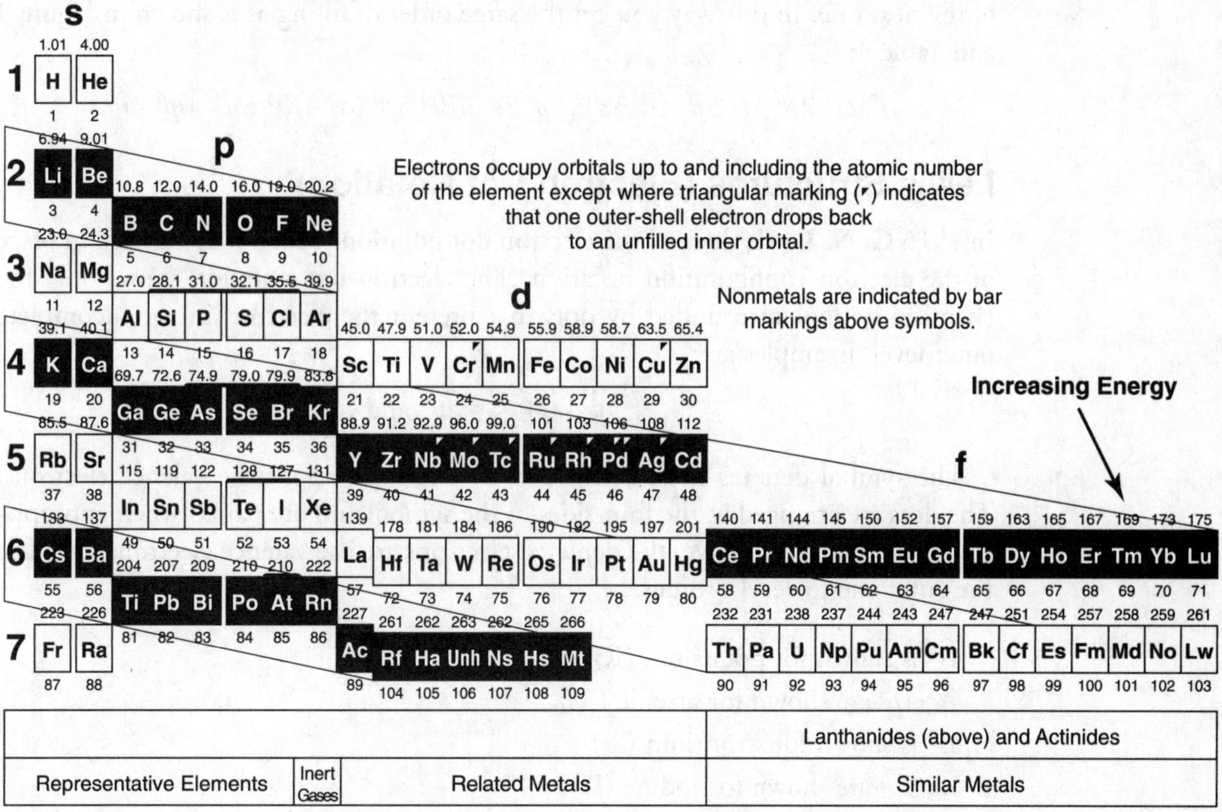

Note: Follow order of the atomic numbers to ascertain the order of filling.

By following the atomic numbers in numerical order in Table 3 you can plot the order of filling of the orbitals for every element shown.

A simplified method of showing the order in which the orbitals are filled is to use the following diagram. It works for all the naturally occurring elements through lanthanum, atomic number 88.

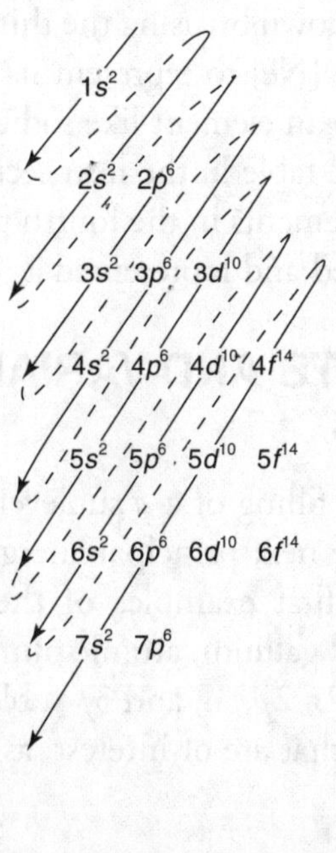

> **TIP**
>
> This is a simple way to remember the order of filling orbitals.

Start by drawing the diagonal arrows through the diagram as shown. The order of filling can be charted by following each arrow from tail to head and then to the tail of the next one. In this way you get the same order of filling as is shown in Figure 9 and Table 3:

$$1s^2\ 2s^2\ 2p^6\ 3s^2\ 3p^6\ 4s^2\ 3d^{10}\ 4p^6\ 5s^2\ 4d^{10}\ 5p^6\ 6s^2\ 4f^{14}\ 5d^{10}\ 6p^6\ 7s^2$$

Lewis Structures (Electron Dot Notation)

In 1916 **G. N. Lewis** devised the electron dot notation, which may be used in place of the electron configuration notation. The electron dot notation shows only the chemical symbol surrounded by dots to represent the electrons in the incomplete outer level. Examples are:

K, ·As·, Sr, :I:, and :Rn:

The symbol denotes the nucleus and all electrons except the valence electrons. The dots are arranged at the four sides of the symbol and are paired when appropriate. In the examples above, the depicted electrons are the valence electrons found in the outer energy level orbitals.

$4s^1$ is shown for potassium (K)
$4s^2\ 4p^3$ are shown for arsenic (As)
$5s^2$ is shown for strontium (Sr)
$5s^2\ 5p^5$ are shown for iodine (I)
$6s^2\ 6p^6$ are shown for radon (Rn)

Noble Gas Notation

Another method of simplifying the electron distribution to the orbitals is called the noble gas notation. In this method you represent all of the lower filled orbitals up to the closest noble gas. By enclosing its symbol in brackets, it represents all of the complete noble gas configuration. Then the remaining orbitals are written in the usual way. An example of this can be shown by using the third period of elements. By using neon as the noble gas, you write [Ne] to represent its orbital structure, which is $1s^2\ 2s^2\ 2p^6$. This allows you to write an element like sodium as [Ne] $3s^1$, which is called sodium's noble gas notation. The table in the next section shows the noble gas notations of some of the transition elements in the fourth period of elements. Notice that the base structure of argon is used and represented as [Ar].

TRANSITION ELEMENTS AND VARIABLE OXIDATION NUMBERS

The elements involved with the filling of a *d* sublevel with electrons after two electrons are in the *s* sublevel of the next principal energy level are often referred to as the **transition elements**. The first examples of these are the elements between calcium, atomic number 20, and gallium, atomic number 31. Their electron configurations are the same in the 1*s*, 2*s*, 2*p*, 3*s*, and 3*p* sublevels. It is the filling of the 3*d* and changes in the 4*s* sublevels that are of interest, as shown in the following table.

Electron Configuration of Some Elements in the Fourth Period

Element	Symbol	Atomic No.	Electron Configuration $1s^2\ 2s^2\ 2p^6\ 3s^2\ 3p^6$	$3d^{10}$	$4s$	Noble Gas Notation
Scandium	Sc	21		1	2	[Ar] $d^1\ s^2$
Titanium	Ti	22	All	2	2	[Ar] $d^2\ s^2$
Vanadium	V	23		3	2	[Ar] $d^3\ s^2$
Chromium	Cr	24	the	5	1*	[Ar] $d^5\ s^1$
Manganese	Mn	25		5	2	[Ar] $d^5\ s^2$
Iron	Fe	26	same	6	2	[Ar] $d^6\ s^2$
Cobalt	Co	27		7	2	[Ar] $d^7\ s^2$
Nickel	Ni	28		8	2	[Ar] $d^8\ s^2$
Copper	Cu	29		10	1*	[Ar] $d^{10}\ s^1$
Zinc	Zn	30		10	2	[Ar] $d^{10}\ s^2$

The asterisk (*) shows where a $4s$ electron is promoted into the $3d$ sublevel. This is because the $3d$ and $4s$ sublevels are very close in energy and that there is a state of greater stability in half-filled and filled sublevels. Therefore, chromium gains stability by the movement of an electron from the $4s$ sublevel into the $3d$ sublevel to give a half-filled $3d$ sublevel. It then has one electron in each of the five orbitals of the $3d$ sublevel. In copper, the movement of one $4s$ electron into the $3d$ sublevel gives the $3d$ sublevel a completely filled configuration.

The fact that the electrons in the $3d$ and $4s$ sublevels are so close in energy levels leads to the possibility of some or all the $3d$ electrons being involved in chemical bonding. With the variable number of electrons available for bonding, it is not surprising that transition elements can exhibit variable oxidation numbers. An example is manganese with possible oxidation numbers of +2, +3, +4, +6, and +7, which correspond, respectively, to the use of no, one, two, four, and five electrons from the $3d$ sublevel.

The transition elements in the other periods of the table show this same type of anomaly, as they have *d* sublevels filling in the same manner.

Transition elements have several common characteristic properties.

- They often form colored compounds.
- They can have a variety of oxidation states.
- At least one of their compounds has an incomplete *d* electron subshell.
- They are often good catalysts.
- They are silvery blue at room temperature (except copper and gold).
- They are solids at room temperature (except mercury).
- They form complex ions.
- They are often paramagnetic.

PERIODIC TABLE OF THE ELEMENTS

History

The history of the development of a systematic pattern for the elements includes the work of a number of scientists such as John Newlands, who, in 1863, proposed the idea of repeating octaves of properties.

Dimitri I. Mendeleev in 1869 proposed a table containing 17 columns and is usually given credit for the first periodic table since he arranged elements in groups according to their atomic weights and properties. It is interesting to note that Lothar Meyer proposed a similar arrangement about the same time. In 1871 Mendeleev rearranged some elements and proposed a table of eight columns, obtained by splitting each of the long periods across into a period of seven elements, an eighth group containing the three central elements (such as Fe, Co, Ni), and a second period of seven elements. The first and second periods of seven across were later distinguished by use of the letters A and B attached to the group symbols, which were Roman numerals. This nomenclature of periods (IA, IIA, etc.) has been revised in the present Periodic Table, even in the extended form of assigning Arabic numbers from 1–18 as shown in Table 4.

TIP

Mendeleev is given credit for the first Periodic Table. It was based on placement by properties.

TIP

Periods are the horizontal rows 1–7.

Groups are the vertical columns 1–18.

TIP

Know these relationships across the table.

Table 4. Periodic Table Properties

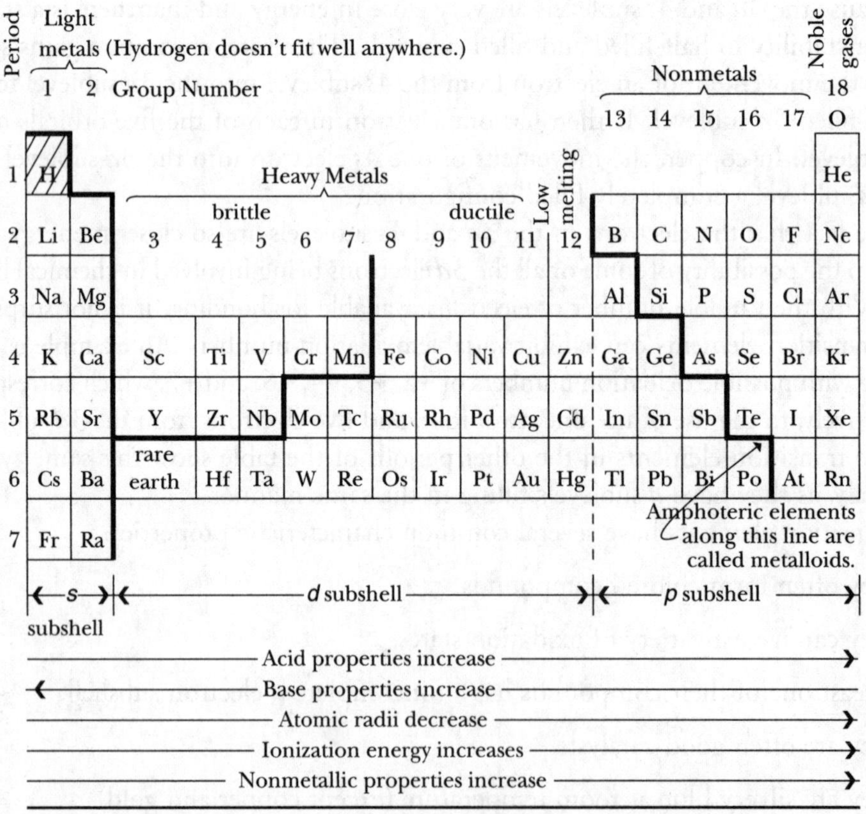

Mendeleev's table had the elements arranged by atomic weights with recurring properties in a periodic manner. Where atomic weight placement disagreed with the properties that should occur in a particular spot in the table, Mendeleev gave preference to the element with the correct properties. He even predicted elements for places that were not yet occupied in the table. These predictions proved to be amazingly accurate and led to wide acceptance of his table.

Periodic Law

Henry Moseley stated, after his work with x-ray spectra in the early 1900s, that the properties of elements are a periodic function of their atomic numbers, thus changing the basis of the periodic law from atomic weight to atomic number. This is the present statement of the periodic law.

The Table

The horizontal rows of the periodic table are called **periods** or **rows**. There are seven periods, each of which begins with an atom having only one valence electron and ends with a complete outer shell structure of an inert gas. The first three periods are short, consisting of 2, 8, and 8 elements, respectively. Periods 4 and 5 are longer, with 18 each, while period 6 has 32 elements, and period 7 is incomplete with 22 elements, most of which are radioactive and do not occur in nature.

In Table 4, you should note the relationship of the length of the periods to the orbital structure of the elements. In the first period, the $1s^2$ orbital is filled with the noble gas helium, He. The second period begins with the $2s^1$ orbital and ends with the filling of the $2p^6$ orbital, again with a noble gas, neon, Ne. The same pattern is repeated in period three, going from $3s^1$ to $3p^6$. The eight elements from sodium, Na, to argon, Ar, complete the filling of the $n = 3$ energy level with $3s^2$ and $3p^6$. In the fourth period, the first two elements fill the $4s^2$ orbital. Beyond calcium, Ca, the pattern becomes more complicated. As discussed in the section "Order of Filling and Notation," the next orbitals to be filled are the five $3d$ orbitals whose elements represent transition elements. Then the three $4p$ orbitals are filled, ending with the noble gas krypton, Kr. The fifth period is similar to the fourth period. The $5s^2$ orbital filling is represented by rubidium, Rb, and strontium, Sr, both of which resemble the elements directly above them on the table. Next come the transition elements that fill the five $4d$ orbitals before the next group of elements, from indium, In, to xenon, Xe, complete the three $5p$ orbitals. (Table 3 should be consulted for the irregularities that occur as the d orbitals fill.) The sixth period follows much the same pattern and has the filling order $6s^2$, $4f^{14}$, $5d^{10}$, $6p^6$. Here, again, irregularities occur and can best be followed by using Table 3.

The vertical columns of the Periodic Table are called **groups** or **families**. The elements in a group exhibit similar or related properties. In 1984 the IUPAC agreed that the groups would be numbered 1 through 18.

PROPERTIES RELATED TO THE PERIODIC TABLE

Metals are found on the left of the chart (see Table 4) with the most active metal in the lower left corner. Nonmetals are found on the right side with the most active nonmetal in the upper right-hand corner. The noble or inert gases are on the far right. Since the most active metals react with water to form bases, the Group 1 metals are called alkali metals. As you proceed to the right, the base-forming property decreases and the acid-forming properties increase. The metals in the first two groups are the light metals, and those toward the center are heavy metals.

The elements found along the dark line in the Periodic Table (Table 4) are called **metalloids**. These elements have certain characteristics of metals and other characteristics of nonmetals. Some examples of metalloids are boron, silicon, arsenic, and tellurium.

Here are some important general summary statements about the Periodic Table:

TIP

These are important trends to remember.

- Acid-forming properties increase from left to right on the table.
- Base-forming properties are high on the left side and decrease to the right.
- The atomic radii of elements decrease from left to right across a period.
- First ionization energies increase from left to right across a period.
- Metallic properties are greatest on the left side of the table and decrease to the right.
- Nonmetallic properties are greatest on the right side of the table and decrease to the left.

Study Table 4 carefully because it summarizes many of these properties. For a more detailed description of metals, alloys, and metalloids see pages 297–301 in Chapter 13.

Radii of Atoms

The size of an atom is difficult to describe because atoms have no definite outer boundary. Unlike a volleyball, an atom does not have a definite circumference.

To overcome this problem, the size of an atom is estimated by describing its radius. In metals, this is done by measuring the distance between two nuclei in the solid state and dividing this distance by 2. Such measurements can be made with x-ray diffraction. For a nonmetallic element that exists in pure form as a molecule, such as chlorine, measurements can be made of the distance between nuclei for two atoms covalently bonded together. Half of this distance is referred to as the **covalent radius**. The method for finding the covalent radius of the chlorine atom is illustrated in the following diagram.

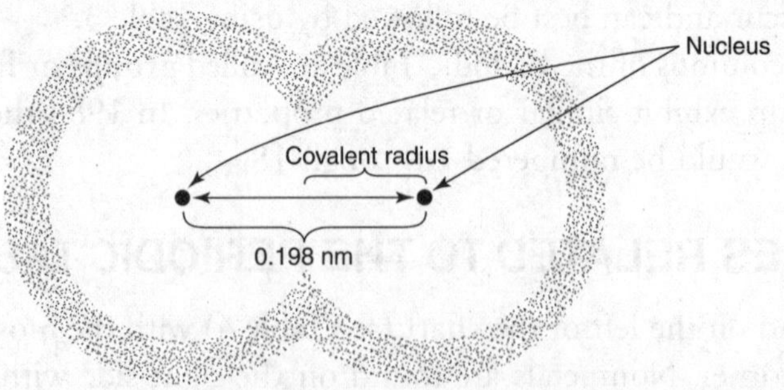

Covalent radius of Cl = $\frac{1}{2}$(0.198) nm = 0.099 nm

Figure 10 shows the relative atomic and ionic radii for some elements. As you review this chart, you should note two trends:

1. Atomic radii decrease from left to right across a period in the Periodic Table (until the noble gases).
2. Atomic radii increase from top to bottom in a group or family.

The reason for these trends will become clear in the following discussions.

> **TIP**
> Know these trends in atomic radii.

Atomic Radii in Periods

> **TIP**
> This is the explanation of these trends in periods.

Since the number of electrons in the outer principal energy level increases as you go from left to right in each period, the corresponding increase in the nuclear charge because of the additional protons pulls the electrons more tightly around the nucleus. This attraction more than balances the repulsion between the added electrons and the other electrons, and the radius is generally reduced. The inert gas at the end of the period has a slight increase in radius because of the electron repulsion in the filled outer principal energy level. For example, lithium's atomic radius in Figure 10 is 0.152 nm at the one end of period 2 whereas fluorine has a radius of only 0.064 nm at the far end of the period. This trend can be seen in Figure 10 across every period.

Atomic Radii in Groups

> **TIP**
> . . . and in group.

For a group of elements, the atoms of each successive member have another outer principal energy level in the electron configuration and the electrons there are held less tightly by the nucleus. This is so because of their increased distance from the nuclear positive charge and the shielding of this positive charge by all the intermediate electrons. Therefore the atomic radius increases down a group. For example, oxygen's atomic radius in Figure 10 is 0.066 nm at the top of group 16, whereas polonium has a radius of 0.167 nm at the bottom of the same group. This trend can be seen in Figure 10 down every group.

Ionic Radius Compared with Atomic Radius

> **TIP**
> Know this relationship between the atomic radius and the ionic radius.

Metals tend to lose electrons in forming positive ions. With this loss of negative charge, the positive nuclear charge pulls in the remaining electrons closer and thus reduces the ionic radius below that of the atomic radius.

Nonmetals tend to gain electrons in forming negative ions. With this added negative charge, which increases the inner electron repulsion, the ionic radius is increased beyond the atomic radius. See Figure 10 for relative atomic and ionic radii values.

Electronegativity

The **electronegativity** of an element is a number that measures the relative strength with which the atoms of the element attract valence electrons in a chemical bond. This electronegativity number is based on an arbitrary scale going from 0 to 4. In general, a value of less than 2 indicates a metal.

Notice in Table 5 that the electronegativity decreases down a group and increases across a period. The inert gases can be ignored. The lower the electronegativity number, the more electropositive an element is said to be. The most electronegative element is in the upper right corner—F, fluorine. The most electropositive is in the lower left corner of the chart—Fr, francium.

Table 5. First Ionization Energies and Electronegativities

Key: upper number = First Ionization Energy (kcal/mol of atoms); lower number = Electronegativity*

1	2	13	14	15	16	17	18
H 313 / 2.2							He 567
Li 125 / 1.0	Be 215 / 1.5	B 191 / 2.0	C 260 / 2.6	N 336 / 3.1	O 314 / 3.5	F 402 / 4.0	Ne 497
Na 119 / 0.9	Mg 176 / 1.2	Al 138 / 1.5	Si 188 / 1.9	P 242 / 2.2	S 239 / 2.6	Cl 300 / 3.2	Ar 363
K 100 / 0.8	Ca 141 / 1.0	Ga 138 / 1.6	Ge 182 / 1.9	As 226 / 2.0	Se 225 / 2.5	Br 273 / 2.9	Kr 323
Rb 96 / 0.8	Sr 131 / 1.0	In 133 / 1.7	Sn 169 / 1.8	Sb 199 / 2.1	Te 208 / 2.3	I 241 / 2.7	Xe 280
Cs 90 / 0.7	Ba 120 / 0.9	Tl 141 / 1.8	Pb 171 / 1.8	Bi 168 / 1.9	Po 194 / 2.0	At 2.2	Rn 248
Fr 0.7	Ra 122 / 0.9						

*Arbitrary scale based on fluorine = 4.0

TIP

The most electronegative element is F, in the upper right corner.

TIP

The least electronegative element is Fr, in the lower left corner.

Electron Affinity

Neutral atoms can either gain or lose electrons. When either of these occur, there is an energy change. The energy change that occurs when an electron is gained by a neutral atom is called the atom's **electron affinity**. Most atoms release energy when they acquire an electron.

$$A + e^- \rightarrow A^- + \text{energy}$$

As is the general custom for the release of energy, the quantity of energy released is represented by a negative number. If an atom is forced to gain an electron by the addition of energy, this would be represented as a positive quantity. A much more common measure of atomic activity is electronegativity.

Ionization Energy

Atoms hold their valence electrons, then, with different amounts of energy. If enough energy is supplied to one outer electron to remove it from its atom, this

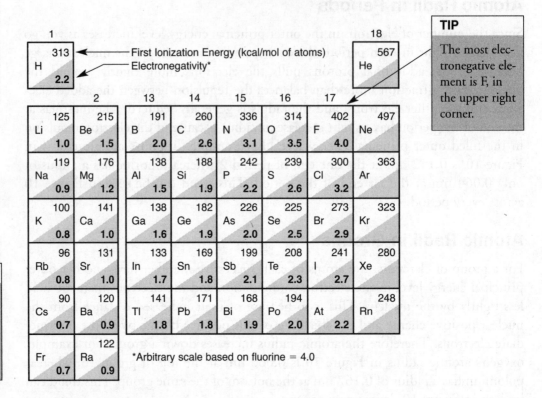

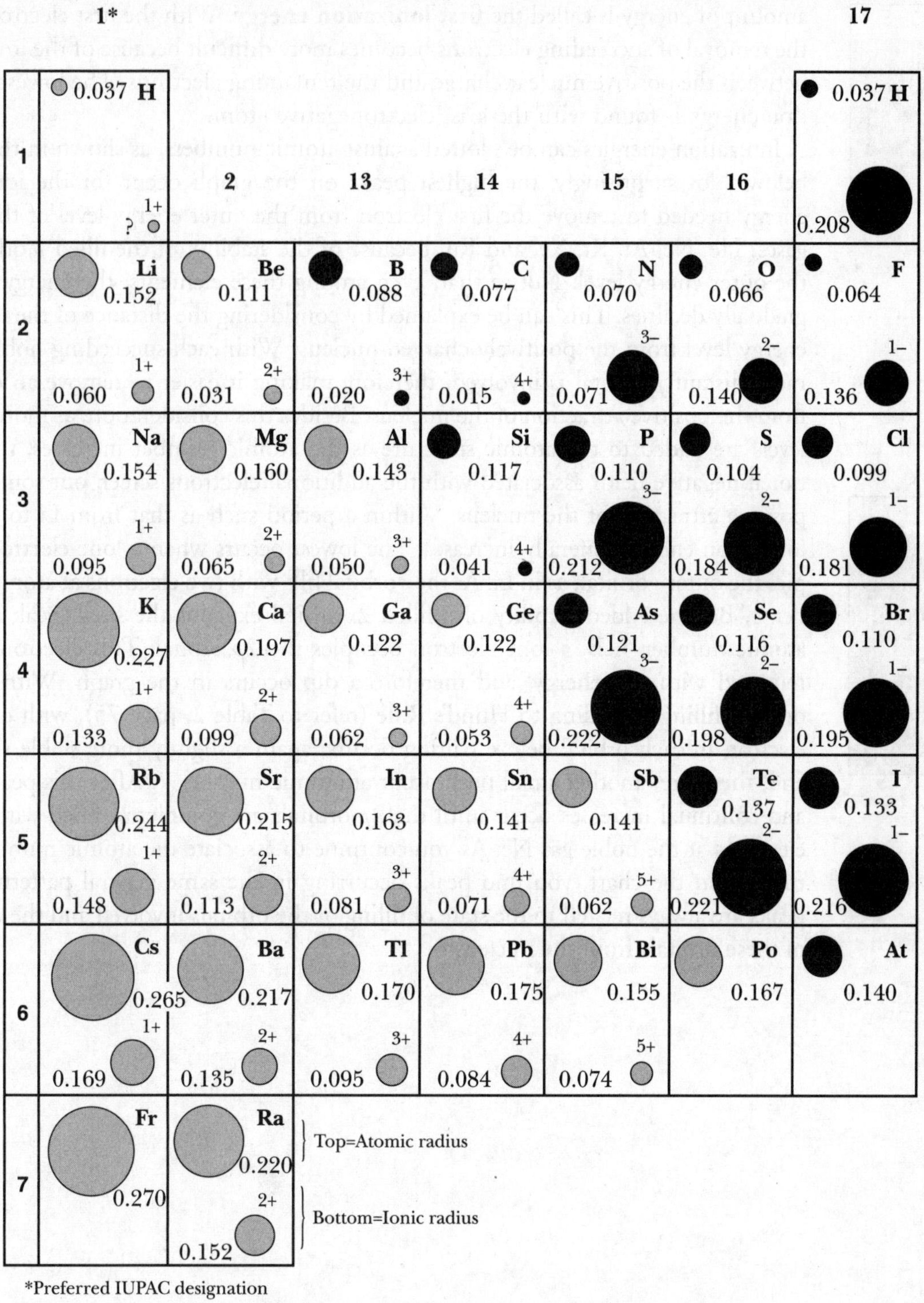

FIGURE 10. Radii of Some Atoms and Ions (in nanometers)

Notes: The atomic radius is usually given for metal atoms, which are shown in gray, and the covalent radius is usually given for atoms of nonmetals, which are shown in black.

TIP

Know this definition of the first ionization energy.

amount of energy is called the **first ionization energy**. With the first electron gone, the removal of succeeding electrons becomes more difficult because of the imbalance between the positive nuclear charge and the remaining electrons. The lowest ionization energy is found with the least electronegative atom.

Ionization energies can be plotted against atomic numbers, as shown in the graph below. Not surprisingly, the highest peaks on the graph occur for the ionization energy needed to remove the first electron from the outer energy level of the noble gases, He, Ne, Ar, Kr, Xe, and Rn, because of the stability of the filled *p* orbitals in the outer energy level. Notice that, even among these elements, the energy needed gradually declines. This can be explained by considering the distance of the involved energy level from the positively charged nucleus. With each succeeding noble gas, a more distant *p* orbital is involved, therefore making it easier to remove an electron from the positive attraction of the nucleus. Besides this consideration, as more energy levels are added to the atomic structure as the atomic number increases, the additional negative fields associated with the additional electrons screen out some of the positive attraction of the nucleus. Within a period such as that from Li to Ne, the ionization energy generally increases. The lowest occurs when a lone electron occupies the outer *s* orbital, as in Li. As the *s* orbital fills with two electrons at atomic number 4, Be, the added stability of a filled 2*s* orbital explains the small peak at 4. At atomic number 5, B, a lone electron occupies the 2*p* orbital. This electron can be removed with less energy, and therefore a dip occurs in the graph. With the 2*p* orbitals filling according to Hund's Rule (refer to Table 2, page 75), with only one electron in each orbital before pairing occurs, again a slightly more stable situation and, therefore, another small peak occur at atomic number 7. After this peak, a dip and continual increases occur until the 2*p* orbitals are completely filled with paired electrons at the noble gas Ne. As you continue to associate the atomic number with the line in the chart, you find peaks occurring in the same general pattern. These peaks are always related to the state of filling of the orbitals involved and the distance of these orbitals from the nucleus.

TIP

Know how this trend relates to the chart on page 87.

Can you explain the peaks?

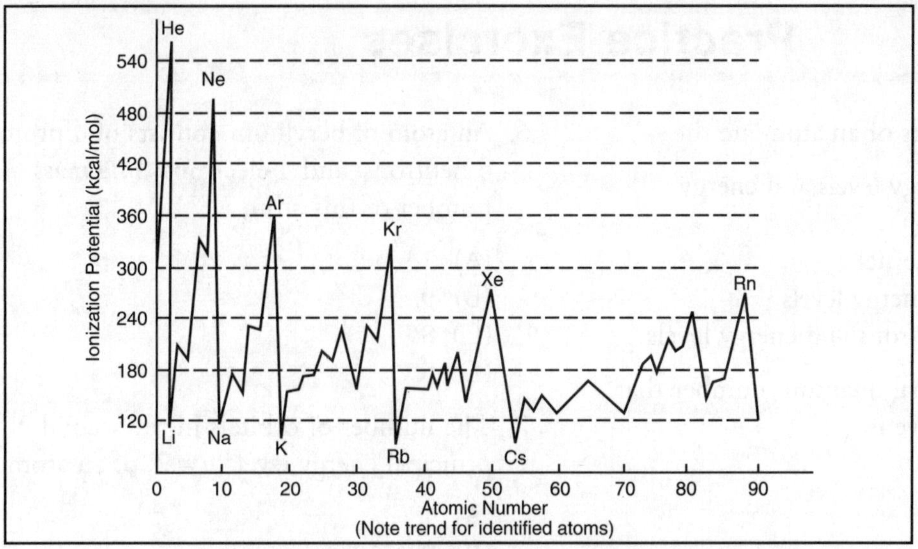

	Element	Atomic Number	First Ionization Energy (eV)	Second Ionization Energy (eV)	Third Ionization Energy (eV)
Group 1 or IA	Li	3	5.39	75.7	
	K	19	4.34	31.8	
Group 2 or IIA	Be	4	9.32	18.2	154
	Mg	12	7.64	15.1	80.3

Sample Ionization Energies for Second and Third Electron Removal

Chapter Summary

The following terms summarize all the concepts and ideas that were introduced in this chapter. You should be able to explain their meaning and how you would use them in chemistry. They appear in boldface type in this chapter to draw your attention to them. The boldface type also makes it easier for you to look them up if you need to. You could also use the search or "Google" action on your computer to get a quick and expanded explanation of these terms, laws, and formulas.

atomic mass
atomic number
atomic radii
Aufbau Principle
Bohr model
covalent radius
Dalton's atomic theory
electron
electron affinity
electronegativity
group or family
Hund's Rule
inert atoms
ionic radii
ionization energy
isotopes
Lewis structure
Mendeleev
metallic atoms
metalloids
Moseley
neutron
nonmetallic atoms
nucleus
oxidation number
Pauli Exclusion Principle
periodic law
period or row
quantum numbers
quantum theory
s, p, d, f orbitals
transition elements
uncertainty principle
valence electrons
wave-mechanical model

Practice Exercises

1. The two main parts of an atom are the
   (A) principal energy levels and energy sublevels
   (B) nucleus and kernel
   (C) nucleus and energy levels
   (D) planetary electrons and energy levels

2. The lowest principal quantum number that an electron can have is
   (A) 0
   (B) 1
   (C) 2
   (D) 3

3. The sublevel that has only one orbital is identified by the letter
   (A) *s*
   (B) *p*
   (C) *d*
   (D) *f*

4. The sublevel that can be occupied by a maximum of 10 electrons is identified by the letter
   (A) *d*
   (B) *f*
   (C) *p*
   (D) *s*

5. An orbital may never be occupied by
   (A) 1 electron
   (B) 2 electrons
   (C) 3 electrons
   (D) 0 electrons

6. An atom of beryllium consists of 4 protons, 5 neutrons, and 4 electrons. The mass number of this atom is
   (A) 13
   (B) 9
   (C) 8
   (D) 5

7. The number of orbitals in the second principal energy level, $n = 2$, of an atom is
   (A) 1
   (B) 9
   (C) 16
   (D) 4

8. Lewis structure consists of the symbol representing the element and an arrangement of dots that usually shows
   (A) the atomic number
   (B) the atomic mass
   (C) the number of neutrons
   (D) the electrons in the outermost energy level

9. Chlorine is represented by the Lewis structure $:\dot{\underset{\cdot\cdot}{\text{Cl}}}:$. The atom that would be represented by an identical electron-dot arrangement has the atomic number
   (A) 7
   (B) 9
   (C) 15
   (D) 19

The following questions are in the format that is used on the **SAT Subject Test in Chemistry**. If you are not familiar with these types of questions, study pages xii – xv before doing the remainder of the review questions.

Questions 10–14

Use this abbreviated periodic table to answer the following questions.

Period ↓	Group 1	Group 2		Group 14	Group 15	Group 16	Group 17
2	(A) **Li**	**Be**		(D) **C**	**N**	**O**	(E) **F**
3	(B) **Na**	**Mg**	....				
4	(C) **K**						

10. Which neutral atom has an outer energy level configuration of $3s^1$?

11. An atom of which element shows the greatest affinity for an additional electron?

12. Which element is the most active metal of this group?

13. Which element has the lowest electronegativity of this group?

14. Which element has an outer orbital configuration of $2s^2\ 2p^2$?

Directions:

(Further explanation and the examples for this type of question can be found on pages xiv–xv.) Every question below contains two statements, I in the left-hand column and II in the right-hand column. For each question, decide if statement I is true or false and whether statement II is true or false, and fill in the corresponding T or F ovals in the answer spaces. *Fill in oval CE only if statement II is a correct explanation of statement I.

I		II
15. Si, with an atomic number of 14, will probably exhibit an oxidation number of +4 in a compound	BECAUSE	silicon is an element that has amphoteric properties.
16. Nonmetallic atoms have larger ionic radii than their atomic radii	BECAUSE	nonmetallic atoms generally gain electrons to form the ionic state and increase the size of the electron cloud.
17. Elements in the upper right corner of the Periodic Table form acid anhydrides	BECAUSE	nonmetallic oxides react with water to form acid solutions.

*Fill in oval CE only if II is a correct explanation of I.

	I	II	CE*
15.	(T) (F)	(T) (F)	◯
16.	(T) (F)	(T) (F)	◯
17.	(T) (F)	(T) (F)	◯

Answers and Explanations

1. **(C)** The two main parts of the atom are the nucleus and its energy levels.

2. **(B)** The principal quantum numbers start with the value of 1 to represent the first level.

3. **(A)** The letter *s* is used to represent the first orbital that can hold two electrons.

4. **(A)** The sublevel d has five orbitals that each can hold two electrons, totaling ten electrons.

5. **(C)** Each orbital can only hold two electrons.

6. **(B)** The mass number is the total of the number of protons and neutrons, which in this case is nine.

7. **(D)** The second principal energy level has an *s* orbital and three *p* orbitals, making a total of four.

8. **(D)** The Lewis electron-dot notation shows the symbol and the outermost energy-level electrons, which are referred to as the valence electrons.

9. **(B)** Chlorine is a member of the halogen family found in group 17. The other element in the same family would be the element with the atomic number 19.

10. **(B)** Because Na (sodium) has the position shown, it has the atomic number 11. The electron configuration is $1s^2\ 2s^2\ 2p^6\ 3s^1$.

11. **(E)** Fluorine has seven electrons in its outer energy level and needs only one more to complete its outer energy level octet. It therefore has the greatest affinity for one more electron.

12. **(C)** The most active metal of the group is found in the lower left-hand corner. Because its outer energy level is furthest from the nucleus and these are the most loosely held electrons, it is the most likely to lose an electron.

13. **(C)** The element that has the lowest electronegativity will be in the lower left corner. For this group, it is K.

14. **(D)** The element that has this configuration is carbon, which has six electrons. The first two are in the first level, and the next four are in the second level, as $2s^2\ 2p^2$.

15. **(T, T)** Both statements are true, but statement II does not explain statement I.

16. **(T, T, CE)** Because nonmetallic atoms gain electrons to form ions, the additional negative charge of the added electrons increases the size of the ion. This occurs because of the increased repulsion of the additional negative charge(s) and additional shielding from the positively charged nucleus.

17. **(T, T, CE)** Elements found in the upper right corner of the Periodic Table are nonmetals that form oxides and react with water to form acids. An example is sulfur that forms sulfur trioxide, and this reacts with water to form sulfuric acid.

Bonding

CHAPTER 3

- Define ionic and covalent bonds, and explain how they form.
- Identify the differences in the continuum that exists between ionic and covalent bonding.
- Describe the implications of the type of bond on the structure of the compound.
- Explain the implications of intermolecular forces and van der Waals forces.
- Explain how VSEPR and hybridization solve the need to comply with known molecular shapes.

Some elements show no tendency to combine with either like atoms or other kinds of elements. These elements are said to be monoatomic molecules; three examples are helium, neon, and argon. A **molecule** is defined as the smallest particle of an element or a compound that retains the characteristics of the original substance. Water is a triatomic molecule since two hydrogen atoms and one oxygen atom must combine to form the substance water with its characteristic properties. When atoms do combine to form molecules, there is a shifting of valence electrons, that is, the electrons in the outer energy level of each atom. Usually, this results in completion of the outer energy level of each atom. This more stable form may be achieved by the gain or loss of electrons or the sharing of pairs of electrons. The resulting attraction of the atoms involved is called a **chemical bond**. When a chemical bond forms, energy is released; when this bond is broken, energy is absorbed.

This relationship of bonding to the valence electrons of atoms can be further explained by studying the electron structures of the atoms involved. As already mentioned, the noble gases are monoatomic molecules. The reason can be seen in the electron distributions of these noble gases as shown in the following table.

Noble Gas	Electron Distribution	Electrons in Valence Energy Level
Helium	$1s^2$	2
Neon	$1s^2\ 2s^2\ 2p^6$	8
Argon	$1s^2\ 2s^2\ 2p^6\ 3s^2\ 3p^6$	8
Krypton	$1s^2\ 2s^2\ 2p^6\ 3s^2\ 3p^6\ 3d^{10}\ 4s^2\ 4p^6$	8
Xenon	$1s^2\ 2s^2\ 2p^6\ 3s^2\ 3p^6\ 3d^{10}\ 4s^2\ 4p^6\ 4d^{10}\ 5s^2\ 5p^6$	8
Radon	$1s^2\ 2s^2\ 2p^6\ 3s^2\ 3p^6\ 3d^{10}\ 4s^2\ 4p^6\ 4d^{10}\ 4f^{14}\ 5s^2\ 5p^6\ 5d^{10}\ 6s^2\ 6p^6$	8

TIP

Notice the reccurrence of the octet (8) of electrons in noble gases.

The distinguishing factor in these very stable configurations is the arrangement of two *s* electrons and six *p* electrons in the valence energy level in five of the six atoms. (Note that helium, He, has only a single *s* valence energy level, which is filled with two electrons, making He a very stable atom.) This arrangement is called a **stable octet.** All other elements, other than the noble gases, have one to seven electrons in their outer energy levels. These elements are reactive to varying degrees. When they do react to form chemical bonds, usually the electrons shift in such a way that stable octets form. In other words, in bond formation, atoms usually attain the stable electron structure of one of the noble gases. The type of bond formed is directly related to whether this structure is achieved by gaining, losing, or sharing electrons.

TYPES OF BONDS

Ionic Bonds

TIP

A 1.7 or greater electronegativity difference between atoms will form an ionic bond.

When the electronegativity values of two kinds of atoms differ by 1.7 or more (especially differences greater than 1.7), the more electronegative atom will borrow the electrons it needs to fill its energy level, and the other atom will lend electrons until it, too, has a complete energy level. Because of this exchange, the borrower becomes negatively charged and the lender becomes positively charged. They are now referred to as **ions**, and the bond or attraction between them is called an **ionic bond**. These ions do not retain the properties of the original atoms. An example can be seen in Figure 11.

These ions do not form an individual molecule in the liquid or solid phase but are arranged into a crystal lattice or giant ion-molecule containing many such ions. Ionic solids of this type tend to have high melting points and will not conduct a current of electricity until they are in the molten state.

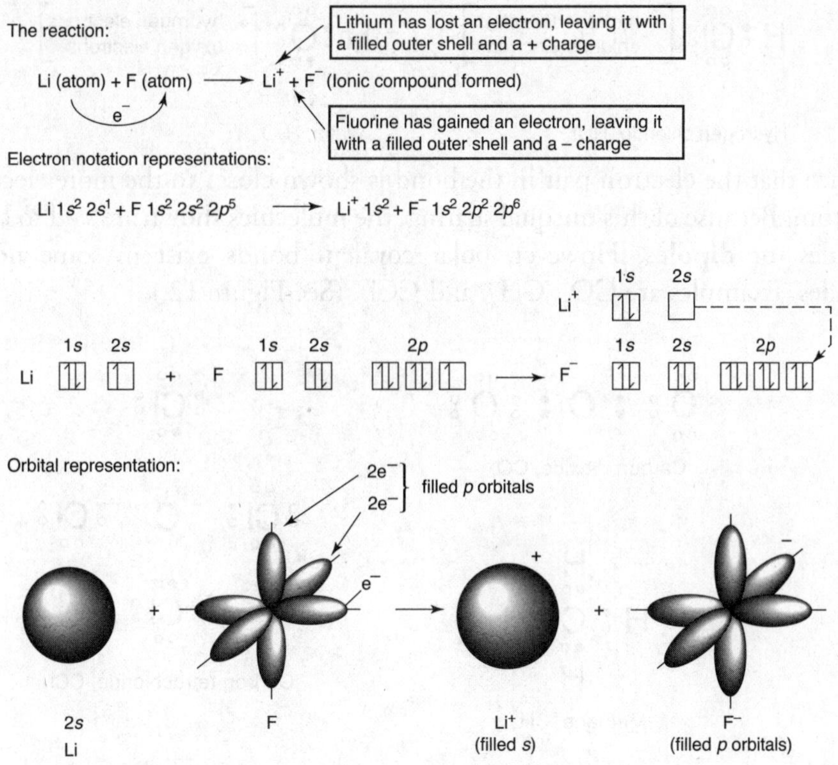

Figure 11. Three Representations of the Ionic Bonding of LiF

Covalent Bonds

When the electronegativity difference between two or more atoms is 0 or very small (not greater than about 0.5), the atoms tend to share the valence electrons in their respective outer energy levels. This attraction is called a **nonpolar covalent bond.** Here is an example using electron-dot notation and orbital notation:

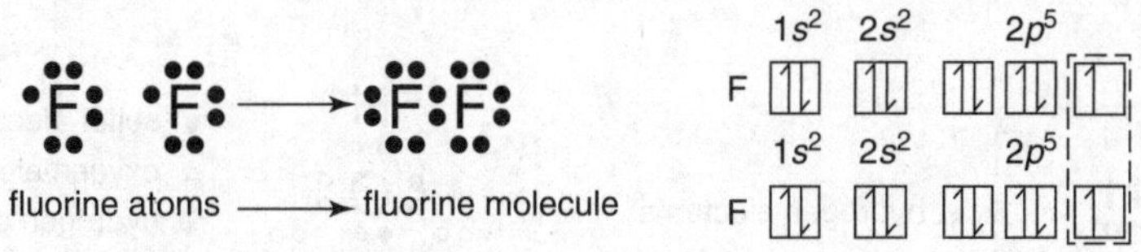

These covalent bonded molecules do not have electrostatic charges like those of ionic bonded substances. In general, covalent compounds are gases, liquids having fairly low boiling points, or solids that melt at relatively low temperatures. Unlike ionic compounds, they do not conduct electric currents.

When the electronegativity difference is between 0.4 and 1.6, there will not be an equal sharing of electrons between the atoms involved. The shared electrons will be more strongly attracted to the atom of greater electronegativity. As the difference in the electronegativities of the two elements increases above 0.4, the polarity or degree of ionic character increases. At a difference of 1.6 or more, the bond has more than 50% ionic character. However, when the difference is between 0.4 and 1.6, the bond is called a **polar covalent bond**. An example:

TIP

Covalent bonds involve a sharing of electrons between atoms. Their electronegativity difference is between 0 to 0.5.

> **TIP**
>
> Polar covalent bonds have unequal sharing of electrons. Their electronegativity difference is between 0.4 and 1.6.

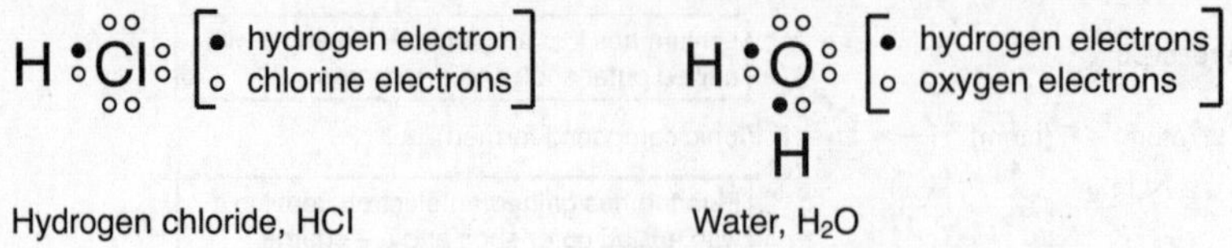

Notice that the electron pair in the bond is shown closer to the more electronegative atom. Because of this unequal sharing, the molecules shown are said to be polar molecules, or **dipoles**. However, polar covalent bonds exist in some nonpolar molecules. Examples are CO_2, CH_4, and CCl_4. (See Figure 12.)

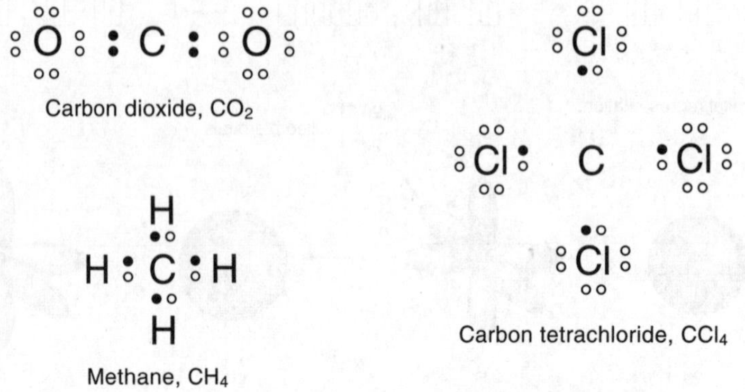

Figure 12. Polar Covalent Bonds in Nonpolar Molecules

In all the examples in Figure 12 the bonds are polar covalent bonds, but the important thing is that they are symmetrically arranged in the molecule. The result is a nonpolar molecule.

In the **covalent** bonds described so far, the shared electrons in the pair were contributed one each from the atoms bonded. In some cases, however, both electrons for the shared pair are supplied by only one of the atoms. Two examples are the bonds in NH_4^+ and H_2SO_4. (See Figure 13.)

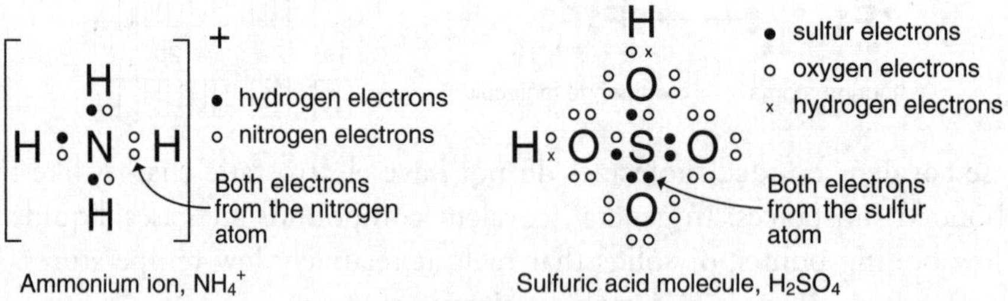

Figure 13. Covalent Bonds (both electrons supplied by one atom)

The formation of a covalent bond can be described in graphic form and related to the potential energies of the atoms involved. Using the formation of the hydrogen molecule as an example, we can show how the potential energy changes as the two atoms approach and form a covalent bond. In the illustration that follows, frames (1), (2), and (3) show the effect on potential energy as the atoms move closer to each other. In frame (3), the atoms have reached the condition of lowest potential energy, but the inertia of the atoms pulls them even closer, as shown in frame

(4). The repulsion between them then forces the two nuclei to a stable position, as shown in frame (5).

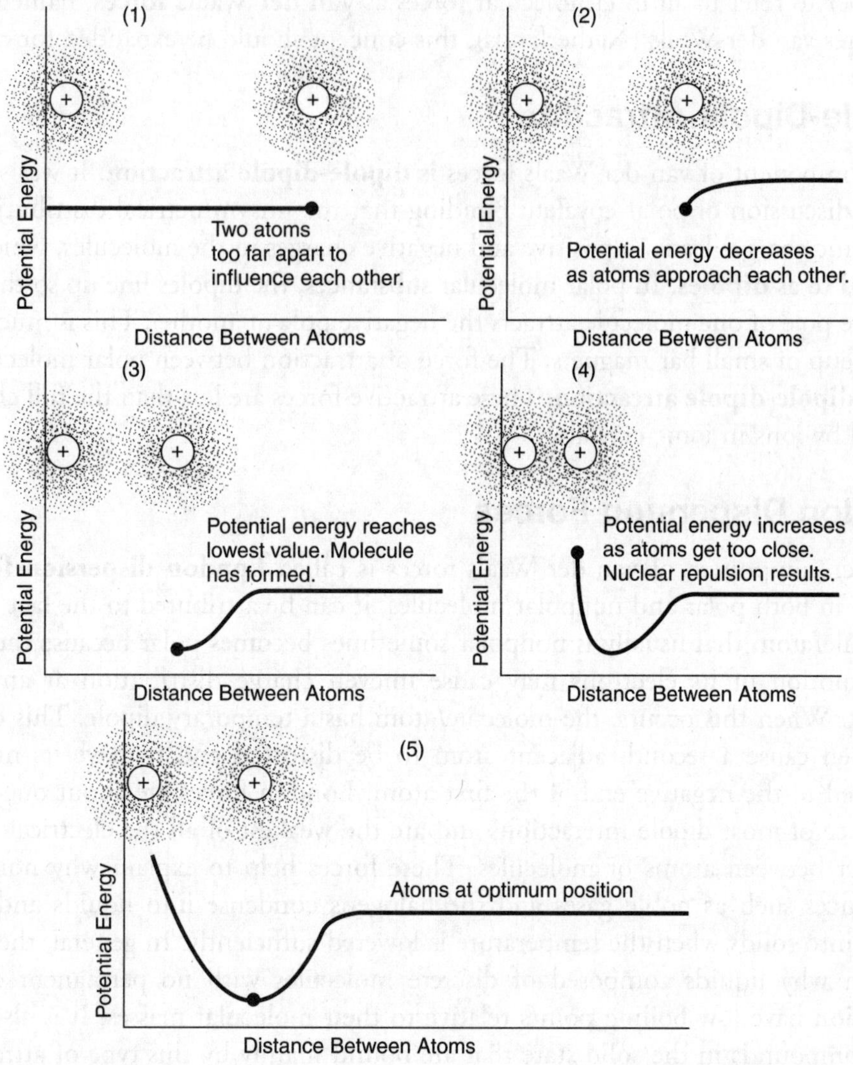

Metallic Bonds

In most metals, one or more of the valence electrons become detached from the atom and migrate in a "sea" of free electrons among the positive metal ions. The attractive force strength varies with the nuclear positive charge of the metal atoms and the number of electrons in this electron sea. Both of these factors are reflected in the amount of heat required to vaporize the metal. The strong attraction between these differently charged particles forms a **metallic bond**. Because of this firm bonding, metals usually have high melting points, show great strength, and are good conductors of electricity.

TIP

Metallic bonds are like positive ions in a "sea" of electrons.

INTERMOLECULAR FORCES OF ATTRACTION

The term **intermolecular forces** refers to attractions *between* molecules. Although it is proper to refer to all intermolecular forces as **van der Waals forces**, named after Johannes van der Waals (Netherlands), this concept should be expanded for clarity.

Dipole-Dipole Attraction

One component of van der Waals forces is **dipole-dipole attraction**. It was shown in the discussion of polar covalent bonding that the unsymmetrical distribution of electronic charges leads to positive and negative charges in the molecules, which are referred to as **dipoles**. In polar molecular substances, the dipoles line up so that the positive pole of one molecule attracts the negative pole of another. This is much like the lineup of small bar magnets. The force of attraction between polar molecules is called **dipole-dipole attraction**. These attractive forces are less than the full charges carried by ions in ionic crystals.

London Dispersion Forces

TIP

Weakest of all, the London dispersion forces are one-tenth the force of most dipole attractions.

Another component of van der Waals forces is called **London dispersion forces**. Found in both polar and nonpolar molecules, it can be attributed to the fact that a molecule/atom that usually is nonpolar sometimes becomes polar because the constant motion of its electrons may cause uneven charge distribution at any one instant. When this occurs, the molecule/atom has a temporary dipole. This dipole can then cause a second, adjacent atom to be distorted and to have its nucleus attracted to the negative end of the first atom. London forces are about one-tenth the force of most dipole interactions and are the weakest of all the electrical forces that act between atoms or molecules. These forces help to explain why nonpolar substances such as noble gases and the halogens condense into liquids and then freeze into solids when the temperature is lowered sufficiently. In general, they also explain why liquids composed of discrete molecules with no permanent dipole attraction have low boiling points relative to their molecular masses. It is also true that compounds in the solid state that are bound mainly by this type of attraction have rather soft crystals, are easily deformed, and vaporize easily. Because of the low intermolecular forces, the melting points are low and evaporation takes place so easily that it may occur at room temperature. Examples of such solids are iodine crystals and moth balls (paradichlorobenzene and naphthalene).

Hydrogen Bonds

A proton or hydrogen nucleus has a high concentration of positive charge. When a hydrogen atom is bonded to a highly electronegative atom, its positive charge will have an attraction for neighboring electron pairs. This special kind of dipole-dipole attraction is called a **hydrogen bond**. The more strongly polar the molecule is, the more effective the hydrogen bonding is in binding the molecules into a larger unit. As a result the boiling points of such molecules are higher than those of similar nonpolar molecules. Good examples are water and hydrogen fluoride.

Figure 14 shows that in the formation of H_2O, H_2S, H_2Se, and H_2Te an unusual rise in the boiling point of H_2O occurs that is not in keeping with the typical slow increase of boiling point as molecular mass increases. Instead of the expected slope of the line between H_2O and H_2S, which is shown in Figure 14 as a dashed line, the actual boiling point of H_2O is quite a bit higher—100°C. The explanation is that hydrogen bonding occurs in H_2O but not to any significant degree in the other compounds.

TIP

Notice how hydrogen bonding elevates the boiling point of H_2O above the expected slope.

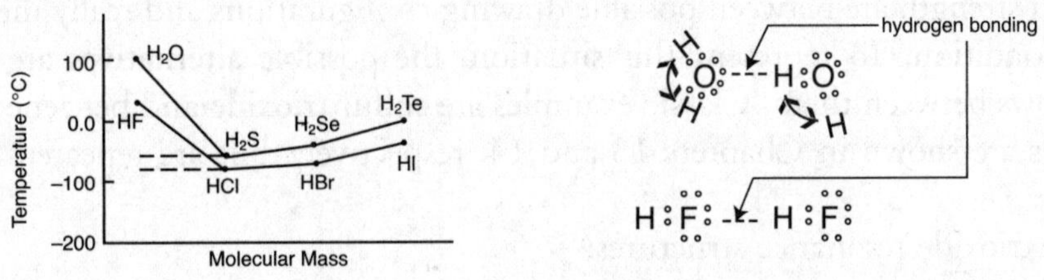

Figure 14. Boiling Points of Hydrogen Compounds with Similar Electron Dot Structures

This same phenomenon occurs with the hydrogen halides (HF, HCl, HBr, and HI). Note in Figure 14 that hydrogen fluoride, HF, which has strong hydrogen bonding, shows an unexpectedly high boiling point.

Hydrogen bonding also explains why some substances have unexpectedly low vapor pressures, high heats of vaporization, and high melting points. In order for vaporization or melting to take place, molecules must be separated. Energy must be expended to break hydrogen bonds and thus break down the larger clusters of molecules into separate molecules. As with the boiling point, the melting point of H_2O is abnormally high when compared with the melting points of the hydrogen compounds of the other elements having six valence electrons, which are chemically similar but which have no apparent hydrogen bonding. The hydrogen bonding effect in water is discussed on pages 189 and 190.

DOUBLE AND TRIPLE BONDS

To achieve the **octet** structure, which is an outer energy level resembling the noble gas configuration of eight electrons, it is necessary for some atoms to share two or even three pairs of electrons. Sharing two pairs of electrons produces a **double bond**. An example:

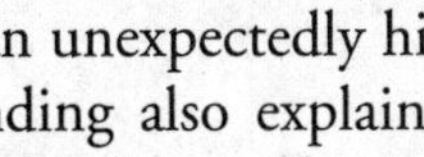, and by a line formula O=C=O

carbon dioxide

In the line formula, only the shared pair of electrons is indicated by a bond (—). The sharing of three electron pairs results in a **triple bond**. An example:

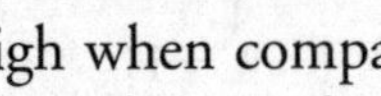, and by a line formula H−C≡C−H

acetylene

It can be assumed from these structures that there is a greater electron density between the nuclei involved and hence a greater attractive force between the nuclei and the shared electrons. Experimental data verify that greater energy is required to break double bonds than single bonds, and triple bonds than double bonds. Also, since these stronger bonds tend to pull atoms closer together, the atoms joined by double and triple bonds have smaller interatomic distances and greater bond strengths, respectively.

TIP

Resonance structure is a **hybrid** of the possible drawings

RESONANCE STRUCTURES

It is not always possible to represent the bonding structure of a molecule by either the Lewis dot structure or the line drawing because data about the bonding distance and bond strength are between possible drawing configurations and really indicate a hybrid condition. To represent this situation, the possible alternatives are drawn with arrows between them. Classic examples are sulfur trioxide and benzene. These structures are shown in Chapters 13 and 14, respectively, but are repeated here as examples.

Sulfur trioxide resonance structures:

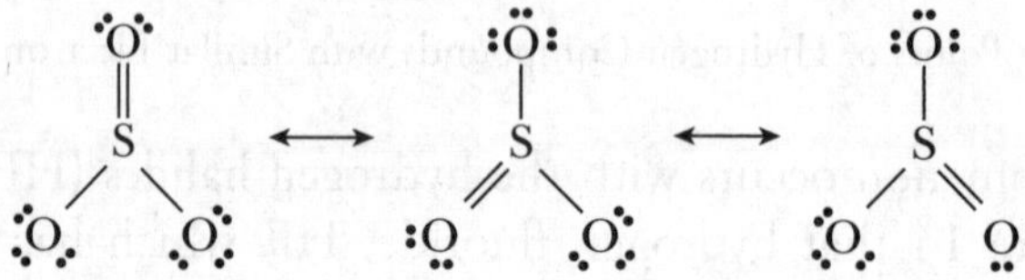

Benzene resonance structures:

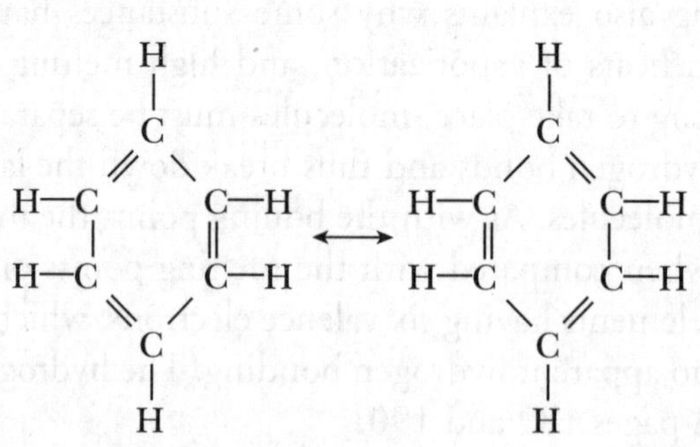

TIP

Two theories explain molecular structure:

VSEPR theory uses valence shell electron pair repulsion.

Hybridization theory uses changes in the orbitals of the valence electrons.

MOLECULAR GEOMETRY—VSEPR—AND HYBRIDIZATION

VSEPR—Electrostatic Repulsion

Properties of molecules depend not only on the bonding of atoms but also on the **molecular geometry**—the three-dimensional arrangement of the molecule's atoms in space. The combination of the polarity of the bonds and the geometry of the molecule determine the molecular polarity. This can be defined as the uneven distribution of the molecular charge. The chemical formula reveals little information about a molecule's geometry. It is only after doing many tests designed to reveal the shapes of the various molecules that chemists developed two different yet equally successful theories to explain certain aspects of their findings. One theory accounts structurally for molecular bond angles. The other is used to describe changes in the orbitals that contain the valence electrons of a molecule's atoms. The structural theory that deals with the bond angles is called the **VSEPR theory**, whereas the one that describes changes in the orbitals that contain the valence electrons is called the **hybridization theory**. (VSEPR represents Valence Shell Electron Pair Repulsion.)

VSEPR uses as its basis the fact that like charges will orient themselves in such a way as to diminish the repulsion between them.

1. Mutual repulsion of two electron clouds forces them to the opposite sides of a sphere. This is called a **linear arrangement**.

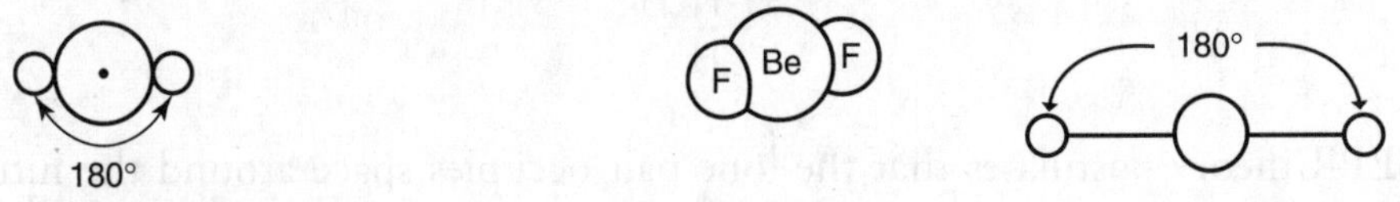

EXAMPLE: BeF_2, berylium fluoride

2. Minimum repulsion between three electron pairs occurs when the pairs are at the vertices of an equilateral triangle inscribed in a sphere. This is called a **trigonal-planar arrangement**.

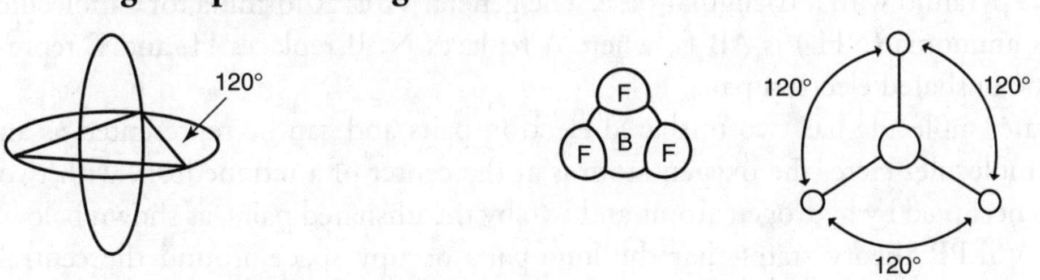

TIP

These basic arrangements are important to learn!

EXAMPLE: BF_3, boron trifluoride

3. Four electron pairs are farthest apart at the vertices of a tetrahedron inscribed in a sphere. This is called a **tetrahedral** shape.

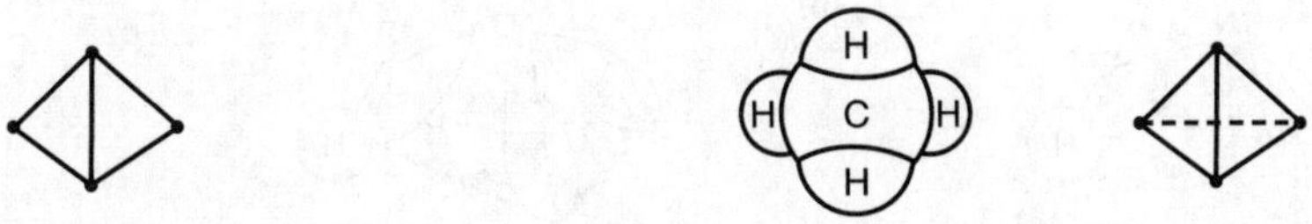

TIP

Configurations often appear as questions on the SAT test.

EXAMPLE: CH_4, methane

4. Mutual repulsion of six identical electron clouds directs them to the corners of an inscribed regular octahedron. This is said to have **octahedral** geometry.

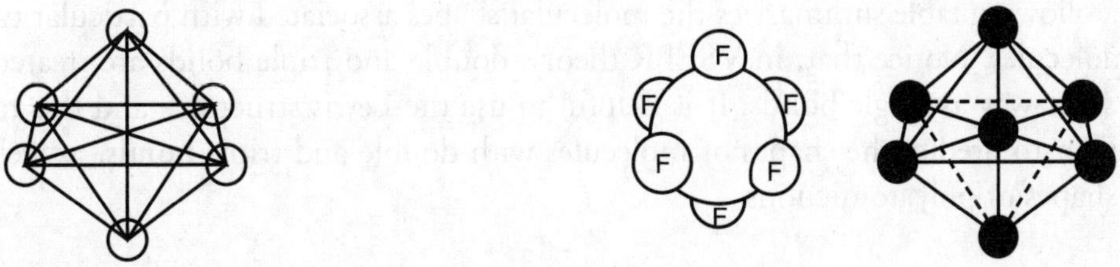

EXAMPLE: SF_6, sulfur hexafluoride

VSEPR and Unshared Electron Pairs

Ammonia, NH_3, and water, H_2O, are examples of molecules in which the central atom has both shared and unshared electron pairs. Here is how the VSEPR theory accounts for the geometries of these molecules.

The Lewis structure of ammonia shows that, in addition to the three electron pairs the central nitrogen atom shares with the three hydrogen atoms, it also has one unshared pair of electrons:

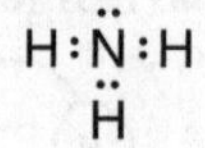

VSEPR theory postulates that the lone pair occupies space around the nitrogen atom just as the bonding pairs do. Thus, as in the methane molecule shown in the preceding section, the electron pairs maximize their separation by assuming the four corners of a tetrahedron. Lone pairs do occupy space, but our description of the observed shape of a molecule refers to the positions of atoms only. Consequently, as shown in the drawing below, the molecular geometry of an ammonia molecule is that of a pyramid with a triangular base. The general VSEPR formula for a molecule such as ammonia (NH_3) is AB_3E, where A replaces N, B replaces H, and E represents the unshared electron pair.

A water molecule has two unshared electron pairs and can be represented as an AB_2E_2 molecule. Here, the oxygen atom is at the center of a tetrahedron, with two corners occupied by hydrogen atoms and two by the unshared pairs, as shown below. Again, VSEPR theory states that the lone pairs occupy space around the central atom but that the actual shape of the molecule is determined only by the positions of the atoms. In the case of water, this results in a "bent," or angular, molecule.

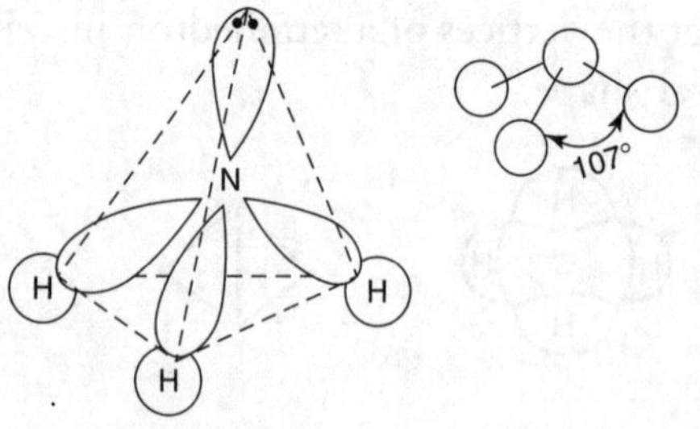

Ammonia, NH_3

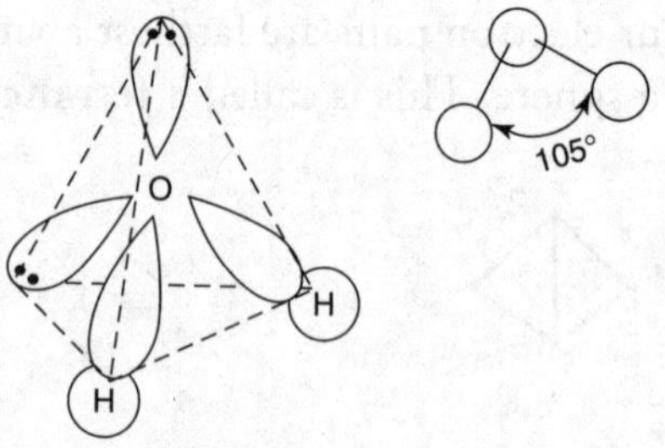

Water, H_2O

VSEPR and Molecular Geometry

The following table summarizes the molecular shapes associated with particular types of molecules. Notice that, in VSEPR theory, double and triple bonds are treated in the same way as single bonds. It is helpful to use the Lewis structures and this table together to predict the shapes of molecules with double and triple bonds, as well as the shapes of polyatomic ions.

Summary of Molecular Shapes

Type of Molecule	Molecular Shape	Atoms Bonded to Central Atom	Lone Pairs of Electrons	Formula Example	Lewis Structure
Linear	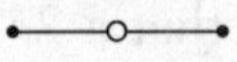	2	0	BeF_2	:F–Be–F:
Bent	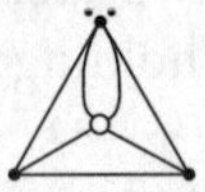	2	1	$SnCl_2$	Sn bonded to :Cl: and :Cl:
Trigonal-planar	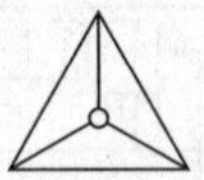	3	0	BF_3	B bonded to :F:, :F:, :F:
Tetrahedral	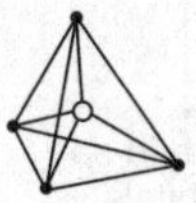	4	0	CH_4	H–C–H with H above and below
Trigonal-pyramidal	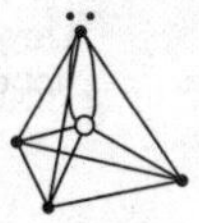	3	1	NH_3	N̈ bonded to H H H
Bent	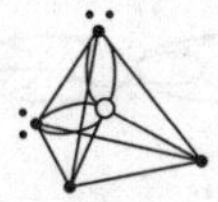	2	2	H_2O	Ö bonded to H H
Trigonal-bipyramidal	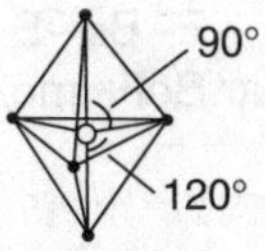	5	0	PCl_5	P bonded to five :Cl:
Octahedral	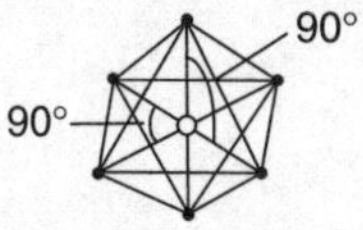	6	0	SF_6	S bonded to six :F:

Hybridization

The molecular configurations derived by VSEPR can also be arrived at through the concept of **hybridization**. Briefly stated, this means that two or more pure atomic orbitals (usually *s*, *p*, and *d*) can be mixed to form two or more new hybrid atomic orbitals that are identical. Hybridization can be illustrated as follows:

1. *sp* Hybrid Orbitals
Spectroscopic measurements of beryllium fluoride, BeF_2, reveal a bond angle of 180° and equal bond lengths.

F—BE—F
180°

The ground state of beryllium is:

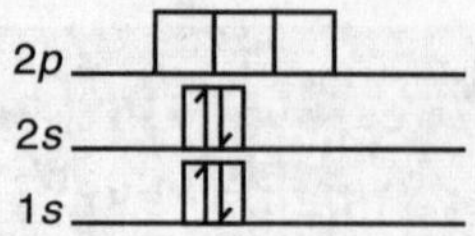

To accommodate the experimental data, we theorize that a 2*s* electron is excited to a 2*p* orbital; then the two orbitals hybridize to yield two identical orbitals called *sp* orbitals. Each contains one electron but is capable of holding two electrons.

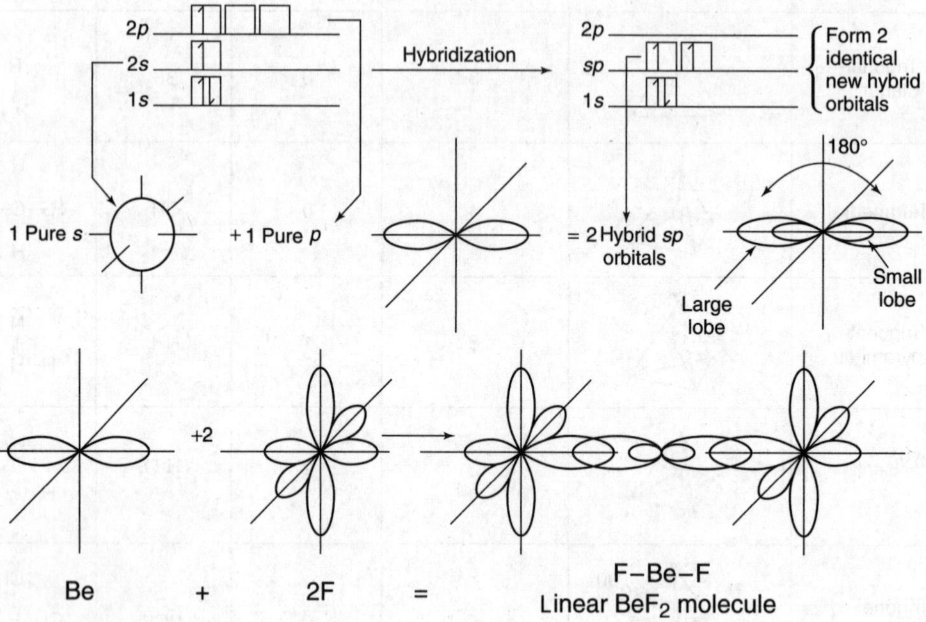

2. sp^2 Hybrid Orbitals

Boron trifluoride, BF_3, has bond angles of 120° of equal strength. To accommodate these data, the boron atom hybridizes from its ground state of $1s^22s^22p^1$ to:

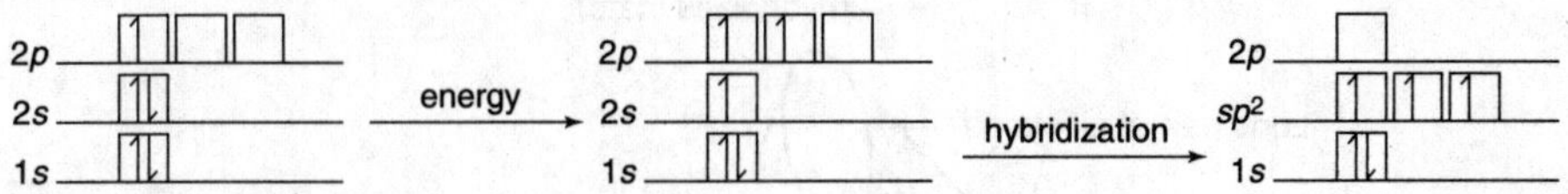

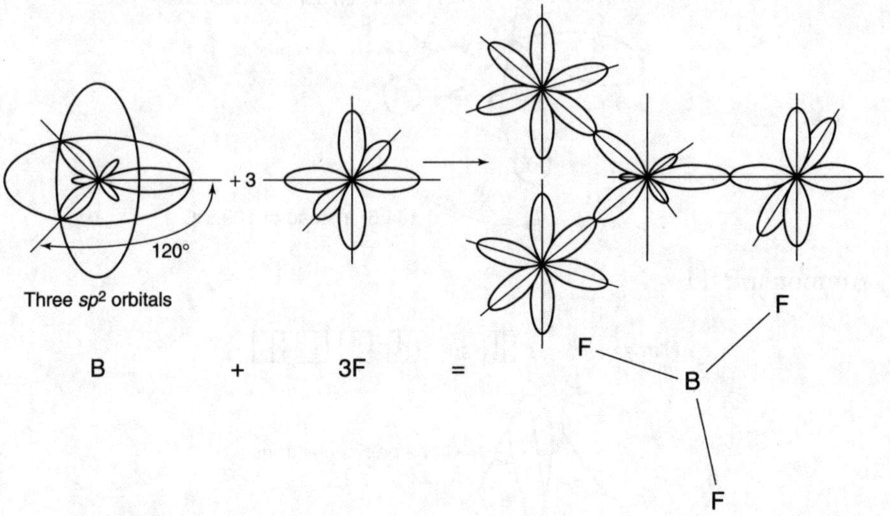

3. sp^3 Hybrid Orbitals

Methane, CH_4, can be used to illustrate this hybridization. Carbon has a ground state of $1s^22s^22p^2$. One $2s$ electron is excited to a $2p$ orbital, and the four involved orbitals then form four new identical sp^3 orbitals.

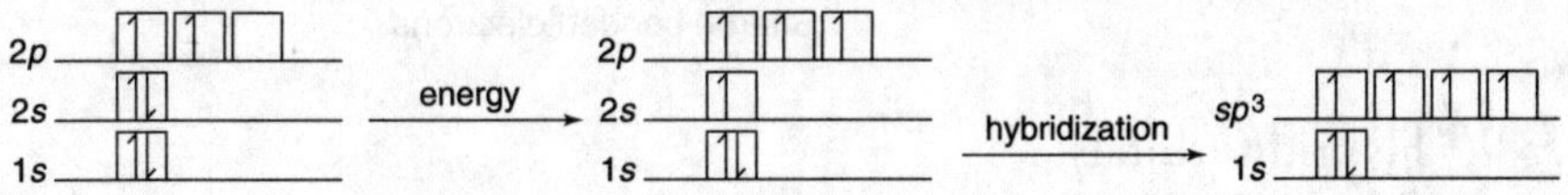

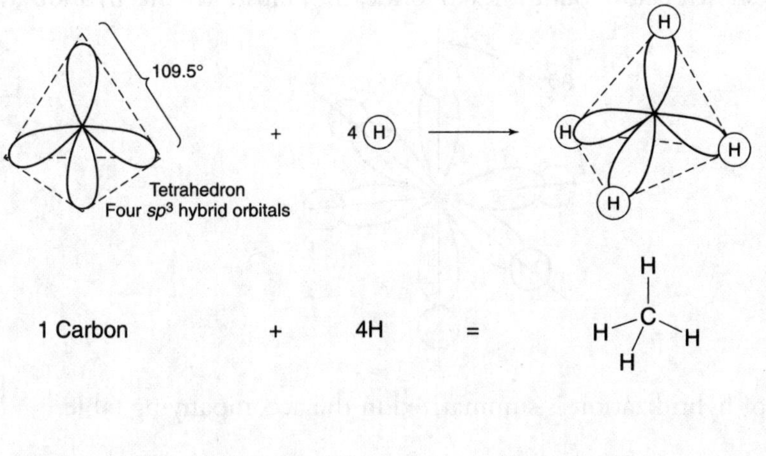

In some compounds where only certain sp^3 orbitals are involved in bonding, distortion in the bond angle occurs because of unbonded electron repulsion. Examples:

a. Water, H_2O

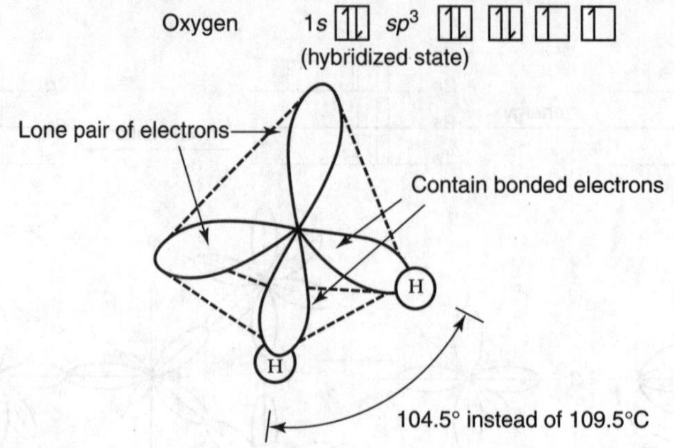

b. Ammonia, NH_3

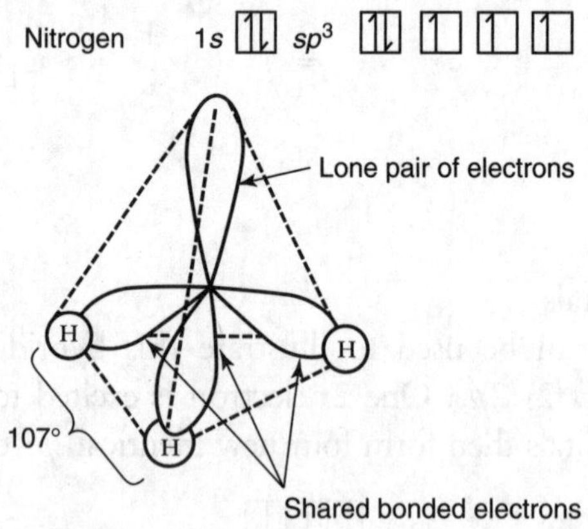

4. sp^3d^2 Hybrid Orbitals

These orbitals are formed from the hybridization of an *s* and a *p* electron promoted to *d* orbitals and transformed into six equal sp^3d^2 orbitals. The spatial form is shown below. Sulfur hexafluoride, SF_6, illustrates this hybridization.

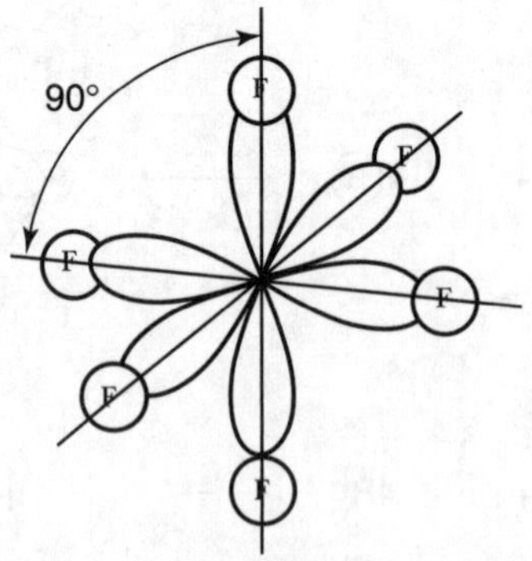

The concept of hybridization is summarized in the accompanying table.

Summary of Hybridization

Number of Bonds	Number of Unused Electron Pairs	Type of Hybrid Orbital	Angle Between Bonded Atoms	Geometry	Example
2	0	*sp*	180°	Linear	BeF_2
3	0	sp^2	120°	Trigonal-planar	BF_3
4	0	sp^3	109.5°	Tetrahedral	CH_4
3	1	sp^3	90° to 109.5°	Pyramidal	NH_3
2	2	sp^3	90° to 109.5°	Angular	H_2O
6	0	sp^3d^2	90°	Octahedral	SF_6

SIGMA AND PI BONDS

When bonding occurs between *s* and *p* orbitals, each bond is identified by a special term. A *sigma bond* is a bond between *s* orbitals, or an *s* orbital and another orbital such as a *p* orbital. It includes bonding between hybrids of *s* orbitals such as *sp*, sp^2, and sp^3.

TIP

Know these hybrid orbitals designations and their corresponding shapes.

In the methane molecule, the sp^3 orbitals are each bonded to hydrogen atoms. These are sigma bonds.

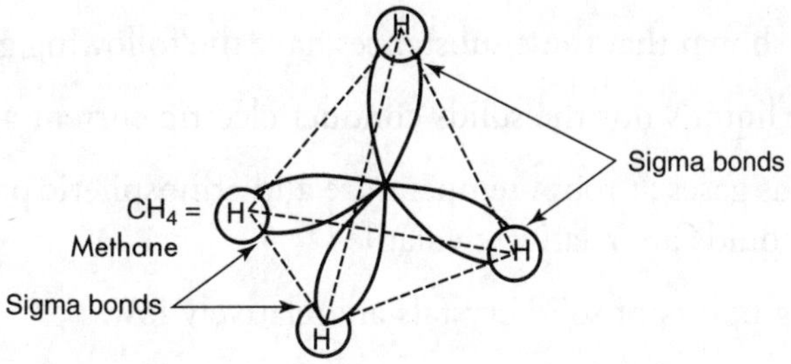

When two *p* orbitals share electrons in a covalent bond, the result is a **pi bond**. Here is an example:

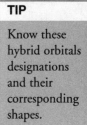

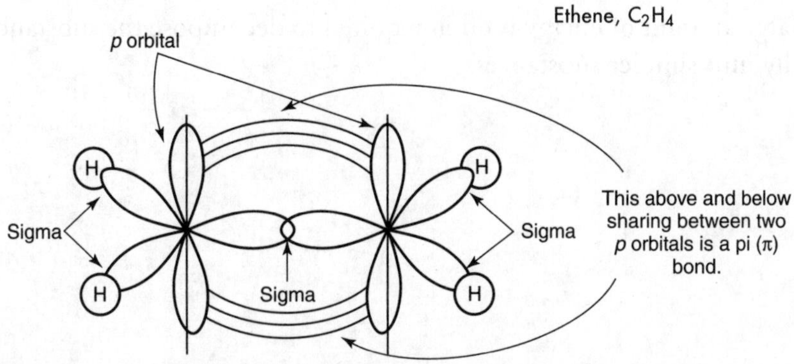

Chapter 14 gives more examples of sigma and pi bonding.

PROPERTIES OF IONIC SUBSTANCES

Laboratory experiments reveal that, in general, ionic substances are characterized by the following properties:

1. In the solid phase at room temperature they do not conduct appreciable electric current.
2. In the liquid phase they are relatively good conductors of electric current. The conductivity of ionic substances is much smaller than that of metallic substances.
3. They have relatively high melting and boiling points. There is a wide variation in the properties of different ionic compounds. For example, potassium iodide (KI) melts at 686°C and boils at 1,330°C, while magnesium oxide (MgO) melts at 2,800°C and boils at 3,600°C. Both KI and MgO are ionic compounds.
4. They have relatively low volatilities and low vapor pressures. In other words, they do not vaporize readily at room temperature.
5. They are brittle and easily broken when stress is exerted on them.
6. Those that are soluble in water form electrolytic solutions that are good conductors of electricity. There is, however, a wide range in the solubilities of ionic compounds. For example, at 25°C, 92 grams of sodium nitrate ($NaNO_3$) dissolves in 100 grams of water, while only 0.0002 grams of barium sulfate ($BaSO_4$) dissolves in the same mass of water.

PROPERTIES OF MOLECULAR CRYSTALS AND LIQUIDS

Experiments have shown that these substances have the following general properties:

1. Neither the liquids nor the solids conduct electric current appreciably.
2. Many exist as gases at room temperature and atmospheric pressure, and many solids and liquids are relatively volatile.
3. The melting points of solid crystals are relatively low.
4. The boiling points of the liquids are relatively low.
5. The solids are generally soft and have a waxy consistency.
6. A large amount of energy is often required to decompose the substance chemically into simpler substances.

Chapter Summary

The following terms summarize all the concepts and ideas that were introduced in this chapter. You should be able to explain their meaning and how you would use them in chemistry. They appear in boldface type in this chapter to draw your attention to them. The boldface type also makes it easier for you to look them up if you need to. You could also use the search or "Google" action on your computer to get a quick and expanded explanation of these terms, laws, and formulas.

covalent bond	ionic bond	stable octet
dipole-dipole attraction	London dispersion forces	sigma bond
electrostatic repulsion	metallic bond	van der Waals forces
hybridization	pi bond	VSEPR
hydrogen bond	resonance structure	

Practice Exercises

The following questions are in the format that is used on the **SAT Subject Area Test in Chemistry**. If you are not familiar with these types of questions, study pages xii – xv before doing these review questions.

Directions: The following set of lettered choices refers to the numbered questions immediately below it. For each numbered item, choose the one lettered choice that fits it best. Then fill in the corresponding oval on the answer sheet. Each choice in the set may be used once, more than once, or not at all.

Questions 1–7

(A) ionic
(B) covalent
(C) polar covalent
(D) metallic
(E) hydrogen bonding

1. When the electronegativity difference between two atoms is 2, what type of bond can be predicted?
2. If two atoms are bonded in such a way that both members of the pair equally share one electron with the other, what is the bond called?
3. Which of the five choices is considered the weakest bond in the group?
4. Which of the above bonds explains water's abnormally high boiling point?
5. If the sharing of an electron pair is unequal and the atoms have an electronegativity difference of 1.4 to 1.6, what is this type of sharing called?
6. If an electron is lost by one atom and completely captured by another, what is this type of bond called?
7. If one or more valence electrons become detached from the atoms and migrate in a "sea" of free electrons among the positive metal ions, what is this type of bonding called?

Directions:

(Further explanation and the examples for this type of question can be found on pages xiv–xv.) Every question below contains two statements, I in the left-hand column and II in the right-hand column. For each question, decide if statement I is true or false and whether statement II is true or false, and fill in the corresponding T or F ovals in the answer spaces. *Fill in oval CE only if statement II is a correct explanation of statement I.

	I		II
8.	Minimum repulsion between two electron pairs in a molecular compound will result in a linear structure	BECAUSE	the VSEPR model says that like charges will orient themselves so as to diminish the repulsion between them.
9.	Sodium chloride is an example of ionic bonding	BECAUSE	sodium and chlorine have the same electronegativity.
10.	Ammonia has a trigonal pyramidal molecular structure	BECAUSE	ammonia has a tetrahedral electron pair geometry with three atoms bonded to the central atom.

*Fill in oval CE only if II is a correct explanation of I.

	I	II	CE*
8.	(T) (F)	(T) (F)	( )
9.	(T) (F)	(T) (F)	( )
10.	(T) (F)	(T) (F)	( )

Answers and Explanations

1. **(A)** When the electronegativity difference between the two atoms is greater than 1.7, the bond between them is considered more than 50% ionic.

2. **(B)** When two atoms equally share a pair of electrons, the bond is considered covalent. With an electronegativity difference of 0 to 0.3, the bond is still considered covalent.

3. **(E)** Hydrogen bonding is the weakest because it is the weak attraction of the hydrogen end of a polar molecule to the partial negative charge of an adjacent molecule. Especially strong in water, it is responsible for many of water's properties.

4. **(E)** Because of the strong hydrogen bonding in water, it takes much more energy to cause the molecules to break away from each other in the liquid state and change to steam.

5. **(C)** When the electronegativity difference between the two bonding atoms is between 0.3 and 1.7, the electrons are not equally shared and result in a polar covalent bond.

6. **(A)** This is the definition of a pure ionic bond.

7. **(D)** Metallic bonds are defined as a "sea" of free electrons that migrate through the metal.

8. **(T, T, CE)** The mutual repulsion of two electron clouds forces them to the opposite sides of a sphere. An example is BeF_2, which forms a linear molecular structure like this, F – Be – F.

9. **(T, F)** The bond between sodium and chlorine in sodium chloride is ionic, because the difference between their electronegativity is greater than 1.7.

10. **(T, T, CE)** The molecular structure of NH_3 is a trigonal pyramidal structure like this.

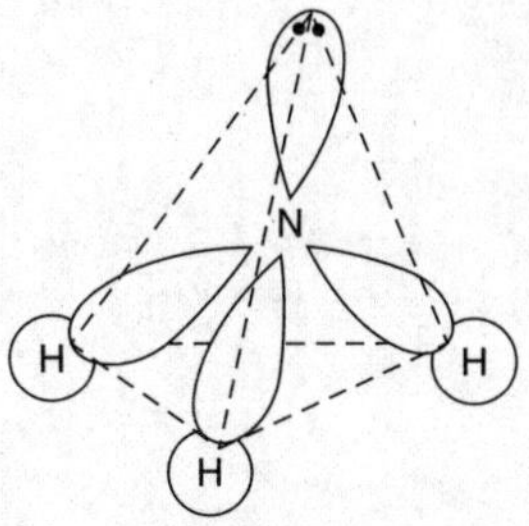

Ammonia, NH_3

Chemical Formulas

CHAPTER 4

- Recall and use the basic rules about oxidation numbers to write correct formulas with the ions listed on page 115. This includes writing formulas with polyatomic ions.
- Name compounds (acids, bases, and salts) using the Stock system and the prefix system, and write their formulas.
- Calculate the oxidation number of an element in any formula.
- Calculate the formula mass of a compound and the percent composition of each element.
- Calculate the empirical formula when given the percent composition of each element. When given the formula mass, you should be able to find the true formula.
- Write a simple balanced equation, indicating the phase (or state) of the reactants and products.

From a knowledge of oxidation numbers and valence and an understanding of atomic structure, it is possible to write chemical formulas.

An oxidation number is a small, whole-number superscript preceded by a plus or minus sign. An ionic charge, on the other hand, is shown by a small, whole-number superscript followed by a plus or minus sign that indicates the charge of the ion. Table 6 is a list of oxidation numbers often encountered in a first-year chemistry course.

OXIDATION STATES AND FORMULA WRITING

1. The symbols of metals have + signs, while those of nonmetals and all the polyatomic ions *except* ammonium have – signs.
2. When an element exhibits two possible oxidation states, the lower state can be indicated by the suffix *-ous* and the higher one by *-ic*. Another method, now preferred, of indicating this difference is to use the Roman numeral of the oxidation state in parentheses after the name of the element. For example: ferrous iron or iron(II) for the +2 oxidation state of iron.
3. A **polyatomic ion** is a group of elements that act like a single atom in the formation of a compound. The bonds within these polyatomic ions are

predominantly covalent, but the groups of atoms as a whole have an excess of electrons when combined, and thus are usually negative ions.

When you attempt to write a formula, it is important to know whether the substance actually exists. For example, one can easily write the formula of carbon nitrate, but no chemist has ever prepared this compound. Here are the basic rules for writing formulas, with three examples carried through each step:

1. Write the symbols of the components, placing the positive part first, and then the negative part.

Sodium chloride	Calcium oxide	Ammonium sulfate
NaCl	CaO	NH_4SO_4

2. Indicate the oxidation number above and to the right of each symbol. (Enclose polyatomic ions in parentheses for the time being.)

$Na^{+1}Cl^{-1}$	$Ca^{+2}O^{-2}$	$(NH_4)^{+1}(SO_4)^{-2}$

3. For each element, write a subscript equal to the oxidation number of the other element or polyatomic ion. This is the same as the mechanical criss-cross method.

$Na_1{}^{+1}Cl_1{}^{-1}$	$Ca_2{}^{+2}O_2{}^{-2}$	$(NH_4)_2{}^{+1}(SO_4)_1{}^{-2}$

Because the positive oxidation number shows the number of electrons that may be lost or shared and the negative oxidation number shows the number of electrons that may be gained or shared, you must have just as many electrons lost (or partially lost in sharing) as are gained (or partially gained in sharing).

4. Now rewrite the formulas, omitting the subscript 1, the parentheses of the polyatomic ions that have the subscript 1, and the plus and minus numbers.

NaCl	Ca_2O_2	$(NH_4)_2SO_4$

5. As a general rule, reduce the subscript numbers in the final formula to their lowest terms. There are, however, certain exceptions, such as hydrogen peroxide (H_2O_2), and acetylene (C_2H_2). For these exceptions, you must have more specific information about the compound. The only one that needs to be changed is Ca_2O_2 to CaO.

The only way to become proficient at writing formulas is to memorize the oxidation numbers of common elements and to practice writing formulas. It is especially important to memorize the polyatomic ions and their oxidation numbers.

MORE ABOUT OXIDATION NUMBERS

Because it is important to know how to assign oxidation numbers in order to write formulas correctly, the basic rules concerning the assignment of oxidation numbers need to be listed. The oxidation number concept was introduced in Chapter 2,

Table 6. Table of Oxidation Numbers

	Monovalent I		Bivalent II		Trivalent III		Tetravalent IV		V	
Metals Cations (+)	Hydrogen	H	Barium	Ba	Aluminum	Al	Carbon	C	Arsenic(V)*	As
	Potassium	K	Calcium	Ca	Gold(III)*	Au	Silicon	Si	Phosphorus(V)*	P
	Sodium	Na	Cobalt	Co	Arsenic (III)*	As	Manganese(IV)*	Mn	Antimony(V)*	Sb
	Silver	Ag	Magnesium	Mg	Chromium	Cr	Tin(IV)*	Sn	Bismuth(V)*	Bi
	Mercury(I)*	Hg	Lead	Pb	Iron(III)*	Fe	Platinum	Pt		
	Copper(I)*	Cu	Zinc	Zn	Phosphorus(III)*	P	Sulfur	S		
	Gold(I)*	Au	Mercury(II)*	Hg	Antimony(III)*	Sb				
	Ammonium†	(NH_4)	Copper(II)*	Cu	Bismuth(III)*	Bi				
			Iron(II)*	Fe						
			Manganese(II)*	Mn						
			Tin(II)*	Sn						
Nonmetals‡ Anions(−)	Fluorine	F	Oxygen	O	Nitrogen	N	Carbon	C		
	Chlorine	Cl	Sulfur	S	Phosphorus	P				
	Bromine	Br								
	Iodine	I								
Polyatomic Ions (−)	Hydroxide	(OH)	Carbonate	(CO_3)	Borate	(BO_3)	Ferrocyanide	$[Fe(CN)_6]$		
	Bicarbonate	(HCO_3)	Sulfite	(SO_3)	Phosphate	(PO_4)				
	Nitrite	(NO_2)	Sulfate	(SO_4)	Phosphite	(PO_3)				
	Nitrate	(NO_3)	Tetraborate	(B_4O_7)	Ferricyanide	$[Fe(CN)_6]$				
	Hypochlorite	(ClO)	Silicate	(SiO_3)						
	Chlorate	(ClO_3)	Chromate	(CrO_4)						
	Chlorite	(ClO_2)	Oxalate	(C_2O_4)						
	Perchlorate	(ClO_4)								
	Acetate	$(C_2H_3O_2)$								
	Permanganate	(MnO_4)								
	Bisulfate	(HSO_4)								

*Use of the Roman numeral, instead of the suffix, is now preferred; for example, iron(II) oxide instead of ferrous oxide.

†Polyatomic ion

‡Last syllable of nonmetal name is changed to *-ide* in binary compound.

which deals with atomic structure. Keep in mind that oxidation numbers are written with the sign preceding the number while ions have the sign following the number.

TIP

Learn these rules!

The basic rules follow:

1. The oxidation number of a monoatomic ion is the same as the charge on the ion. Therefore, in NaCl, the oxidation number assigned to sodium is +1, and the oxidation number of chlorine is −1.
2. The oxidation number of any element is zero when the atoms of that element are in the elemental state. An atom is in its elemental state when it is not chemically combined with atoms of other elements.
3. In a covalent compound of two elements, the more electronegative element has a negative oxidation number and the other element has a positive oxidation number. For example, in ammonia, NH_3, the nitrogen is more electronegative. Each of the valence electrons from the three hydrogen atoms is drawn nearer to the nitrogen atom. The oxidation number of the nitrogen in ammonia is therefore −3. Since each hydrogen atom partially loses one electron, the oxidation number of hydrogen is +1.
4. The oxidation number of oxygen in most of its compounds is −2. In peroxides an exception occurs, and oxygen is assigned −1.
5. The oxidation number of hydrogen in most compounds is +1. In hydrides an exception occurs, and hydrogen is assigned −1.
6. The oxidation number of fluorine in all compounds is −1 because it is the most electronegative element.
7. In neutral compounds, for all the atoms represented by the formula, the quantitative sum of the positive oxidation numbers must equal the quantitative sum of the negative oxidation numbers, and their sum must equal 0. This can best be illustrated in an example.

TIP

The most important rule is number 7.

TIP

Know how to solve this type of question.

In Na_2SO_4, what is the oxidation number of sulfur?

$$Na_2SO_4$$

We know the oxidation number of one atom of Na is +1. There are two atoms of Na.	We know one atom of O is assigned −2. There are four atoms of O.
$2 \times +1 = +2$	$4 \times -2 = -8$

Since the positive sum plus the negative sum must equal 0,

$$(+2) + x + (-8) = 0$$
$$\underline{x = +6}$$

The sulfur must be using a +6 oxidation state.

Another example: What is the oxidation number of chromium in $K_2Cr_2O_7$?

$$
\begin{aligned}
K &= 2 \times (+1) = +2 \\
Cr &= 2 \times (x) = 2x \\
O &= 7 \times (-2) = -14 \\
(+2) + (2x) &+ (-14) = 0 \\
2x &= +12 \\
\underline{x} &\underline{= +6}
\end{aligned}
$$

Chromium is using a +6 oxidation number.

8. In a polyatomic ion, the algebraic sum of the positive and negative oxidation numbers of all the atoms in the formula must equal the charge on the ion.

Example: What is the oxidation number of sulfur in SO_4^{2-}?

$$
\begin{aligned}
S &= 1 \times (x) = 1x \\
O &= 4 \times (-2) = -8 \\
(1x) + (-8) &= -2 \text{ (the charge of the ion)} \\
x &= +6
\end{aligned}
$$

The sulfur is using a +6 oxidation state.

Another example: In the NO_3^- ion, find the oxidation number of nitrogen.

$$
\begin{aligned}
&NO_3^- \\
(x) + 3(-2) &= -1 \text{ (the charge of the ion)} \\
x &= +5
\end{aligned}
$$

Nitrogen is using an oxidation number of +5.

NAMING COMPOUNDS

A **binary compound** consists of two elements. The name of the compound also consists of the two elements, but the second name has its ending changed to *-ide*. For example, the compound formula from sodium and chlorine, NaCl, is named sodium chloride; AgCl = silver chloride. If a metal has two different oxidation numbers, this can be indicated by the use of the suffix *-ous* for the lower one and *-ic* for the higher one. As previously stated, however, the preferred way, called the **Stock system**, is to use a Roman numeral after the name to indicate the oxidation state.

> **TIP**
>
> The Stock system uses Roman numerals after the name to indicate the oxidation state.

EXAMPLES: $FeCl_2$ = iron(II) chloride
$FeCl_3$ = iron(III) chloride

If elements combine in varying proportions, thus forming two or more compounds of varying compositions, the name of the second element may be preceded by a prefix, such as *mon(o)-* (one), *di-* (two), *tri-* (three), or *pent(a)-* (five). (The prefix does not use the letter in parentheses when preceding a vowel in the word to which it is attached.) Here are some examples: carbon dioxide, CO_2; carbon monoxide, CO; phosphorous trioxide, P_2O_3; phosphorous pentoxide, P_2O_5. Notice that, when these prefixes are used, it is not necessary to indicate the oxidation state of the first element in the name since it is given indirectly by the prefixed second element.

The following are examples of names in both the Stock system and the prefix system:

	Stock system	Prefix system
PCl_3	phosphorus(III) chloride	phosphorus trichloride
PCl_5	phosphorus(V) chloride	phosphorus pentachloride
N_2O	nitrogen(I) oxide	dinitrogen monoxide
NO	nitrogen(II) oxide	nitrogen monoxide
PbO_2	lead(IV) oxide	lead dioxide
Mo_2O_3	molybdenum(III) oxide	dimolybdenum trioxide

A binary acid is named by placing the prefix *hydro-* in front of the stem or full name of the nonmetallic element, and adding the ending *-ic.* Examples are *hydro*chlor*ic* acid (HCl) and *hydro*sulfur*ic* acid (H_2S).

A **ternary compound** consists of three elements, usually an element and a polyatomic ion. To name the compound, you merely name each component in the order of positive first and negative second.

Ternary acids usually contain hydrogen, a nonmetal, and oxygen. Because the amount of oxygen often varies, the name of the most common form of the acid in the series consists of merely the stem of the nonmetal with the ending *-ic.* The acid containing one less atom of oxygen than the most common acid is designated by the ending *-ous.* The name of the acid containing one more atom of oxygen than the most common acid has the prefix *per-* and the ending *-ic*; that of the acid containing one less atom of oxygen than the *-ous* acid has the prefix *hypo-* and the ending *-ous.* This is evident in Table 7 with the acids containing H, Cl, and O.

You can remember the names of the common acids and their salts by learning the following simple rules:

TIP

Learn these rules.

Rule	Example
-ic acids form *-ate* salts.	Sulfuric acid forms sulfate salts.
-ous acids form *-ite* salts.	Sulfurous acid forms sulfite salts.
hydro-(stem)-*ic* acids form *-ide* salts. salts.	Hydrochloric acid forms chloride

When the name of the ternary acid has the prefix *hypo-* or *per-*, that prefix is retained in the name of the salt (hypochlorous acid = sodium hypochlorite).

The names and formulas of some comon acids and bases are listed in Table 7.

Table 7. Formulas of Common Acids and Bases

ACIDS, BINARY		ACIDS, TERNARY	
Name	*Formula*	*Name*	*Formula*
Hydrofluoric	HF	Nitric	HNO_3
Hydrochloric	HCl	Nitrous	HNO_2
Hydrobromic	HBr	Hypochlorous	$HClO$
Hydriodic	HI	Chlorous	$HClO_2$
Hydrosulfuric	H_2S	Chloric	$HClO_3$
		Perchloric	$HClO_4$
BASES		Sulfuric	H_2SO_4
Sodium hydroxide	$NaOH$	Sulfurous	H_2SO_3
Potassium ″	KOH	Phosphoric	H_3PO_4
Ammonium ″	NH_4OH	Phosphorous	H_3PO_3
Calcium ″	$Ca(OH)_2$	Carbonic	H_2CO_3
Magnesium ″	$Mg(OH)_2$	Acetic	$HC_2H_3O_2$
Barium ″	$Ba(OH)_2$	Oxalic	$H_2C_2O_4$
Aluminum ″	$Al(OH)_3$	Boric	H_3BO_3
Iron(II) ″	$Fe(OH)_2$	Silicic	H_2SiO_3
Iron(III) ″	$Fe(OH)_3$		
Zinc ″	$Zn(OH)_2$		
Lithium ″	$LiOH$		

CHEMICAL FORMULAS

The chemical formula is an indication of the makeup of a compound in terms of the kinds of atoms and their relative numbers. It also has some quantitative applications. By using the atomic masses assigned to the elements, we can find the **formula mass** of a compound. If we are sure that the formula represents the actual makeup of one molecule of the substance, the term **molecular mass** may be used as well. In some cases the formula represents an ionic lattice and no discrete molecule exists, as in the case of table salt, NaCl, or the formula merely represents the simplest ratio of the combined substances and not a molecule of the substance. For example, CH_2 is the simplest ratio of carbon and hydrogen united to form the actual compound ethylene, C_2H_4. This simplest ratio formula is called the **empirical formula**, and the actual formula is the **true formula**. The formula mass is determined by multiplying the atomic mass units (as a whole number) by the subscript for that element and then adding these values for all the elements in the formula. For example:

$Ca(OH)_2$ (one calcium amu + two hydrogen and two oxygen amu = formula mass).

$$\begin{array}{l} 1\text{Ca (amu} = 40) = 40 \\ 2\text{O (amu} = 16) = 32 \\ \underline{2\text{H (amu} = 1) = -2} \end{array}$$

$$\text{Formula mass } Ca(OH)_2 = 74 \text{ amu (or } \mu)$$

In Chapter 6, the concept of a mole is introduced. If you have 6.02×10^{23} atoms of an element, then the atomic mass units can be expressed in grams, and then the formula mass can be called the **molar mass**. This is further explained on page 156. Another example is Fe_2O_3.

$$\begin{array}{l} 2\text{Fe (amu} = 56) = 112 \\ \underline{3\text{O (amu} = 16) = -48} \end{array}$$

$$\text{Formula mass } Fe_2O_3 = 160 \text{ amu}$$

TIP

Know how to compute the percentage composition of an element in a compound.

It is sometimes useful to know what percent of the total weight of a compound is made up of a particular element. This is called finding the **percentage composition**. The simple formula for this is:

$$\frac{\text{Total amu of the element in the compound}}{\text{Total formula amu}} \times 100\% = \begin{array}{c}\text{Percentage composition} \\ \text{of that element}\end{array}$$

To find the percent composition of calcium in calcium hydroxide in the example above, we set the formula up as follows:

$$\frac{\text{Ca} = 40 \text{ amu}}{\text{Formula mass} = 74 \text{ amu}} \times 100\% = 54\% \text{ Calcium}$$

To find the percent composition of oxygen in calcium hydroxide:

$$\frac{\text{O} = 32 \text{ amu}}{\text{Formula mass} = 74 \text{ amu}} \times 100\% = 43\% \text{ Oxygen}$$

To find the percent composition of hydrogen in calcium hydroxide:

$$\frac{\text{H} = 2 \text{ amu}}{\text{Formula mass} = 74 \text{ amu}} \times 100\% = 2.7\% \text{ Hydrogen}$$

ANOTHER EXAMPLE: Find the percent compositions of Cu and H_2O in the compound $CuSO_4 \cdot 5H_2O$ (the dot is read "with").

First, we calculate the formula mass:

$$\begin{aligned} &1\ Cu &&= 64\ \text{amu} \\ &1\ S &&= 32\ \text{amu} \\ &4\ O &&= 64\ (4 \times 16)\ \text{amu} \\ &5\ H_2O &&= 90\ (5 \times 18)\ \text{amu} \\ \hline & && 250\ \text{amu} \end{aligned}$$

and then find the percentages:

Percentage Cu:

$$\frac{Cu = 64\ \text{amu}}{\text{Formula mass} = 250\ \text{amu}} \times 100\% = 26\%$$

Percentage H_2O:

$$\frac{5H_2O = 90\ \text{amu}}{\text{Formula mass} = 250\ \text{amu}} \times 100\% = 36\%$$

When you are given the percentage of each element in a compound, you can find the empirical formula as shown with the following example:

Given that a compound is composed of 60.0% Mg and 40.0% O, find the empirical formula of the compound.

1. It is easiest to think of 100 mass units of this compound. In this case, the 100 mass units are composed of 60 amu of Mg and 40 amu of O. Because you know that 1 unit of Mg is 24 amu (from its atomic mass) and, likewise, 1 unit of O is 16, you can divide 60 by 24 to find the number of units of Mg in the compound and divide 40 by 16 to find the number of units of O in the compound.

Mg	O
24)60	16)40
2.5 units Mg	2.5 units O

2. Now, because we know formulas are made up of whole-number units of the elements, which are expressed as subscripts, we must manipulate these numbers to get whole numbers. This is usually accomplished by dividing these numbers by the smallest quotient. In this case they are equal, so we divide by 2.5.

Mg	O
2.5)2.5	2.5)2.5
1	1

3. So, the empirical formula is MgO.

ANOTHER EXAMPLE: Given: Ba = 58.81%, S = 13.73%, and O = 27.46%.
Find the empirical formula.

1. Divide each percent by the amu of the element.

Ba	S	O
137)58.8	32)13.7	16)27.5
0.43	0.43	1.72

2. Manipulate numbers to get small whole numbers. Try dividing them all by the smallest first.

Ba	S	O
0.43)0.43	0.43)0.43	0.43)1.72
1	1	4

3. The formula is $BaSO_4$.

In some cases you may be given the true formula mass of the compound. To check if your empirical formula is correct, add up the formula mass of the empirical formula and compare it with the given formula mass. If it is *not* the same, multiply the empirical formula by the small whole number that gives you the correct formula mass. For example, if your empirical formula is CH_2 (which has a formula mass of 14) and the true formula mass is given as 28, you can see that you must double the empirical formula by doubling all the subscripts. The true formula is C_2H_4.

LAWS OF DEFINITE COMPOSITION AND MULTIPLE PROPORTIONS

In the problems involving percent composition, we have depended on two things: each unit of an element has the same atomic mass, and every time the particular compound forms, it forms in the same percent composition. That this latter statement is true no matter the source of the compound is the **Law of Definite Composition.** There are some compounds formed by the same two elements in which the mass of one element is constant, but the mass of the other varies. In every case, however, the mass of the other element is present in a small-whole-number ratio to the weight of the first element. This is called the **Law of Multiple Proportions**. An example is H_2O and H_2O_2.

In H_2O the proportion of H : O = 2 : 16 or 1 : 8
In H_2O_2 the proportion of H : O = 2 : 32 or 1 : 16

The ratio of the mass of oxygen in each is 8 : 16 or 1 : 2 (a small-whole-number ratio).

WRITING AND BALANCING SIMPLE EQUATIONS

An equation is a simplified way of recording a chemical change. Instead of words, chemical symbols and formulas are used to represent the reactants and the products.

Here is an example of how this can be done. The following is the word equation of the reaction of burning hydrogen with oxygen:

Hydrogen + oxygen yields water.

Replacing the words with the chemical formulas, we have

$$H_2 + O_2 \rightarrow H_2O$$

We replaced hydrogen and oxygen with the formulas for their diatomic molecular states and wrote the appropriate formula for water based on the respective oxidation (valence) numbers for hydrogen and oxygen. Note that the word **yields** was replaced with the arrow.

Although the chemical statement tells what happened, it is not an equation because the two sides are not equal. While the left side has two atoms of oxygen, the right side has only one. Knowing that the Law of Conservation of Matter dictates that matter cannot easily be created or destroyed, we must get the number of atoms of each element represented on the left side to equal the number on the right. To do this, we can only use numbers, called **coefficients**, in front of the formulas. It is important to note that in attempting to balance equations THE SUBSCRIPTS IN THE FORMULAS MAY NOT BE CHANGED.

TIP

You cannot change subscripts of formulas to attempt to balance an equation!

Looking again at the skeleton equation, we notice that if 2 is placed in front of H_2O the numbers of oxygen atoms represented on the two sides of the equation are equal. However, there are now four hydrogens on the right side with only two on the left. This can be corrected by using a coefficient of 2 in front of H_2. Now we have a balanced equation:

$$2H_2 + O_2 \rightarrow 2H_2O$$

This equation tells us more than merely that hydrogen reacts with oxygen to form water. It has quantitative meaning as well. It tells us that two molecular masses of hydrogen react with one molecular mass of oxygen to form two molecular masses of water. Because molecular masses are indirectly related to grams, we may also relate the masses of reactants and products in grams.

$2H_2$	+	O_2	$\rightarrow$	$2H_2O$
2(2)		32		2(18)
4 units	+	32 units	=	36 units

4 grams of H_2 + 32 grams of O_2 = 36 grams of water

This aspect will be important in solving problems related to the masses of substances in a chemical equation.

Here is another, more difficult example: Write the balanced equation for the burning of butane (C_4H_{10}) in oxygen. First, we write the skeleton equation:

$$C_4H_{10} + O_2 \text{ yields } CO_2 + H_2O.$$

TIP

First deal with multiatomic reactants

Looking at the oxygens, we see that there are an even number on the left but an odd number on the right. This is a good place to start. If we use a coefficient of 2 for H_2O, that will even out the oxygens but introduce four hydrogens on the right while there are ten on the left. A coefficient of 5 will give us the right number of hydrogens but introduces an odd number of oxygens. Therefore we have to go to the next even multiple of 5, which is 10. Ten gives us 20 hydrogen atoms on the right. By placing another coefficient of 2 in front of C_4H_{10}, we also have 20 hydrogen atoms on the left. Now the carbons need to be balanced. By placing an 8 in front of CO_2, we have eight carbons on both sides. The remaining step is to balance the oxygens. We have 26 on the right side, so we need a coefficient of 13 in front of the O_2 on the left to give us 26 oxygens on both sides. Our balanced equation is:

TIP

Be sure you have included all sources of a particular element since it may occur in two or more compounds.

$$2C_4H_{10} + 13O_2 \rightarrow 8CO_2 + 10H_2O$$

For more practice in balancing equations, see page 175. (The balancing of equations using the electron exchange method is introduced at the end of Chapter 12.)

SHOWING PHASES IN CHEMICAL EQUATIONS

Once an equation is balanced, you may choose to give additional information in the equation. This can be done by indicating the phases of substances, telling whether each substance is in the liquid phase (ℓ), the gaseous phase (g), or the solid phase (s). Since many solids will not react to any appreciable extent unless they are dissolved in water, the notation (aq) is used to indicate that the substance exists in a water (aqueous) solution. Information concerning phase is given in parentheses following the formula for each substance. Several illustrations of this notation are given below:

TIP

(g) = gaseous state

(ℓ) = liquid state

(s) = solid state

Formula with Phase Notation	Meaning
$Cl_2(g)$	Chlorine gas
$H_2O(\ell)$	Water in the liquid state as opposed to ice or steam
NaCl(s)	Sodium chloride as a solid
NaCl(aq)	A water solution of dissolved sodium chloride

An example of phase notation in an equation:

$$2HCl(aq) + Zn(s) \rightarrow ZnCl_2(aq) + H_2(g)$$

In words, this says that a water solution of hydrogen chloride (called hydrochloric acid) reacts with solid zinc to produce zinc chloride dissolved in water plus hydrogen gas.

WRITING IONIC EQUATIONS

At times, chemists choose to show only the substances that react in the chemical action. These equations are called **ionic** equations because they stress the reaction and production of ions. If we look at the preceding equation, we see the complete cast of "actors":

Reactants

$2HCl(aq)$ releases	$\rightarrow$	$2H^+(aq) + 2Cl^-(aq)$ in solution
$Zn(s)$ stay as		$Zn(s)$ particles

Products

$ZnCl_2(aq)$ releases	$\rightarrow$	$Zn^{2+}(aq) + 2Cl^-(aq)$ in solution
$H_2(g)$ stay as		$H_2(g)$ molecules

Writing the complete reaction using these results, we have:

$$2H^+(aq) + 2Cl^-(aq) + Zn(s) \rightarrow Zn^{2+}(aq) + 2Cl^-(aq) + H_2(g)$$

Notice that nothing happened to the chloride ion. It appears the same on both sides of the equation. It is referred to as a spectator ion. In writing the net ionic equation, spectator ions are omitted, so the net ionic equation is:

$$2H^+(aq) + Zn(s) \rightarrow Zn^{2+}(aq) + H_2(g)$$

TIP

In net ionic equations, do not show "spectator" ions that do not change.

Chapter Summary

The following terms summarize all the concepts and ideas that were introduced in this chapter. You should be able to explain their meaning and how you would use them in chemistry. They appear in boldface type in this chapter to draw your attention to them. The boldface type also makes it easier for you to look them up if you need to. You could also use the search or "Google" action on your computer to get a quick and expanded explanation of these terms, laws, and formulas.

binary compound
coefficient
empirical formula
formula mass
molecular mass
percentage composition
polyatomic ion
Stock system
ternary compound
true formula

Law of Definite Composition
Law of Multiple Proportions

Practice Exercises

Write the formula or Stock name in questions 1 through 10:

1. $AgCl$ ______________________
2. $CaSO_4$ ______________________
3. $Al_2(SO_4)_3$ ______________________
4. NH_4NO_3 ______________________
5. $FeSO_4$ ______________________
6. Potassium chromate ______________________
7. Sodium fluoride ______________________
8. Magnesium sulfite ______________________
9. Copper(II) sulfate ______________________
10. Iron(III) chloride ______________________

Directions: The following set of lettered choices refers to the numbered questions immediately below it. For each numbered item, choose the one lettered choice that fits it best. Then fill in the corresponding oval on the answer sheet. Each choice in the set may be used once, more than once, or not at all.

Questions 11–15

Use the following choices to indicate the oxidation number of the underlined symbol in the given formulas.

(A) +1
(B) +2
(C) +4
(D) +5
(E) +6

11. $K_2\underline{Cr}O_4$
12. $Na_2\underline{S}_2O_3$
13. $\underline{P}O_4^{3-}$
14. $\underline{Ca}CO_3$
15. $Mg(\underline{H}CO_3)_2$

---

16. Find the percentage of sulfur in H_2SO_4.
17. What are the empirical formula and the true formula of a compound composed of 85.7% C and 14.3% H with a true formula mass of 42?

Directions:

(Further explanation and the examples for this type of question can be found on pages xiv–xv.)

Every question below contains two statements, I in the left-hand column and II in the right-hand column. For each question, decide if statement I is true or false and whether statement II is true or false, and fill in the corresponding T or F ovals in the answer spaces. *Fill in oval CE only if statement II is a correct explanation of statement I.

I		II
18. In the formula of a compound, the algebraic sum of the oxidation numbers must be 0	BECAUSE	oxygen's oxidation number in most compounds is −2.
19. Fluorine is assigned an oxidation number of −1 in all compounds	BECAUSE	fluorine is the most electronegative element.
20. Balanced equations have the same number of reactant atoms as the product atoms	BECAUSE	the conservation of matter must apply in all regular chemical equations.

*Fill in oval CE only if II is a correct explanation of I.

	I	II	CE*
18.	(T) (F)	(T) (F)	( )
18.	(T) (F)	(T) (F)	( )
20.	(T) (F)	(T) (F)	( )

Answers and Explanations

1. Silver chloride
2. Calcium sulfate
3. Aluminum sulfate
4. Ammonium nitrate
5. Iron(II) sulfate or ferrous sulfate
6. K_2CrO_4
7. NaF
8. $MgSO_3$

9. $CuSO_4$

10. $FeCl_3$

11. $\overset{K^{+1}}{}\ \overset{Cr^{x}}{}\ \overset{O^{-2}}{}$
+6 because $2(+1) + 1(x) + 4(-2) = 0,\ x = +6$

12. $\overset{Na^{+1}}{}\ \overset{S^{x}}{}\ \overset{O^{-2}}{}$
+2 because $2(+1) + 2(x) + 3(-2) = 0,\ x = +2$

13. $P^{x}\ \ O^{-2}$
+5 because $1(x) + 4(-2) = -3,\ x = +5$

14. $Ca^{x}\ \ C^{+4}\ \ O^{-2}$
(B) +2, because $1(x) + 1(+4) + 3(-2) = 0,\ x = +2$

15. $Mg^{+2}\ \ H^{x}\ \ C^{+4}\ \ O^{-2}$
(A) +1, because $1(+2) + 2(x) + 2(+4) + 6(-2) = 0,\ x = 1$

16. H_2SO_4 is composed of:

2H = 2	Percentage of S:
1S = 32	$\frac{S = 32}{Total = 98} \times 100\% = \underline{\underline{32.65}}$ or $\underline{\underline{33\%}}$
4O = 64	
Total = 98	

17. **(C)** $= 12)\ 85.7\%\ C \quad H = 1)\ 14.3\%\ H$
$\quad 7.14 \qquad\qquad 14.3$

To find the lowest ratio of the whole numbers:

$7.14)\underline{7.14} \quad 7.14)\underline{14.3}$
$\quad 1.0 \qquad\qquad 2.0$

The empirical formula is $\underline{\underline{CH_2}}$.

Since the formula mass is given as 42, the empirical formula CH_2, which represents a mass of 14, divides into 42 three times. Therefore, the true formula is $\underline{\underline{C_3H_6}}$.

18. **(T, T)** The sum of all the oxidation totals for a compound must equal 0, and oxygen in most compounds is assigned a –2 oxidation number, but this does not explain the I statement.

19. **(T, T, CE)** Statement II does explain statement I.

20. **(T, T, CE)** The Law of Conservation of Matter does require that there are equal numbers of respective atoms on both sides of a regular (not atomic) equation. Because the formula of each substance is determined by oxidation numbers, the equation is balanced by using coefficients.

Gases and the Gas Laws

CHAPTER 5

- Describe the physical and chemical properties of oxygen and hydrogen and the electronic makeup of their diatomic molecules.
- Explain how atmospheric pressure is measured, how to read the pressure in a manometer, and the units used to measure pressure.
- Read and explain a graphic distribution of the number of molecules versus kinetic energy at different temperatures.
- Know and use the following laws to solve gas problems: Graham's, Charles's, Boyle's, Dalton's, the Combined Gas Law, and the Ideal Gas Law.

When we discuss gases today, the most pressing concern is the gases in our atmosphere. These are the gases that are held against Earth by the gravitational field. The mixture of gases in the air today has had 4.5 billion years in which to evolve. The earliest atmosphere must have consisted of volcanic discharges alone. Gases erupted by volcanoes today are mostly a mixture of water vapor, carbon dioxide, sulfur dioxide, and nitrogen, with almost no oxygen. If this is the same mixture that existed in the early atmosphere, various processes would have had to operate to produce the mixture we have today. One of these processes was condensation. As it cooled, much of the volcanic water vapor condensed to fill the earliest oceans. Chemical reactions would also have occurred. Some carbon dioxide would have reacted with the rocks of Earth's crust to form carbonate minerals, and some would have dissolved in the new oceans. Later, as primitive life capable of photosynthesis evolved in the seas, new marine organisms began producing oxygen. Almost all the free oxygen in the air today is believed to have formed this way. About 570 million years ago, the oxygen content of the atmosphere and oceans became high enough to permit marine life capable of respiration. Later, about 400 million years ago, the atmosphere contained enough oxygen for air-breathing animals to emerge from the seas.

The principal constituents of the atmosphere of Earth today are nitrogen (78%) and oxygen (21%). The gases in the remaining 1% are argon (0.9%), carbon dioxide (0.03%), varying amounts of water vapor, and trace amounts of hydrogen, ozone, methane, carbon monoxide, helium, neon, krypton, and xenon. Oxides and other pollutants added to the atmosphere by factories and automobiles have become a major concern because of their damaging effects in the form of acid rain. In addition, a strong possibility exists that the steady increase in atmospheric carbon dioxide, mainly attributed to fossil fuel combustion over the past century, may

TIP

The major components of Earth's atmosphere:

78% nitrogen

21% oxygen

affect Earth's climate by causing a greenhouse effect, resulting in a steady rise in temperatures worldwide.

Studies of air samples show that up to 55 miles above sea level the composition of the atmosphere is substantially the same as at ground level; continuous stirring produced by atmospheric currents counteracts the tendency of the heavier gases to settle below the lighter ones. In the lower atmosphere, ozone is normally present in extremely low concentrations. The atmospheric layer 12 to 30 miles up contains more ozone that is produced by the action of ultraviolet radiation from the Sun. In this layer, however, the percentage of ozone is only 0.001 by volume. Human activity adds to the ozone concentration in the lower atmosphere where it can be a harmful pollutant.

The ozone layer became a subject of concern in the early 1970s when it was found that chemicals known as fluorocarbons, or chlorofluoromethanes, were rising into the atmosphere in large quantities because of their use as refrigerants and as propellants in aerosol dispensers. The concern centered on the possibility that these compounds, through the action of sunlight, could chemically attack and destroy stratospheric ozone, which protects Earth's surface from excessive ultraviolet radiation. As a result, U.S. industries and the Environmental Protection Agency phased out the use of certain chlorocarbons and fluorocarbons as of the year 2000. There is still ongoing concern about both these environmental problems: the greenhouse effect and the deterioration of the ozone layer.

SOME REPRESENTATIVE GASES

Oxygen

Of the gases that occur in the **atmosphere**, the most important one to us is oxygen. Although it makes up only approximately 21% of the atmosphere, by volume, the oxygen found on Earth is equal in weight to all the other elements combined. About 50% of Earth's crust (including the waters on Earth, and the air surrounding it) is oxygen. (Note Figure 15.)

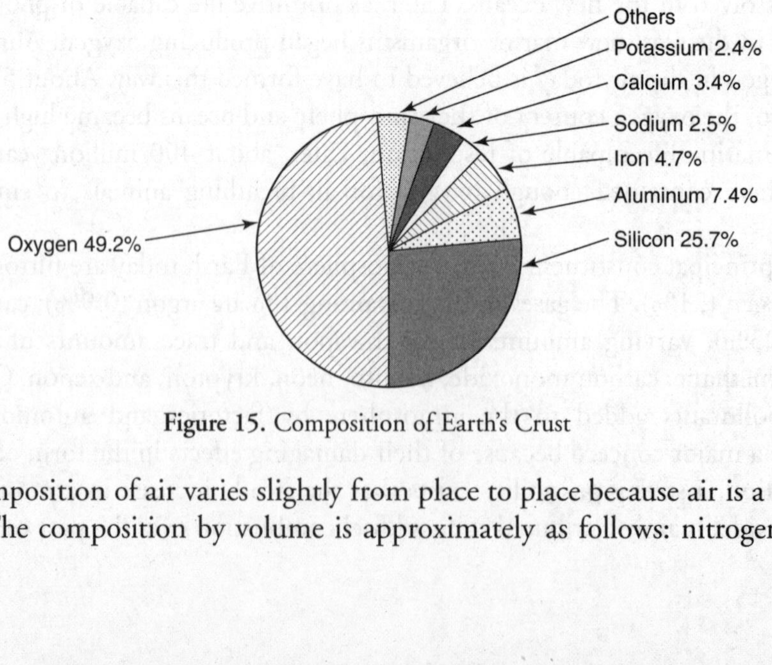

Figure 15. Composition of Earth's Crust

The composition of air varies slightly from place to place because air is a mixture of gases. The composition by volume is approximately as follows: nitrogen, 78%;

oxygen, 21%; argon, 1%. There are also small amounts of carbon dioxide, water vapor, and trace gases.

PREPARATION OF OXYGEN. In 1774, an English scientist named Joseph Priestley discovered oxygen by heating mercuric oxide in an enclosed container with a magnifying glass. That mercuric oxide decomposes into oxygen and mercury can be expressed in an equation: $2HgO \rightarrow 2Hg + O_2$. After his discovery, Priestley visited one of the greatest of all scientists, Antoine Lavoisier, in Paris. As early as 1773 Lavoisier had carried on experiments concerning burning, and they had caused him to doubt the phlogiston theory (that a substance called phlogiston was released when a substance burned; the theory went through several modifications before it was finally abandoned). By 1775, Lavoisier had demonstrated the true nature of burning and called the resulting gas "oxygen."

Today oxygen is usually prepared in the lab by heating an easily decomposed oxygen compound such as potassium chlorate ($KClO_3$). The equation for this reaction is:

$$2KClO_3 + MnO_2 \rightarrow 2KCl + 3O_2(g) + MnO_2$$

A possible laboratory setup is shown in Figure 16.

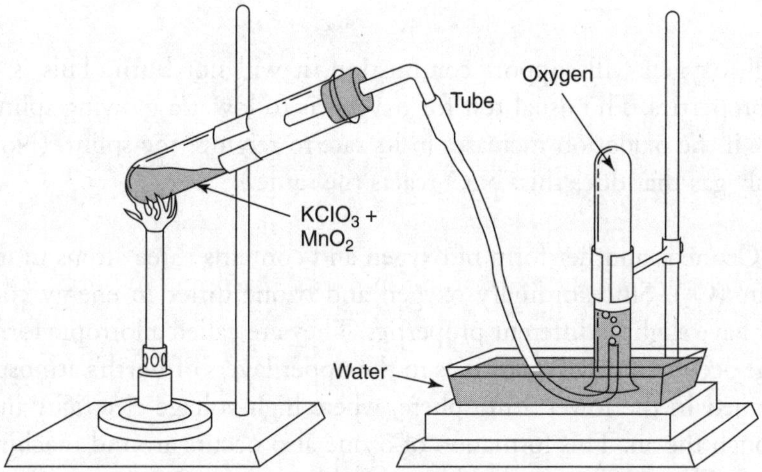

Figure 16. A Possible Laboratory Preparation of Oxygen

TIP

A catalyst speeds up the rate of reaction by lowering the activation energy needed for the reaction. A catalyst is not consumed.

In this preparation manganese dioxide (MnO_2) is often used. This compound is not used up in the reaction and can be shown to have the same composition as it had before the reaction occurred. The only effect it seems to have is that it lowers the temperature needed to decompose the $KClO_3$, and thus speeds up the reaction. Substances that behave in this manner are referred to as **catalysts**. The mechanism by which a catalyst acts is not completely understood in all cases, but it is known that in some reactions the catalyst does change its structure temporarily. Its effect is shown graphically in the reaction graphs in Figure 17.

TIP

Graphic representation of how a catalyst lowers the required activation energy

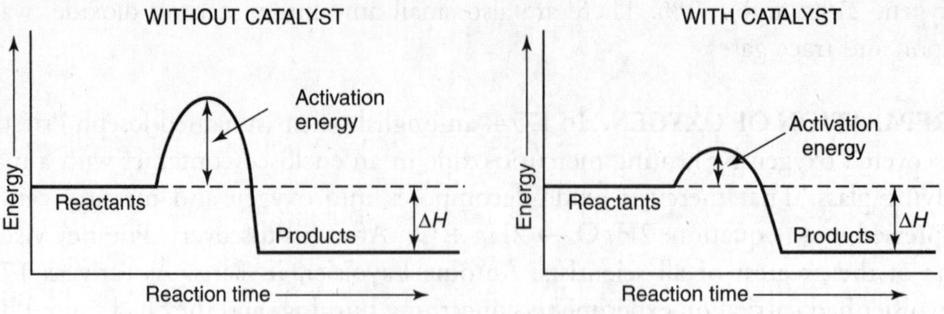

Figure 17. Effect of Catalyst on Reaction

PROPERTIES OF OXYGEN. Oxygen is a gas under ordinary conditions of temperature and pressure, and it is a gas that is colorless, odorless, tasteless, and slightly heavier than air; all these physical properties are characteristic of this element. Oxygen is only slightly soluble in water, thus making it possible to collect the gas over water, as shown in Figure 16. Oxygen at −183°C changes to a pale blue liquid with magnetic properties. This attraction by a magnet (paramagnetism) is believed to be due to an unpaired group of electrons found in the *p* orbitals. The Lewis structure for oxygen is:

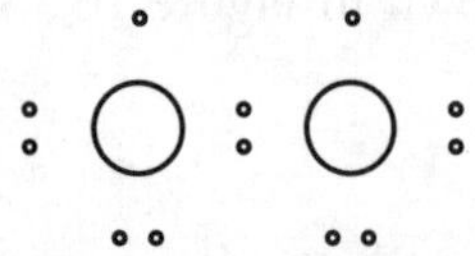

Although oxygen will support combustion, it will not burn. This is one of its chemical properties. The usual test for oxygen is to lower a glowing splint into the gas and see if the oxidation increases in its rate to reignite the splint. (Note: This is not the only gas that does this. N_2O reacts the same.)

OZONE. Ozone is another form of oxygen and contains three atoms in its molecular structure (O_3). Since ordinary oxygen and ozone differ in energy content and form, they have slightly different properties. They are called allotropic forms of oxygen. Ozone occurs in small quantities in the upper layers of Earth's atmosphere, and can be formed in the lower atmosphere, where high-voltage electricity in lightning passes through the air. This formation of ozone also occurs around machinery using high voltage. The reaction can be shown by this equation:

TIP

The ozone layer protects us from UV rays from the sun.

$$3O_2 + \text{elec.} \rightarrow 2O_3$$

Because of its higher energy content, ozone is more reactive chemically than oxygen.

The ozone layer prevents harmful wavelengths of ultraviolet (UV) light from passing through Earth's atmosphere. UV rays have been linked to biological consequences such as skin cancer.

Concern is growing that the ozone layer is slowly depleting due to gaseous emissions from Earth in spite of some mandated restrictions. The rate of depletion is claimed to be 4% per decade.

Hydrogen

PREPARATION OF HYDROGEN. Although there is evidence of the preparation of hydrogen before 1766, Henry Cavandish was the first person to recognize this gas as a separate substance. He observed that, whenever it burned, it produced water. Lavoisier named it **hydrogen**, which means "water former."

Electrolysis of water, which is the process of passing an electric current through water to cause it to decompose, is one method of obtaining hydrogen. This is a widely used commercial method, as well as a laboratory method.

Another method of producing hydrogen is to displace it from the water molecule by using a metal. To choose the metal you must be familiar with its activity with respect to hydrogen. The acitivities of the common metals are shown in Table 8.

TABLE 8. Activity Chart of Metals Compared to Hydrogen

Increasing Activity ↑

Potassium	In cold water
Calcium	
Sodium	
Magnesium	In hot water
Aluminum	In most dilute acids
Zinc	
Iron	
Tin	
HYDROGEN	. . . .
Copper	Not active enough
Mercury	
Silver	
Gold	

TIP

Know the relative activity of metals.

As noted in Table 8, any of the first three metals will react with cold water; the reaction is as follows:

Very active metal + Water = Hydrogen + Metal hydroxide

Using sodium as an example:

$$2Na + 2HOH \rightarrow H_2(g) + 2NaOH$$

With the metals that react more slowly, a dilute acid reaction is needed to produce hydrogen in sufficient quantities to collect in the laboratory. This general equation is:

Active metal + Dilute acid → Hydrogen + Salt of the acid

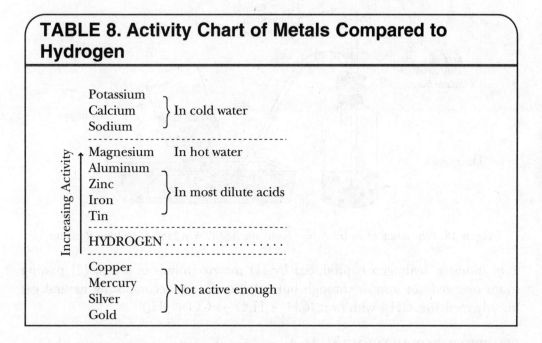

An example:

$$Zn + \text{dil. } H_2SO_4 \rightarrow H_2(g) + ZnSO_4$$

This equation shows the usual laboratory method of preparing hydrogen. Mossy zinc is used in a setup as shown in Figure 18. The acid is introduced down the thistle tube after the zinc is placed in the reacting bottle. In this sort of setup, you would not begin collect-ing the gas that bubbles out of the delivery tube for a few minutes so that the air in the system has a chance to be expelled and you can collect a rather pure volume of the gas generated.

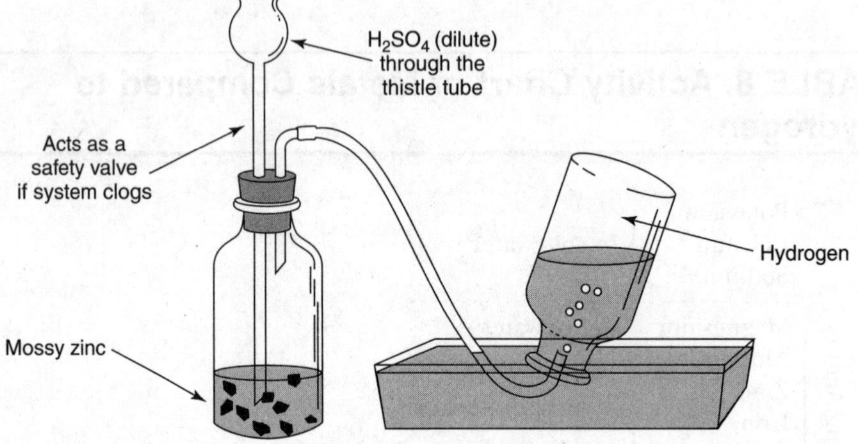

Figure 18. Preparation of an Insoluble Gas by the Addition of Liquid to Other Reactant

In industry, hydrogen is produced by (1) the electrolysis of water, (2) passing steam over red-hot iron or through hot coke, or (3) by decomposing natural gas (mostly methane, CH_4) with heat ($CH_4 + H_2O \rightarrow CO + 3H_2$).

PROPERTIES OF HYDROGEN. Hydrogen has the following important physical properties:

1. It is ordinarily a gas; colorless, odorless, tasteless when pure.
2. It weighs 0.9 gram per liter at 0°C and 1 atmosphere pressure. This is $\frac{1}{14}$ as dense as air.
3. It is slightly soluble in water.
4. It becomes a liquid at a temperature of −240°C and a pressure of 13 atmospheres.
5. It diffuses (moves from place to place in gases) more rapidly than any other gas. This property can be demonstrated as shown in Figure 19.

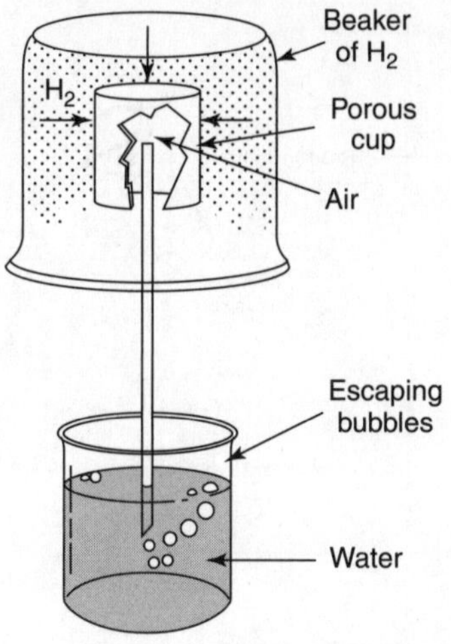

Here the H_2 in the beaker that is placed over the porous cup diffuses faster through the cup than the air can diffuse out. Consequently, there is a pressure buildup in the cup, which pushes the gas out through the water in the lower beaker.

Figure 19. Diffusion of Hydrogen

The chemical properties of hydrogen are:

1. It burns in air or in oxygen, giving off large amounts of heat. Its high heat of combustion makes it a good fuel.
2. It does not support ordinary combustion.
3. It is a good reducing agent in that it withdraws oxygen from many hot metal oxides.

GENERAL CHARACTERISTICS OF GASES

Measuring the Pressure of a Gas

Pressure is defined as force per unit area. With respect to the atmosphere, pressure is the result of the weight of a mixture of gases. This pressure, which is called **atmospheric pressure**, **air pressure**, or **barometric pressure**, is approximately equal to the weight of a kilogram mass on every square centimeter of surface exposed to it. This weight is about 10 newtons.

The pressure of the atmosphere varies with altitude. At higher altitudes, the weight of the overlying atmosphere is less, so the pressure is less. Air pressure also varies somewhat with weather conditions as low- and high-pressure areas move with weather fronts. On the average, however, the air pressure at sea level can support a column of mercury 760 millimeters in height. This average sea-level air pressure is known as **normal atmospheric pressure**, also called **standard pressure**.

The instrument most commonly used for measuring air pressure is the **mercury barometer**. The diagram below shows how it operates. Atmospheric pressure is exerted on the mercury in the dish, and this in turn holds the column of mercury up in the tube. This column at standard pressure will measure 760 millimeters above the level of the mercury in the dish below.

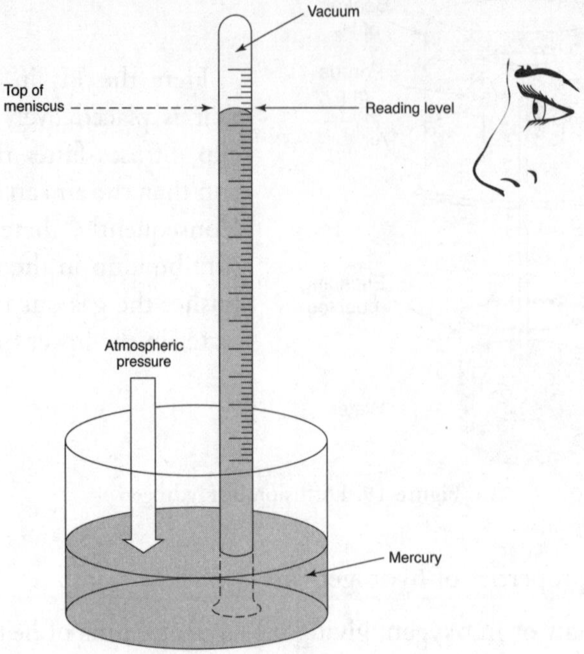

Mercury Barometer

TIP

Read the top of the meniscus for mercury but the bottom of the meniscus for water.

In gas-law problems pressure may be expressed in various units. One standard atmosphere (1 atm) is equal to 760 millimeters of mercury (760 mm Hg) or 760 **torr**, a unit named for Evangelista Torricelli. In the SI system, the unit of pressure is the **pascal** (Pa), named in honor of the scientist of the same name, and standard pressure is 101,325 pascals or 101.325 kilopascals (kPa). One pascal (Pa) is defined as the pressure exerted by the force of one newton (1N) acting on an area of one square meter. In many cases, as in atmospheric pressure, it is more convenient to express pressure in kilopascals (kPa).

Summary of Units of Pressure

Unit	Abbreviation	Unit Equivalent to 1 atm
Atmosphere	atm	1 atm
Millimeters of Hg	mm Hg	760 mm Hg
Torr	torr	760 torr
Pascal	Pa	101,325 Pa
Kilopascal	kPa	101.3 kPa

A device similar to the barometer can be used to measure the pressure of a gas in a confined container. This apparatus, called a **manometer**, is illustrated below. A manometer is basically a U-tube containing mercury or some other liquid. When both ends are open to the air, as in (1) in the diagram, the level of the liquid will be the same on both sides since the same pressure is being exerted on both ends of the tube. In (2) and (3), a vessel is connected to one end of the U-tube. Now the height of the mercury column serves as a means of reading the pressure inside the vessel if the atmospheric pressure is known. When the pressure inside the vessel is the same as the atmospheric pressure outside, the levels of liquid are the same. When the pressure inside is greater than outside, the column of liquid will be higher on the side that is exposed to the air, as in (2). Conversely, when the pressure inside the vessel is less than the outside atmospheric pressure, the additional pressure will force the liquid to a higher level on the side near the vessel, as in (3).

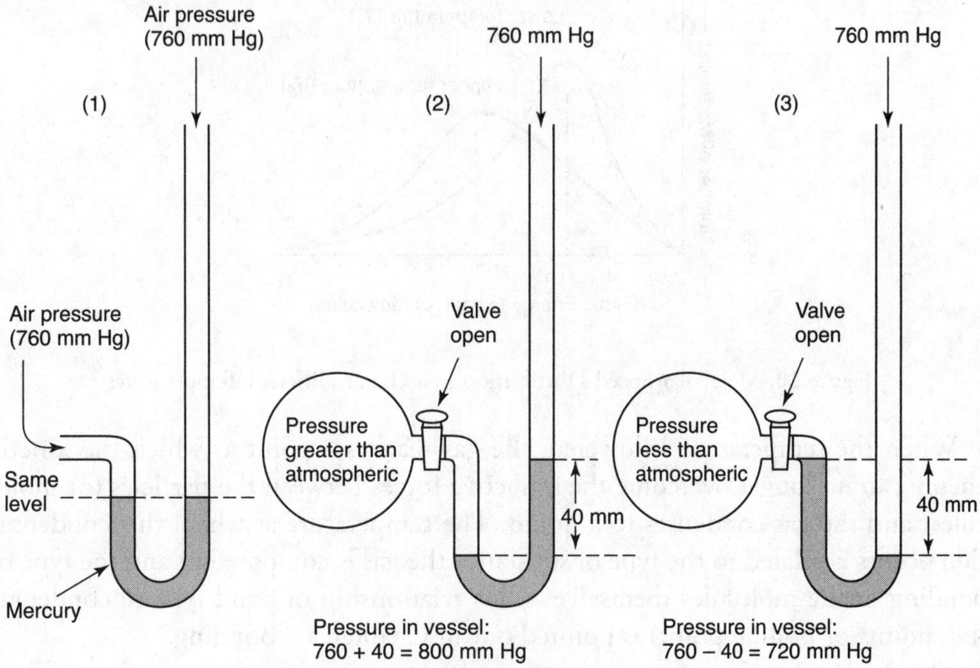

Manometer

TIP

Know how to calculate the pressure in a closed vessel like in the manometer shown.

Kinetic-Molecular Theory

By indirect observations, the Kinetic-Molecular Theory has been arrived at to explain the forces between molecules and the energy the molecules possess. There are three basic assumptions to the Kinetic-Molecular Theory:

1. Matter in all its forms (solid, liquid, and gas) is composed of extremely small particles. In many cases these are called molecules. The space occupied by the gas particles themselves is ignored in comparison with the volume of the space in which they are contained.
2. The particles of matter are in constant motion. In solids, this motion is restricted to a small space. In liquids, the particles have a more random pattern but still are restricted to a kind of rolling over one another. In a gas, the particles are in continuous, random, straight-line motion.
3. When these particles collide with each other or with the walls of the container, there is no loss of energy.

TIP

Know these basic assumptions of the Kinetic-Molecular Theory.

Some Particular Properties of Gases

As the temperature of a gas is increased, its kinetic energy is increased, thereby increasing the random motion. At a particular temperature not all the particles have the same kinetic energy, but the temperature is a measure of the average kinetic energy of the particles. A graph of the various kinetic energies resembles a normal bell-shaped curve with the average found at the peak of the curve (see Figure 20).

> **TIP**
> When you read the temperature of a substance, you are measuring its average kinetic energy.

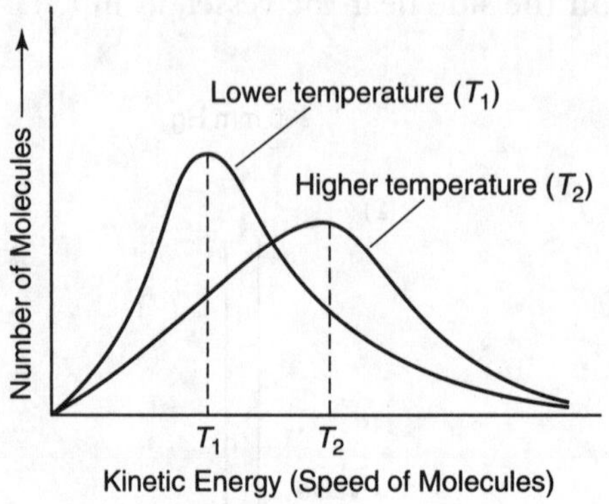

Figure 20. Molecular Speed Distribution in a Gas at Different Temperatures

> **TIP**
> *Diffusion* means spreading out.

> **TIP**
> *Effusion* means passing of a gas through an orifice (like through the neck of a balloon).

When the temperature is lowered, the gas reaches a point at which the kinetic energy can no longer overcome the attractive forces between the particles (or molecules) and the gas condenses to a liquid. The temperature at which this condensation occurs is related to the type of substance the gas is composed of and the type of bonding in the molecules themselves. This relationship of bond type to condensation point (or boiling point) is pointed out in Chapter 3, "Bonding."

The random motion of gases in moving from one position to another is referred to as **diffusion**. You know that, if a bottle of perfume is opened in one corner of a room, the perfume, that is, its molecules, will move or diffuse to all parts of the room in time. The rate of diffusion is the rate of the mixing of gases.

Effusion is the term used to describe the passage of a gas through a tiny orifice into an evacuated chamber. The rate of effusion measures the speed at which the gas is transferred into the chamber.

GAS LAWS AND RELATED PROBLEMS

Graham's Law of Effusion (Diffusion)

This law relates the rate at which a gas diffuses (or effuses) to the type of molecule in the gas. It can be expressed as follows:

> The Rate of Effusion of a Gas Is Inversely Proportional to the Square Root of its Molecular Mass.

Hydrogen, with the lowest molecular mass, can diffuse more rapidly than other gases under similar conditions.

TYPE PROBLEM: Compare the rate of diffusion of hydrogen to that of oxygen under similar conditions.

The formula is

$$\frac{\text{Rate A}}{\text{Rate B}} = \frac{\sqrt{\text{Molecular mass of B}}}{\sqrt{\text{Molecular mass of A}}}$$

Let A be H_2 and B be O_2.

$$\frac{\text{Rate } H_2}{\text{Rate } O_2} = \frac{\sqrt{32}}{\sqrt{2}} = \frac{\sqrt{16}}{\sqrt{1}} = \frac{4}{1}$$

Therefore hydrogen diffuses four times as fast as oxygen.

In dealing with the gas laws, a student must know what is meant by standard conditions of temperature and pressure (abbreviated as STP). The standard pressure is defined as the height of mercury that can be held in an evacuated tube by 1 atmosphere of pressure (14.7 lb/in.^{2}). This is usually expressed as 760 millimeters of Hg or 101.3 pascals. **Standard temperature** is defined as 273 Kelvin or absolute (which corresponds to 0° Celsius).

Charles's Law $\left(\frac{V}{T} = k\right)$

Jacques Charles, a French chemist of the early nineteenth century, discovered that, when a gas under constant pressure is heated from 0°C to 1°C, it expands 1/273 of its volume. It contracts this amount when the temperature is dropped 1 degree to −1°C. Charles reasoned that, if a gas at 0°C was cooled to −273°C (actually found to be −273.15°C), its volume would be zero. Actually, all gases are converted into liquids before this temperature is reached. By using the Kelvin scale to rid the problem of negative numbers, we can state Charles's Law as follows:

> If the Pressure Remains Constant, the Volume of a Gas Varies Directly as the Absolute Temperature. Then
>
> $$\text{Initial } \frac{V_1}{T_1} = \text{Final} \frac{V_2}{T_2} \text{ at constant pressure or } \frac{V}{T} = k$$

TIP

Charles's Law

$$\frac{V_1}{T_1} = \frac{V_2}{T_2}$$

at constant pressure is a direct proportion.

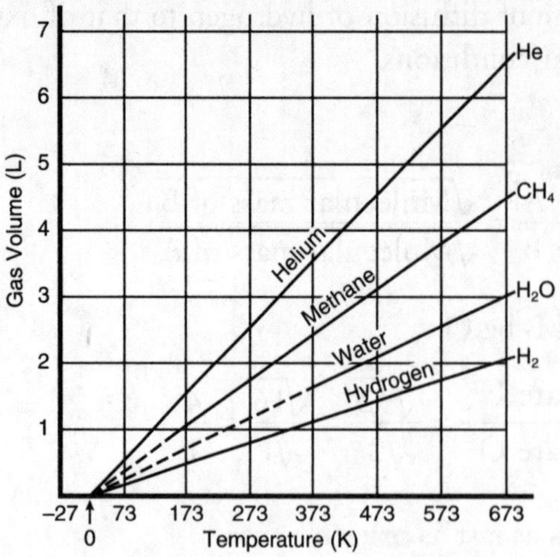

Graphic relationship—Charles's Law. The dashed lines represent extrapolation of the data into regions where the gas would become liquid or solid. Extrapolation shows that each gas, if it remained gaseous, would reach zero volume at 0 K or –273°C.

Plots of *V* versus *T* for representative gases.

TIP

Assume dry gases unless otherwise stated.

TYPE PROBLEM: The volume of a gas at 20°C is 500 mL. Find its volume at standard temperature if pressure is held constant.

Convert temperatures:

$$20°C = 20° + 273° = 293\text{ K}$$
$$0°C = 0° + 273° = 273\text{ K}$$

TIP

STP = standard temperature of 273K

standard pressure of 760 mm Hg or 1 atmosphere (atm) or 10^5 pascals.

If you know that cooling a gas decreases its volume, then you know that 500 mL will have to be multiplied by a fraction (made up of the Kelvin temperatures) that has a smaller numerator than the denominator. So

$$500\text{ mL} \times \frac{273}{293} = 465\text{ mL}$$

Or you can use the formula and substitute known values:

$$\frac{V_1}{T_1} = \frac{V_2}{T_2}$$

$$\frac{500\text{ mL}}{293} = \frac{x\text{ mL}}{273}$$

$$x\text{ mL} = \mathbf{465\text{ mL}}$$

ANOTHER EXAMPLE: A sample of gas occupies 24 L at 175.0 K. What volume would the gas occupy at 400.0 K?

The temperature of the gas is increased. Charles's Law predicts that the gas volume will also increase. So

$$\frac{V_1}{T_1} = \frac{V_2}{T_2}$$

$$V_2 = \frac{V_1 T_2}{T_2}$$

$$V_2 = 24\ \text{L} \times \frac{400.0\ \cancel{\text{K}}}{175.0\ \cancel{\text{K}}}$$

$$V_2 = 55\ \text{L}$$

The final volume has increased as predicted.

Boyle's Law ($PV = k$)

Robert Boyle, a seventeenth century English scientist, found that the volume of a gas decreases when the pressure on it is increased, and vice versa, when the temperature is held constant. Boyle's Law can be stated as follows:

> **TIP**
>
> **Boyle's Law**
>
> $P_1V_1 = P_2V_2$
>
> temperature is held constant

> If the Temperature Remains Constant, the Volume of a Gas Varies Inversely as the Pressure Changes. Then
>
> $P_1V_1 = P_2V_2$ at a constant temperature
>
> or
>
> $PV = k$

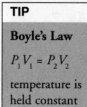

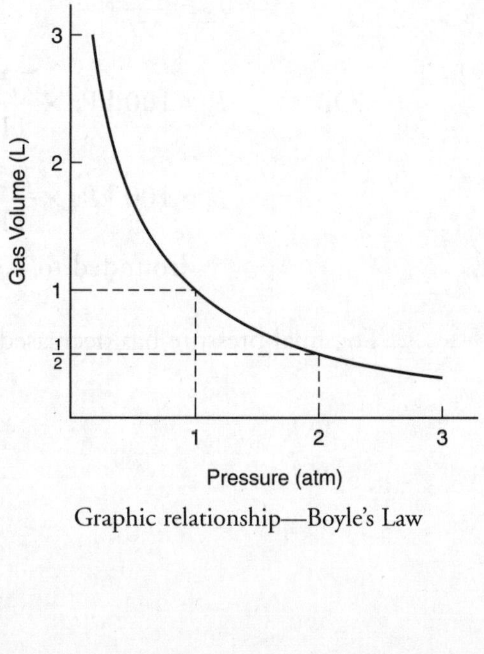

Graphic relationship—Boyle's Law

> **TIP**
>
> Volume vs. pressure for a gas at constant temperature. This is an **inverse** proportion. As the pressure increases by 2, the volume drops by 1/2.

TYPE PROBLEM: Given the volume of a gas as 200. mL at 1.05 atm pressure, calculate the volume of the same gas at 1.01 atm. Temperature is held constant.

If you know that this decrease in pressure will cause an increase in the volume, then you know 200. mL must be multiplied by a fraction (made up of the two pressures) that has a larger numerator than the denominator. So

$$200.\ \text{mL} \times \frac{1.05\ \cancel{\text{atm}}}{1.01\ \cancel{\text{atm}}} = 208\ \text{mL}$$

Or you can use the formula:

$$P_1V_1 = P_2V_2$$

$$V_2 = V_1 \times \frac{P_1}{P_2}$$

$$= 200.\ \text{mL} \times \frac{1.05\ \cancel{\text{atm}}}{1.01\ \cancel{\text{atm}}} = 208\ \text{mL}$$

ANOTHER EXAMPLE: The gas in a balloon has a volume of 7.5 L at 100 kPa. The balloon is released into the atmosphere, and the gas in it expands to a volume of 11 L. Assuming a constant temperature, what is the pressure on the balloon at the new volume?

The volume of the gas has increased. Boyle's Law predicts that the gas pressure will decrease. So

$$P_1V_1 = P_2V_2$$

$$P_2 = \frac{P_1V_1}{V_2}$$

Or $$P_2 = 100\ \text{kPa} \times \frac{7.5\ \cancel{\text{L}}}{11\ \cancel{\text{L}}}$$

$$P_2 = 100\ \text{kPa} \times \frac{7.5\ \cancel{\text{L}}}{11\ \cancel{\text{L}}} = 68 = 70\ \text{kPA}$$

(rounded to 1 significant figure)

The final pressure has decreased as predicted.

Combined Gas Law

This is a combination of the two preceding gas laws. The formula is as follows:

$$\frac{P_1V_1}{T_1} = \frac{P_2V_2}{T_2}$$

> **TIP**
>
> In the combined gas law, all subscripts are 1s on the left and all subscripts are 2s on the right.

TYPE PROBLEM: The volume of a gas at 780 mm pressure and 30°C is 500. mL. What volume would the gas occupy at STP?

You again can use reasoning to determine the kind of fractions the temperatures and pressures must be to arrive at your answer. Since the pressure is going from 780 mm to 760 mm, the volume should increase. The fraction must then be $\frac{780}{760}$. Also, since the temperature is going from 30°C (303 K) to 0°C (273 K), the volume should decrease; this fraction must be $\frac{273}{303}$. So

$$500.\text{ mL} \times \frac{780}{760} \times \frac{273}{303} = 462\text{ mL}$$

Or you can use the formula:

$$\frac{P_1V_1}{T_1} = \frac{P_2V_2}{T_2}$$

Solve for $V_2 = V_1 \times \frac{P_1}{P_2} \times \frac{T_2}{T_1}$

$$V_2 = 500.\text{ mL} \times \frac{780\ \cancel{\text{mm Hg}}}{760\ \cancel{\text{mm Hg}}} \times \frac{273\ \cancel{\text{K}}}{303\ \cancel{\text{K}}} = 462\text{ mL}$$

ANOTHER EXAMPLE: The volume of a gas is 27.5 mL at 22.0°C and 0.974 atm. What will the volume be at 15.0°C and 0.993 atm?

Using the combined gas law, $\frac{P_1V_1}{T_1} = \frac{P_2V_2}{T_2}$

and solving for $V_2 = V_1 \times \frac{P_1 T_2}{P_2 T_1}$

$$V_2 = 27.5\text{ mL} \times \frac{0.974\ \cancel{\text{atm}}}{0.993\ \cancel{\text{atm}}} \times \frac{(15.0°\text{C} + 273 = 288\ \cancel{\text{K}})}{(22.0°\text{C} + 273 = 295\ \cancel{\text{K}})}$$

$V_2 = 26.3$ mL

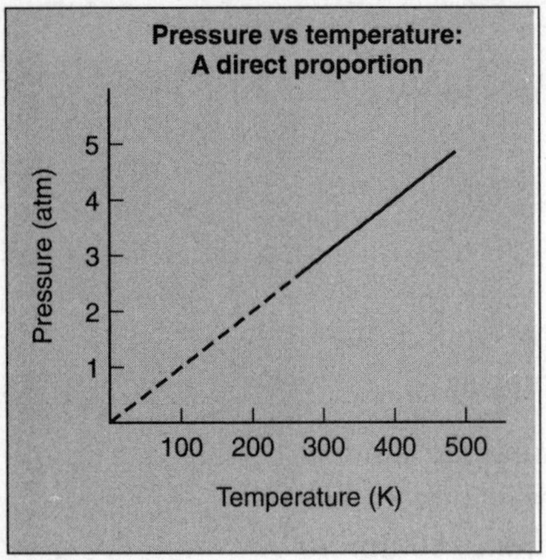

Pressure Versus Temperature (Gay-Lussac's Law)

At Constant Volume, the Pressure of a Given Mass of Gas Varies Directly with the Absolute Temperature. Then

$$\frac{P_1}{T_1} = \frac{P_2}{T_2} \text{ at constant volume or } \frac{P_1}{V_1} = k$$

TYPE PROBLEM: A steel tank contains a gas at 27°C and a pressure of 12. atms. Determine the gas pressure when the tank is heated to 100.°C.

Reasoning that an increase in temperature will cause an increase in pressure at constant volume, you know the pressure must be multiplied by a fraction that has a larger numerator than denominator. The fraction must be $\frac{373\text{ K}}{300\text{ K}}$. So

$$12.\text{ atm} \times \frac{373\text{ }\cancel{K}}{300\text{ }\cancel{K}} = 14.9\text{ atm or } 15.\text{ atm}$$

Or you can use the formula:

$$\frac{P_1}{T_1} = \frac{P_2}{T_2}$$

$$P_2 = P_1 \times \frac{T_2}{T_1}$$

$$P_2 = 12\text{ atm} \times \frac{373\ K}{300\ K} = 14.9\text{ atm}$$

ANOTHER EXAMPLE: At 120.°C, the pressure of a sample of nitrogen is 1.07 atm. What will the pressure be at 205°C, assuming constant volume?

Using the relationship of: $\frac{P_1}{T_1} = \frac{P_2}{T_2}$

and solving for $P_2 = \frac{P_1 T_2}{T_1}$

$$\text{So } P_2 = \frac{1.07 \text{ atm} \times (205° + 273° = 478 \not{K})}{(120.° + 273° = 393 \not{K})}$$

$$P_2 = 1.30 \text{ atm}$$

Dalton's Law of Partial Pressures

When a Gas Is Made Up of a Mixture of Different Gases, the Total Pressure of the Mixture Is Equal to the Sum of the Partial Pressures of the Components; That Is, the Partial Pressure of the Gas Would Be the Pressure of the Individual Gas If It Alone Occupied the Volume. The Formula Is:

$$P_{total} = P_{gas\ 1} + P_{gas\ 2} + P_{gas\ 3} + \cdots$$

TIP

Know Dalton's Law of Partial Pressures.

TYPE PROBLEM: A mixture of gases at 760. mm Hg pressure contains 65% nitrogen, 15% oxygen, and 20% carbon dioxide by volume. What is the partial pressure of each gas?

$0.65 \times 760. = 494$ mm pressure (N_2)
$0.15 \times 760. = 114$ mm pressure (O_2)
$0.20 \times 760. = 152$ mm pressure (CO_2)

If the pressure was given as 1.0 atm, you would substitute 1.0 atm for 760. mm Hg. The answers would be:

0.65×1.0 atm $= 0.65$ atm (N_2)
0.15×1.0 atm $= 0.15$ atm (O_2)
0.20×1.0 atm $= 0.20$ atm (CO_2)

Corrections of Pressure

CORRECTION OF PRESSURE WHEN A GAS IS COLLECTED OVER WATER. When a gas is collected over a volatile liquid, such as water, some of the water vapor is present in the gas and contributes to the total pressure. Assuming that the gas is saturated with water vapor at the given temperature, you can find the partial pressure due to the water vapor in a table of such water vapor values. This vapor

> **TIP**
>
> When a gas is collected over water, subtract the water vapor pressure at the given temperature from the atmospheric pressure to find the partial pressure of the gas.
>
> $P_{gas} = P_{atm} - P_{H_2O}$

pressure, which depends only on the temperature, must be subtracted from the total pressure to find the partial pressure of the gas being measured.

CORRECTION OF DIFFERENCE IN THE HEIGHT OF THE FLUID. When gases are collected in eudiometers (glass tubes closed at one end), it is not always possible to get the level of the liquid inside the tube to equal the level on the outside. This deviation of levels must be taken into account when determining the pressure of the enclosed gas. There are then two possibilities: (1) When the level inside is higher than the level outside the tube, the pressure on the inside is less, by the height of fluid in excess, than the outside pressure. If the fluid is mercury, you simply subtract the difference from the outside pressure reading (also in height of mercury and in the same units) to get the corrected pressure of the gas. If the fluid is water, you must first convert the difference to an equivalent height of mercury by dividing the difference by 13.6 (since mercury is 13.6 times as heavy as water, the height expressed in terms of Hg will be 1/13.6 the height of water). This is shown pictorially in Figure 21. Again, care must be taken that this equivalent height of mercury is in the same units as the expression for the outside pressure before it is subtracted to obtain the corrected pressure for the gas in the eudiometer. (2) When the level inside is lower than the level outside the tube, a correction must be added to the outside pressure. If the difference in height between the inside and the outside is expressed in terms of water, you must take 1/13.6 of this quantity to correct it to millimeters of mercury. This quantity is then added to the expression of the outside pressure, which must also be in millimeters of mercury. If the tube contains mercury, then the difference between the inside and outside levels is merely added to the outside pressure to get the corrected pressure for the enclosed gas.

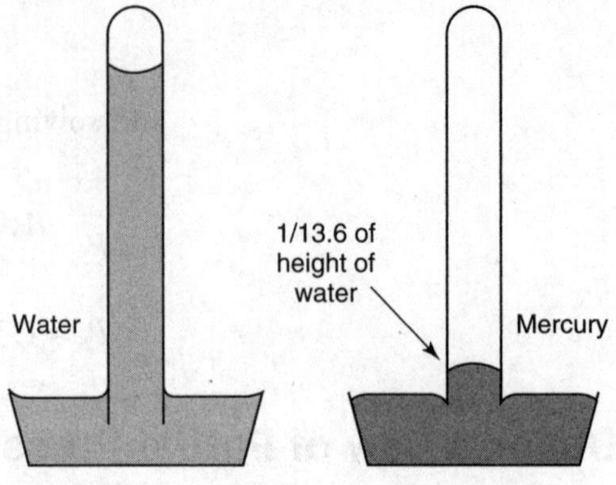

Figure 21. Same Pressure Exerted on Both Liquids

TYPE PROBLEM: Hydrogen gas was collected in a eudiometer tube over water. It was impossible to level the outside water with that in the tube, so the water level inside the tube was 40.8 mm higher than that outside. The barometric pressure was 730 mm of Hg. The water vapor pressure at the room temperature of 29°C was found in a handbook to be 30.0 mm. What is the pressure of the dry hydrogen?

Step 1: To find the true pressure of the gas, we must first subtract the water-level difference expressed in mm of Hg:

$$\frac{40.8}{13.6} = 3 \text{ mm of Hg}$$

Then 730 mm − 3 mm = 727 mm total pressure of gases in the eudiometer

Step 2: Correcting for the partial pressure due to water vapor in the hydrogen, we subtract the vapor pressure (30.0 mm) from 727 mm and get our answer: 697 mm.

Ideal Gas Law

The preceding laws do not include the relationship of number of moles of a gas to the pressure, volume, and temperature of the gas. A law derived from the Kinetic-Molecular Theory relates these variables. It is called the Ideal Gas Law and is expressed as

> **TIP**
>
> Know how to use the ideal gas law, $PV = nRT$

$$PV = nRT$$

P, V, and T retain their usual meanings, but n stands for the number of moles of the gas and R represents the ideal gas constant.

Boyle's Law and Charles's Law are actually derived from the Ideal Gas Law. Boyle's Law applies when the number of moles and the temperature of the gas are constant. Then in $PV = nRT$, the number of moles, n, is constant; the gas constant (R) remains the same; and by definition T is constant. Therefore, $PV = k$. At the initial set of conditions of a problem, P_1V_1 = a constant (k). At the second set of conditions, the terms on the right side of the equation are equal to the same constant, so $P_1V_1 = P_2V_2$. This matches the Boyle's Law equation introduced earlier.

The same can also be done with Charles's Law, because $PV = nRT$ can be expressed with the variables on the left and the constants on the right:

$$\frac{V}{T} = \frac{nR}{P}$$

In Charles's Law the number of moles and the pressure are constant. Substituting k for the constant term, $\frac{nR}{P}$, we have

$$\frac{V}{T} = k$$

The expression relating two sets of conditions can be written as

$$\frac{V_1}{T_1} = \frac{V_2}{T_2}$$

TIP

Know the ideal gas constant

$$R = 0.0821 \frac{\text{L atm}}{\text{mol K}}$$

To use the Ideal Gas Law in the form $PV = nRT$, the gas constant, R, must be determined. This can be done mathematically as shown in the following example.

One mole of oxygen gas was collected in the laboratory at a temperature of 24.0°C and a pressure of exactly 1 atmosphere. The volume was 24.38 liters. Find the value of R.

$$PV = nRT$$

Rearranging the equation to solve for R gives

$$R = \frac{PV}{nT}$$

Substituting the known values on the right, we have

$$R = \frac{1 \text{ atm} \times 24.38 \text{ L}}{1 \text{ mol} \times 297 \text{ K}}$$

Calculating R, we get

$$R = 0.0821 \frac{\text{L atm}}{\text{mol K}}$$

Once R is known, the Ideal Gas Law can be used to find any of the variables, given the other three.

For example, calculate the pressure, at 16.0°C, of 1.00 gram of hydrogen gas occupying 2.54 liters.

Rearranging the equation to solve for P, we get

$$P = \frac{nRT}{V}$$

The molar mass of hydrogen is 2.00 g/mol, so the number of moles in this problem would be

TIP

Remember to use appropriate units:

moles (mol)

liters (L)

atmosphere (atm)

$$\frac{1.00 \text{ g}}{2.00 \text{ g/mol}} = 0.500 \text{ mol}$$

Substituting the known values, we have

$$P = \frac{(0.500 \cancel{\text{mol}})\left(0.0821 \frac{\cancel{\text{L}} \text{ atm}}{\cancel{\text{mol}}\ \cancel{\text{K}}}\right) 289 \cancel{\text{K}}}{2.54 \text{ L}}$$

Calculating the value, we get

$$P = 4.66 \text{ atm}$$

Another use of the ideal gas law is to find the number of moles of a gas when P, T, and V are known.

For example, how many moles of nitrogen gas are there in 0.38 liter of gas at 0°C and 380 mm Hg pressure?

Rearranging the equation to solve for n gives

$$n = \frac{PV}{RT}$$

Changing temperature to kelvins and pressure to atmospheres gives

$$T = 0° + 273 = 273 \text{ K}$$

$$P = \frac{380 \text{ mm Hg}}{760 \text{ mmHg/atm}} = 0.50 \text{ atm}$$

Substituting in the equation, we have

$$n = \frac{(0.50 \cancel{\text{atm}})(0.38 \cancel{\text{L}})}{\left(0.0821 \frac{\cancel{\text{L}}\ \cancel{\text{atm}}}{\text{mol } \cancel{\text{K}}}\right)(273 \cancel{\text{K}})} = 0.0085 \text{ mol}$$

$$= 0.0085 \text{ mol of nitrogen gas}$$

Ideal Gas Deviations

In the use of the gas laws, we have assumed that the gases involved were "ideal" gases. This means that the molecules of the gas were not taking up space in the gas volume and that no intermolecular forces of attraction were serving to pull the molecules closer together. You will find that a gas behaves like an ideal gas at low pressures and high temperatures, which move the molecules as far as possible from conditions that would cause condensation. In general, pressures below a few atmospheres will cause most gases to exhibit sufficiently ideal properties for the application of the gas laws with a reliability of a few percent or better.

If, however, high pressures are used, the molecules will be forced into closer proximity with each other as the volume decreases until the attractive force between molecules becomes a factor. This factor decreases the volume, and therefore the PV values at high pressure conditions will be less than those predicted by the Ideal Gas Law, where PV remains a constant.

Examining what occurs at very low temperatures creates a similar situation. Again, the molecules, because they have slowed down at low temperatures, come into closer proximity with each other and begin to feel the attractive force between them. This tends to make the gas volume smaller and, therefore, causes the PV to be lower than that expected in the ideal gas situation. Thus, under conditions of very high pressures and low temperatures, deviations from the expected results of the ideal gas law will occur.

TIP

Least deviations occur at low pressures and high temperatures.

High deviations occur at high pressures and very low temperatures.

Chapter Summary

The following terms summarize all the concepts and ideas that were introduced in this chapter. You should be able to explain their meaning and how you would use them in chemistry. They appear in boldface type in this chapter to draw your attention to them. The boldface type also makes it easier for you to look them up if you need to. You could also use the search or "Google" action on your computer to get a quick and expanded explanation of these terms, laws, and formulas.

atmosphere	mercury barometer	standard pressure
atmospheric pressure	ozone	standard temperature
manometer	pascal	torr

Boyle's Law ($PV = k$) constant T

Charles's Law $\frac{V}{T} = k$

Combined Gas Law $\frac{P_1V_1}{T_1} = \frac{P_2V_2}{T_2}$

Dalton's Law of Partial Pressures

$\frac{P_1}{V_1} = k$ (constant T)

Graham's Law
Ideal Gas Law
Kinetic-Molecular Theory

Practice Exercises

1. The most abundant element in Earth's crust is
   (A) sodium
   (B) oxygen
   (C) silicon
   (D) aluminum

2. A compound that can be decomposed to produce oxygen gas in the lab is
   (A) MnO_2
   (B) NaOH
   (C) CO_2
   (D) $KClO_3$

3. In the usual laboratory preparation equation for the reaction in question 2, what is the coefficient of O_2?
   (A) 1
   (B) 2
   (C) 3
   (D) 4

4. In the graphic representation of the energy contents of the reactants and the resulting products in an exothermic reaction, the energy content would be

(A) higher for the reactants
(B) higher for the products
(C) the same for both
(D) impossible to determine

5. The process of separating components of a mixture by making use of the difference in their boiling points is called

(A) destructive distillation
(B) displacement
(C) fractional distillation
(D) filtration

6. When oxygen combines with an element to form a compound, the resulting compound is called

(A) a salt
(B) an oxide
(C) oxidation
(D) an oxalate

7. According to the activity chart of metals, which metal would react most vigorously in a dilute acid solution?

(A) zinc
(B) iron
(C) aluminum
(D) magnesium

8. Graham's Law refers to

(A) boiling points of gases
(B) gaseous diffusion
(C) gas compression problems
(D) volume changes of gases when the temperature changes

9. When 200 milliliters of a gas at constant pressure is heated, its volume

(A) increases
(B) decreases
(C) remains unchanged

10. When 200 milliliters of a gas at constant pressure is heated from 0°C to 100°C, the volume must be multiplied by

(A) 0/100
(B) 100/0
(C) 273/373
(D) 373/273

11. If you wish to find the corrected volume of a gas that was at 20°C and 1 atmosphere pressure and conditions were changed to 0°C and 0.92 atmosphere pressure, by what fractions would you multiply the original volume?

(A) $\frac{293}{273} \times \frac{1}{0.92}$

(B) $\frac{273}{293} \times \frac{0.92}{1}$

(C) $\frac{273}{293} \times \frac{1}{0.92}$

(D) $\frac{293}{273} \times \frac{0.92}{1}$

12. When the level of mercury inside a gas tube is higher than the level in the reservoir, you find the correct pressure inside the tube by taking the outside pressure reading and ? the difference in the height of mercury.

(A) subtracting
(B) adding
(C) dividing by 13.6
(D) doing both (c) and (a)

13. If water is the liquid in question 12 instead of mercury, you can change the height difference to an equivalent mercury expression by

(A) dividing by 13.6
(B) multiplying by 13.6
(C) adding 13.6
(D) subtracting 13.6

14. Standard conditions are
    (A) 0°C and 14.7 mm
    (B) 273 K and 760 mm Hg
    (C) 273°C and 760 mm Hg
    (D) 4°C and 7.6 mm Hg

15. When a gas is collected over water, the pressure is corrected by
    (A) adding the vapor pressure of water
    (B) multiplying by the vapor pressure of water
    (C) subtracting the vapor pressure of water at that temperature
    (D) subtracting the temperature of the water from the vapor pressure.

16. At 5 atmospheres pressure and 343 K how many moles are present in 1.5 liters of O_2 gas?
    (A) 0.036
    (B) 0.267
    (C) 0.536
    (D) 0.103

Directions: Every set of the given lettered choices below refers to the numbered questions immediately below it. For each numbered item, choose the one lettered choice that fits it best. Every choice in a set may be used once, more than once, or not at all.

Questions 17–20 refer to the following graphs:

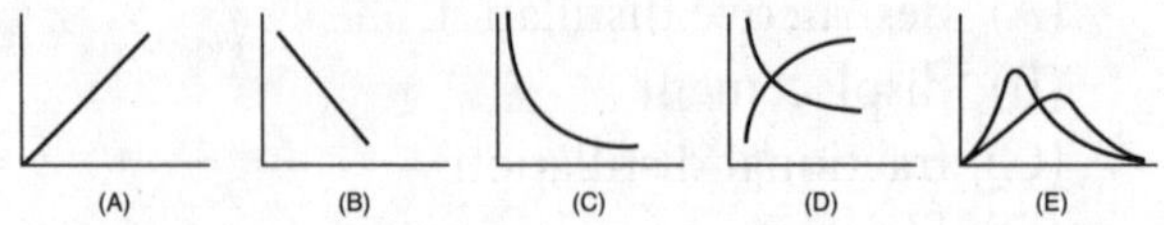

17. Which is a graphic depiction of Boyle's Law?

18. Which is a graphic depiction of Charles's Law?

19. Which is a graphic depiction of the relationship of the pressure of a given volume of gas with the absolute temperature?

20. Which graph shows the distribution of molecules with respect to their kinetic energy at different temperatures?

Answers and Explanations

1. **(B)** Oxygen is the most abundant element in Earth's crust, with 21%.

2. **(D)** $KClO_3$ can be decomposed with heat to form KCl and O_2.

3. **(C)** The equation is: $2KClO_3 \rightarrow 2KCl + 3O_2\uparrow$

4. **(A)** Because energy is given off in an exothermic reaction, the heat content of the reactants would be higher than the products.

5. **(C)** Fractional distillation can separate liquids of different boiling points.

6. **(B)** Oxygen compounds formed in combination reactions are called oxides.

7. **(D)** Magnesium, which is the most active of the given metals, will react to release hydrogen.

8. **(B)** Graham's Law refers to gaseous diffusion or effusion.

9. **(A)** When confined gases at a constant pressure are heated, they expand.

10. **(D)** The temperature must be changed to absolute temperature by adding 273 to the centigrade readings.

11. **(C)** Because the temperature is decreasing, the volume must decrease with the temperature fraction, and, because the pressure is decreasing, the volume must increase. The correct answer does this.

12. **(A)** The level is higher because the pressure inside the tube is less than outside; therefore, you must subtract the height inside the tube from the outside pressue.

13. **(A)** If water is the liquid, you must divide by 13.6 to change the height to the equivalent height of mercury.

14. **(B)** The only correct indication of standard conditions is (B). It is usually given as 273 K and 760 mm Hg.

15. **(C)** Because the vapor pressure of water is a part of the pressure reading, it must be subtracted to get the atmospheric pressure.

16. **(B)** Using the general gas law, $PV = nRT$, convert 70°C to K by adding 273 = 343 K, and solving for n, you get

$$n = \frac{PV}{RT} = \frac{5\ \cancel{\text{atm}} \times 1.5\ \cancel{\text{L}}}{0.0821 \frac{\cancel{\text{L}}\ \cancel{\text{atm}}}{\text{mol}\ \cancel{\text{K}}} 343\ \cancel{\text{K}}} = 0.267 \text{ mol}$$

17. **(C)** Boyle's Law is an inverse relationship. As the pressure increases, the volume decreases, as shown by the graph.

18. **(A)** Charles's Law is a direct relationship. As the temperature increases, the volume increases.

19. **(A)** The relationship of pressure to absolute temperature while the volume is held constant is a direct one. As the temperature on a given volume increases, the pressure will increase.

20. **(D)** The graphs in (D) show the distributions of the kinetic energy of molecules at two different temperatures.

Stoichiometry (Chemical Calculations) and the Mole Concept

CHAPTER 6

- Use the mole concept to find the molar mass of different formulas and monoatomic and diatomic molecules, and how gas volumes are related to molar mass.
- Solve stoichiometric problems involving Gay-Lussac's Law, density = mass/volume equation, mass–volume relation, mass–mass problems, volume–volume problems, problems with a limiting reactant (excess of one reactant), and finding the percent yield.

This chapter deals with the solving of a variety of chemistry problems, which is often referred to as **stoichiometry**. Solving problems should be done in an organized manner, and it would be to your advantage to go back to the Introduction to this book and review the section called, "How Can You Improve Your Problem Solving Skills?" It describes a well-planned method for attacking the process of solving problems that you will find helpful in this chapter.

Two or three methods of solving the problems in this chapter are usually given. It would be wise on your part to learn the mole method, the proportion method, and the factor-label method, so that you have more tools available for solving problems. Remember that no matter which method you use, you should still use the estimation process to verify the plausibility of the answer.

THE MOLE CONCEPT

Providing a name for a quantity of things taken as a whole is common in everyday life. Some examples are a dozen, a gross, and a ream. Each of these represents a specific number of items and is not dependent on the commodity. A dozen eggs, oranges, or bananas will always represent 12 items.

In chemistry we have a unit that decribes a quantity of particles. It is called the **mole** (sometimes abbreviated as **mol**). A mole is 6.02×10^{23} particles. The

> **TIP**
>
> Just like a dozen is 12 units of an item, a mole is 6.02×10^{23} units of an item. It is called Avogadro's number.

particles can be atoms, molecules, ions, electrons, and so forth. Because particles are so small in chemistry, the mole is a very convenient unit. The number 6.02×10^{23} is often referred to as **Avogadro's number** in honor of the Italian scientist whose hypothesis led to its determination.

In 1811, Amedeo Avogadro made a far-reaching scientific assumption (hypothesis) that also bears his name. He stated that equal volumes of different gases contain equal numbers of molecules at the same pressure and temperature. The statement is called Avogadro's Law. It means that, under the same conditions, the number of molecules of hydrogen in a 1-liter container is exactly the same as the number of molecules of carbon dioxide or any other gas in a 1-liter container even though the individual molecules of the different gases have different masses and sizes. Because of the substantiation of this hypothesis by many data since its inception, it is often referred to as **Avogadro's Law**.

MOLAR MASS AND MOLES

When the formula mass of an ionic compound is determined by the addition of its component relative atomic masses and expressed in grams, it is called the **molar mass.**

EXAMPLE: Find the molar mass of $CaCO_3$.

1Ca = 40
1C = 12
3O = 48 = (3 × 16)

$CaCO_3$ = 100 formula mass
100 g/mol is the molar mass

The **molar mass** is equivalent to the mass of 1 **mole** of that compound expressed in grams.

ANOTHER EXAMPLE: Find the mass of 1 mole of $KAl(SO_4)_2 \cdot 12H_2O$.

1K = 39
1Al = 27
$2(SO_4)$ = 192 = 2(32 + 16 × 4)
$12H_2O$ = 216 = 12(2 + 16)

1 mole = 474 g

TRY THESE PROBLEMS: Find the molecular mass or formula mass of each of the following:

1. HNO_3 Ans. 63 g/mol
2. $C_6H_{10}O_5$ Ans. 162 g/mol
3. H_2SO_4 Ans. 98 g/mol
4. KCl Ans. 74.5 g/mol
5. $C_{12}H_{22}O_{11}$ Ans. 342 g/mol

In some cases, there may be a question of the mass of 1 mole of an element if you are not told specifically what is referred to—the single atoms or a molecular state. An example of this situation may arise with hydrogen and other elements that form diatomic molecules. If you are asked the mass of 1 mole of hydrogen, you could say it is 1 gram if you are dealing with single atoms of hydrogen. This 1 mole of hydrogen atoms would contain 6.02×10^{23} atoms. In a similar fashion, 1 mole of hydrogen molecules (H_2) would have a molar mass of 2 grams. This 1 mole of hydrogen molecules would contain 6.02×10^{23} molecules and, since each molecule is composed of two atoms, 12.04×10^{23} atoms.

The other elements that exist as diatomic molecules are oxygen, nitrogen, fluorine, chlorine, bromine, iodine, and astatine.

Try these questions.

TIP

Know these relationships!

What is the mass of 1 mole of oxygen?

Answer: Since the atomic mass is 16 grams and the molecule is diatomic, 2×16 grams = 32 grams.

What is the mass of one molecule of oxygen?

Answer: There are 6.02×10^{23} molecules in 1 mole of this gas. 32 grams ÷ 6.02×10^{23} molecules = 5.32×10^{-23} gram.

What is the mass of 1 mole of oxygen atoms?

Answer: Without the diatomic molecule structure, the mass of one mole of oxygen atoms expressed in grams is the answer. That is 16 g/mol.

What is the mass of one atom of oxygen?

Answer: The atomic molar mass, 16 grams, is divided by 6.02×10^{23}. This gives 2.66×10^{-23} gram.

If you gave the same answer to the first two questions, you are confused about the mole concept. To say that a molecule of oxygen gas has the same mass as a mole of oxygen gas is the same as saying that an apple has the same mass as a dozen apples.

MOLE RELATIONSHIPS

Because the mole is used often in chemistry to quantify the number of atoms, molecules, and several other items, it is important to know the relationships that exist and how to move from one to another. The following summarizes a number of these relationships:

TIP

Know how to use these mole relationships.

When dealing with elements—

Moles of an element × molar mass (atomic) = mass of the element

Mass of an element/molar mass (atomic) = moles of the element

When dealing with compounds—

Moles of a compound × molar mass (molecular) = mass of the compound

Mass of a compound/molar mass (molecular) = moles of the compound

When dealing with the molecules of a compound—
Moles of molecules × 6.02×10^{23} = number of molecules
Number of molecules/6.02×10^{23} = moles of molecules

When dealing with the atoms of elements—
Moles of atoms × 6.02×10^{23} = number of atoms
Number of atoms/6.02×10^{23} = moles of atoms

GAS VOLUMES AND MOLAR MASS

Because the volume of a gas may vary depending on the conditions of temperature and pressure, a standard is set for comparing gases. As stated in Chapter 5, the standard conditions of temperature and pressure (abbreviated STP) are 273 K and 760 mm of mercury pressure.

The molecular mass of a gas expressed in grams and under standard conditions occupies 22.4 liters. This is an important relationship to remember! The 22.4 liters is referred to as the gram-molecular volume (gmv) or molar volume. Two scientists are associated with this relationship.

> **TIP**
> This is Gay-Lussac's Law of combining gases.

Gay-Lussac's Law states that, when only gases are involved in a reaction, the volumes of the reacting gases and the volumes of the gaseous products are in a small-whole-number ratio to each other. This law may be illustrated by the following cases:

Example 1: $H_2(g) + Cl_2(g) \rightarrow 2HCl(g)$

This balanced equation shows that

1 vol. hydrogen + 1 vol. chlorine = 2 vols. hydrogen chloride

Example 2: $2H_2(g) + O_2(g) \rightarrow 2H_2O(g)$

This balanced equation shows that

2 vols. hydrogen + 1 vol. oxygen = 2 vols. steam

Avogadro's Law, which explains Gay-Lussac's, states that equal volumes of gases under the same conditions of temperature and pressure contain equal numbers of molecules. This means that 1 mole of any gas at STP occupies 22.4 liters, so:

32 g O_2 at STP occupies 22.4 liters
2 g H_2 at STP occupies 22.4 liters
44 g CO_2 at STP occupies 22.4 liters
$2O_2$ (2 moles O_2) = 64 g = 44.8 liters at STP
$3H_2$ (3 moles H_2) = 6 g = 67.2 liters at STP

DENSITY AND MOLAR MASS

Because the density of a gas is usually given in grams/liter of gas at STP, we can use the molar volume to molar mass relationship to solve the following types of problems.

> **TIP**
>
> For a dry gas at STP, the mass of 1 liter × 22.4 = the molar mass of the gas.

EXAMPLE: Find the molar mass of a gas when the density is given as 1.25 grams/liter.

Because it is known that 1 mole of a gas occupies 22.4 liters at STP, we solve this problem by multiplying the mass of 1 liter by 22.4 liters/mole.

$$\frac{1.25\text{ g}}{\cancel{\text{L}}} \times \frac{22.4\ \cancel{\text{L}}}{1\text{ mol}} = 28\text{ g/mol}$$

Even if the weight given is not for 1 liter, the same setup can be used.

PROBLEM: If 3 liters of a gas weighs 2 grams, find the molar mass.

Solution:

$$\frac{2\text{ g}}{3\ \cancel{\text{L}}} \times \frac{22.4\ \cancel{\text{L}}}{1\text{ mol}} = 14.9\text{ g/mol}$$

You can also find the density of a gas if you know the molar mass. Since the molar mass occupies 22.4 liters at STP, dividing the molar mass by 22.4 liters will give you the mass per liter, or the density.

EXAMPLE: Find the density of oxygen at STP.

> **TIP**
>
> For a dry gas at STP,
>
> $$\text{density} = \frac{\text{molar mass}}{22.4\text{ L}}$$

Solution:

Oxygen is diatomic in its molecular structure.
O_2 = molar mass of 2 × 16 or 32 g/mol
32 g/mol occupies 22.4 L.

Therefore

$$\frac{32\text{ g}}{1\text{ mol}} \div \frac{22.4\text{ L}}{1\text{ mol}} = \frac{32\text{ g}}{1\ \cancel{\text{mol}}} \times \frac{1\ \cancel{\text{mol}}}{22.4\text{ L}} = 1.43\text{ g/L}$$

You can find the density of a gas, then, by dividing its molar mass by 22.4 liters.

MASS–VOLUME RELATIONSHIPS

A typical **mass-volume problem**: How many liters of oxygen (STP) can you prepare from the decomposition of 42.6 grams of sodium chlorate?

1st step. Write the balanced equation for the reaction.

$$2NaClO_3 \xrightarrow{\Delta} 2NaCl + 3O_2(g)$$

2nd step. Write the given quantity and the unknown quantity above the appropriate substances.

$$\overset{42.6\text{ g}}{2NaClO_3} \xrightarrow{\Delta} 2NaCl + \overset{x\text{ L}}{3O_2}$$

3rd step. Calculate the equation mass or volume of each substance that has something indicated above it, and write the results under the substances. Note that the units above and below *must* match.

$$\underset{213\text{ g}}{\overset{42.6\text{ g}}{2NaClO_3}} \xrightarrow{\Delta} 2NaCl + \underset{67.2\text{ L}}{\overset{x\text{ L}}{3O_2}}$$

(2 × molar mass of $NaClO_3$) (3 × 22.4 L)

4th step. Form the proportion.

$$\frac{42.6\text{ g}}{213\text{ g}} = \frac{x\text{ L}}{67.2\text{ L}}$$

5th step. Solve for *x*.

$$x = 13.4\text{ L } O_2$$

This problem can also be solved using methods other than the **proportion method** shown above.

Another method to proceed from step 3 is called the **factor-label method**. The reasoning is this: Because the equation shows that 213 grams of reactant produces 67.2 liters of the required product, multiplying the given amount by this equality (so that the units of the answer will be correct) will give the same answer as above. Then step 4 would be:

4th step. $$42.6\ \cancel{\text{g } NaClO_3} \times \frac{67.2\text{ L } O_2}{213\ \cancel{\text{g } NaClO_3}} = 13.4\text{ L } O_2$$

Still another method of solving this problem is called the **mole method.** Steps 1 and 2 are the same. Then step 3 is as follows:

3rd step. Determine how many moles of substance are given.

$$42.6\text{ g} \div \frac{106.5\text{ g}}{1\text{ mol NaClO}_3} = 0.4\text{ mol NaClO}_3$$

The equation shows that 2 moles $NaClO_3$ makes 3 moles O_2. Then 0.4 mole $NaClO_3$ will yield

$$0.4\text{ mol NaClO}_3 \times \frac{3\text{ mol O}_2}{2\text{ mol NaClO}_3} = 0.6\text{ mol O}_2$$

4th step. Convert the moles of O_2 to liters.

$$0.6\ \cancel{\text{mol O}_2} \times \frac{22.4\text{ L O}_2}{1\ \cancel{\text{mol O}_2}} = 13.4\text{ L O}_2$$

ANOTHER EXAMPLE: Find the weight of $CaCO_3$ needed to produce 11.2 L CO_2 when the calcium carbonate is reacted with hydrochloric acid.

1st step. Write the balanced equation for the reaction.

$$CaCO_3 + 2HCl \rightarrow CaCl_2 + H_2O + CO_2$$

2nd step. Write the given quantity and the unknown quantity above the appropriate substances.

$$\overset{x\text{ g}}{CaCO_3} + 2HCl \rightarrow CaCl_2 + H_2O + \overset{11.2\text{ L}}{CO_2}$$

3rd step. Calculate the equation mass or volume for each substance that has something indicated above it, and write the results under the substances. Note that the units above and below *must* match.

x g 11.2 L

$$CaCO_3 + 2HCl \rightarrow CaCl_2 + H_2O + CO_2$$

100. g 22.4 L

PROPORTION METHOD:

4th step. Form the proportion.

$$\frac{x\text{ g}}{100.\text{ g}} = \frac{11.2\text{ L}}{22.4\text{ L}}$$

5th step. Solve for *x*.

$x = 50.0\text{ g } CaCO_3$

FACTOR-LABEL METHOD:

4th step. $$11.2\ \cancel{\text{L } CO_2} \times \frac{100.\text{ g } CaCO_3}{22.4\ \cancel{\text{L } CO_2}} = 50.0\text{ g } CaCO_3$$

MOLE METHOD:

$$11.2\text{ L } CO_2 \div \frac{22.4\text{ L}}{1\text{ mol}} = 0.50\text{ mol } CO_2$$

Equation shows 1 mole $CaCO_3$ yields 1 mole CO_2, so

$$0.50\ \cancel{\text{mol } CO_2} \times \frac{1\text{ mol } CaCO_3}{1\ \cancel{\text{mol } CO_2}} = 0.50\text{ mol } CaCO_3$$

Converting moles to grams:

$$0.50\ \cancel{\text{mol } CaCO_3} \times \frac{100.\text{ g } CaCO_3}{1\ \cancel{\text{mol } CaCO_3}} = 50.0\text{ g } CaCO_3$$

TRY THESE PROBLEMS:

1. What mass of water must be electrolyzed to obtain 20.0 liters of oxygen at STP?

   Ans. 32.2 g*

2. How many grams of aluminum will be completely oxidized by 44.8 liters of oxygen at STP?

   Ans. 72 g*

*Answers explained at the end of this chapter, pages 170–171.

MASS–MASS PROBLEMS

A typical problem concerning just mass relationships is as follows:

What mass of oxygen can be obtained from heating 100. grams of potassium chlorate?

1st step. Write the balanced equation for the reaction.

$$2KClO_3 \rightarrow 2KCl + 3O_2$$

2nd step. Write the given quantity and the unknown quantity above the appropriate substances.

$$\overset{100.\ g}{2KClO_3} \rightarrow 2KCl + \overset{x\ g}{3O_2}$$

3rd step. Calculate the equation mass for each substance that has something indicated above it, and write the results under the substances. Note that the units above and below *must* match.

$$\underset{245\ g}{\overset{100.\ g}{2KClO_3}} \rightarrow 2KCl + \underset{96\ g}{\overset{x\ g}{3O_2}}$$

USING THE PROPORTION METHOD:

4th step. Form the proportion.

$$\frac{100.\ g}{245\ g} = \frac{x\ g}{96\ g}$$

5th step. Solve for x.

$$x = 39.3\ g\ of\ O_2$$

USING THE FACTOR-LABEL METHOD:

From the 3rd step on you would proceed as follows:

4th step. The equation indicates that 245 g $KClO_3$ yields 96 g of O_2. Therefore multiplying the given quantity by a factor made up of these two quantities arranged appropriately so that the units of the answer remain uncanceled will give the answer to the problem.

$$100.\ \cancel{g\ KClO_3} \times \frac{96\ g\ O_2}{245\ \cancel{g\ KClO_3}} = 39.3\ g\ O_2$$

TIP

Put what is given above the equation.

Put the calculated amount below.

The **units must match!**

TIP

After you cancel all similar units above and below, you should have remaining the units for your answer.

USING THE MOLE METHOD:

To solve this problem you would proceed as follows from the 2nd step:

3rd step. Determine how many moles of substance are given.

$$100.\ \text{g}\ \cancel{KClO_3} \times \frac{1\ \text{mol}\ KClO_3}{122.5\ \text{g}\ \cancel{KClO_3}} = 0.815\ \text{mol}\ KClO_3$$

The equation shows that 2 moles of $KClO_3$ yields 3 moles of O_2, so 0.815 mole of $KClO_3$ will yield

$$0.815\ \cancel{\text{mol}\ KClO_3} \times \frac{3\ \text{mol}\ O_2}{2\ \cancel{\text{mol}\ KClO_3}} = 1.22\ \text{mol}\ O_2$$

4th step. Convert the moles of O_2 to grams of O_2.

$$1.22\ \cancel{\text{mol}\ O_2} \times \frac{32\ \text{g}\ O_2}{1\ \cancel{\text{mol}\ O_2}} = 39\ \text{g}\ O_2$$

ANOTHER EXAMPLE: What mass of potassium hydroxide is needed to neutralize 20.0 grams of sulfuric acid?

1st step. $2KOH + H_2SO_4 \rightarrow K_2SO_4 + 2H_2O$

x g 20.0 g

2nd step. $2KOH + H_2SO_4 \rightarrow K_2OS_4 + 2H_2O$

x g 20.0 g

3rd step. $2KOH + H_2SO_4 \rightarrow K_2SO_4 + 2H_2O$

112 g 98 g

USING THE PROPORTION METHOD:

4th step. $\frac{x\ \text{g}}{112\ \text{g}} = \frac{20.0\ \text{g}}{98\ \text{g}}$

5th step. $x = 22.8$ g KOH

USING THE FACTOR-LABEL METHOD:

4th step. $20.0\ \cancel{\text{g}\ H_2SO_4} \times \frac{112\ \text{g}\ KOH}{98\ \cancel{\text{g}\ H_2SO_4}} = 22.8\ \text{g}\ KOH$

USING THE MOLE METHOD:

3rd step. $20.0\ \cancel{g\ H_2SO_4} \times \dfrac{1\ mol\ H_2SO_4}{98\ \cancel{g\ H_2SO_4}} = 0.204\ mol\ H_2SO_4$

$0.204\ \cancel{mol\ H_2SO_4} \times \dfrac{2\ mol\ KOH}{1\ \cancel{mol\ H_2SO_4}} = 0.408\ mol\ KOH$

4th step. $0.408\ \cancel{mol\ KOH} \times \dfrac{56\ g\ KOH}{1\ \cancel{mol\ KOH}} = 22.8\ g\ KOH$

TRY THESE PROBLEMS:

3. What mass of manganese dioxide is needed to react with an excess of hydrochloric acid so that 100. grams of chlorine is liberated?

   Ans.: 122.5 g*

4. A 20.0 gram sample of Mg is burned in 20.0 grams of O_2. How much MgO is formed? (Hint: Determine which is in excess.)

   Ans. 33.3 g*

VOLUME–VOLUME PROBLEMS

TIP

Know Gay-Lussac's Law.

This type of problem involves only volume units and therefore can make use of **Gay-Lussac's Law**: "When gases combine, they combine in simple whole-number ratios." These simple numbers are the coefficients of the equation.

A typical problem concerning just volume relationships is as follows:

What volume of NH_3 is produced when 22.4 liters of nitrogen are made to combine with a sufficient quantity of hydrogen under the appropriate conditions?

1st step. Write the balanced equation for the reaction.

$N_2 + 3H_2 \rightarrow 2NH_3$

2nd step. Write the given quantity and the unknown quantity above the respective substances.

$$\overset{22.4\ L}{N_2} + 3H_2 \rightarrow \overset{x\ L}{2NH_3}$$

3rd step. Set up a proportion using the coefficient of the substances that have something indicated above them as denominators.

$$\frac{22.4\ L}{1\ L} = \frac{x\ L}{2\ L}$$

*Answers explained at the end of this chapter, pages 171–172.

4th step. Solve for *x*.

$x = 44.8$ L

USING THE FACTOR-LABEL METHOD:

From the 2nd step on you would proceed as follows:

3rd step. The equation shows that 1 volume of N_2 will yield 2 volumes of NH_3. Therefore multiplying the given quantity by a factor made up of these two quantities appropriately arranged so that the units of the answer remain uncanceled will solve the problem.

$$22.4\ \cancel{L\ N_2} \times \frac{2\ L\ NH_3}{1\ \cancel{L\ N_2}} = 44.8\ L\ NH_3$$

USING THE MOLE METHOD:

To solve this problem you would proceed as follows:

3rd step. The given quantity is converted to moles.

$$\cancel{22.4\ L\ N_2} \times \frac{1\ mol\ N_2}{\cancel{22.4\ L\ N_2}} = 1\ mol\ N_2$$

The equation shows that 1 mole of N_2 will yield 2 moles of NH_3. Therefore using this relationship will yield

$$1\ \cancel{mol\ N_2} \times \frac{2\ mol\ NH_3}{1\ \cancel{mol\ N_2}} = 2\ mol\ NH_3$$

Convert the moles of NH_3 to liters of NH_3.

$$2\ \cancel{mol\ NH_3} \times \frac{22.4\ L\ NH_3}{1\ \cancel{mol\ NH_3}} = 44.8\ L\ NH_3$$

ANOTHER EXAMPLE: What volume of SO_2 will result from the complete burning of pure sulfur in 8 liters of oxygen until all the oxygen is used?

1st step. $S + O_2 \rightarrow SO_2$

$$\quad 8\ L \qquad x\ L$$

2nd step. $S + O_2 \rightarrow SO_2$

USING THE PROPORTION METHOD:

3rd step. $\frac{8\text{ L}}{1\text{ L}} = \frac{x\text{ L}}{1\text{ L}}$

4th step. $x = 8\text{ L of }SO_2$

USING THE FACTOR-LABEL METHOD:

3rd step. $8\ \cancel{\text{L}\,O_2} \times \frac{1\text{ L }SO_2}{1\ \cancel{\text{L}\,O_2}} = 8\text{ L }SO_2$

USING THE MOLE METHOD (NOT REALLY PRACTICAL IN THIS CASE):

3rd step. $8\ \cancel{\text{L}\,O_2} \times \frac{1\text{ mol }O_2}{22.4\ \cancel{\text{L}\,O_2}} = 0.358\text{ mol }O_2$

$0.358\ \cancel{\text{mol}\,O_2} \times \frac{1\text{ mol }SO_2}{1\ \cancel{\text{mol}\,O_2}} = 0.358\text{ mol }SO_2$

4th step. $0.358\ \cancel{\text{mol}\,SO_2} \times \frac{22.4\text{ L }SO_2}{1\ \cancel{\text{mol}\,SO_2}} = 8\text{ L }SO_2$

TRY THESE PROBLEMS:

5. How many liters of hydrogen are necessary to react with sufficient chlorine to produce 12 liters of hydrogen chloride gas?

   Ans. 6 L H_2*

6. How many liters of oxygen will be needed to burn 100 liters of carbon monoxide?

   Ans. 50 L O_2*

PROBLEMS WITH AN EXCESS OF ONE REACTANT

It will not always be true that the amounts given in a particular problem are exactly in the proportion required for the reaction to use up all of the reactants. In other words, at times some of one reactant will be left over after the other has been used up. This is similar to the situation in which two eggs are required to mix with one cup of flour in a particular recipe, and you have four eggs and four cups of flour. Since two eggs require only one cup of flour, four eggs can use only two cups of flour and two cups of flour will be left over.

TIP

The paragraph describes a practical example of a "limiting reactant."

*Answers explained at the end of this chapter, page 173.

TIP

Remember this recipe analogy to solve limiting-reactant questions!

A chemical equation is very much like a recipe. Consider the following equation:

$$Zn + 2HCl \rightarrow ZnCl_2 + H_2$$

If you are given 65 grams of zinc and 65 grams of HCl, how many grams of hydrogen gas can be produced? Which reactant will be left over? How many grams of this reactant will not be consumed?

1st step. Set up the problem.

$$\begin{array}{cccc} 65\text{ g} & 65\text{ g} & & x\text{ g} \\ Zn + & 2HCl \rightarrow & ZnCl_2 + & H_2 \\ 65\text{ g} & 73\text{ g} & & 2\text{ g} \end{array}$$

The given quantities are above the equation, and the equation masses are given beneath. To solve for the grams of hydrogen gas, the limiting reactant must be determined.

2nd step. Compare the given quantities with the equation requirements.

Knowing that it takes 65 grams of zinc to react with 73 grams of hydrochloric acid, it is reasonable to surmise that, since there are only 65 grams of hydrochloric acid, not all of the 65 grams of zinc can be used. The limiting factor will be the amount of hydrochloric acid.

3rd step. Solve for the quantity of product that will be produced using the amount of the limiting factor.

Now that we know that the limiting factor is the amount of hydrochloric acid (65 g), the equation can be solved. This means that the proportion is set up ignoring the 65 grams of zinc.

$$\frac{65\text{ g}}{73\text{ g}} = \frac{x\text{ g}}{2\text{ g}}$$

Solving for x, we get $x = 1.78$ grams or 1.8 grams of hydrogen produced.

4th step. To find the number of grams of zinc that were not consumed, a separate problem is necessary. The equation would be set up as follows:

$$\begin{array}{cccc} y\text{ g} & 65\text{ g} & & \\ Zn + & 2HCl \rightarrow & ZnCl_2 + & H_2 \\ 65\text{ g} & 73\text{ g} & & \end{array}$$

This gives the proportion:

$$\frac{y\text{ g}}{65\text{ g}} = \frac{65\text{ g}}{73\text{ g}}$$

$$y = \frac{65 \times 65}{73} = 57.8 \text{ or } 58\text{ g}$$

Subtracting the 58 grams of zinc that will be used from the original 65 grams leaves 7 grams of zinc that will not be consumed.

PERCENT YIELD OF A PRODUCT

In most stoichiometric problems, we assume that the results are exactly what we would theoretically expect. In reality, the resulting yield is rarely the actual yield. Why the actual yield of a reaction may be less than the theoretical yield occurs for many reasons. Some of the product is often lost during the purification or collection process.

Chemists are usually interested in the efficiency of a reaction. The efficiency is expressed by comparing the actual and the theoretical yields.

The **percent yield** is the ratio of the actual yield to the theoretical yield, multiplied by 100%.

$$\text{percent yield} = \frac{\text{actual yield}}{\text{theoretical yield}} \times 100$$

EXAMPLE: Quicklime, CaO, is to be prepared by roasting 2.00×10^3 g of limestone, $CaCO_3$, according to the following reaction. However, the actual yield of CaO is 1.05×10^3 g. What is the percent of yield?

The reaction is:

$$CaCO_3(s) \rightarrow CaO(s) + CO_2(g)$$

The percent yield is found by using the above formula and inserting the actual and theoretical yields.

To find the theoretical yield, you solve the following:

2.00×10^3 g x g

$CaCO_3(s) \rightarrow CaO(s) + CO_2$ (g)

100 g/mol   56 g/mol

$x = 1.12 \times 10^3$ g of CaO, the theoretical yield

$$\text{percent yield} = \frac{1.05 \times 10^3 \text{ g CaO (actual yield)}}{1.12 \times 10^3 \text{ g of CaO (theoretical yield)}} \times 100$$

$= 93.7\%$, percent yield

Answers to Chapter Problems

After each type of problem in this chapter there are several problems to solve; therefore the chapter review does not include problems. The answer explanations for the problems previously presented are given below.

PAGE 162 **1.** Equation: $\overset{x\text{ g}}{2H_2O} \rightarrow 2H_2 + \overset{20.\text{ L}}{O_2}$

Calculate the equation mass and volume so that the units match the ones above the equation.

$$\overset{x\text{ g}}{2H_2O} \rightarrow 2H_2 + \overset{20.\text{ L}}{O_2}$$

(2 × molecular mass of water) (1 × gram-molecular volume)

36 g 22.4 L

Then

$$\frac{x\text{ g}}{36\text{ g}} = \frac{20.\text{ L}}{22.4\text{ L}}$$

$$x = 32\text{ g or }32\text{ g}$$

Using the mole method:

$$\frac{20.\ \cancel{L}}{22.4\ \cancel{L}/\text{mol}} = .89\text{ mol }O_2$$

The mole relationship in the equation is:

2 moles $H_2O \rightarrow 2H_2$ + 1 mole O_2

Then $\dfrac{2\text{ mol }H_2O}{x\text{ mol }H_2O} = \dfrac{1\text{ mol }O_2}{.89\text{ mol }O_2}$

$x = 1.78\text{ mol }H_2O$

Because 1 mol H_2O = 18 g /mol

$$18\text{ g}/\cancel{\text{mol}} \times 1.78\ \cancel{\text{mol}} = \underline{\underline{32\text{ g }H_2O}}$$

PAGE 162 **2.** x g 44.8 L

$4Al + 3O_2 \rightarrow 2Al_2O_3$

Calculate the equation mass and volume so that the units match the ones above the equation.

$$\begin{array}{cccc} x\text{ g} & & 44.8\text{ L} & \\ 4Al & + & 3O_2 & \rightarrow 2Al_2O_3 \\ (4 \times 27\text{ g}) & & (3 \times 22.4\text{ L}) & \\ 108\text{ g} & & 67.2\text{ L} & \end{array}$$

Then

$$\frac{x\text{ g}}{108\text{ g}} = \frac{44.8\text{ L}}{67.2\text{ L}}$$

$$x = \underline{\underline{72\text{ g}}}$$

Using the mole method:

$$\frac{44.8\ \cancel{\text{L}}}{22.4\ \frac{\cancel{\text{L}}}{\text{mol}}} = 2\text{ mol }O_2$$

The balanced equation shows that 4 mol Al needed 3 mol O_2.

Then

$$\frac{4\text{ mol Al}}{3\text{ mol }O_2} = \frac{x\text{ mol Al}}{2\text{ mol }O_2}$$

$$x = 2.67\text{ mol Al}$$

Because there is 27 g in 1 mol Al,

$$2.67\ \cancel{\text{mol Al}} \times 27\text{ g}/\cancel{\text{mol Al}} = \underline{\underline{72\text{ g Al}}}$$

PAGE 165 **3.** The equation:

x g 100. g

$MnO_2 + 4HCl \rightarrow MnCl_2 + Cl_2 + 2H_2O$

Calculate the equation masses for the substances involved.

$$\begin{array}{ccccc} x\text{ g} & & & 100.\text{ g} & \\ MnO_2 & + 4HCl \rightarrow MnCl_2 + & & Cl_2 & + 2H_2O \\ (1 \times 87\text{ g}) & & & (1 \times 71\text{ g}) & \\ 87\text{ g} & & & 71\text{ g} & \end{array}$$

Then

$$\frac{x\text{ g}}{87\text{ g}} = \frac{100.\text{ g}}{71\text{ g}}$$
$$x = 122.5\text{ g or } 123\text{ g}$$

Using the mole method:

$$\frac{100.\ \cancel{\text{g Cl}_2}}{\frac{71\ \cancel{\text{g Cl}_2}}{\text{mol}}} = 1.41\text{ mol Cl}_2$$

The balanced equation shows that 1 mol MnO_2 needed 1 mol Cl_2.

Then

$$\frac{1\text{ mol MnO}_2}{1\text{ mol Cl}_2} = \frac{x\text{ mol MnO}_2}{1.41\text{ mol Cl}_2}$$
$$x = 1.41\text{ mol MnO}_2$$

Because there is (54.9 + 32.0) or 86.9 g in 1 mol MnO_2,

$$86.9\text{ g}/\cancel{\text{mol MnO}_2} \times 1.41\ \cancel{\text{mol MnO}_2} = 122.5\text{ g MnO}_2.$$

PAGE 165 **4.** The equation:

$$\begin{array}{ccccc} 20.\text{g} & & 20.\text{ g} & & x\text{ g} \\ 2\text{Mg} & + & \text{O}_2 & \rightarrow & 2\text{MgO} \\ 48\text{ g} & & 32\text{ g} & & 80\text{ g} \end{array}$$

Comparing the amounts needed below the equation, you see that you need more Mg than O_2. Therefore, the amount of Mg will be the limiting factor because it will be used up first. Because the equation masses indicate you must have more Mg than O_2, there is an excess of oxygen.

$$\frac{20.\text{g Mg}}{48\text{ g Mg}} = \frac{x\text{ g MgO}}{80\text{ g MgO}}$$
$$x = 33.3\text{ g MgO}$$

Using the mole method:

$$\frac{20.\ \cancel{\text{g Mg}}}{\frac{24\ \cancel{\text{g Mg}}}{\text{mol Mg}}} = .83\text{ mol Mg}$$

The balanced equation shows that 2 mol Mg forms 2 mol MgO.

Then

$$\frac{2 \text{ mol Mg}}{2 \text{ mol MgO}} = \frac{.83 \text{ mol Mg}}{x \text{ mol MgO}}$$

$$x = .83 \text{ mol Mg}$$

Because there is (24 + 16) or 40 g of MgO in 1 mol MgO,

$$\frac{40 \text{ g/}\cancel{\text{MgO}}}{\cancel{\text{mol/MgO}}} \times .83 \ \cancel{\text{mol MgO}} = 33 \text{ g MgO.}$$

PAGE 167 **5.** The equation:

x L 12 L
$H_2 + Cl_2 \rightarrow 2HCl$

Set up the proportion using the coefficients of the substances that have something indicated above them as denominators. Gay-Lussac's Law says that with reacting gases you can use the coefficients of the balanced equation to solve the problem.

$$\frac{x \text{ L } H_2}{1 \text{ L } H_2} = \frac{12 \text{ L HCL}}{2 \text{ L HCL}}$$

$$\underline{\underline{x = 6 \text{ L } H_2}}$$

PAGE 167 **6.** The equation:

100 L x L
$2CO + O_2 \rightarrow 2CO_2$

Set up the proportion using the coefficients of the substances that have something indicated above them as denominators. Here again, because the reaction is of gases, you can use the coefficients of the balance equation to set up the proportion.

$$\frac{100 \text{ L CO}}{2 \text{ L CO}} = \frac{x \text{ L } O_2}{1 \text{ L } O_2}$$

$$\underline{\underline{x = 50 \text{ L } O_2}}$$

Chapter Summary

The following terms summarize all the concepts and ideas that were introduced in this chapter. You should be able to explain their meaning and how you would use them in chemistry. They appear in boldface type in this chapter to draw your attention to them. The boldface type also makes it easier for you to look them up if you need to. You could also use the search or "Google" action on your computer to get a quick and expanded explanation of these terms, laws, and formulas.

molar mass
molar volume
$$\text{percent yield} = \frac{\text{actual yield}}{\text{theoretical yield}} \times 100$$
mole
STP
Avogadro's Law
Avogadro's number = (6.02×10^{23})
Gay-Lussac's Law
Density of gases

Practice Exercises

Often there are some equations on a chemistry test to either complete or complete and balance. The following list includes representative equations that you should be capable of writing and balancing:

1. zinc + sulfuric acid →
2. iron(III) chloride + sodium hydroxide → iron(III) hydroxide(s) + sodium chloride
3. aluminum hydroxide + sulfuric acid → aluminum sulfate + water
4. potassium + water →
5. magnesium + oxygen →
6. silver nitrate + copper → copper(II) nitrate + silver(s)
7. magnesium bromide + chlorine → magnesium chloride + bromine

Answers

Below the word equations are the complete balanced equations.

1. zinc + sulfuric acid → zinc sulfate + hydrogen(g)
   $Zn + H_2SO_4 \rightarrow ZnSO_4 + H_2(g)$
2. $FeCl_3 + 3NaOH \rightarrow Fe(OH)_3(s) + 3NaCl$
3. $2Al(OH)_3 + 3H_2SO_4 \rightarrow Al_2(SO_4)_3 + 6H_2O$
4. potassium + water → potassium hydroxide + hydrogen(g)
   $2K + 2H_2O \rightarrow 2KOH + H_2(g)$
5. magnesium + oxygen → magnesium oxide
   $2Mg + O_2 \rightarrow 2MgO$
6. $2AgNO_3 + Cu \rightarrow Cu(NO_3)_2 + 2Ag(s)$
7. $MgBr_2 + Cl_2 \rightarrow MgCl_2 + Br_2$

Liquids, Solids, and Phase Changes

CHAPTER 7

- Explain, using a graph, the distribution of the kinetic energy of molecules of a liquid at different temperatures.
- Describe the states of matter and what occurs when a substance changes state.
- Define critical temperature and pressure.
- Analyze a phase diagram and the triple point.
- Solve water calorimetry problems that include changes of state.
- Explain the polarity of the water molecule and hydrogen bonding.
- Solve solubility problems, concentration problems, and changes in boiling point/freezing point of water problems.
- Describe the continuum of water mixtures including solutions, colloids, and suspensions.

Liquids and solids each have their own properties, including intermolecular interactions, surface tension, and more. In fact, one very important compound—water—has distinct properties necessary for life to exist on this planet.

LIQUIDS

Importance of Intermolecular Interaction

A liquid can be described as a form of matter that has a definite volume and takes the shape of its container. In a liquid, the volume of the molecules and the intermolecular forces between them are much more important than in a gas. When you consider that in a gas the molecules constitute far less than 1% of the total volume, while in the liquid state the molecules constitute 70% of the total volume, it is clear that in a liquid the forces between molecules are more important. Because of this decreased volume and increased intermolecular interaction, a liquid expands and contracts only very slightly with a change in temperature and lacks the compressibility typical of gases.

TIP

An increase in temperature increases the average kinetic energy of the molecules.

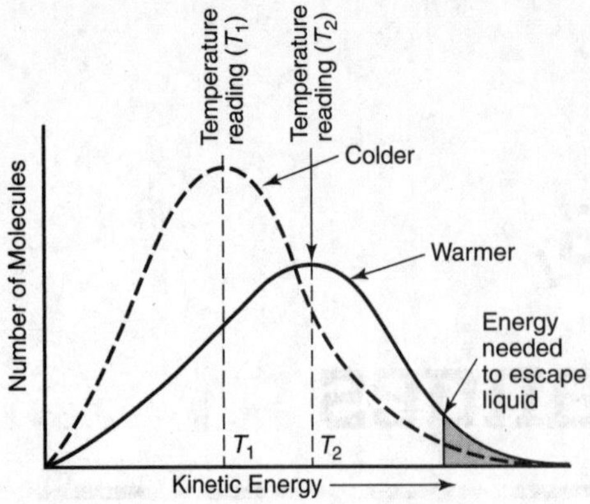

Figure 22. Distribution of the Kinetic Energy of Molecules

Kinetics of Liquids

Even though the volume of space between molecules has decreased in a liquid and the mutual attraction forces between neighboring molecules can have great effects on the molecules, they are still in motion. This motion can be verified under a microscope when colloidal particles are suspended in a liquid. The particles' zigzag path, called **Brownian movement**, indicates molecular motion and supports the **Kinetic-Molecular Theory**.

Increases in temperature increase the average kinetic energy of molecules and the rapidity of their movement. This is shown graphically in Figure 22. The molecules in the sample of cold liquid have, on the average, less kinetic energy than those in the warmer sample. Hence the temperature reading T_1 will be less than the temperature reading T_2. If a particular molecule gains enough kinetic energy when it is near the surface of a liquid, it can overcome the attractive forces of the liquid phase and escape into the gaseous phase. This is called a **change of phase**. When fast-moving molecules with high kinetic energy escape, the average energy of the remaining molecules is lower; hence the temperature is lowered.

TIP

More viscous liquids move more slowly.

Viscosity

Viscosity is the friction or resistance to motion that exists between the molecules of a liquid when they move past each other. It is logical that the stronger the attraction between the molecules of a liquid, the greater its resistance to flow—and thus the greater its viscosity. The viscosity of a liquid depends on its intermolecular forces. Because hydrogen bonds are such strong intermolecular forces, liquids with hydrogen bonds tend to have high viscosities. Water, for example, is strongly hydrogen bonded and has a relatively high viscosity. You may have noticed how fast liquids with low viscosity, such as alcohol and gasoline, flow.

Surface Tension

Molecules at the surface of a liquid experience attractive forces downward, toward the inside of the liquid, and sideways, along the surface of the liquid. On the other hand, molecules in the center of the liquid experience uniformly distributed attractive forces. This imbalance of forces at the surface of a liquid results in a property called **surface tension.** The uneven forces make the surface behave as if it had a tight film stretched across it. Depending on the magnitude of the surface tension of the liquid, the film is able to support the weight of a small object, such as a razor blade or a needle. Surface tension also explains the beading of raindrops on the shiny surface of a car.

Capillary Action

Capillary action, the attraction of the surface of a liquid to the surface of a solid, is a property closely related to surface tension. A liquid will rise quite high in a very narrow tube if a strong attraction exists between the liquid molecules and the molecules that make up the surface of the tube. This attraction tends to pull the liquid molecules upward along the surface against the pull of gravity. This process continues until the weight of the liquid balances the gravitational force. Capillary action can occur between water molecules and paper fiber, causing the water molecules to rise up the paper. When a water soluble ink is placed on the paper, the ink moves up the paper and separates into its various colored components. This separation occurs because the water and the paper attract the molecules of the ink components differently. These phenomena are used in the separation process of **paper chromatography,** as shown in the paper chromatography experiment on page 333. Capillary action is at least partly responsible for the transportation of water from the roots of a plant to its leaves. The same process is responsible for the concave liquid surface, called a **meniscus**, that forms in a test tube or graduated cylinder. (See the drawing on page 51.)

PHASE EQUILIBRIUM

Figure 23 shows water in a container enclosed by a bell jar. Observation of this closed system would show an initial small drop in the water level, but after some time the level would become constant. The explanation is that, at first, more energetic molecules near the surface are escaping into the gaseous phase faster than some of the gaseous water molecules are returning to the surface and possibly being caught by the attractive forces that will retain them in the liquid phase. After some time the rates of evaporation and condensation equalize. This is known as **phase equilibrium**.

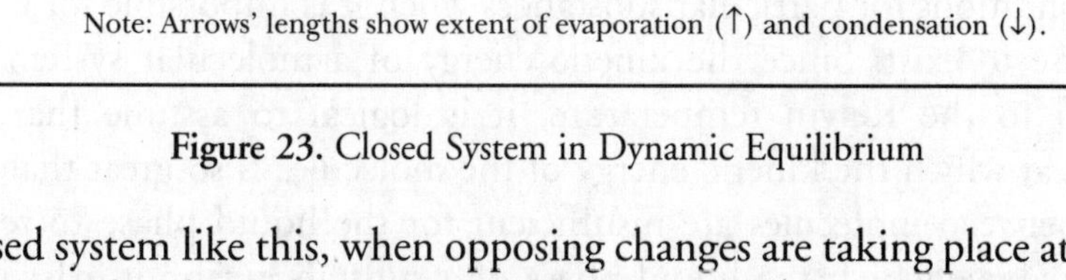

Note: Arrows' lengths show extent of evaporation (↑) and condensation (↓).

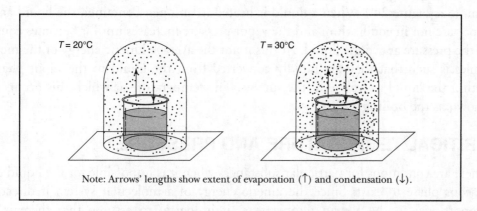

Figure 23. Closed System in Dynamic Equilibrium

In a closed system like this, when opposing changes are taking place at equal rates, the system is said to have **dynamic equilibrium**. At higher temperatures, since the number of molecules at higher energies increases, the number of molecules in the liquid phase will be reduced and the number of molecules in the gaseous phase will

TIP

More about equilibrium is discussed in Chapter 9.

be increased. The rates of evaporation and condensation, however, will again become equal.

The behavior of the system described above illustrates what is known as **Le Châtelier's Principle.** It is stated as follows: When a system at equilibrium is disturbed by the application of a stress (a change in temperature, pressure, or concentration), it reacts so as to minimize the stress and attain a new equilibrium position.

In the discussion above, if the 20°C system is heated to 30°C, the number of gas molecules will be increased while the number of liquid molecules will be decreased:

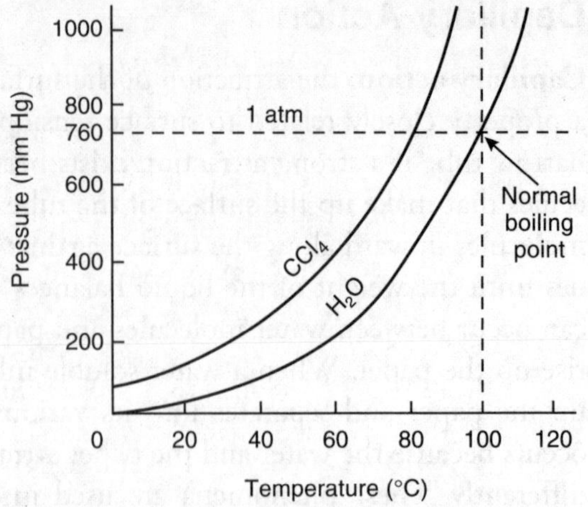

Figure 24. Vapor Pressure-Temperature Relationship for Carbon Tetrachloride and Water

$$\text{Heat} + H_2O(\ell) \rightleftharpoons H_2O(g)$$

The equation shifts to the right (any similar system that is endothermic shifts to the right when temperature is increased) until equilibrium is reestablished at the new temperature.

The molecules in the vapor that are in equilibrium with the liquid at a given temperature exert a constant pressure. This is called the **equilibrium vapor pressure** at that temperature.

TIP

Boiling point is defined as the point at which the liquid's vapor pressure equals the atmospheric pressure.

BOILING POINT

The vapor pressure-temperature relation can be plotted on a graph for a closed system. (See Figure 24.) When a liquid is heated in an open container, the liquid and vapor are not in equilibrium and the vapor pressure increases until it becomes equal to the pressure above the liquid. At this point the average kinetic energy of the molecules is such that they are rapidly converted from the liquid to the vapor phase within the liquid as well as at the surface. The temperature at which this occurs is known as the **boiling point**.

CRITICAL TEMPERATURE AND PRESSURE

There are conditions for particular substances when it is impossible for the liquid or gaseous phase to exist. Since the kinetic energy of a molecular system is directly proportional to the Kelvin temperature, it is logical to assume that there is a temperature at which the kinetic energy of the molecules is so great that the attractive forces between molecules are insufficient for the liquid phase to remain. The temperature above which the liquid phase of a substance cannot exist is called its **critical temperature**. Above its critical temperature, no gas can be liquefied regardless of the pressure applied. The minimum pressure required to liquefy a gas at its critical temperature is called its **critical pressure**.

SOLIDS

Whereas particles in gases have the highest degree of disorder, the solid state has the most ordered system. Particles are fixed in rather definite positions and maintain definite shapes. Because of their variation in packing, solids can be divided into three categories: **Crystalline solids** have a three-dimensional representation much like a brick wall. They have a regular structure, in which the particles pack in a repeating pattern from one edge of the solid to the other. **Amorphous solids** (literally, "solids without form") have a random structure, with little if any long-range order. **Polycrystalline** solids are an aggregate of a large number of small crystals or grains in which the structure is regular, but the crystals or grains are arranged in a random fashion.

Particles in solids do vibrate in position, however, and may even diffuse through the solid. (Example: Gold clamped to lead shows diffusion of some gold atoms into the lead over long periods of time.) Other solids do not show diffusion because of strong ionic or covalent bonds in network solids. (Examples: NaCl and diamond, respectively.)

When heated at certain pressures, some solids vaporize directly without passing through the liquid phase. This is called **sublimation**. Solids like solid carbon dioxide and solid iodine exhibit this property because of unusually high vapor pressure.

The temperature at which atomic or molecular vibrations of a solid become so great that the particles break free from fixed positions and begin to slide freely over each other in a liquid state is called the **melting point**. The amount of energy required at the melting point temperature to cause the change of phase to occur is called the **heat of fusion**. The amount of this energy depends on the nature of the solid and the type of bonds present.

PHASE DIAGRAMS

The simplest way to discuss a phase diagram is by an example, such as Figure 25.

A phase diagram ties together the effects of temperature and also pressure on the phase changes of a substance. In Figure 25 the line *BD* is essentially the vapor-pressure curve for the liquid phase. Notice that at a pressure of 760 millimeters of mercury (1 atmosphere) the water will boil (change to the vapor phase) at 100°C (point *F*). However, if the pressure is raised, the boiling point temperature increases; and, if the pressure is less than 760 millimeters, the boiling point decreases along the *BD* curve down to point *B*.

TIP

Know the significance of each gray area and boundary line.

TIP

The triple point is the only temperature and pressure at which three phases of a substance can exist.

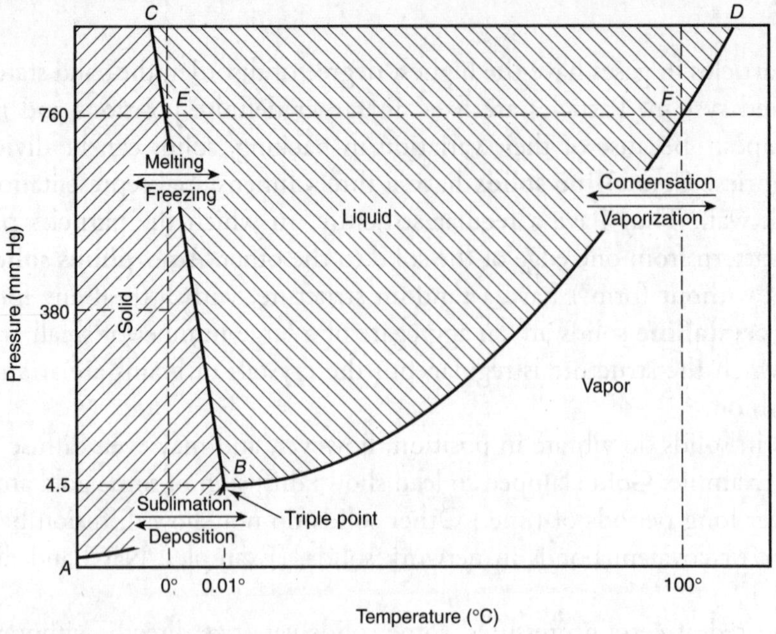

Figure 25. Partial Phase Diagram for Water (distorted somewhat to distinguish the triple point from the freezing point)

At 0°C the freezing point of water is found along the line *BC* at point *E* for pressure at 1 atmosphere or 760 millimeters. Again, this point is affected by pressure along the line *BC* so that, if pressure is decreased, the freezing point is slightly higher up to point *B* or 0.01°C.

Point *B* represents the point at which the solid, liquid, and vapor phases may all exist at equilibrium. This point is known as the **triple point**. It is the only temperature and pressure at which three phases of a pure substance can exist in equilibrium with one another in a system containing only the pure substance.

WATER

History of Water

Ancient philosophers like Aristotle regarded water as a basic element that typified all liquid substances. Scientists did not discard this view until the latter half of the eighteenth century. In 1781 the British chemist Henry Cavendish synthesized water by detonating a mixture of hydrogen and air. However, the results of his experiments were not clearly interpreted until two years later when the French chemist Antoine Laurent Lavoisier proved that water was not an element but a compound of oxygen and hydrogen. In a scientific paper presented in 1804, the French chemist Joseph Gay-Lussac and the German naturalist Alexander von Humboldt demonstrated jointly that water consisted of two volumes of hydrogen to one of oxygen, as expressed by the well-known formula H_2O.

It was not until 1932 that the American chemist Harold Clayton Urey discovered the presence of a small amount (1 part in 6000 parts) of so-called heavy water, or deuterium oxide (D_2O). Deuterium is the hydrogen isotope with an atomic number of 2. In 1951 the American chemist Aristid Grosse discovered that naturally occurring water also contained minute traces of tritium oxide (T_2O). Tritium is the hydrogen isotope with an atomic weight of 3.

Purification of Water

Water is so often involved in chemistry that it is important to have a rather complete understanding of this compound and its properties. Pure water has become a matter of national concern. Although commercial methods of purification will not be discussed here, the usual laboratory method of obtaining pure water, distillation, will be covered.

The process of distillation involves the evaporation and condensation of the water molecules. The usual apparatus for the distillation of any liquid is shown in Figure 26.

This method of purification will remove any substance that has a boiling point higher than that of water. It cannot remove dissolved gases or liquids that boil off before water. These substances will be carried over into the condenser and subsequently into the distillate.

TIP

In distillation, first boil and then condense.

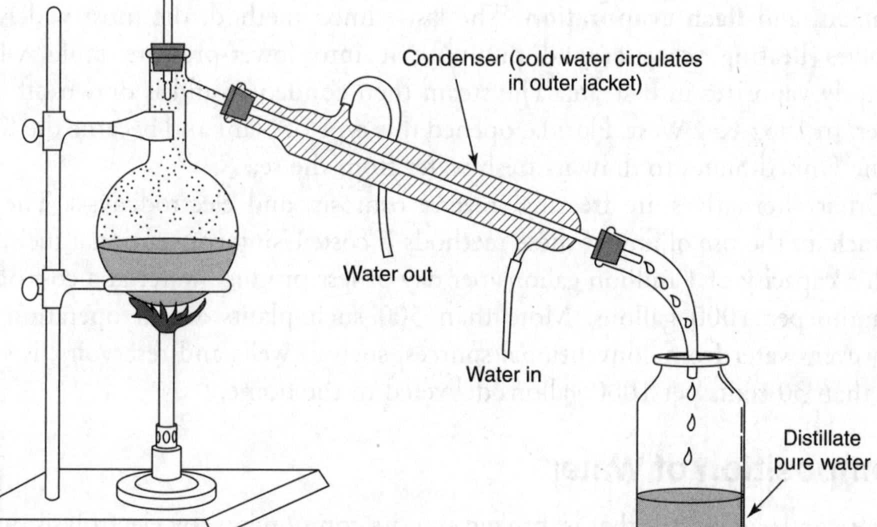

Figure 26. Distillation of water.

Another method of water purification is **aeration**, or the saturation of water with air. In this method, water is brought into contact with air in such a manner as to produce maximum diffusion, usually by spraying the water into the air in fountains. Aeration removes odors and taste caused by decomposing organic matter, as well as industrial wastes such as phenols and volatile gases such as chlorine. It also converts dissolved iron and manganese compounds to insoluble hydrated oxides of the metals, which then settle out as precipitates.

When water contains bicarbonates of Ca, Mg, or Fe it is said to have carbonate hardness, formerly called "temporary hardness." Such hardness may be removed by boiling, since heat causes the bicarbonate to decompose to a carbonate precipitate. This precipitate can cause detrimental scales in industrial boilers if not removed.

Natural waters are also apt to contain varying amounts of calcium sulfate, magnesium sulfate, calcium chloride, and magnesium chloride. Water in which these chemicals are dissolved is said to have noncarbonate hardness, formerly known as "permanent hardness." One way of softening water is to pass it through an artificial zeolite, which replaces the "hard" positive ions (called **cations**) with "soft" ions of sodium. It is the "hard" cations that are responsible for "bathtub ring" by reacting with the soap to form an insoluble compound. Another effective method that gives pure water like that obtained by distillation is the use of organic deionizers (e.g., Dowex, Amberlite, and zeolite). These organic resins exchange H+ ions for the cations, and OH^- ions for the anions.

With the ever-increasing demands for fresh water, especially in arid and semiarid areas or very densely populated areas, much research has gone into finding efficient methods of removing salt from seawater and brackish water. The state of California is seriously considering desalinization as a possible solution to its growing water problems. In the United States, desalinization research is directed by the Bureau of Reclamation, Department of the Interior. Several processes are being developed to produce fresh water cheaply.

Three of the processes involve evaporation followed by condensation of the resultant steam and are known as **multiple-effect evaporation**, **vapor-compression distillation**, and **flash evaporation**. The last-named method, the most widely used, involves heating seawater and pumping it into lower-pressure tanks where it abruptly vaporizes into steam. The steam then condenses and is drawn off as pure water. In 1967 Key West, Florida, opened this type of plant and became the first city in the United States to draw its fresh water from the sea.

Other alternatives are freezing, reverse osmosis, and electrodialysis. The major obstacle to the use of any of these methods is cost. Using conventional fuels, plants with a capacity of 1 million gallons per day or less produce water at a cost of $1.00 or more per 1000 gallons. More than 500 such plants are in operation today. However, water from conventional sources, such as wells and reservoirs, is sold for less than 30 cents per 1000 gallons delivered to the home.

TIP

Water ratios of combination:

H_2 to O_2

2 vol : 1 vol in gaseous state

1 g : 8 g by mass

Composition of Water

Water can be analyzed, that is, broken into its components, by electrolysis, which is shown on page 210. This process shows that the composition of water by volume is 2 parts of hydrogen to 1 part of oxygen. Water composition can also be arrived at by synthesis. **Synthesis** is the formation of a compound by uniting its components. Water can be made by mixing hydrogen and oxygen in a eudiometer over mercury and passing an igniting spark through the mixture. Again the ratio of combination is found to be 2 parts hydrogen: 1 part oxygen. In a steam-jacketed eudiometer, which keeps the water formed in the gas phase, 2 volumes of hydrogen combine with 1 volume of oxygen to form 2 volumes of steam.

Another interesting method is the Dumas experiment pictured in Figure 27. Data obtained show that hydrogen and oxygen combine to form water in a ratio of 1 : 8 by mass. This means that 1 gram of hydrogen combines with 8 grams of oxygen to form 9 grams of water.

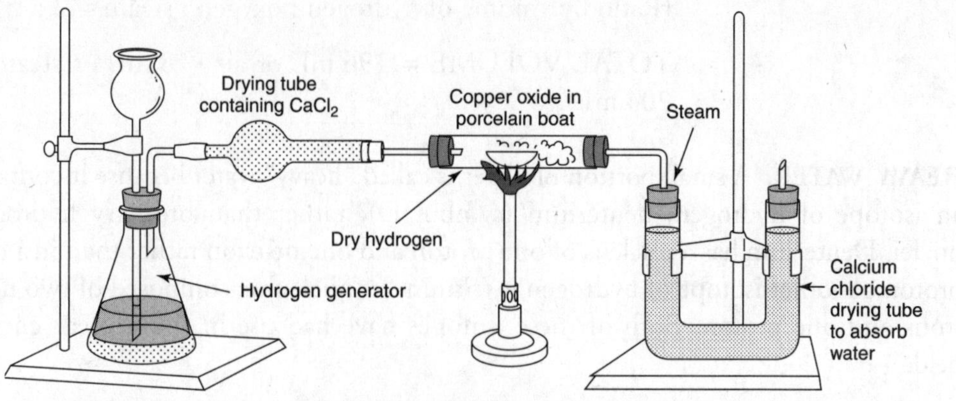

$CuO + H_2 \rightarrow Cu + H_2O$

Figure 27. Synthesis of Water

Some sample problems involving the composition of water are shown below.

TYPE PROBLEM (BY MASS): An electric spark is passed through a mixture of 12 grams of hydrogen and 24 grams of oxygen in the eudiometer setup shown. Find the number of grams of water formed and the number of grams of gas left uncombined.

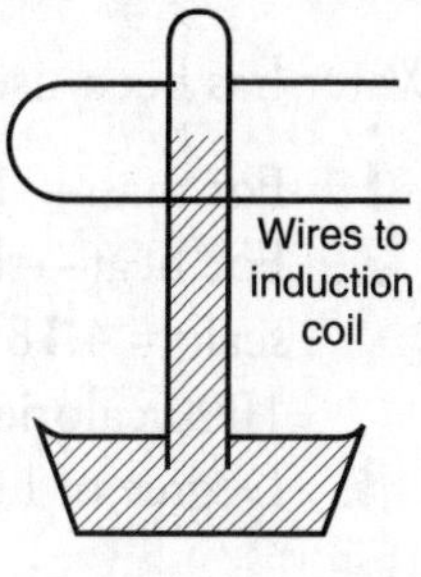

Since water forms in a ratio of 1 : 8 by mass, to use up the oxygen (which by inspection will be the limiting factor since it has enough hydrogen present to react completely) we need only 3 grams of hydrogen.

3 g of hydrogen + 24 g of oxygen = 27 g of water

This leaves 12 g − 3 g = 9 g of hydrogen uncombined.

TYPE PROBLEM (BY VOLUME): A mixture of 8 milliliters of hydrogen and 200 milliliters of air is placed in a steam-jacketed eudiometer, and a spark is passed through the mixture. What will be the total volume of gases in the eudiometer?

Because this is a combination by volume, 8 mL of hydrogen require 4 mL of oxygen. (Ratio $H_2 : O_2$ by volume = 2 : 1.)

The 200 mL of air is approximately 21% oxygen. This will more than supply the needed oxygen and leave 196 mL of the air uncombined.

The 8 mL of hydrogen and 4 mL of oxygen will form 8 mL of steam since the eudiometer is steam-jacketed and keeps the water formed in the gaseous state.

(Ratio by volume of hydrogen : oxygen : steam = 2 : 1: 2)

TOTAL VOLUME = 196 mL of air + 8 mL of steam = 204 mL

HEAVY WATER. A small portion of water is called "heavy" water because it contains an isotope of hydrogen, deuterium (symbol D), rather than ordinary hydrogen nuclei. Deuterium has a nucleus of one proton and one neutron rather than just one proton. Another isotope of hydrogen is tritium. Its nucleus is composed of two neutrons and one proton. Both of these isotopes have had use in the nuclear energy field.

HYDROGEN PEROXIDE. The prefix *per-* indicates that this compound contains more than the usual oxide. Its formula is H_2O_2. It is a well-known bleaching and oxidizing agent. Its electron-dot formula is shown in Figure 28.

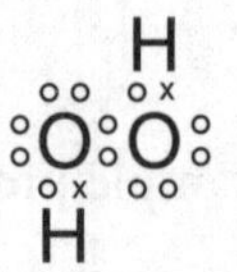

Figure 28. Hydrogen Peroxide

Properties and Uses of Water

Water has been used in the definition of various standards.

1. For mass—1 milliliter (or 1 cc) of water at 277 K is 1 gram.
2. For heat—the heat needed to raise 1 gram of water 1 degree on the Kelvin scale = 4.18 joules (J)
   1000 calories = 1 kilocalorie (kcal)
3. Degree of heat—the freezing point of water = 0°C, 273 K
   —the boiling point of water = 100°C, 373 K

Water Calorimetry Problems

A calorimeter is a container well insulated from outside sources of heat or cold so that most of its heat is contained in the vessel. If a very hot object is placed in a calorimeter containing some ice crystals, we can find the final temperature of the mixture mathematically and check it experimentally. To do this, however, certain behaviors must be understood. Ice changing to water and then to steam does not represent a continuous and constant change of temperature as time progresses. In fact, the chart would look as shown in Figure 29.

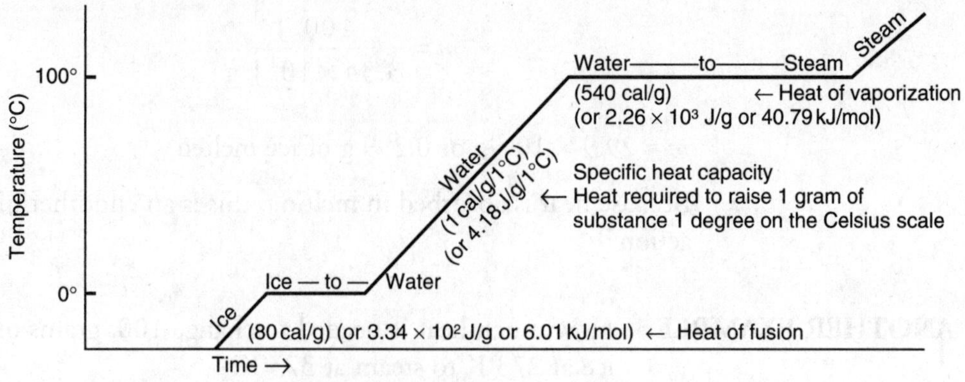

Figure 29. Changing Ice to Steam

TIP

Water's heat of vaporization = 40.79 kJ/mol

TIP

Water's heat of fusion = 6.01 kJ/mol

From this graph, you see that heat is being used at 0°C and 100°C to change the state of water, but not its temperature. One gram of ice at 0°C needs 80 calories or 3.34×10^2 joules to change to water at 0°C. This is called its **heat of fusion**. Likewise, energy is being used at 100°C to change water to steam, not to change the temperature. One gram of water at 100°C absorbs 540 calories or 2.26×10^3 joules of heat to change to 1 gram of steam at 100°C. This is called its **heat of vaporization**. This energy absorbed at the plateaus in the curve is being used to break up the bonding forces between molecules by increasing the potential energy content of the molecules so that a specific change of state can occur.

The amount of heat energy required to melt one mole of solid at its melting point is called its **molar heat of fusion**. If the quantity of ice melted is one mole (18 grams), then, for ice, it is 6.01 kJ/mole. Likewise, the amount of heat energy required to vaporize one mole of liquid at its boiling point is called its **molar heat of vaporization**. If the quantity of water vaporized is one mole, then its molar heat of vaporization is 40.79 kJ/mole.

The following two problems are samples of calorimetry problems. The first is solved by using the SI units of joules. The second is done by using calories but is finally converted to joules.

TYPE PROBLEM: What quantity of ice at 273 K can be melted by 100 joules of heat?

Heat to fuse (melt) a substance = heat of fusion of the substance × mass of the substance.

This quantity can be expressed by the following formula, where q denotes the heat measurement made in a calorimeter:

$$q = m(\text{mass}) \times C(\text{heat of fusion})$$

Solving for m, we get

$$m = \frac{q}{C}$$

$$m = \frac{100.\text{ J}}{3.34 \times 10^2 \text{ J/g}}$$

$m = 29.9 \times 10^{-2}$ g or 0.299 g of ice melted

Because heat is absorbed in melting, this is an endothermic action.

ANOTHER EXAMPLE: How much heat is needed to change 100. grams of ice at 273 K to steam at 373 K?

To melt 100. grams of ice at 273 K:

Use: m(mass) × C(heat of fusion) = q(quantity of heat)

$$100.\ \cancel{g} \times \frac{80 \text{ cal}}{1\ \cancel{g}} = 8{,}000 \text{ cal} = 8 \text{ kcal}$$

To heat 100. grams of water from 273 K to 373 K:

Use: $m \times \Delta T \times$ specific heat = q

$$100.\ \cancel{g} \times \overbrace{(373-273)\text{ k } \cancel{C}}^{\text{temperature change}} \times \frac{1 \text{ cal}}{1\ \cancel{g} \times 1 \text{ K}} = 10{,}000 \text{ cal} = 10 \text{ kcal}$$

To vaporize 100. grams of water at 100.°C to steam at 100.°C:

Use: $m \times$ heat of vaporization = q

$$100.\ \cancel{g} \times \frac{540 \text{ cal}}{1\ \cancel{g}} = 54{,}000 \text{ cal} = 54 \text{ kcal}$$

Total heat = 8 + 10 + 54 = 72 kcal

Using the factor-label system to express the answer in joules, we have

$$72\ \cancel{\text{kcal}} \times \frac{1{,}000\ \cancel{\text{cal}}}{1\ \cancel{\text{kcal}}} \times \frac{4.18 \text{ J}}{1\ \cancel{\text{cal}}} = 3.01 \times 10^5 \text{ J}$$

Water's Reactions with Anhydrides

Anhydrides are certain oxides that react with water to form two classes of compounds—acids and bases.

Many metal oxides react with water to form bases such as sodium hydroxide, potassium hydroxide, and calcium hydroxide. For this reason, they are called **basic anhydrides**. The common bases are water solutions which contain the hydroxyl (OH^-) ion. Some common examples are:

$Na_2O + H_2O \rightarrow 2NaOH$, sodium hydroxide
$CaO + H_2O \rightarrow Ca(OH)_2$, calcium hydroxide

In general then: Metal oxide + $H_2O \rightarrow$ Metal hydroxide

In a similar manner, nonmetallic oxides react with water to form an acid such as sulfuric acid, carbonic acid, or phosphoric acid. For this reason, they are referred to as **acid anhydrides**. The common acids are water solutions containing hydrogen ions (H^+). Some common examples are:

$CO_2 + H_2O \rightarrow H_2CO_3$, carbonic acid
$SO_3 + H_2O \rightarrow H_2SO_4$, sulfuric acid
$P_2O_5 + 3H_2O \rightarrow 2H_3PO_4$, phosphoric acid

In general, then: Nonmetallic oxide + $H_2O \rightarrow$ Acid

POLARITY AND HYDROGEN BONDING

Water is different from most liquids in that it reaches its greatest density at 4°C and then its volume begins to expand. By the time water freezes at 0°C, its volume has expanded by about 9 percent. Most other liquids contract as they cool and change state to a solid because their molecules have less energy, move more slowly, and are closer together. This abnormal behavior of water can be explained as follows. X-ray studies of ice crystals show that H_2O molecules are bound into large molecules in which each oxygen atom is connected through **hydrogen bonds** to four other oxygen atoms as shown in Figure 30.

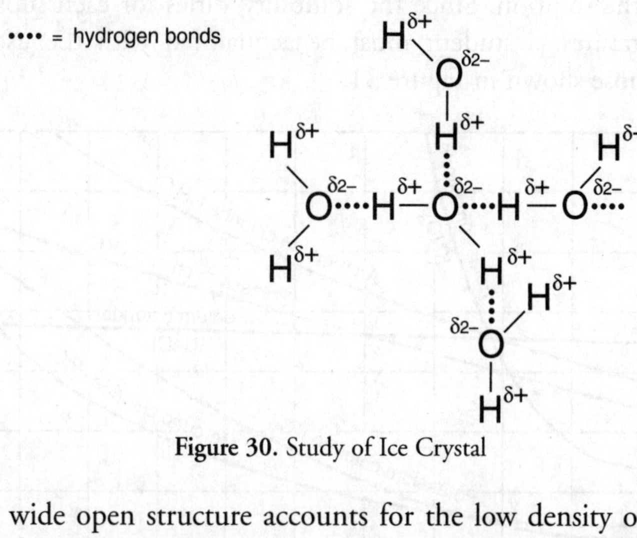

Figure 30. Study of Ice Crystal

This rather wide open structure accounts for the low density of ice. As heat is applied and melting begins, this structure begins to collapse but not all the hydrogen bonds are broken. The collapsing increases the density of the water, but the remaining bonds keep the structure from completely collapsing. As heat is absorbed, the kinetic energy of the molecules breaks more of these bonds as the temperature rises from 0° to 4°C. At the same time this added kinetic energy tends to distribute the molecules farther apart. At 4°C these opposing forces are in balance—thus the greatest density. Above 4°C the increasing molecular motion again causes a decrease in density since it is the dominate force and offsets the breaking of any more hydrogen bonds.

This behavior of water can be explained by studying the water molecule itself. The water molecule is composed of two hydrogen atoms bonded by a polar covalent bond to one oxygen atom.

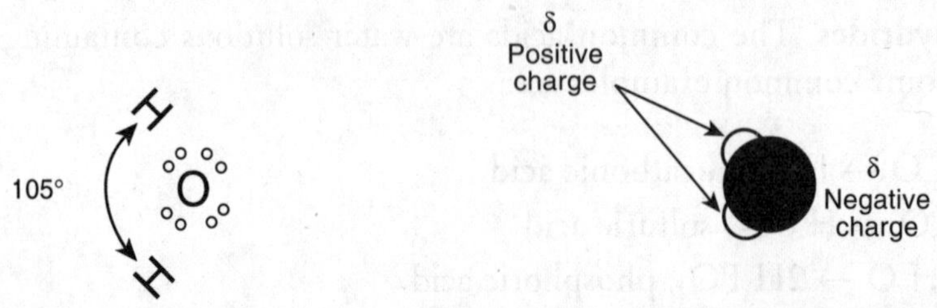

Because of the polar nature of the bond, the molecule exhibits the charges shown in the above drawing. It is this polar charge that causes the polar bonding discussed in Chapter 3 as the hydrogen bond. This bonding is stronger than the usual molecular attraction called **van der Waals forces** or dipole-dipole attractions.

SOLUBILITY

Water is often referred to as "the universal solvent" because of the number of common substances that dissolve in water. When substances are dissolved in water to the extent that no more will dissolve at that temperature, the solution is said to be **saturated**. The substance dissolved is called the **solute** and the dissolving medium is called a **solvent**. To give an accurate statement of a substance's solubility, three conditions are mentioned: the amount of solute, the amount of solvent, and the temperature of the solution. Since the solubility varies for each substance and for different temperatures, a student must be acquainted with the use of solubility curves such as those shown in Figure 31.

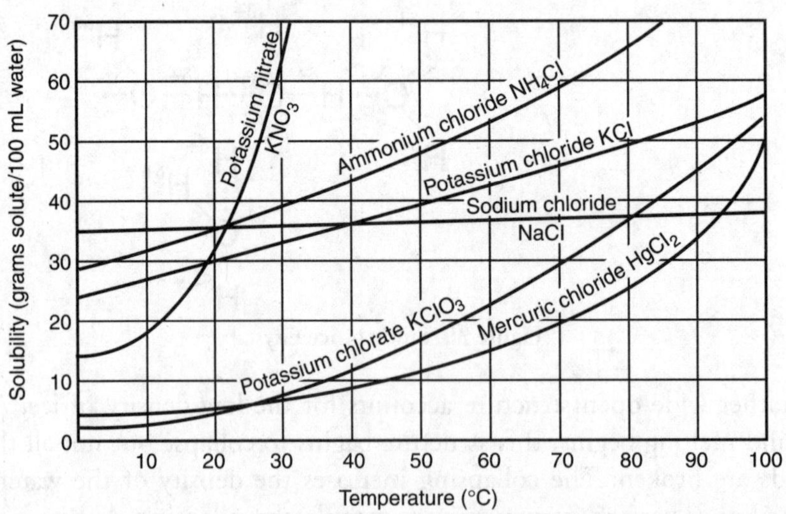

Figure 31. Solubility Curves of Some Common Salts

These curves show the number of grams of solute that will dissolve in 100 grams (milliliters) of water over a temperature range of 0°C to 100°C. Take, for example, the very lowest curve at 0°C. This curve shows the number of grams of $KClO_3$ that will dissolve in 100 grams of water over a temperature range of 0°C to 100°C. To find the solubility at any particular temperature, for example at 50°C, you follow the vertical line up from 50°C until it crosses the curve. At that point you place a ruler horizontally across the page and take the reading on the vertical axis. This point happens to be slightly below the 20 gram mark, or 18 grams. This means that 18 grams of $KClO_3$ will dissolve in 100 grams (milliliters) of water at 50°C.

As a soluble solute is added to water at a given temperature, the solute will continue to go into solution until the water cannot dissolve anymore. This is shown graphically in the figure below. It is the sloped portion of the graph. From that point on, the solute will fall to the bottom and does not dissolve, because an equilibrium has been established between molecules leaving and entering the solid phase. This is the condition when the line on the graph becomes horizontal. The solution is holding the maximum amount of dissolved solute in this amount of water and at 20°C. This is a **saturated solution.** If more water is added to the saturated solution, then more sodium acetate will dissolve in it. A solution that contains less solute than a saturated solution under the existing conditions is described as an **unsaturated solution**.

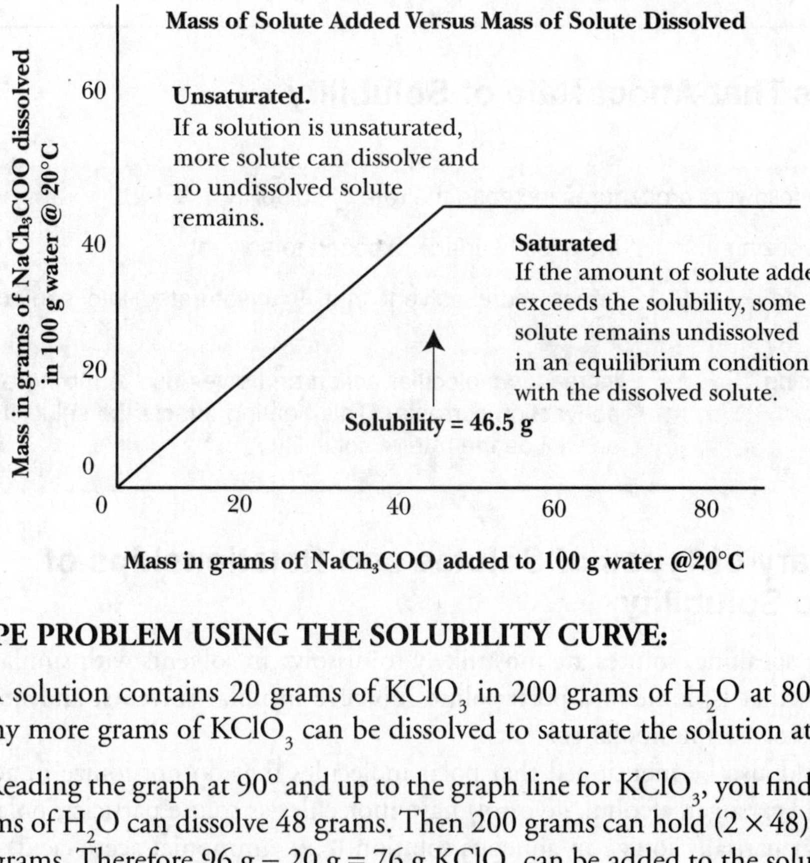

TYPE PROBLEM USING THE SOLUBILITY CURVE:

A solution contains 20 grams of $KClO_3$ in 200 grams of H_2O at 80°C. How many more grams of $KClO_3$ can be dissolved to saturate the solution at 90°C?

Reading the graph at 90° and up to the graph line for $KClO_3$, you find that 100 grams of H_2O can dissolve 48 grams. Then 200 grams can hold (2×48) grams or 96 grams. Therefore 96 g − 20 g = 76 g $KClO_3$ can be added to the solution.

General Rules of Solubility

All nitrates, acetates, and chlorates are soluble.

All common compounds of sodium, potassium, and ammonium are soluble.

All chlorides are soluble except those of silver, mercury(I), and lead. (Lead chloride is noticeably soluble in hot water.)

All sulfates are soluble except those of lead, barium, strontium, and calcium. (Calcium sulfate is slightly soluble.)

The normal carbonates, phosphates, silicates, and sulfides are insoluble except those of sodium, potassium, and ammonium.

All hydroxides are insoluble except those of sodium, potassium, ammonium, calcium, barium, and strontium.

Some general trends of solubility are shown in the chart below.

	Temperature Effect	Pressure Effect
Solid	Solubility usually increases with temperature increase.	Little effect
Gas	Solubility usually decreases with temperature increase.	Solubility varies in direct proportion to the pressure applied to it: **Henry's Law.**

Factors That Affect Rate of Solubility

The following procedures increase the rate of solubility

Pulverizing	increases surface exposed to solvent.
Stirring	brings more solvent that is unsaturated into contact with solute.
Heating	increases molecular action and gives rise to mixing by convection currents. (This heating affects the solubility as well as the rate of solubility.)

Summary of Types of Solutes and Relationships of Type to Solubility

Generally speaking, solutes are most likely to dissolve in solvents with similar characteristics; that is, ionic and polar solutes dissolve in polar solvents, and nonpolar solutes dissolve in nonpolar solvents.

It should also be mentioned that polar molecules that do not ionize in aqueous solution (e.g., sugar, alcohol, glycerol) have molecules as solute particles; polar molecules that partially ionize in aqueous solution (e.g., ammonia, acetic acid) have a mixture of molecules and ions as solute particles; and polar molecules that completely ionize in aqueous solution (e.g., hydrogen chloride, hydrogen iodide) have ions as solute particles.

WATER SOLUTIONS

To make molecules or ions of another substance go into solution, water molecules must overcome the forces that hold these molecules or ions together. The mechanism of the actual process is complex. To make sugar molecules go into solution, the water molecules cluster around the sugar molecules, pull them off, and disperse, forming the solution.

For an ionic crystal such as salt, the water molecules orient themselves around the ions (which are charged particles) and again must overcome the forces holding the ions together. Since the water molecule is polar, this orientation around the ion is an attraction of the polar ends of the water molecule. For example:

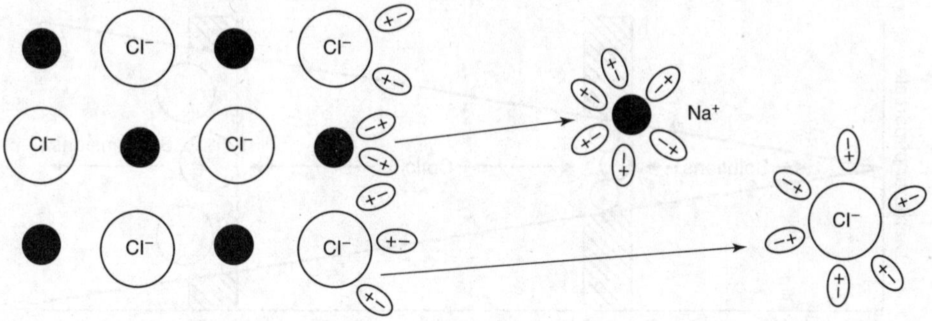

Once surrounded, the ion is insulated to an extent from other ions in solution because of the **dipole** property of water. The water molecules that surround the ion differ in number for various ions, and the whole group is called a **hydrated ion.** This is referred to as **dissociation.**

Some molecular compounds actually form ions when they go into solution. Usually they are polar compounds. Ions are formed from solute molecules by the action of the solvent in a process called **ionization.** The term means to create ions where there were none. Ionization is different from dissociation in that when an ionic compound dissolves, the ions that were already present separate from one another. When a molecular compound dissolves and ionizes in a polar solvent, ions are formed where none existed in the undissolved compound. The ions formed by such a molecular solvent are hydrated.

In general, the extent of the ionization depends on the strength of the bonds within the molecules of the solute and the strength of attraction between the solute and solvent molecules. If the strength of a bond within the solute molecule is weaker than the attractive forces of the solvent molecules, then the covalent bond of the solute breaks and the molecule is separated into ions. Hydrogen chloride, HCl, is a molecular compound that ionizes in an aqueous solution. The attraction between a polar HCl molecule and the polar water molecules is strong enough to break the HCl bond, forming the hydrogen and chloride ions.

$$HCl \xrightarrow{H_2O} H^+(aq) + Cl^-(aq)$$

In general, as stated in the preceding section, polar substances and ions dissolve in polar solvents and nonpolar substances such as fats dissolve in nonpolar solvents such as gasoline. The process of going into solution is **exothermic** if energy is released in the process, and **endothermic** if energy from the water is used up to a greater extent than energy is released in freeing the particle.

When two liquids are mixed and they dissolve in each other, they are said to be completely **miscible**. If they separate and do not mix, they are said to be **immiscible**.

Two molten metals may be mixed and allowed to cool. This gives a "solid solution" called an **alloy**.

CONTINUUM OF WATER MIXTURES

Figure 32 shows the general sizes of the particles found in a water mixture.

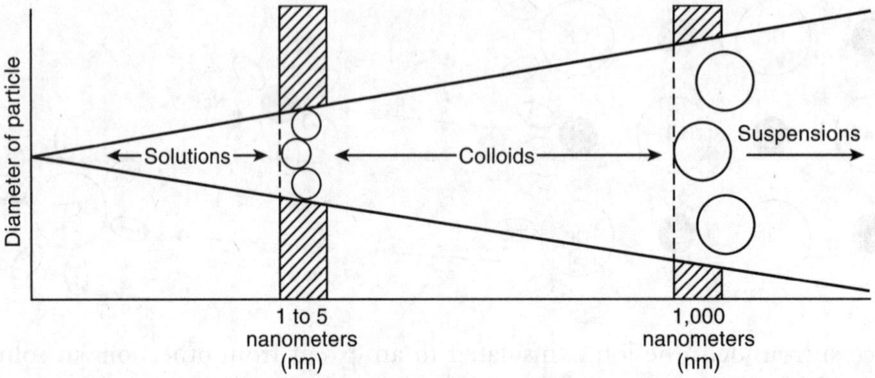

Figure 32. Size of Particles in Water Mixture

The basic difference between a colloid and a suspension is the diameter of the particles dispersed. All the boundaries marked in Figure 32 indicate only the general ranges in which the distinctions between solutions, colloids, and suspensions are usually made.

The characteristics of water mixtures are as follows:

Solutions	Colloids	Suspensions
. . . . 1 nm	. . . . 1,000 nm	. . . .
Clear; may have color		Cloudy; opaque color
Particles do not settle.		Settle on standing
Particles pass through ordinary filter paper.		Do not pass through ordinary filter paper
Particles pass through membranes.	Do not pass through semipermeable membranes such as animal bladders, cellophane, and parchment, which have very small pores*	
Particles are not visible.	Visible in ultramicroscope	Visible with microscope or naked eye
	Show Brownian movement	No Brownian movement

**Separation of a solution from a colloidal dispersion through a semipermeable membrane is called dialysis.*

When a bright light is directed at right angles to the stage of an ultramicroscope, the individual reflections of colloidal particles can be observed to be following a random zigzag path. This is explained as follows: The molecules in the dispersing medium are in motion and continuously bumping into the colloidal particles, causing them to change direction in a random fashion. This motion is called **Brownian movement** after the Scottish botanist Robert Brown, who first observed it.

EXPRESSIONS OF CONCENTRATION

There are general terms and very specific terms used to express the concentration of a solution. The general terms and their definitions are:

Dilute	= small amount of solute is dispersed in the solvent.
Concentrated	= large amount of solute is dissolved in the solvent.
Saturated	= the solution is holding all the solute possible at that temperature. This is not a static condition; that is, some solute particles are exchanging places with some of the undissolved particles, but the total solute in solution remains the same. This is an example of equilibrium.
Unsaturated	= more solute can go into solution at that temperature. The solvent has further capacity to hold more solute.
Supersaturated	= sometimes a saturated solution at a higher temperature can be carefully cooled so that the solute does not get a chance to come out of solution. At a lower temperature, then, the solution will be holding more solute in solution than it should for saturation and is said to be supersaturated. As soon as the solute particles are jarred or a "seed" particle is added to the solution to act as a nucleus, they rapidly come out of solution so that the solution reverts to the saturated state.

It is interesting to note that the words *saturated* and *concentrated* are NOT synonymous. In fact, it is possible to have a saturated dilute solution when the solute is only slightly soluble and a small amount of it makes the solution saturated, but the concentration of the solution is still dilute.

The more specific terms used to describe concentration are mathematically calculated.

Percentage concentration is based on the percent of solute in the solution by mass. The general formula is:

$$\frac{\text{No. of grams of solute}}{\text{No. of grams of solution}} \times 100\% = \%\text{ concentration}$$

TYPE PROBLEM: How many grams of NaCl are needed to prepare 200 grams of a 10% salt solution?

10% of 200 grams = 20 grams of salt

You could also solve the problem using the above formula and solving for the unknown quantity.

$$\frac{x \text{ g solute}}{200 \text{ g solution}} \times 100\% = 10\%$$

$$x \text{ g solute} = \frac{10}{100} \times 200 = 20 \text{ g of solute}$$

Specific gravity can also be used to describe the concentration.

Specific gravity is the ratio of the mass of a substance to the mass of an equal volume of water. Often chemistry handbooks give the specific gravity of solutions. These, in turn, can be used to determine the mass of a solute in solution. Specific gravity has no units. However, since the density of water is 1.0 gram/milliliter, specific gravity is numerically equal to the density in grams per milliliter. For example, 7.9 is the specific gravity of iron. This means that the density of iron is 7.9 gram/milliliter.

This relationship can be used in solving problems of concentration that use percentage composition and a known specific gravity.

TYPE PROBLEM: Suppose that a solution of hydrogen chloride gas dissolved in water (hydrochloric acid) has a concentration of 30.0% HCl by mass. How much solute is there in 100. milliliters of this solution if the specific gravity of the solution is 1.15? (This means that its density is 1.15 g/mL.)

First, determine the mass of 100. mL of the solution:

$$\text{mass} = \text{density} \times \text{volume}$$
$$\text{mass} = 1.15 \text{ g/mL} \times 100.\text{mL} = 115 \text{ g}$$

Then, determine what mass of HCl is dissolved in 115 g of the solution:

$$\text{mass HCl} = 30.0\% \times \text{mass of solution}$$
$$\text{mass HCl} = 0.30 \times 115 \text{ g} = 34.5 \text{ g}$$

Thus, 34.5 g of HCl gas are dissolved in 100. mL of this solution.

The next two expressions depend on the fact that, if the formula mass of a substance is expressed in grams, it is called a **gram-formula mass** (gfm), **molar mass**, or 1 **mole**. **Gram-molecular mass** can be used in place of gram-formula mass when the substance is really of molecular composition, and not ionic like NaCl or NaOH. The definitions and examples are:

MOLARITY (abbreviated M) is defined as the number of moles of a substance dissolved in 1 liter of solution.

EXAMPLE: A 1 molar H_2SO_4 solution has 98 grams of H_2SO_4 (its molar mass) in 1 liter of the solution.

This can be expressed as a formula:

$$\text{Molarity} = \frac{\text{No. of moles of solute}}{1\,\text{L of solution}}$$

If the molarity (M) and the volume of a solution are known, the mass of the solute can be determined. First, use the above equation and solve for the number of moles of solute. Then, multiply this number by the molar mass.

TYPE PROBLEM: How many grams of NaOH are dissolved in 200. milliliters of solution if its concentration is 1.50 M?

$$\text{Because M} = \frac{\text{No. of moles of solute}}{1\,\text{L of solution}}$$

Solving for number of moles gives:

No. of moles of solute = M × volume in liters of solution

No. of moles of NaOH = 1.50 M × volume in liters of solution

$$\text{No. of moles of NaOH} = 1.50\frac{\text{mol}}{\text{L}} \times 0.200\,\text{L} = 0.30\,\text{mol of NaOH}$$

The molar mass of NaOH = 23 + 16 + 1 = 40 g of NaOH

$$0.30\,\cancel{\text{mol NaOH}} \times \frac{40\,\text{g of NaOH}}{1\,\cancel{\text{mol NaOH}}} = 12\,\text{g of NaOH}$$

MOLALITY (abbreviated *m*) is defined as the number of moles of the solute dissolved in 1,000 grams of solvent.

EXAMPLE: A 1 molal solution of H_2SO_4 has 98 grams of H_2SO_4 dissolved in 1,000 grams of water. This, you will notice, gives a total volume greater than 1 liter, whereas the molar solution had 98 grams in 1 liter of solution.

$$\text{Molality} = \frac{\text{Moles of solute}}{1{,}000\,\text{g of solvent}}$$

TYPE PROBLEM: Suppose that 0.25 mole of sugar is dissolved in 500 grams of water. What is the molality of this solution?

$$m = \frac{\text{No. of moles of solute}}{1{,}000 \text{ g of solvent}}$$

If 0.25 mole is in 500 grams of H_2O, then 0.50 mole is in 1,000 grams of H_2O. Thus:

$$m = \frac{0.50 \text{ mol of sugar}}{1{,}000 \text{ g of } H_2O}$$

$$m = 0.50$$

NORMALITY (abbreviated N) is used less frequently to express concentration and depends on knowledge of what a **gram-equivalent mass** is. This can be defined as the amount of a substance that reacts with or displaces 1 gram of hydrogen or 8 grams of oxygen. A simple method of determining the number of equivalents in a formula is to count the number of hydrogens or find the total positive oxidation numbers since each +1 charge can be replaced by a hydrogen.

In H_2SO_4 (gram-formula mass = 98 g), there are 2 hydrogens, so the gram-equivalent mass will be:

$$\frac{98 \text{ g}}{2} = 49 \text{ g}$$

In Al_2SO_4 (molar mass = 150 g), there are 2 aluminums, each of which has a +3 oxidation number, making a total of +6. The gram-equivalent mass will be:

$$\frac{150 \text{ g}}{6} = 49 \text{ g}$$

This idea of gram-equivalent mass is used in the definition of this means of expressing concentration, namely, the normality of a solution.

Normality is defined as the number of gram-equivalent masses of solute in 1 liter of solution.

EXAMPLE: If 49 grams of H_2SO_4 is mixed with enough H_2O to make 500 milliliters of solution, what is the normality?

Because normality is expressed for a liter, we must double the expression to 98 g of H_2SO_4 in 1,000 mL of solution. To find the normality we find the number of gram-equivalents in a liter of solution.

$$\frac{98 \text{ g (no. of grams in 1 L of solution)}}{49 \text{ g (gram-equivalent mass)}} = 2 \text{ N (normal)}$$

MOLE FRACTION is another way of indicating the concentration of a component in a solution. It is simply the number of moles of that component divided by the total moles of all the components. The mole fraction of component *i* is written as

X_i. For a solution consisting of n_A moles of component *A*, n_B moles of component *B*, n_C moles of component *C*, and so on, then the mole fraction of component *A* is given by:

$$X_A = \frac{n_A}{n_A + n_B + n_C + \ldots}$$

As an example, if a mixture is obtained by dissolving 10 moles of NaCl in 90 moles of water, the mole fraction of NaCl in that mixture is 10 (moles of NaCl) divided by (10 + 90) or 100 moles, giving an answer of 0.1, the mole fraction of NaCl.

DILUTION

Because the expression of molarity gives the quantity of solute per volume of solution, the amount of solute dissolved in a given volume of solution is equal to the product of the concentration times the volume. Hence, 0.5 liter of 2 M solution contains

$$M \times V = \text{amount of solute (in moles)}$$

$$\frac{2\text{ mol}}{\text{L}} \times 0.5\text{ L} = 1\text{ mol (of solute in 0.5 L)}$$

Notice that volume units must be identical.

If you dilute a solution with water, the amount or number of moles of solute present remains the same, but the concentration changes. You can use the expression:

Before *After*

$$M_1V_1 = M_2V_2$$

This expression is useful in solving problems involving dilution.

TYPE PROBLEM: If you wish to make 1 liter of solution that is 6 M into 3 M solution, how much water must be added?

$$M_1V_1 = M_2V_2$$
$$6\text{ M} \times 1\text{ L} = 3\text{ M} \times ?\text{ L}$$

Solving this expression:

? L = 2 L. This is the total volume of the solution after dilution and means that 1 liter of water had to be added to the original volume of 1 liter to get a total of 2 liters for the dilute solution volume.

An important use of the molarity concept is in the solution of **titration** problems, which are covered in Chapter 11, along with pH expressions of concentration for acids.

COLLIGATIVE PROPERTIES OF SOLUTIONS

Colligative properties are properties that depend primarily on the concentration of particles and not the type of particle. There is usually a direct relationship between the concentration of particles and the effect recorded.

The vapor pressure of an aqueous solution is always lowered by the addition of more solute. From the molecular standpoint, it is easy to see that there are fewer molecules of water per unit volume in the liquid, and therefore fewer molecules of water in the vapor phase are required to maintain equilibrium. The concentration in the vapor drops and so does the pressure that molecules exert. This is shown graphically below.

Notice that the effects of this change in vapor pressure are registered in the freezing point and the boiling point. The freezing point is lowered, and the boiling point is raised, in direct proportion to the number of particles of solute present. For water solutions, the concentration expression that expresses this relationship is molality (*m*), that is, the number of moles of solute per kilogram of solvent. For molecules that do not dissociate, it has been found that a 1 *m* solution freezes at −1.86°C (271.14 K) and boils at 100.51°C (373.51 K). A 2 *m* solution would then freeze at twice this lowering, or −3.72°C (269.28 K), and boil at twice the 1 molal increase of 0.51°C, or 101.02°C (374.02 K).

Vapor Pressure Versus Temperature for Water and a Solution

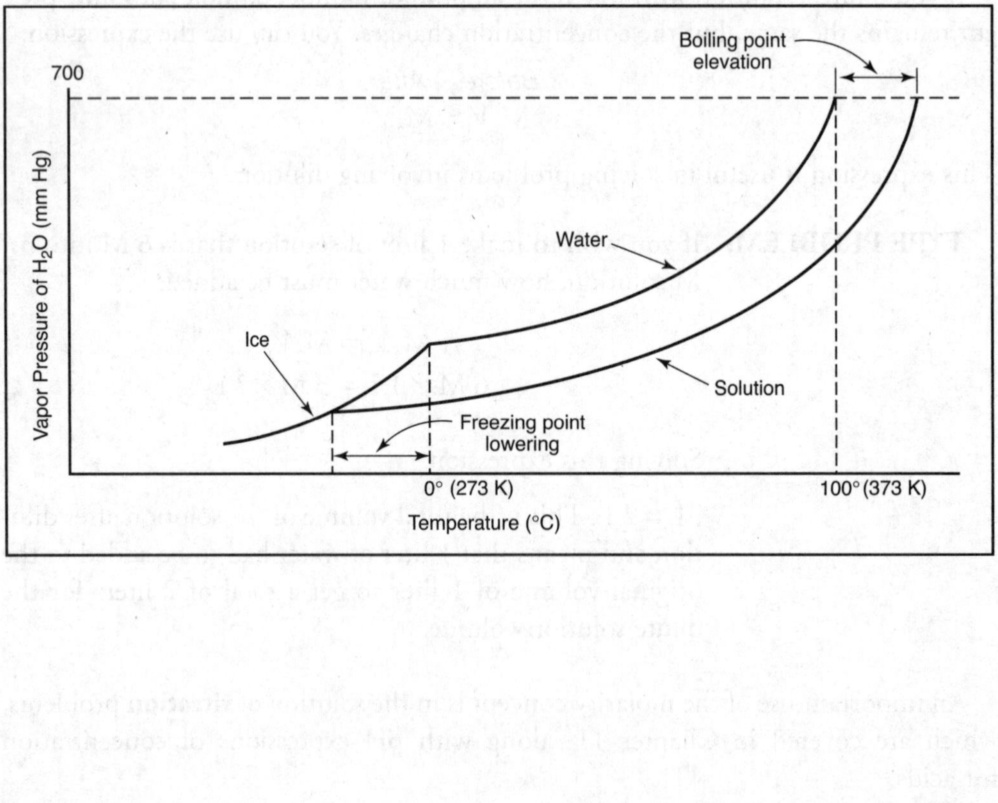

The chart below summarizes the colligative effect for aqueous solutions.

Molal Freezing Point and Boiling Point Comparisons

Type	Concentration (m)	Example	Moles of Particles	Freezing Point (°C)	Boiling Point (°C)
Molecular, nonionizing	1	Sugar or urea	1	−1.86	100.51
Molecular, completely ionized	1	HCl	2	−3.72	101.02
Ionic, completely dissociated	1	NaCl	2	−3.72	101.02
Ionic, completely dissociated	1	$CaCl_2$ or $Cu(NO_3)_2$	3	−5.58	101.53

SAMPLE PROBLEMS: 1. A 1.50-gram sample of urea is dissolved in 105.0 grams of water and produces a solution that boils at 100.12°C. From these data, what is the molecular mass of urea?

Because this property is related to the molality, then

$$\frac{1.5\text{ g}}{105.0\text{ g}} = \frac{x\text{ g}}{1{,}000\text{ g}}$$

$$x = 14.28\text{ g in } 1{,}000\text{ g of water}$$

The boiling point change is 0.12°C, and since each mole of particles causes a 0.51° increase, then

$$0.12°\,\not{C} \div \frac{0.51°\not{C}}{\text{mol}} = 0.235\text{ mol}$$

Then 14.28 g = 0.235 mol

and

$$\frac{14.28\text{ g}}{0.235\text{ mol}} = \frac{x\text{ g}}{1\text{ mol}}$$

$$x = 60.8\text{ g/mol}$$

2. Suppose that there are two water solutions, one of glucose (molar mass = 180), the other of sucrose (molar mass = 342). Each contains 50. grams of solute in 1,000 grams (1 kg) of water. Which has the higher boiling point? The lower freezing point?

   The molality of each of these nonionizing substances is found by dividing the number of grams of solute by the molecular mass.

$$\text{Glucose: } \frac{50.0\text{ g}}{180\text{ g/mol}} = 0.278\text{ mol}$$

$$\text{Sucrose: } \frac{50.0\text{ g}}{342\text{ g/mol}} = 0.146\text{ mol}$$

   Therefore their respective molalities are 0.278 *m* and 0.146 *m*. Since the freezing point and boiling point are colligative properties, the effect depends only on the concentration. Because the glucose has a higher concentration, it will have a higher boiling point and a lower freezing point. The respective boiling points would be:

$$0.278\ \cancel{m} \times \frac{0.51°\text{C rise}}{1\ \cancel{m}} = 0.14°\text{C rise or } 100.14°\text{C boiling point for glucose.}$$

$$0.146\ \cancel{m} \times \frac{0.51°\text{C rise}}{1\ \cancel{m}} = 0.07°\text{C rise in boiling point or } 100.07°\text{C for sucrose}$$

   The lowering of the freezing point would be:

$$0.278\ \cancel{m} \times \frac{-1.86°\text{C drop}}{1\ \cancel{m}} = -0.52°\text{C drop below } 0°\text{C for glucose.}$$

$$0.146\ \cancel{m} \times \frac{-1.86°\text{C drop}}{1\ \cancel{m}} = -0.27°\text{C drop below } 0°\text{C for sucrose}$$

Using a solute that is an ionic solid and that completely ionizes in an aqueous solution introduces consideration of the number of particles present in the solution. Notice in the preceding chart that a 1 molal solution of NaCl yields a solution with 2 moles of particles because:

NaCl(aq)		Na^+		Cl^-
1 mole of ionic	=	1 mole of	+	1 mole of
sodium chloride salt		Na^+ ions		Cl^- ions

Thus a 1 molal solution of NaCl has 2 moles of ion particles in 1,000 grams of solvent. The colligative property of lowering the freezing point and raising the boiling point depends primarily on the concentration of particles and not the type of particles.

In a 1 molal solution of NaCl, 1 mole of sodium chloride salt would be dissolved in 1,000 grams of water so that a total of 2 moles of ions were released—1 mole of Na^+ ions and 1 mole of Cl^- ions. Because this provides 2 moles of particles, it will cause a $2 \times -1.86°C$ (drop caused by 1 mole) $= -3.72°C$ drop in the freezing point.

Likewise, it will cause a 2 × 0.51°C (rise caused by 1 mole) = +1.02°C rise in the boiling point or a boiling point of 101.02°C.

The chart also shows that $CaCl_2$ releases 3 moles of particles from 1 mole of $CaCl_2$ dissolved in 1,000 grams of water. Note that its effect is to lower the freezing point to 3 × −1.86°C or to −5.58°C. The boiling point rise is also 3 times the molal rise of 0.51°C, resulting in a boiling point of 101.53°C.

This explains the use of salt on icy roads in the winter and the increased effectiveness of calcium chloride per mole of solute. The use of glycols in antifreeze solutions in automobile radiators is also based on this same concept.

CRYSTALLIZATION

Many substances form a repeated pattern structure as they come out of solution. The structure is bounded by plane surfaces that make definite angles with each other to produce a geometric form called a **crystal**. The smallest portion of the crystal lattice that is repeated throughout the crystal is called the **unit cell**. The kinds of unit cells are shown in Figure 33.

The crystal structure can also be classified by its internal axis, as shown in Figure 34.

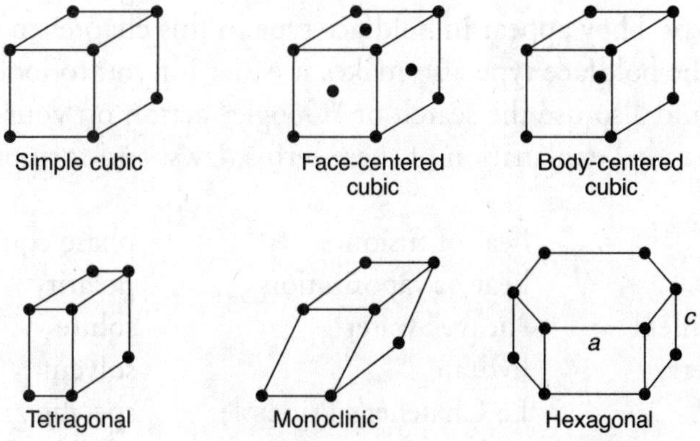

Figure 33. Kinds of Unit Cells

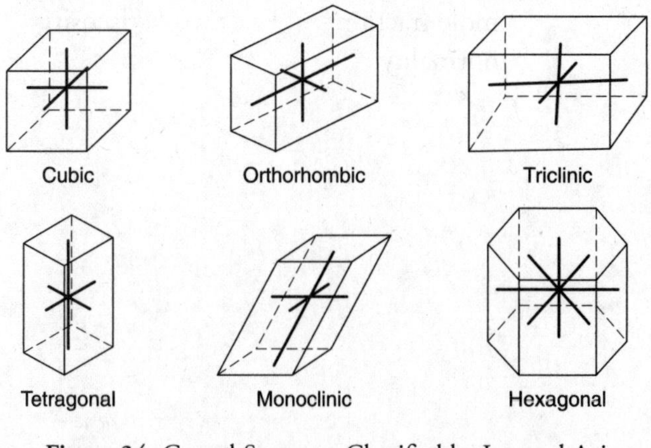

Figure 34. Crystal Structure Classified by Internal Axis

TIP

A triclinic crystal is the least symmetrical of those shown. All of the axes are of different lengths. None of the axes are perpendicular to each other.

A substance that holds a definite proportion of water in its crystal structure is called a **hydrate**. The formulas of hydrates show this water in the following manner: $CuSO_4 \cdot 5H_2O$; $CaSO_4 \cdot 2H_2O$; and $Na_2CO_3 \cdot 10H_2O$. (The · is read as "with.") When these crystals are heated gently, the water of hydration can be forced out of the crystal and the structure collapses into an anhydrous (without water) powder. The dehydration of hydrated $CuSO_4$ serves as a good example since the hydrated crystals are deep blue because of the copper ions present with water molecules. When this water is removed, the structure crumbles into the anhydrous white powder. Some hydrated crystals, such as magnesium sulfate (Epsom salt), lose the water of hydration on exposure to air at ordinary temperatures. They are said to be **efflorescent**. Other hydrates, such as magnesium chloride and calcium chloride, absorb water from the air and become wet. They are said to be **deliquescent** or **hydroscopic**. This property explains why calcium chloride is often used as a drying agent in laboratory experiments.

Chapter Summary

The following terms summarize all the concepts and ideas that were introduced in this chapter. You should be able to explain their meaning and how you would use them in chemistry. They appear in boldface type in this chapter to draw your attention to them. The boldface type also makes it easier for you to look them up if you need to. You could also use the search or "Google" action on your computer to get a quick and expanded explanation of these terms, laws, and formulas.

anhydride
boiling point
Brownian movement
colligative property
critical pressure
critical temperature
crystal
dynamic equilibrium
endothermic
exothermic
heat of fusion
heat of vaporization
"heavy" water
hydrate
Le Châtelier's Principle
melting point
molality
molarity
mole fraction
normality
phase equilibrium
polarity
solute
solvent
specific gravity
sublimation
surface tension
van der Waals forces
viscosity

Practice Exercises

1. Distillation of water cannot remove
   (A) volatile liquids like alcohol
   (B) dissolved salts
   (C) suspensions
   (D) precipitates

2. The ratio in water of hydrogen to oxygen by mass is
   (A) 1 : 9
   (B) 2 : 1
   (C) 1 : 2
   (D) 1 : 8

3. Decomposing water by an electric current will give a ratio of hydrogen to oxygen by volume equal to
   (A) 1 : 9
   (B) 2 : 1
   (C) 1 : 2
   (D) 1 : 8

4. If 10 grams of ice melts at 0°C, the total quantity of heat absorbed is
   (A) 10 cal
   (B) 80 cal
   (C) 800 cal
   (D) 800 kilocal

5. To heat 10 grams of water from 4°C to 14°C requires
   (A) 10 cal
   (B) 14 cal
   (C) 100 cal
   (D) 140 cal

6. The abnormally high boiling point of water in comparison to similar compounds is due primarily to
   (A) van der Waals forces
   (B) polar covalent bonding
   (C) dipole insulation
   (D) hydrogen bonding

7. A metallic oxide placed in water would most likely yield
   (A) an acid
   (B) a base
   (C) a metallic anhydride
   (D) a basic anhydride

8. A solution can be *both*
   (A) dilute and concentrated
   (B) saturated and dilute
   (C) saturated and unsaturated
   (D) supersaturated and saturated

9. The solubility of a solute must indicate
   (A) the temperature of the solution
   (B) the quantity of solute
   (C) the quantity of solvent
   (D) all of these

10. A foam is an example of
    (A) a gas dispersed in a liquid
    (B) a liquid dispersed in a gas
    (C) a solid dispersed in a liquid
    (D) a liquid dispersed in a liquid

11. When another crystal was added to a water solution of the same substance, the crystal seemed to remain unchanged. Its particles were
    (A) going into an unsaturated solution
    (B) exchanging places with others in the solution
    (C) causing the solution to become supersaturated
    (D) not going into solution in this static condition

12. A 10% solution of NaCl means that in 100 grams of solution there is
    (A) 5.85 g NaCl
    (B) 58.5 g NaCl
    (C) 10 g NaCl
    (D) 94 g of H_2O

13. The gram-equivalent mass of $AlCl_3$ is its gram-formula mass divided by

(A) 1
(B) 3
(C) 4
(D) 6

14. The molarity of a solution made by placing 98 grams of H_2SO_4 in sufficient water to make 500 milliliters of solution is

(A) 0.5
(B) 1
(C) 2
(D) 3

15. If 684 grams of sucrose (molecular mass = 342 g/mol) is dissolved in 2,000 grams of water (essentially 2 L), what will be the freezing point of this solution?

(A) −.51°C
(B) −1.86°C
(C) −3.72°C
(D) −6.58°C

The following questions are in the format that is used on the **SAT Subject Test in Chemistry.** If you are not familiar with these types of questions, study pages xii–xv before doing the remainder of the review questions.

Directions: Each of the following sets of lettered choices refers to the numbered questions immediately below it. For each numbered item, choose the one lettered choice that fits it best. Every choice in a set may be used once, more than once, or not at all.

Questions 16–18 refer to the following graphs:

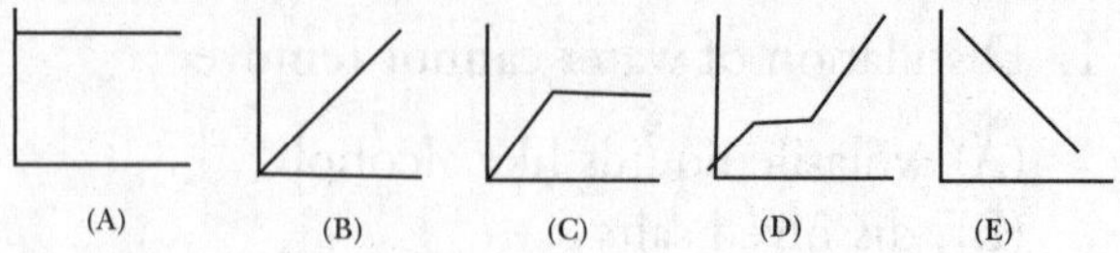

16. Which of the graphs represents the solubility curve of a substance that has no change in its solubility as the temperature increases?

17. Which of the graphs represents the amount of a solute in a solution as the solute is added even after the saturation point is reached and no more can go into solution?

18. Which of the graphs represents the temperature of a sample of ice over time as it is heated at an even rate from below its freezing point to a temperature where it is in the liquid state above the freezing point?

Questions 19–20

(A) 1 molar
(B) 1 molal
(C) .5 molar
(D) .5 molal
(E) 25 molar

19. The concentration of a solution of $Ca(OH)_2$ when 74 grams are completely dissolved in a container holding 2 liters of water?

20. The concentration of a solution of sucrose, $C_{12}H_{22}O_{11}$, when 684 grams are completely dissolved in 2,000 grams of water?

Answers and Explanations

1. **(A)** Volatile liquids cannot be removed by distillation because they would come over with the steam in this process.

2. **(D)** In the molar mass of water, there are 2 hydrogen = 2 g and 1 oxygen = 16 g. So the H/O ratio by mass is 2 g/16 g or 1/8.

3. **(B)** In the balanced reaction, $2H_2O + 2H_2\ (g) + O_2\ (g)$, 2 volumes of hydrogen are released to 1 volume of oxygen, so the ratio is 2:1.

4. **(C)** It takes 80 cal to melt 1 g (heat of fusion for 1 gram) so 10 g × 80 cal/g = 800 cal to melt the ice.

5. **(C)** m (mass) × ΔT (change of temperature) × specific heat = q (quantity of heat) So 10 g × (14° – 4°= 10) × 1 cal/°/g = 100 cal.

6. **(D)** Because hydrogen bonding is so strong in water, it takes more heat energy to reach its boiling point.

7. **(B)** Metal oxides are acid anhydrides, because they react with water to form acids. Example: $CaO + H_2O \rightarrow Ca(OH)_2$

8. **(B)** Saturated means that the solution is holding all the solute it can at that temperature. Dilute means that there is a small amount of solute in solution compared with the amount of solvent. So a substance that is only slightly soluble can form a saturated dilute solution.

9. **(D)** Solubility of a solute must give the amount of solute dissolved in a given amount of solvent at a particular temperature.

10. **(A)** A foam is an example of a colloidal dispersion of a gas in a liquid.

11. **(B)** This situation describes a saturated solution where an equilibrium exists between the undissolved solute and the solute particles in solution.

12. **(C)** A 10% solution contains 10 g of solute/100 g of solution because the percentage is by mass/100 g of solution.

13. **(B)** The gram-equivalent mass is equal to gram-formula mass divided by 3, because Al has a +3 oxidation number that is equivalent to 3 hydrogens of +1 each.

14. **(C)** 98 g H_2SO_4 = 1 mol H_2SO_4
So, if there are 500 mL (which is 0.5 L) then 1 mol in 0.5 L would be equivalent to 2 mol in 1 L or a 2 molar solution.

15. **(B)** 684 g of sucrose $= \dfrac{684\ \text{g}}{342\ \text{g/mol}} = 2$ mol of sucrose

So $\dfrac{2\ \text{mol}}{2{,}000\ \text{g}\ H_2O} = 1$ molal solution

The freezing point is lowered to −1.86°C

16. **(A)** The graph of the solubility of a substance that has no change in its solubility as the temperature is increased would be a straight line, as shown in A.

17. **(C)** The graph that shows the amount of solute in solution as it is added up through its saturation point is C. The point at which saturation occurs is where the line flattens out to a horizontal straight line.

18. **(D)** Graph D shows the ice getting warmer until it starts to melt. This is where the temperature stays the same until it is all melted. The heat being added is being used for the change in state from a solid to a liquid and the line stays horizontal, indicating no change in temperature during this time. The line then starts to ascend as liquid water is absorbing the heat and increasing in temperature.

19. **(C)** The molar mass of $Ca(OH)_2$ is Ca = 40, 2 O = 32, and 2 H = 2, which, added together equals 74 g/mol. If this is dissolved to make a 2 L solution, there is only .5 mol in 1 L. So the molarity (M), the moles of solute in 1 L = .5 M

20. **(B)** The molar mass of $C_{12}H_{22}O_{11}$ is 12 C = 144, 22 H = 22, and 11 O = 176, which added together equals 342 g/mol. Because 684 g is given, then 684 g divided by 342 g/mol = 2 mol of sucrose. If this is dissolved in 2,000 g of water, it gives 1 mol /1,000 g of water or a 1 molal solution.

CHAPTER 3

Chemical Reactions and Thermochemistry

CHAPTER 8

- Identify the driving force for these four major types of chemical reactions and write balanced equations for each: combination (or synthesis), decomposition (or analysis), single replacement, and double replacement.
- Explain hydrolysis with a balanced equation.
- Identify and explain graphically enthalpy changes in exothermic and endothermic reactions.
- Use Hess's Law to show the additivity of heats of reactions.
- Calculate enthalpy from bond energies.

The many kinds of reactions you may encounter can be placed into four basic categories: combination, decomposition, single replacement, and double replacement.

The first type, **combination**, can also be called **synthesis**. This means the formation of a compound from the union of its elements. Some examples of this type are:

$$Zn + S \rightarrow ZnS$$
$$2H_2 + O_2 \rightarrow 2H_2O$$
$$C + O_2 \rightarrow CO_2$$

The second type of reaction, **decomposition**, can also be referred to as **analysis**. This means the breakdown of a compound to release its components as individual elements or other compounds. Some examples of this type are:

$$2H_2O \rightarrow 2H_2 + O_2 \text{ (electrolysis of water)}$$
$$C_{12}H_{22}O_{11} \rightarrow 12C + 11H_2O$$
$$2HgO \rightarrow 2Hg + O_2$$

The third type of reaction is called **single replacement** or **single displacement**. This type can best be shown by examples in which one substance is displacing another. Some examples are:

$$Fe + CuSO_4 \rightarrow FeSO_4 + Cu$$
$$Zn + H_2SO_4 \rightarrow ZnSO_4 + H_2(g)$$
$$CU + 2AgNO_3 \rightarrow Cu(NO_3)_2 + 2Ag$$

TIP

Know the four types of chemical reactions:

1. Combination or synthesis
2. Decomposition or analysis
3. Single replacement
4. Double replacement

The last type of reaction is called **double replacement** or **double displacement** because there is an actual exchange of "partners" to form new compounds. Some examples are:

$$AgNO_3 + NaCl \rightarrow AgCl + NaNO_3$$
$$H_2SO_4 + 2NaOH \rightarrow Na_2SO_4 + 2H_2O \text{ (neutralization)}$$
$$CaCO_3 + 2HCl \rightarrow H_2CO_3 + CaCl_2$$

(unstable)

$$\rightarrow H_2O + CO_2(g)$$

PREDICTING REACTIONS

One of the most important topics of chemistry deals with the reasons why reactions take place. Taking each of the above types of reactions, let us see how a prediction can be made concerning how the reaction gets the driving force to make it occur.

TIP

$-\Delta H$ = exothermic reaction

$+\Delta H$ = endothermic reaction

1. Combination (Also Known as Synthesis)

The best source of information to predict a chemical combination is the heat of formation table. A **heat of formation** table gives the number of calories evolved or absorbed when a mole (gram-formula mass) of the compound in question is formed by the direct union of its elements. In this book a positive number indicates that heat is absorbed and a negative number that heat is evolved. It makes some difference whether the compounds formed are in the solid, liquid, or gaseous state. Unless otherwise indicated (g = gas, ℓ = liquid), the compounds are in the solid state. The values given are in kilojoules, 4.18 joules is the amount of heat needed to raise the temperature of 1 gram of water 1 degree on the Kelvin scale. The symbol ΔH is used to indicate the heat of formation.

If the heat of formation is a large number preceded by a minus sign, the combination is likely to occur spontaneously and the reaction is exothermic. If, on the other hand, the number is small and negative or is positive, heat will be needed to get the reaction to proceed at any noticeable rate. Some examples are:

$$Zn + S \rightarrow ZnS + 202.7 \text{ kJ} \qquad \Delta H = -202.7 \text{ kJ}$$

This means that 1 mole of zinc (65 grams) reacts with 1 mole of sulfur (32 grams) to form 1 mole of zinc sulfide (97 grams) and releases 202.7 kilojoules of heat.

SIMILARLY:

$$Mg + \tfrac{1}{2}O_2 \rightarrow MgO + 601.6 \text{ kJ} \qquad \Delta H = -601.6 \text{ kJ/mol}$$

indicates that the formation of 1 mole of magnesium oxide requires 1 mole of magnesium and $\frac{1}{2}$ mole of oxygen with the release of 601.6 kJ of heat. Notice the use of the fractional coefficient for oxygen. If the equation had been written with the usual whole-number coefficients, 2 moles of magnesium oxide would have been released.

$$2Mg + O_2 \rightarrow 2MgO + 2(+601.6) \text{ kJ}$$

Since, by definition, the heat of formation is given for the formation of *1* mole, this latter thermal equation shows $2 \times (-601.6)$ kJ released.

ANOTHER EXAMPLE:

$$H_2(g) + \tfrac{1}{2}O_2(g) \rightarrow H_2O(\ell) + 285.8\text{ kJ} \qquad \Delta H = -285.8\text{ kJ}$$

In combustion reactions the heat evolved when 1 mole of a substance is completely oxidized is called the **heat of combustion** of the substance, so, in this equation:

$$C(g) + O_2(g) \rightarrow CO_2(g) + 393.5\text{ kJ} \qquad \Delta H = -393.5\text{ kJ}$$

ΔH is the **heat of combustion** of carbon. Because the energy of a system is conserved during chemical activity, the same equation could be arrived at by adding the following equations:

$$\begin{array}{ll} C(s) + \tfrac{1}{2}O_2(g) \rightarrow \cancel{CO}(g) & \Delta H = 110.5\text{ kJ} \\ \cancel{CO}(g) + \tfrac{1}{2}O_2(g) \rightarrow CO_2(g) & \Delta H = -283.0\text{ kJ} \\ \hline C(s) + O_2(g) \rightarrow CO_2(g) & \Delta H = -393.5\text{ kJ} \end{array}$$

2. Decomposition (Also Known as Analysis)

The prediction of decomposition reactions uses the same source of information, the heat of formation table. If the heat of formation is a high exothermic (ΔH is negative) value, the compound will be difficult to decompose since this same quantity of energy must be returned to the compound. A low heat of formation indicates decomposition would not be difficult, such as the decomposition of mercuric oxide with $\Delta H = -90.8$ kJ/mole:

$$2HgO \rightarrow 2Hg + O_2 \text{ (Prestley's method of preparation)}$$

A high positive heat of formation indicates extreme instability of a compound, which can explosively decompose.

3. Single Replacement

A prediction of the feasibility of this type of reaction can be based on a comparison of the heat of formation of the original compound and that of the compound to be formed. For example, in a reaction of zinc with hydrochloric acid, the 2 moles of HCl have $\Delta H = 2 \times -92.3$ kJ.

$$\begin{array}{ll} Zn + 2HCl \rightarrow ZnCl_2 + H_2(g) & \textit{Note: } \Delta H = 0 \text{ for elements} \\ 2 \times -92.3\text{ kJ} & \\ -184.6\text{ kJ} \neq -415.5\text{ kJ} & \end{array}$$

and the zinc chloride has $\Delta H = -415.5$ kJ. This comparison leaves an excess of 230.9 kJ of heat given off, so the reaction would occur.

In this next example, $-928.4 - (-771.4) = -157.0$ kJ, which is the excess to be given off as the reaction occurs:

$$\begin{array}{cccc} Fe + & CuSO_4 & \rightarrow \; FeSO_4 & + Cu \\ & -771.4\text{ kJ} & -928.4\text{ kJ} & \end{array}$$

Another simple way of predicting single replacement reactions is to check the relative positions of the two elements in the activity series below. If the element that is to replace the other in the compound is higher on the chart, the reaction will occur. If it is below, there will be no reaction.

Some simple examples of this are the following reactions.

In predicting the replacement of hydrogen by zinc in hydrochloric acid, reference to the activity series shows that zinc will replace hydrogen. This reaction would occur:

$$Zn + 2HCl \rightarrow ZnCl_2 + H_2(g)$$

In fact, most metals in the activity series would replace hydrogen in an acid solution. If a metal such as copper were chosen, no reaction would occur.

$$Cu + HCl \rightarrow \text{no reaction}$$

The determination of these replacements using a quantative method is covered in Chapter 12.

Activity Series of Common Elements

	Activity of metals		Activity of halogen nonmetals
Most active	Li Rb K Ba Sr Ca Na	React with cold water and acids, replacing hydrogen. React with oxygen, forming oxides.	F_2 Cl_2 Br_2 I_2
	Mg Al Mn Zn Cr Fe Cd	React with steam (but not cold water) and acids, replacing hydrogen. React with oxygen, forming oxides.	
	Co Ni Sn Pb	Do not react with water. React with acids, replacing hydrogen. React with oxygen, forming oxides.	
	H_2 Sb Bi Cu Hg	React with oxygen, forming oxides.	
Least active	Ag Pt Au	Fairly unreactive, forming oxides only indirectly

4. Double Replacement

For double replacement reactions to go to completion, that is, proceed until the supply of one of the reactants is exhausted, one of the following conditions must be present: an insoluble precipitate is formed, a nonionizing substance is formed, or a gaseous product is given off.

To predict the formation of an insoluble precipitate, you should have some knowledge of the solubilities of compounds. Table 9 gives some general solubility rules.

Table 9. Solubilities of Compounds

SOLUBLE	EXCEPT
Na^+, NH_4^+, K^+ compounds	
Acetates	
Bicarbonates	
Chlorates	
Chlorides	Ag^+, Hg^-, Pb ($PbCl_2$, sol. in hot water)
Nitrates	
Sulfates	Ba, Ca (slight), Pb
INSOLUBLE	
Carbonates, phosphates	Na, NH_4, K compounds
Sulfides, hydroxides	Na, NH_4, K, Ba, Ca

TIP

Know the general solubility rules.

(A table of solubilities could also be used as reference.)

An example of this type of reaction is given in its complete ionic form.

$$(K^+ + Cl^-) + (Ag^+ + NO_3^-) \rightarrow AgCl(s) + (K^+ + NO_3^-)$$

TIP

Notice that the ions "switch" partners.

The silver ions combine with the chloride ions to form an insoluble precipitate, silver chloride. If the reaction had been like this:

$$(K^+ + Cl^-) + (Na^+ + NO_3^-) \rightarrow K^+ + NO_3^- + Na^+ + Cl^-$$

merely a mixture of the ions would have been shown in the final solution.

Another reason for a reaction of this type to go to completion is the formation of a nonionizing product such as water. This weak electrolyte keeps its component ions in molecular form and thus eliminates the possibility of reversing the reaction. All neutralization reactions are of this type.

$$(H^+ + Cl^-) + (Na^+ + OH^-) \rightarrow H_2O + Na^+ + Cl^-$$

This example shows the ions of the reactants, hydrochloric acid and sodium hydroxide, and the nonelectrolyte product water with sodium and chloride ions in solution. Since the water does not ionize to any extent, the reverse reaction cannot occur.

The third reason for double displacement to occur is the evolution of a gaseous product. An example of this is calcium carbonate reacting with hydrochloric acid:

$$CaCO_3 + 2HCl \rightarrow CaCl_2 + H_2O + CO_2(g)$$

Another example of a compound that evolves a gas in sodium sulfite with an acid is:

$$Na_2SO_3 + 2HCl \rightarrow 2NaCl + H_2O + SO_2(g)$$

In general, acids with carbonates or sulfites are good examples of this type of equation.

Hydrolysis Reactions

TIP

Salt interacts with water.

Hydrolysis reactions are the opposite of neutralization reactions. In hydrolysis the salt and water react to form an acid and a base. For example, if sodium chloride is placed in solution, this reaction occurs to some degree:

$$(Na^+ + Cl^-) + H_2O \rightarrow (Na^+ + OH^-) + (H^+ + Cl^-)$$

In this hydrolysis reaction the same number of hydrogen ions and of hydroxide ions is released so that the solution is neutral. This occurs because sodium hydroxide is a strong base and hydrochloric acid is a strong acid. (There are charts of relative acid and base strengths on pages 248 and 249 to use as references.) Because they are both classified as strong, sodium hydroxide and hydrochloric acid essentially exist as ions in solutions. Therefore, the NaCl solution has an excess of neither hydrogen nor hydroxide ions, and it will test neutral. Thus, the salt of a strong acid and a strong base forms a neutral solution when dissolved in water. However, if Na_2CO_3 is dissolved, we have:

$$(2Na^+ + CO_3^{2-}) + 2H_2O \rightarrow (2Na^+ + 2OH^-) + H_2CO_3$$

The H_2CO_3 is written as a single entity because it is a slightly ionized acid, or, in other words, a weak acid. Since the hydroxide ions are free in the solution, the solution is basic. Notice that here it was the salt of a strong base and a weak acid that formed a basic solution. This generalization is true for this type of salt.

If we use $ZnCl_2$, which is the salt of a strong acid and a weak base, the reaction will be:

$$(Zn^{2+} + 2Cl^-) + H_2O \rightarrow (2H^+ + 2Cl^-) + Zn(OH)_2$$

In this case the hydroxide ions are held in the weakly ionizing compound while the hydrogen ions are free to make the solution acidic. In general, then, the salt of a strong acid and a weak base forms an acid solution by hydrolysis.

The fourth possibility is that of a salt of a weak acid and weak base dissolving in water. An example would be ammonium carbonate, $(NH_4)_2CO_3$, which is the salt of a weak base and a weak acid. The hydrolysis reaction would be:

$$(NH_4)_2CO_3 + 2H_2O \rightarrow 2NH_4OH + H_2CO_3$$

Both the ammonium hydroxide, NH_4OH, and the carbonic acid, H_2CO_3, are written as nonionized compounds because they are classified as a weak base and a weak acid, respectively. Therefore, a salt of a weak acid and a weak base forms a neutral solution since neither hydrogen ion nor hydroxide ion will be present in excess.

Entropy

In many of the preceding predictions of reactions, we used the concept that reactions will occur when they result in the lowest possible energy state.

There is, however, another driving force to reactions that is related to their state of disorder or of randomness. This state of disorder is called **entropy**. A reaction is also driven, then, by a need for a greater degree of disorder. An example is the intermixing of gases in two connected flasks when a valve is opened to allow the two previously isolated gases to travel between the two flasks. Because temperature remains constant throughout the process, the total heat content cannot have changed to a lower energy level, and yet the gases will become evenly distributed in the two flasks. The system has thus reached a higher degree of disorder or entropy. A quantitative treatment of entropy is given on page 241.

TIP

Entropy is a measure of the degree of disorder.

THERMOCHEMISTRY

In general, all chemical reactions either liberate or absorb heat. The origin of chemical energy lies in the position and motion of atoms, molecules, and subatomic particles. The total energy possessed by a molecule is the sum of all the forms of potential and kinetic energy associated with it.

The energy changes in a reaction are due, to a large extent, to the changes in potential energy that accompany the breaking of chemical bonds in reactants to form new bonds in products.

The molecule may also have rotational, vibrational, and translational energy, along with some nuclear energy sources. All these make up the total energy of molecules. In beginning chemistry, the greatest concern in reactions is the electronic energy involved in the making and breaking of chemical bonds.

Because it is virtually impossible to measure the total energy of molecules, the energy change is usually the experimental data that we deal with in reactions. This change in quantity of energy is known as the change in **enthalpy** (heat content) of the chemical system and is symbolized by ΔH.

CHANGES IN ENTHALPY

Changes in enthalpy for exothermic and endothermic reactions can be shown graphically, as in the examples below.

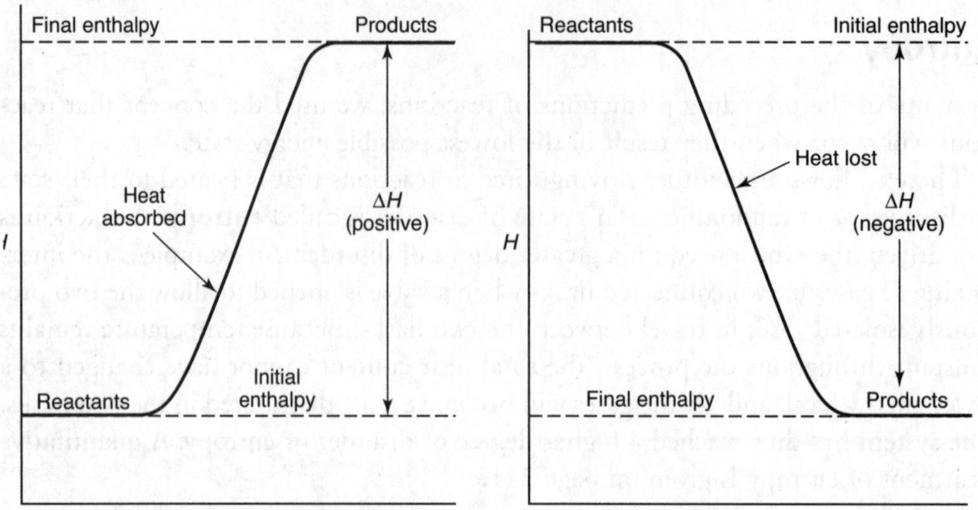

Course of Endothermic Reaction **Course of Exothermic Reaction**

Notice that the ΔH for an endothermic reaction is positive, while that for an exothermic reaction is negative. It should be noted also that changes in enthalpy are always indepen-dent of the path taken to change a system from the initial state to the final state.

> **TIP**
> This paragraph gives the definition of the molar heat of formation.

Because the quantity of heat absorbed or liberated during a reaction varies with the temperature, scientists have adopted 25°C and 1 atmosphere pressure as the **standard state** condition for reporting heat data. A superscript zero on ΔH (i.e., ΔH^0) indicates that the corresponding process was carried out under standard conditions. The **standard enthalpy of formation** (ΔH^0_f) of a compound is defined as the change in enthalpy that accompanies the formation of 1 mole of a compound from its elements with all substances in their standard states at 25°C. This value is called the **molar heat of formation.**

To calculate the enthalpy of a reaction, it is necessary to write an equation for the reaction. The standard enthalpy change, ΔH, for a given reaction is usually expressed in kilocalories and depends on how the equation is written. For example, the following equations express the reaction of hydrogen with oxygen in two ways:

$$H_2(g) + \tfrac{1}{2}O_2(g) \rightarrow H_2O(g) \quad \Delta H^0_f = -241.8 \text{ kJ}$$

$$2H_2(g) + O_2(g) \rightarrow 2H_2O(g) \quad \Delta H^0_f = -285.8 \text{ kJ}$$

Experimentally, ΔH^0_f for the formation of 1 mole of $H_2O(g)$ is −241.8 kJ. Since the second equation represents the formation of 2 moles of $H_2O(g)$, the quantity is twice −241.8, or −485.6 kJ. It is assumed that the initial and final states are measured at 25°C and 1 atmosphere, although the reaction occurs at a higher temperature.

PROBLEM: How much heat is liberated when 40.0 grams of $H_2(g)$ reacts with excess O(g)? The reaction equation is:

$$H_2(g) + \tfrac{1}{2}O_2(g) \rightarrow H_2O(g) \qquad \Delta H^0_f = -241.8 \text{ kJ}$$

This represents 1 mol or 2 g of H(g) forming 1 mol of $H_2O(g)$:

$$40 \text{ g} \times \frac{1 \text{ mol}}{2 \text{ g}} = 20 \text{ mol of hydrogen}$$

Since each mole gives off −241.8 kJ, then

$$20 \text{ mol} \times \frac{-241.8 \text{ kJ}}{1 \text{ mol}} = -4.836 \text{ kJ}$$

Notice that the physical state of each participant must be given since the phase changes involve energy changes. Combustion reactions produce a considerable amount of energy in the form of light and heat when a substance is combined with oxygen. The heat released by the complete combustion of 1 mole of a substance is called the **heat of combustion** of that substance. Heat of combustion is defined in terms of 1 mole of reactant, whereas the heat of formation is defined in terms of 1 mole of product. All substances are in their standard state. The general enthalpy notation, ΔH, applies to heats of reaction, but the addition of a subscript ***c***, $\boldsymbol{\Delta H_c}$, specifically indicates heat of combustion.

ADDITIVITY OF REACTION HEATS AND HESS'S LAW

Chemical equations and ΔH^0 values may be manipulated algebraically. Finding the ΔH for the formation of vapor from liquid water shows how this can be done.

$$H_2(g) + \tfrac{1}{2}O_2(g) \rightarrow H_2O(g) \qquad \Delta H^0_f = -241.8 \text{ kJ}$$

$$H_2(g) + \tfrac{1}{2}O_2(g) \rightarrow H_2O(\ell) \qquad \Delta H^0_f = -285.8 \text{ kJ}$$

Since we want the equation for $H_2O(\ell) \rightarrow H_2O(g)$, we can reverse the second equation. This changes the sign of ΔH.

$$H_2O(\ell) \rightarrow H_2O(g) + \tfrac{1}{2}O_2(g) \qquad \Delta H^0_f = +285.8 \text{ kJ}$$

Adding

$$H_2(g) + \tfrac{1}{2}O_2(g) \rightarrow H_2O(g) \qquad \Delta H^0_f = -241.8 \text{ kJ}$$

yields

$$H_2O(\ell) + H_2(g) + \tfrac{1}{2}O_2(g) \rightarrow H_2(g) + \tfrac{1}{2}O_2(g) + H_2O(g) \qquad \Delta H^0_f = +44.0 \text{ kJ}$$

Simplification gives the net equation:

$$H_2O(\ell) \rightarrow H_2O(g) \qquad \Delta H^0_f = +44.0 \text{ kJ}$$

The principle underlying the preceding calculations is known as **Hess's Law of Heat Summation**. This principle states that, when a reaction can be expressed as the algebraic sum of two or more other reactions, the heat of the reaction is the algebraic sum of the heats of these reactions. This is based upon the **First Law of Thermodynamics**, which, simply stated, says that the total energy of the universe is constant and cannot be created or destroyed.

These laws allow calculations of ΔHs that cannot be easily determined experimentally. An example is the determination of the ΔH of CO from the ΔH^0_f of CO_2.

$$C(s) + \tfrac{1}{2}O_2(g) \rightarrow CO(g) \qquad \Delta H = ?$$

The calculation of ΔH can be done in kilocalorie/mole or kilojoules/mole units. The calculation for the above example is shown using both units.

$$C(s) + O_2(g) \rightarrow CO_2(g) \qquad \Delta H^0_f = -393.5 \text{ kJ/mol}$$

$$CO(g) + \tfrac{1}{2}O_2(g) \rightarrow CO_2(g) \qquad \Delta H^0_c = -283.0 \text{ kJ/mol}$$

The equation wanted is

$$C(s) + \tfrac{1}{2}O_2(g) \rightarrow CO(g)$$

To get this, we reverse the second equation and add it to the first:

$$C(s) + O_2(g) \rightarrow CO_2(g) \qquad \Delta H^0_f = -393.5 \text{ kJ/mol}$$

$$CO_2(g) \rightarrow \tfrac{1}{2}O_2(g) + CO(g) \qquad \Delta H^0_f = +283.0 \text{ kJ/mol}$$

Addition yields

$$C(s) + \tfrac{1}{2}O_2(g) \rightarrow CO(g) \qquad \Delta H^0_f = -110.5 \text{ kJ/mol}$$

This relationship can also be shown schematically as follows:

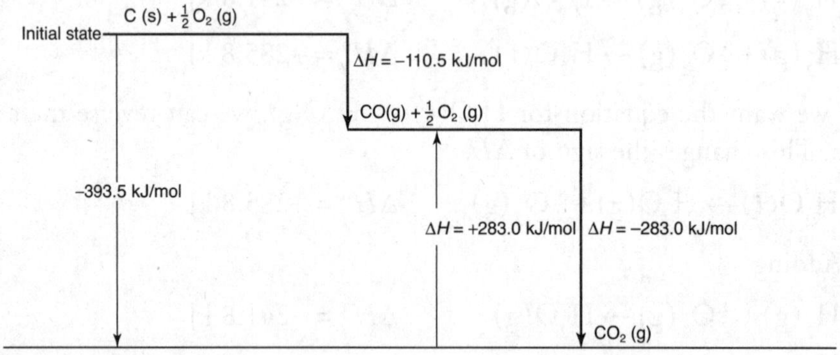

Some commonly used standard heats of formation (enthalpy), designated as ΔH^0_f, are listed in Table Ⓐ in the Tables for Reference section.

An alternative (and easier) method of calculating enthalpies is based on the concept that $\Delta H^0_{\text{reaction}}$ is equal to the difference between the total enthalpy of the reactants and that of the products. This can be expressed as follows:

$$\Delta H^0_{\text{reaction}} = \sum(\text{sum of})\,\Delta H^0_f\,(\text{products}) - \sum(\text{sum of})\,\Delta H^0_f\,(\text{reactants})$$

PROBLEM: Calculate $\Delta H^0_{reaction}$ for the decomposition of sodium chlorate.

$$NaClO_3(s) \rightarrow NaCl(s) + \tfrac{3}{2}O_2(g)$$

Step 1. Obtain ΔH^0_f for all substances.

$$\begin{aligned} NaClO_3(s) &= -358.2 \text{ kJ} \\ NaCl(s) &= -410.5 \text{ kJ} \\ O_2(g) &= 0 \text{ kcal/mol (all elements=0)} \end{aligned}$$

Step 2. Substitute these values in the equation.

$$\begin{aligned} \Delta H^0_{reaction} &= \sum \Delta H^0_f(\text{products}) - \sum \Delta H^0_f(\text{reactants}) \\ \Delta H^0_{reaction} &= -410.5 - (-358.2) \text{ kJ} \\ \Delta H^0_{reaction} &= -52.2 \text{ kJ} \end{aligned}$$

ANOTHER EXAMPLE: Calculate $\Delta H_{reaction}$ for this oxidation of ammonia:

$$4NH_3(g) + 5O_2(g) \rightarrow 6H_2O(g) + 4NO(g)$$

The individual ΔH^0_f's are

$$\begin{aligned} 4NH_3 &= 4(-45.9 \text{ kJ}) &&= -183.9 \text{ kJ} \\ 5O_2(g) &= 0 && \\ 6H_2O(g) &= 6(-241.8 \text{ kJ}) &&= -1450.8 \text{ kJ} \\ 4NO(g) &= 4(90.29 \text{ kJ}) &&= 361.16 \text{ kJ} \end{aligned}$$

Substituting these in the $\Delta H_{reaction}$ equation gives

$$\begin{aligned} \Delta H_{reaction} &= [-1450.8(6H_2O) + 361.16(4NO)] - [-183.9(4NH_3) + 0(5O_2)] \\ \Delta H_{reaction} &= -905.7 \text{ kJ} \end{aligned}$$

BOND DISSOCIATION ENERGY

The same principle of additivity applies to bond energies. Experimentation has yielded average bond energies for particular bonds. Some of the more common are shown in the chart.

Bond	Energy (kJ/mol)	Bond	Energy (kJ/mol)
H—H	432	O=O	495
C—H	413	O—H	467
C—C	347	C—O	358
C=C	614	C=O	745
C≡C	839	H—F	565
O—O	146	H—Cl	427
N—N	160	H—Br	363
Cl—Cl	239	H—I	295

SAMPLE CALCULATION: Find the bond energy, H_{be}, for ethane.

Ethane =
$$
\begin{array}{ccc}
 & \text{H} & \text{H} \\
 & | & | \\
\text{H}- & \!\!\text{C}-\text{C}\!\! & -\text{H} \\
 & | & | \\
 & \text{H} & \text{H}
\end{array}
$$
and has 6 C–H bonds and 1 C–C bond.

$$H_{be} = 6\,E_{C-H} + 1\,E_{C-C} = 6(413\text{ kJ}) + 1(347\text{ kJ}) = 2825\text{ kJ}$$

ENTHALPY FROM BOND ENERGIES

A reaction's enthalpy can be approximated through the summation of bond energies. An example is this reaction:

$$H_2(g) + Cl_2(g) \rightarrow 2HCl(g)$$

Bonds Broken	Heat Energy Absorbed (kJ)	Bonds Formed	Heat Energy Evolved (kJ)
H—H	432		
Cl—Cl	239	2 H—Cl	2(427) = 854
Totals	671 kJ		854 kJ

The difference between heat evolved and heat absorbed is $854 - 671 = 183.0$ kJ. This is for 2 mol of HCl, so -183.0 kJ divided by $2 = -91.5$ kJ/mol. This answer is very close to the experimentally determined ΔH^0_f.

Further discussions of enthalpy (ΔH) and entropy (ΔS) are given on pages 214 and 241.

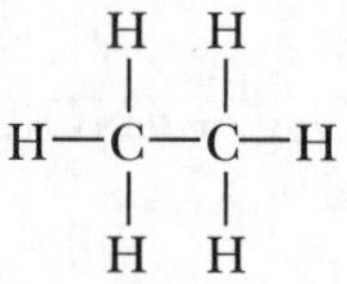

Chapter Summary

The following terms summarize all the concepts and ideas that were introduced in this chapter. You should be able to explain their meaning and how you would use them in chemistry. They appear in boldface type in this chapter to draw your attention to them. The boldface type also makes it easier for you to look them up if you need to. You could also use the search or "Google" action on your computer to get a quick and expanded explanation of these terms, laws, and formulas.

combination reaction
decomposition reaction
double replacement reaction
enthalpy
entropy
heat of combustion
heat of formation
hydrolysis reaction
molar heat of formation
single replacement reaction
standard enthalpy of formation

First Law of Thermodynamics
Hess's Law of Heat Summation

Practice Exercises

1. A synthesis reaction will occur spontaneously after the activation energy is provided if the heat of formation of the product is

   (A) large and negative
   (B) small and negative
   (C) large and positive
   (D) small and positive

2. The reaction of aluminum with dilute H_2SO_4 can be classified as

   (A) synthesis
   (B) decomposition
   (C) single replacement
   (D) double replacement

3. For a metal atom to replace another kind of metallic ion in a solution, the metal atom must be

   (A) a good oxidizing agent
   (B) higher in the activity series than the metal in solution
   (C) lower in the electromotive chart than the metal in solution
   (D) equal in activity to the metal in solution

4. One reason for a double displacement reaction to go to completion is that

   (A) a product is soluble
   (B) a product is given off as a gas
   (C) the products can react with each other
   (D) the products are miscible

5. Hydrolysis will give an acid reaction when which of these is placed in solution with water?

   (A) Na_2SO_4
   (B) K_2SO_4
   (C) $NaNO_3$
   (D) $Cu(NO_3)_2$

6. A salt derived from a strong base and a weak acid will undergo hydrolysis and give a solution that will be

   (A) basic
   (B) acid
   (C) neutral
   (D) volatile

7. Enthalpy is an expression for the

(A) heat content
(B) energy state
(C) reaction rate
(D) activation energy

8. The ΔH_f^0 of a reaction is recorded for

(A) 273 K
(B) 298 K
(C) 373 K
(D) 473 K

9. The property of being able to add enthalpies is based on the

(A) Law of Conservation of Heat
(B) First Law of Thermodynamics
(C) Law of Constants
(D) Einstein equation, $E = mc^2$

10. A $\Delta H_{\text{reaction}}$ of –100 kJ/mole indicates the reaction is

(A) endothermic
(B) unstable
(C) in need of a catalyst
(D) exothermic

Directions:

Before attempting to answer the following questions (11–15), you may want to review the directions for this type of question, found on pages xiv–xv.

Every question below contains two statements, I in the left-hand column and II in the right-hand column. For each question, decide if statement I is true or false and whether statement II is true or false, and fill in the corresponding T or F ovals in the answer spaces. *Fill in oval CE only if statement II is a correct explanation of statement I.

Record your answers here:

	I	II	CE*
11.	(T) (F)	(T) (F)	◯
12.	(T) (F)	(T) (F)	◯
13.	(T) (F)	(T) (F)	◯
14.	(T) (F)	(T) (F)	◯
15.	(T) (F)	(T) (F)	◯

	I		II
11.	If the heat of formation of a compound is a large number preceded by a minus sign, the reaction is exothermic	BECAUSE	the First Law of Thermodynamics states that a negative heat of formation is associated with an exothermic reaction.
12.	The burning of carbon with excess O_2 to form CO_2 will go to completion	BECAUSE	when a reaction results in the release of a gas that is allowed to escape, the reaction will go to completion.

	I		II
13.	The heat of formation of a compound can be calculated by algebraically adding two or more thermal reaction equations	BECAUSE	Hess's Law states that a heat of reaction can be arrived at by the algebraic summation of two or more other thermal reactions.
14.	Entropy can be described as the state of disorder of a system	BECAUSE	when high amounts of energy are released from a reaction, the reaction is said to be exothermic.
15.	The reaction in which HgO is heated to release O_2 is called a decomposition	BECAUSE	in a decomposition reaction, the original compound is broken apart into equal numbers of atoms.

Answers and Explanations

1. **(A)** A large negative heat of formation indicates that the reaction will give off a large amount of energy and will self-perpetuate after getting the activation energy necessary for it to start, like burning paper.

2. **(C)** Because the aluminum replaces the hydrogen to form aluminum sulfate, this is classified as a single-replacement reaction.

3. **(B)** For a metal to replace another metallic ion in a solution, it must be higher in the activity series of elements than the metal in solution.

4. **(B)** For a double displacement to go to completion, a product or products must either deposit as a precipitate or leave the reaction as a gas.

5. **(D)** The only salt that hydrolyzes to a strong acid and a weak base is $Cu(NO_3)_2$. This would give an acid solution. All the others are salts of strong acids and strong bases, so they would hydrolyze to a relatively neutral solution.

6. **(A)** Salts of strong bases and weak acids react with water to give a basic solution. An example would be Na_2CO_3, because it hydrolyzes into NaOH, a strong base, and H_2CO_3, a weak acid.

7. **(A)** Enthalpy is defined as the heat content of a system.

8. **(B)** The ΔH^0 with the superscript zero indicates that the process was carried out under standard conditions, which are 298 K and 1 atmosphere pressure.

9. **(B)** The property of being able to add enthalpies comes from Hess's Law of Heat Summation and is based on the First Law of Thermodynamics, which simply says that the total energy of the universe is constant and cannot be created or destroyed.

10. **(D)** Because the heat of the reaction is negative, it indicates that energy (heat) is given off when this reaction occurs. It is therefore an exothermic reaction.

11. **(T, F)** While it is true that a negative heat of formation is associated with an exothermic reaction, the First Law of Thermodynamics states that the total energy of the universe is constant and cannot be easily created nor destroyed. It is a convention of chemistry that energy given off from a combination reaction is given a negative value.

12. **(T, T, CE)** When a combination reaction results in a gas that is allowed to be released, a precipitate that drops to the bottom of the container, or a nonionizing product, the reaction will go to completion.

13. **(T, T, CE)** Both statements are true and Hess's Law is the basis for arriving at a heat of reaction by algebraically adding two or more thermal equations.

14. **(T, T)** Both statements are true, but the second does not explain the first, and there is no cause and effect relationship. Entropy is a measure of a system's randomness or disorder. Enthalpy is related to the change of heat content in a reaction. Negative enthalpies indicate exothermic reactions.

15. **(T, F)** In the decomposition of HgO, Hg and O_2 are the products. The equation is $2HgO \rightarrow 2Hg + O_2\uparrow$

    It is not always true that the number of atoms of each product is equal. It happens in this reaction because there is an equal number of atoms of Hg and O in the formula. In decomposing water you have: $2H_2O \rightarrow 2H_2\uparrow + O_2\uparrow$ because there are twice as many atoms of hydrogen than oxygen in the original formula.

Rates of Chemical Reactions

CHAPTER 9

- Explain how each of the following factors affect the rate of a chemical reaction: nature of the reactants, surface area exposed, concentrations, temperature, and the presence of a catalyst.
- Describe collision theory.
- Draw reaction diagrams with and without a catalyst.
- Explain the Law of Mass Action.
- Describe the relationship between reaction mechanisms and rates of reaction.

The measurement of reaction rate is based on the rate of appearance of a product or disappearance of a reactant. It is usually expressed in terms of a change in concentration of one of the participants per unit time.

Experiments have shown that for most reactions the concentrations of all participants change most rapidly at the beginning of the reaction; that is, the concentration of the products shows the greatest rate of increase, and the concentrations of the reactants the highest rate of decrease, at this point. This means that the rate of a reaction changes with time. Therefore a rate must be identified with a specific time.

FACTORS AFFECTING REACTION RATES

Five important factors control the rate of a chemical reaction. These are summarized below.

1. **The nature of the reactants.** In chemical reactions, some bonds break and others form. Therefore, the rates of chemical reactions should be affected by the nature of the bonds in the reacting substances. For example, reactions between ions in an aqueous solution may take place in a fraction of a second. Thus, the reaction between silver nitrate and sodium chloride is very fast. The white silver chloride precipitate appears immediately. In reactions where many covalent bonds must be broken, reaction usually takes place slowly at room temperatures. The decomposition of hydrogen peroxide into water and oxygen happens slowly at room temperatures. In fact, about 17 minutes is required for half the peroxide in a 0.50 M solution to decompose.

2. **The surface area exposed.** Since most reactions depend on the reactants coming into contact, the surface exposed proportionally affects the rate of the reaction. Consider what happens to the surface area exposed when a cube is sliced in half.

   Slicing increases by $33\frac{1}{3}$%, the surface area that can become involved in a chemical reaction. This is the reason why reactants that depend on surface reactions are often introduced as powders into reaction vessels. It also explains the danger of a dust explosion if the reactant is flammable.

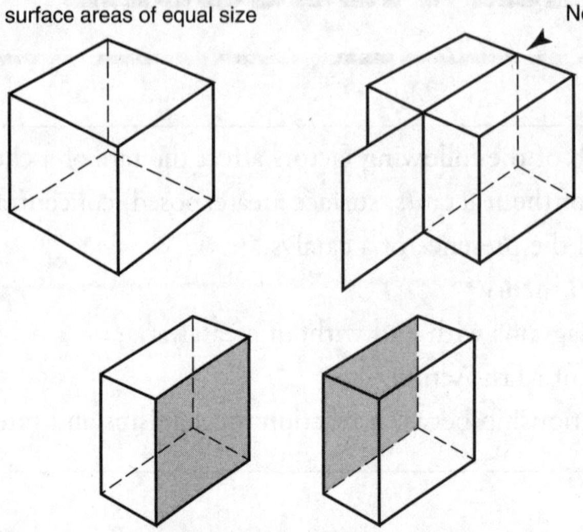

Exposes 2 additional surface areas
or a total of 8 surface areas of equal area.

3. **The concentrations.** The reaction rate is usually proportional to the concentrations of the reactants. The usual dependence of the reaction rate on the concentration of the reactants can simply be explained by theorizing that, if more molecules or ions of the reactant are in the reaction area, then there is a greater chance that more reactions will occur. This idea is further developed in the collision theory discussed below.
4. **The temperature.** A temperature increase of 10°C above room temperature usually causes the reaction rate to double or triple. The basis for this generality is that, as the temperature increases, the average kinetic energy of the particles involved increases. As a result the particles move faster and have a greater probability of hitting other reactant particles. Because the particles have more energy, they can cause an effective collision, resulting in the chemical reaction that forms the product substance.
5. **The presence of a catalyst.** It is a substance that increases or decreases the rate of a chemical reaction without itself undergoing any permanent chemical change. The catalyst provides an alternative pathway by which the reaction can proceed and in which the activation energy is lower. It thus increases the rate at which the reaction comes to completion or equilibrium. Generally, the term is used for a substance that increases reaction rate (a positive catalyst). Some reactions can be slowed down by negative catalysts.

COLLISION THEORY OF REACTION RATES

Collision theory makes the assumption that, for a reaction to occur, there must be collisions between the reacting species. Therefore, the rate of reaction depends on two factors: the number of collisions per unit time, and the fraction of these collisions that are successful because enough energy is involved.

There is a definite relationship between the concentrations of the reactants and the number of collisions. Graphically, the reaction of

$$A + B \rightarrow AB$$

can be examined as the concentrations are changed.

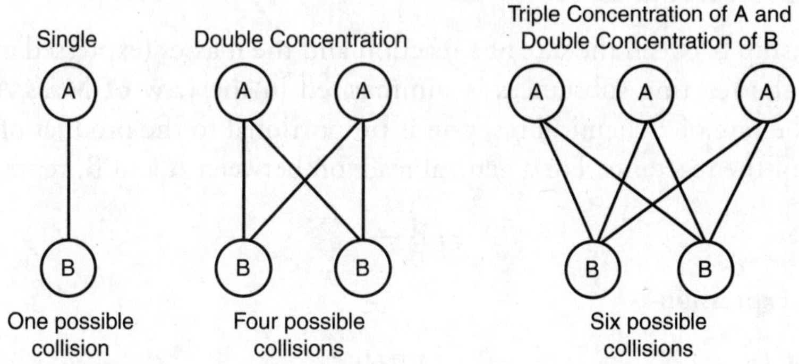

The diagrams show that the number of collisions, and consequently the rate of reaction, are proportional to the product of the concentrations. Simply stated, then, the rate of reaction is directly proportional to the concentrations of the reactants.

ACTIVATION ENERGY

Often a reaction rate may be increased or decreased by affecting the activation energy, that is, the energy necessary to cause a reaction to occur. This is shown graphically below.

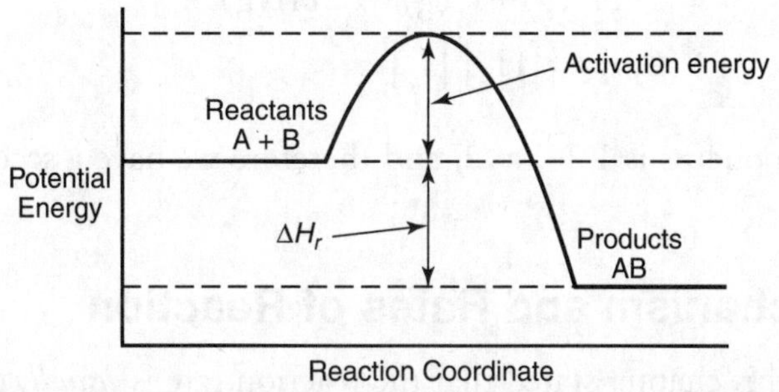

TIP

Activation energy is the energy necessary to cause a reaction to occur.

A **catalyst**, as explained in the preceding section, is a substance that is introduced into a reaction to speed up the reaction by changing the amount of activation energy needed. The effect of a catalyst used to speed up a reaction can be shown as follows:

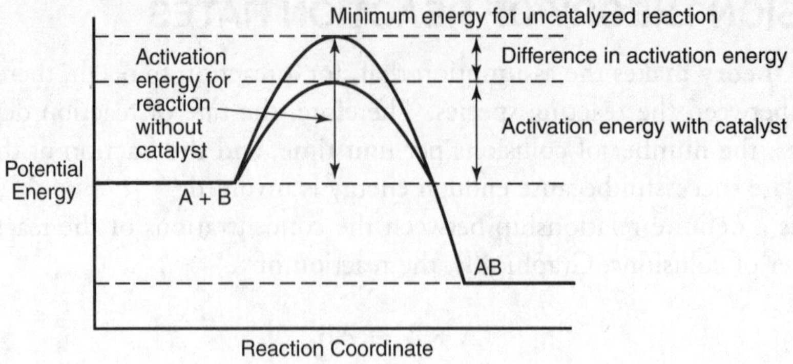

TIP

A **catalyst** speeds up the reaction but is not consumed itself.

REACTION RATE LAW

The relationship between the rate of a reaction and the masses (expressed as concentrations) of the reacting substances is summarized in the **Law of Mass Action**. It states that the rate of a chemical reaction is proportional to the product of the concentrations of the reactants. For a general reaction between A and B, represented by

$$aA + bB \rightarrow \ldots$$

the rate law expression is

$$r \propto [A]^a [B]^b$$

or, inserting a constant of proportionality that mathematically changes the expression to an equality, we have

$$r = k[A]^a [B]^b$$

Here k is called the **specific rate** constant for the reaction at the temperature of the reaction.

The exponents a and b may be added to give the total reaction order. For example:

$$H_2(g) + I_2(g) \rightarrow 2HI(g)$$
$$r = k[H_2]^1[I_2]^1$$

The sum of the exponents is 1 + 1 = 2, and therefore we have a second-order reaction.

Reaction Mechanism and Rates of Reaction

The beginning of this chapter stated that the reaction rate is *usually* proportional to the concentrations of the reactants. This occurs because some reactions do not directly occur between the reactants but may go through intermediate steps to get to the final product. The series of steps by which the reacting particles rearrange themselves to form the products of a chemical reaction is called the **reaction mechanism**. For example:

Step 1	$A + B$	$\rightarrow$	I_1 (fast)
Step 2	$A + I_1$	$\rightarrow$	I_2 (slow)
Step 3	$C + I_2$	$\rightarrow$	D (fast)
Net equation	$2A + B + C$	$\rightarrow$	D

Notice that the reactions of steps 1 and 3 occur relatively fast compared with the reaction of step 2. Now suppose that we increase the concentration of C. This will make the reaction of step 3 go faster, but it will have little effect on the speed of the overall reaction since step 2 is the rate-determining step. If, however, the concentration of A is increased, then the overall reaction rate will increase because step 2 will be accelerated. Knowing the reaction mechanism provides the basis for predicting the effect of a concentration change of a reactant on the overall rate of reaction. Another way of determining the effect of concentration changes is actual experimentation.

Chapter Summary

The following terms summarize all the concepts and ideas that were introduced in this chapter. You should be able to explain their meaning and how you would use them in chemistry. They appear in boldface type in this chapter to draw your attention to them. The boldface type also makes it easier for you to look them up if you need to. You could also use the search or "Google" action on your computer to get a quick and expanded explanation of these terms, laws, and formulas.

catalyst
collision theory
factors affecting reaction rates
Law of Mass Action
reaction mechanism

Practice Exercises

1. List the five factors that affect the rate of a reaction:

1.

2.

3.

4.

5.

2. The addition of a catalyst to a reaction

(A) changes the enthalpy
(B) changes the entropy
(C) changes the nature of the products
(D) changes the activation energy

3. An increase in concentration

(A) is related to the number of collisions directly
(B) is related to the number of collisions inversely
(C) has no effect on the number of collisions

4. At the beginning of a reaction, the reaction rate for the reactants is

(A) largest, then decreasing
(B) largest and remains constant
(C) smallest, then increasing
(D) smallest and remains constant

5. The reaction rate law applied to the reaction $aA + bB \rightarrow AB$ gives the expression

(A) $r \propto [A]^b[B]^a$
(B) $r \propto [AB]^a[A]^b$
(C) $r \propto [B]^a[AB]^b$
(D) $r \propto [A]^a[B]^b$

Answers and Explanations

1. (1) Nature of the reactants
(2) Surface area exposed
(3) Concentrations
(4) Temperature
(5) Presence of a catalyst

2. **(D)** When a catalyst is added to a reaction, it is used to either increase or decrease the activation energy needed for the reaction to occur. When the activation energy is decreased, it allows the reaction to occur at a faster rate; conversely, increasing the activation energy slows down the forward reaction.

3. **(A)** When the concentration of reactants is increased, the number of collisions between reactants is directly affected because there are more reactant units (ions, atoms, or molecules) available.

4. **(A)** At the beginning of a reaction, the reaction rate of the reactants is the highest because their concentration is the highest. As the reaction progresses, the concentration of the reactants decreases while the concentration of the products increases.

5. **(D)** For the reaction $aA + bB \rightarrow AB$, the reaction rate law would give the expression of r as proportional to $[A]^a[B]^b$.

Chemical Equilibrium

CHAPTER 10

- Explain the development of an equilibrium condition and how it is expressed as an equilibrium constant, and use it mathematically.
- Describe Le Châtelier's Principle and how changes in temperature, pressure, and concentrations affect an equilibrium.
- Solve problems dealing with ionization of water, finding the pH, solubility products, and the common ion effect.
- Explain the relationship of enthalpy and entropy as driving forces in a reaction and how they are combined in the Gibbs equation.

In some reactions no product is formed to allow the reaction to go to completion; that is, the reactants and products can still interact in both directions. This can be shown as follows:

$$A + B \rightleftharpoons C + D$$

The double arrow indicates that C and D can react to form A and B, while A and B react to form C and D.

REVERSIBLE REACTIONS AND EQUILIBRIUM

The reaction is said to have reached **equilibrium** when the forward reaction rate is equal to the reverse reaction rate. Notice that this is a dynamic condition, NOT a static one, although in appearance the reaction *seems* to have stopped. An example of an equilibrium is a crystal of copper sulfate in a saturated solution of copper sulfate. Although to the observer the crystal seems to remain unchanged, there is actually an equal exchange of crystal material with the copper sulfate in solution. As some solute comes out of solution, an equal amount is going into solution.

TIP

Equilibrium is reached when the forward and reverse reaction rate are equal.

To express the rate of reaction in numerical terms, we can use the **Law of Mass Action**, discussed in Chapter 9, which states: The rate of a chemical reaction is proportional to the product of the concentrations of the reacting substances. The concentrations are expressed in moles of gas per liter of volume or moles of solute per liter of solution. Suppose, for example, that 1 mole/liter of gas A_2 (diatomic molecule) is allowed to react with 1 mole/liter of another diatomic gas, B_2, and they form gas AB; let R be the rate for the forward reaction forming AB. The bracketed symbols $[A_2]$ and $[B_2]$ represent the concentrations in moles per liter for these diatomic molecules. Then $A_2 + B_2 \rightarrow 2AB$ has the rate expression

$$R \propto [A_2] \times [B_2]$$

where $\propto$ is the symbol for "proportional to." When $[A_2]$ and $[B_2]$ are both 1 mole/liter, the reaction rate is a certain constant value (k_1) at a fixed temperature.

$$R = k_1 \text{ } (k_1 \text{ is called the rate constant})$$

For any concentrations of A and B, the reaction rate is

$$R = k_1 \times [A_2] \times [B_2]$$

If $[A_2]$ is 3 moles/liter and $[B_2]$ is 2 moles/liter, the equation becomes

$$R = k_1 \times 3 \times 2 = 6k_1$$

The reaction rate is six times the value for a 1 mole/liter concentration of each reactant.

At the fixed temperature of the forward reaction, AB molecules are also decomposing. If we designate this reverse reaction as R', then, since

$$2AB \text{ (or } AB + AB) \rightarrow A_2 + B_2$$

two molecules of AB must decompose to form one molecule of A_2 and one of B_2. Thus the reverse reaction in this equation is proportional to the square of the molecular concentration of AB:

$$R' \propto [AB] \times [AB]$$
$$\text{or } R' \propto [AB]^2$$
$$\text{and } R' \propto k_2 \times [AB]^2$$

where k_2 represents the rate of decomposition of AB at the fixed temperature. Both reactions can be shown in this manner:

$$A_2 + B_2 \rightleftharpoons 2AB \text{ (note double arrow)}$$

When the first reaction begins to produce AB, some AB is available for the reverse reaction. If the initial condition is only the presence of A_2 and B_2 gases, then the forward reaction will occur rapidly to produce AB. As the concentration of AB increases, the reverse reaction will increase. At the same time, the concentrations of A_2 and B_2 will be decreasing and consequently the forward reaction rate will decrease. Eventually the two rates will be equal, that is, $R = R'$. At this point, equilibrium has been established, and

$$k_1[A_2] \times [B_2] = k_2[AB]^2$$

or

$$\frac{k_1}{k_2} = \frac{[AB]^2}{[A_2] \times [B_2]} = K_{eq}$$

The convention is that k_1 (forward reaction) is placed over k_2 (reverse reaction) to get this expression. Then k_1/k_2 can be replaced by K_{eq}, which is called the **equilibrium constant** for this reaction under the particular conditions.

In another general example:

$$aA + bB \rightleftharpoons cC + dD$$

the reaction rates can be expressed as

$$R = k_1[A]^a \times [B]^b$$
$$R' = k_2[C]^c \times [D]^d$$

Note that the values of k_1 and k_2 are different, but that each is a constant for the conditions of the reaction. At the start of the reaction, [A] and [B] will be at their greatest values, and R will be large; [C], [D], and R' will be zero. Gradually R will decrease and R' will become equal. At this point the reverse reaction is forming the original reactants just as rapidly as they are being used up by the forward reaction. Therefore no further change in R, R', or any of the concentrations will occur.

If we set R' equal to R, we have:

$$k_2 \times [C]^c \times [D]^d = k_1 \times [A]^a \times [B]^b$$

or

$$\frac{[C]^c \times [D]^d}{[A]^a \times [B]^b} = \frac{k_1}{k_2} = K_{eq}$$

This process of two substances, A and B, reacting to form products C and D and the reverse can be shown graphically to represent what happens as equilibrium is established. The hypothetical equilibrium reaction is described by the following general equation:

$$aA + bB \rightleftharpoons cC + dD$$

At the beginning (time t_0), the concentrations of C and D are zero and those of A and B are maximum. The graph below shows that over time the rate of the forward reaction decreases as A and D are used up. Meanwhile, the rate of the reverse reaction increases as C and D are formed. When these two reaction rates become equal (at time t_1), equilibrium is established. The individual concentrations of A, B, C, and D no longer change if conditions remain the same. To an observer, it appears that all reaction has stopped; in fact, however, both reactions are occurring at the same rate.

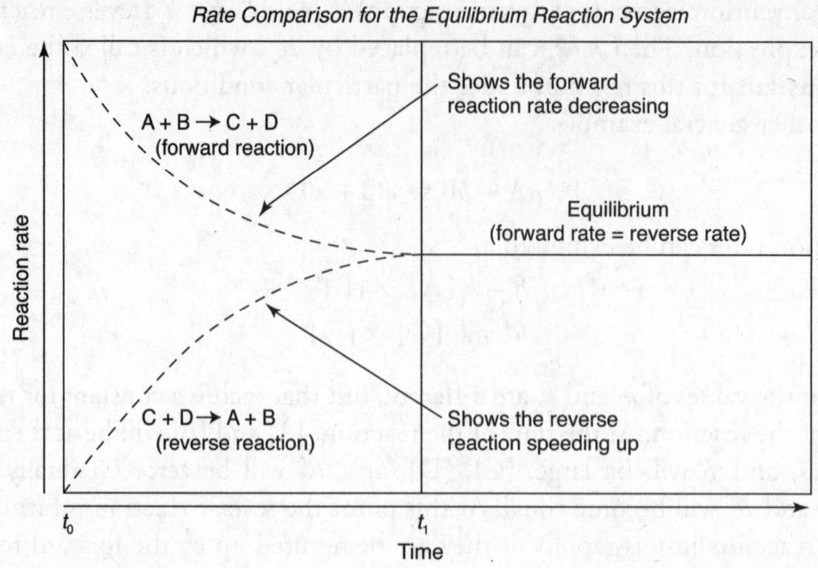

We see that, for the given reaction and the given conditions, K_{eq} is a constant, called the **equilibrium constant**. If K_{eq} is large, this means that equilibrium does not occur until the concentrations of the original reactants are small and those of the products large. A small value of K_{eq} means that equilibrium occurs almost at once and relatively little product is produced.

The equilibrium constant, K_{eq}, has been determined experimentally for many reactions, and the values are given in chemical handbooks.

Suppose we find K_{eq} for reacting H_2 and I_2 at 490°C to be equal to 45.9. Then the equilibrium constant for this reaction

$$H_2 + I_2 \rightleftharpoons 2HI \text{ at } 490°C$$

is

$$K_{eq} = \frac{[HI]^2}{[H_2][I_2]} = 45.9$$

SAMPLE PROBLEM: Three moles of H_2 and 3 moles of I_2 are introduced into a 1-liter box at a temperature of 490°C. Find the concentration of each substance in the box when equilibrium is established.

Initial conditions:

$$[H_2] = 3 \text{ mol/L}$$
$$[I_2] = 3 \text{ mol/L}$$
$$[HI] = 0 \text{ mol/L}$$

The reaction proceeds to equilibrium and

$$K_{eq} = \frac{[HI]^2}{[H_2][I_2]} = 45.9$$

At equilibrium, then,

$[H_2] = (3 - x)$ mol/L (where x is the number of moles of H_2 that are in the form of HI at equilibrium)

$[I_2] = (3 - x)$ mol/L (the same x is used since 1 mol of H_2 requires 1 mol of I_2 to react to form 2 mol of HI)

$[HI] = 2x$ mol/L

Then

$$K_{eq} = \frac{(2x)^2}{(3-x)(3-x)} = 45.9$$

If

$$\frac{(2x)^2}{(3-x)^2} = 45.9$$

then taking the square root of each side gives

$$\frac{2x}{3-x} = 6.77$$

Solving for x:

$$x = 2.32$$

> **Note:** On the SAT Subject Test in Chemistry, calculators are not permitted. So answers to questions like this would be "perfect squares."

Substituting this x value into the concentration expressions at equilibrium we have

$$[H_2] = (3 - x) = 0.68 \text{ mol/L}$$
$$[I_2] = (3 - x) = 0.68 \text{ mol/L}$$
$$[HI] = 2x = 4.64 \text{ mol/L}$$

The crucial step in this type of problem is setting up your concentration expressions from your knowledge of the equation. Suppose that this problem had been:

Find the concentrations at equilibrium for the same conditions as in the preceding example except that only 2 moles of HI are injected into the box.

$$[H_2] = 0 \text{ mol/L}$$
$$[I_2] = 0 \text{ mol/L}$$
$$[HI] = 2 \text{ mol/L}$$

At equilibrium

$$[HI] = (2 - x) \text{ mol/L}$$

(For every mole of HI that decomposes, only 0.5 mole of H_2 and 0.5 mole of I_2 are formed.)

$$[H_2] = 0.5x$$
$$[I_2] = 0.5x$$
$$K_{eq} = \frac{(2-x)^2}{(x/2)^2} = 45.9$$

Solving for x gives:

$$x = 0.456$$

Then, substituting into the equilibrium conditions,

$$[HI] = 2 - x = 1.54 \text{ mol/L}$$
$$[I_2] = \frac{1}{2}x = 0.228 \text{ mol/L}$$
$$[H_2] = \frac{1}{2}x = 0.228 \text{ mol/L}$$

LE CHÂTELIER'S PRINCIPLE

A general law, **Le Châtelier's Principle**, can be used to explain the results of applying any change of condition (stress) on a system in equilibrium. It states that, if a stress is placed upon a system in equilibrium, the equilibrium is displaced in the direction that counteracts the effect of the stress. An increase in concentration of a substance favors the reaction that uses up that substance and lowers its concentration. A rise in temperature favors the reaction that absorbs heat and so tends to lower the temperature. These ideas are further developed below.

EFFECTS OF CHANGING CONDITIONS

TIP

At equilibrium, K_{eq} stays the same at a given temperature.

Effect of Changing the Concentrations

When a system at equilibrium is disturbed by adding or removing one of the substances (thus changing its concentration), all the concentrations will change until a new equilibrium point is reached with the same value of K_{eq}.

If the concentration of a reactant in the forward action is increased, the equilibrium is displaced to the right, favoring the forward reaction. If the concentration of a reactant in the reverse reaction is increased, the equilibrium is displaced to the left. Decreases in concentration will produce effects opposite to those produced by increases.

TIP

Know how each factor affects equilibrium.

Effect of Temperature on Equilibrium

If the temperature of a given equilibrium reaction is changed, the reaction will shift to a new equilibrium point. If the temperature of a system in equilibrium is raised, the equilibrium is shifted in the direction that absorbs heat. Note that the shift in equilibrium as a result of temperature change is actually a change in the value of the equilibrium constant. This is different from the effect of changing the concentration of a reactant; when concentrations are changed, the equilibrium shifts to a condition that maintains the same equilibrium constant.

Effect of Pressure on Equilibrium

A change in pressure affects only equilibria in which a gas or gases are reactants or products. Le Châtelier's Law can be used to predict the direction of displacement. If it is assumed that the total space in which the reaction occurs is constant, the pressure will depend on the total number of molecules in that space. An increase in the number of molecules will increase pressure; a decrease in the number of molecules will decrease pressure. If the pressure is increased, the reaction that will be favored is the one that will lower the pressure, that is, decrease the number of molecules.

An example of the application of these principles is the Haber process of making ammonia. The reaction is

$$N_2 + 3H_2 \rightleftharpoons 2NH_3 + \text{heat (at equilibrium)}$$

If the concentrations of the nitrogen and hydrogen are increased, the forward reaction is increased. At the same time, if the ammonia produced is removed by dissolving it into water, the forward reaction is again favored.

Because the reaction is exothermic, the addition of heat must be considered with care. Increasing the temperature causes an increase in molecular motion and collisions, thus allowing the product to form more readily. At the same time, the equilibrium equation shows that the reverse reaction is favored by the increased temperature, so a compromise temperature of about 500°C is used to get the best yield.

An increase in pressure will cause the forward reaction to be favored since the equation shows that four molecules of reactants are forming two molecules of products. This effect tends to reduce the increase in pressure by the formation of more ammonia.

EQUILIBRIA IN HETEROGENEOUS SYSTEMS

The examples so far have involved systems made up of only gaseous substances. Expression of the K_{eq} values of systems changes when other phases are present.

Equilibrium Constant for Systems Involving Solids

If the experimental data for this reaction are studied:

$$CaCO_3(s) \rightleftharpoons CaO(s) + CO_2(g)$$

it is found that at a given temperature an equilibrium is established in which the concentration of CO_2 is constant. It is also true that the concentrations of the solids have no effect on the CO_2 concentration as long as both solids are present. Therefore the K_{eq}, which would conventionally be written like this:

$$K_{eq} = \frac{[CaO][CO_2]}{[CaCO_3]}$$

TIP

If solids are present, their concentrations are incorporated into $K_{eq} = K$.

can be modified by incorporating the concentrations of the two solids. This can be done since the concentration of solids is fixed by the density. Then K_{eq} becomes a new constant, K:

$$K = [CO_2]$$

Any heterogeneous reaction involving gases does not include the concentrations of pure solids. As another example, K for the reaction

$$NH_4Cl(s) \rightleftharpoons NH_3(g) + HCl(g)$$

is

$$K = [NH_3][HCl]$$

Acid Ionization Constants

When a weak acid does not ionize completely in a solution, an equilibrium is reached between the acid molecule and its ions. The mass action expression can be used to derive an equilibrium constant, called the **acid ionization constant**, for this condition. For example, an acetic acid solution ionizing is shown as

$$HC_2H_3O_2 + H_2O \rightleftharpoons H_3O^+ + C_2H_3O_2^-$$

$$K = \frac{[H_3O^+][C_2H_3O_2^-]}{[HC_2H_3O_2][H_2O]}$$

TIP

K_a incorporates the concentration of water.

The concentration of water in moles/liter is found by dividing the mass of 1 liter of water (which is 1,000 g at 4°C) by its gram-molecular mass, 18 grams, giving H_2O a value of 55.6 moles/liter. Because this number is so large compared with the other numbers involved in the equilibrium constant, it is practically constant and is incorporated into a new equilibrium constant, designated as K_a. The new expression is

$$K_a = \frac{[H_3O^+][C_2H_3O_2^-]}{[HC_2H_3O_2]}$$

Ionization constants have been found experimentally for many substances and are listed in chemical tables. The ionization constants of ammonia and acetic acid are about 1.8×10^{-5}. For boric acid $K_a = 5.8 \times 10^{-10}$, and for carbonic acid $K_a = 4.3 \times 10^{-7}$.

If the concentrations of the ions present in the solution of a weak electrolyte are known, the value of the ionization constant can be calculated. Also, if the value of K_a is known, the concentrations of the ions can be calculated.

A small value for K_a means that the concentration of the un-ionized molecule must be relatively large compared with the ion concentrations. Conversely, a large value for K_a means that the concentrations of ions are relatively high. Therefore the smaller the ionization constant of an acid, the weaker the acid. Thus, of the three acids referred to above, the ionization constants show that the weakest is boric acid, and the strongest, acetic acid. It should be remembered that, in all cases where ionization constants are used, the electrolytes must be weak in order to be involved in ionic equilibria.

Ionization Constant of Water

Because water is a very weak electrolyte, its ionization constant can be expressed as follows:

$$2H_2O \rightleftharpoons H_3O^+ + OH^-$$

$$\text{(Equilibrium constant) } K = \frac{[H_3O^+][OH^-]}{[H_2O]^2} \quad \begin{array}{l}\text{(since } [H_2O]^2 \text{ remains} \\ \text{relatively constant,} \\ \text{it is incorporated into } K_w)\end{array}$$

TIP

K_w incorporates the $[H_2O]^2$.

$$\text{(Ionization constant) } K_w = [H_3O^+][OH^-] = 1 \times 10^{-14} \text{ at } 25°C$$

From this expression, we see that for distilled water $[H_3O^+] = [OH^-] = 1 \times 10^{-7}$. Therefore the pH, which is $-\log [H_3O^+]$, is

$$pH = -\log [1 \times 10^{-7}]$$
$$pH = -[-7] = 7 \text{ for a neutral solution}$$

The pH range of 1 to 6 is acid, and the pH range of 8 to 14 is basic.

SAMPLE PROBLEM: (This sample incorporates the entire discussion of ionization constants, including finding the pH.)

Calculate (a) the $[H_3O^+]$, (b) the pH, and (c) the percentage dissociation for 0.100 M acetic acid at 25°C. The symbol K_a is used for the ionization of acids. K_a for $HC_2H_3O_2$ is 1.8×10^{-5}.

(a) For this reaction

$$H_2O(\ell) + HC_2H_3O_2(\ell) \rightleftharpoons H_3O^+(aq) + C_2H_3O_2^-(aq)$$

and

$$K_a = \frac{[H_3O^+][C_2H_3O_2^-]}{[HC_2H_3O_2]} = 1.8 \times 10^{-5}$$

Let x = number of moles/liter of $HC_2H_3O_2$ that dissociate and reach equilibrium. Then

$$[H_3O^+] = x, \ [C_2H_3O_2^-] = x, \ [HC_2H_3O_2] = 0.1 - x$$

Substituting in the expression for K_a gives

$$K_a = 1.8 \times 10^{-5} = \frac{(x)(x)}{0.10 - x}$$

Because a weak acid, such as acetic, at concentrations of 0.01 M or greater dissociates very little, the equilibrium concentration of the acid is very nearly equal to the original concentration, that is,

$$0.10 - x \cong 0.10$$

Therefore, the expression can be changed to

$$1.8 \times 10^{-5} = \frac{(x)(x)}{0.10}$$
$$x^2 = 1.8 \times 10^{-6}$$
$$x = 1.3 \times 10^{-3} = [H_3O^+]$$

(b) Substituting this result in the pH expression gives

$$pH = -\log [H_3O^+] = -\log [1.3 \times 10^{-3}]$$
$$pH = 3 - \log 1.3$$
$$pH = 2.9$$

(c) The percentage of dissociation of the original acid may be expressed as

$$\% \text{ dissociation} = \frac{\text{moles/liter that dissociate}}{\text{original concentration}}(100)$$
$$\% \text{ dissociation} \times \frac{1.3 \times 10^{-3}}{1.0 \times 10^{-1}}(100) = 1.3\%$$

Solubility Products

A saturated solution of a substance has been defined as an equilibrium condition between the solute and its ions. For example:

> **TIP**
>
> K_{sp} incorporates the concentration of the solute.

$$AgCl(s) \rightleftharpoons Ag^+(aq) + Cl^-(aq)$$

The equilibrium constant would be:

$$K = \frac{[Ag^+][Cl^-]}{[AgCl]}$$

Since the concentration of the solute remains constant for that temperature, the [AgCl] is incorporated into the K to give the K_{sp}, called the **solubility constant:**

$$K_{sp} = [Ag^+][Cl^-] = 1.2 \times 10^{-10} \text{ at } 25°C$$

This setup can be used to solve problems in which the ionic concentrations are given and the K_{sp} is to be found or the K_{sp} is given and the ionic concentrations are to be determined.

TYPE PROBLEM: Finding the K_{sp}.

By experimentation it is found that a saturated solution of $BaSO_4$ at 25°C contains 3.9×10^{-5} mole/liter of Ba^{2+} ions. Find the K_{sp} of this salt.

Since $BaSO_4$ ionizes into equal numbers of Ba^{2+} and SO_4^{2-}, the barium ion concentration will equal the sulfate ion concentration. Then the solution is

$$BaSO_4(s) \rightleftharpoons Ba^{2+}(aq) + SO_4^{2-}(aq)$$

and

$$K_{sp} = [Ba^{2+}][SO_4^{2-}]$$

Therefore

$$K_{sp} = (3.9 \times 10^{-5})(3.9 \times 10^{-5}) = 1.5 \times 10^{-9}$$

ANOTHER TYPE PROBLEM: Finding the solubility.

If the K_{sp} of radium sulfate, $RaSO_4$, is 4×10^{-11}, calculate the solubility of the compound in pure water. Let x = moles of $RaSO_4$ that dissolve per liter of water. Then, in the saturated solution,

$$[Ra^{2+}] = x \text{ mol/L}$$
$$[SO_4^{2-}] = x \text{ mol/L}$$
$$RaSO_4(s) \rightleftharpoons Ra^{2+}(aq) + SO_4^{2-}(aq)$$
$$[Ra^{2+}][SO_4^{2-}] = K_{sp} = 4 \times 10^{-11}$$

Let x = $[Ra^{2+}]$ and $[SO_4^{2-}]$. Then

$$(x)(x) = 4 \times 10^{-11} = 40 \times 10^{-12}$$
$$x = 6 \times 10^{-6} \text{ mol/L}$$

Thus the solubility of $RaSO_4$ is 6×10^{-6} mole/liter of water, for a solution 6×10^{-6} M in Ra^{2+} and 6×10^{-6} M in SO_4^{2-}.

ANOTHER TYPE PROBLEM: Predicting the formation of a precipitate.

In some cases, the solubility products of solutions can be used to predict the formation of a precipitate.

Suppose we have two solutions. One solution contains 1.00×10^{-3} mole of silver nitrate, $AgNO_3$, per liter. The other solution contains 1.00×10^{-2} mole of sodium chloride, NaCl, per liter. If 1 liter of the $AgNO_3$ solution and

1 liter of the NaCl solution are mixed to make a 2-liter mixture, will a precipitate of AgCl form?

In the $AgNO_3$ solution, the concentrations are:

$$[Ag^+] = 1.00 \times 10^{-3} \text{ mol/L and } [NO_3^-] = 1.00 \times 10^{-3} \text{ mol/L}$$

In the NaCl solution, the concentrations are:

$$[Na^+] = 1.00 \times 10^{-2} \text{ mol/L and } [Cl^-] = 1.00 \times 10^{-2} \text{ mol/L}$$

When 1 liter of one of these solutions is mixed with 1 liter of the other solution to form a total volume of 2 liters, the concentrations will be halved.

In the mixture then, the initial concentrations will be:

$$[Ag^+] = 0.50 \times 10^{-3} \quad \text{or} \quad 5.0 \times 10^{-4} \text{ mol/L}$$
$$[Cl^-] = 0.50 \times 10^{-2} \quad \text{or} \quad 5.0 \times 10^{-3} \text{ mol/L}$$

For the K_{sp} of AgCl,

$$[Ag^+][Cl^-] = [5.0 \times 10^{-4}][5.0 \times 10^{-3}]$$
$$[Ag^+][Cl^-] = 25 \times 10^{-7} \quad \text{or} \quad 2.5 \times 10^{-6}$$

This is far greater than 1.7×10^{-10}, which is the K_{sp} of AgCl. These concentrations cannot exist, and Ag^+ and Cl^- will combine to form solid AgCl precipitate. Only enough Ag^+ ions and Cl^- ions will remain to make the product of the respective ion concentrations equal 1.7×10^{-10}.

COMMON ION EFFECT

When a reaction has reached equilibrium, and an outside source adds more of one of the ions that is already in solution, the result is to cause the reverse reaction to occur at a faster rate and reestablish the equilibrium. This is called the **common ion effect**. For example, in this equilibrium reaction:

$$NaCl(s) \rightleftharpoons Na^+ + Cl^-$$

the addition of concentrated HCl (12 M) adds H^+ and Cl^- both at a concentration of 12 M. This increases the concentration of the Cl^- and disturbs the equilibrium. The reaction will shift to the left and cause some solid NaCl to come out of solution.

The "common" ion is the one already present in an equilibrium before a substance is added that increases the concentration of that ion. The effect is to reverse the solution reaction and to decrease the solubility of the original substance, as shown in the above example.

DRIVING FORCES OF REACTIONS

Relation of Minimum Energy (Enthalpy) to Maximum Disorder (Entropy)

Some reactions are said to go to completion because the equilibrium condition is achieved when practically all the reactants have been converted to products. At the other extreme, some reactions reach equilibrium immediately with very little product being formed. These two examples are representative of very large K values and very small K values, respectively. There are essentially two driving forces that control the extent of a reaction and determine when equilibrium will be established. These are the drive to the lowest heat content, or **enthalpy**, and the drive to the greatest randomness or disorder, which is called **entropy**. Reactions with negative ΔH's (enthalpy or heat content) are exothermic, and reactions with positive ΔS's (entropy or randomness) are proceeding to greater randomness.

The **Second Law of Thermodynamics** states that the entropy of the universe increases for any spontaneous process. This means that the entropy of a system may increase or decrease but that, if it decreases, then the entropy of the surroundings must increase to a greater extent so that the overall change in the universe is positive. In other words,

$$\Delta S_{universe} = \Delta S_{system} + \Delta S_{surroundings}$$

The following is a list of conditions in which ΔS is positive for the system:

TIP

When the ΔS is positive for the system, it means greater disorder.

1. When a gas is formed from a solid, for example,

$$CaCO_3(s) \rightarrow CaO(s) + CO_2(g).$$

2. When a gas is evolved from a solution, for example,

$$Zn(s) + 2H^+(aq) \rightarrow H_2(g) + Zn^{2+}(aq).$$

3. When the number of moles of gaseous product exceeds the moles of gaseous reactant, for example,

$$2C_2H_6(g) + 7O_2(g) \rightarrow 4CO_2(g) + 6H_2O(g).$$

4. When crystals dissolve in water, for example,

$$NaCl(s) \rightarrow Na^+(aq) + Cl^-(aq).$$

Looking at specific examples, we find that in some cases endothermic reactions occur when the products provide greater randomness or positive entropy. This reaction is an example:

$$CaCO_3(s) \rightleftharpoons CaO(s) + CO_2(g)$$

The production of the gas and thus greater entropy might be expected to take this reaction almost to completion. However, this does not occur because another force is hampering this reaction. It is the absorption of energy, and thus the increase in enthalpy, as the $CaCO_3$ is heated.

The equilibrium condition, then, at a particular temperature, is a compromise between the increase in entropy and the increase in enthalpy of the system.

The Haber process of making ammonia is another example of this compromise of driving forces that affect the establishment of an equilibrium. In this reaction

$$N_2(g) + 3H_2(g) \rightleftharpoons 2NH_3(g) + \text{heat}$$

the forward reaction to reach the lowest heat content and thus release energy cannot go to completion because the force to maximum randomness is driving the reverse reaction.

TIP

Free energy, ΔG, depends on ΔH (enthalpy) and ΔS (entropy).

Change in Free Energy of a System—the Gibbs Equation

These factors, enthalpy and entropy, can be combined in an equation that summarizes the change of **free energy** in a system. This is designated as ΔG. The relationship is

$$\Delta G = \Delta H - T\Delta S \quad (T \text{ is temperature in kelvins})$$

and is called the **Gibbs free-energy equation.**

The sign of ΔG can be used to predict the spontaneity of a reaction at constant temperature and pressure. If ΔG is negative, the reaction is (probably) spontaneous; if ΔG is positive, the reaction is improbable; and if ΔG is 0, the system is at equilibrium and there is no net reaction.

The ways in which the factors in the equation affect ΔG are shown in this table:

TIP

Know these relationships.

ΔH	ΔS	ΔG	Will It Happen?	Comment
Exothermic (–)	+	Always –	Yes	No exceptions
Exothermic (–)	–	– At lower temperatures	Probably	At low temperature
Endothermic (+)	+	– At higher temperatures	Probably	At high temperature
Endothermic (+)	–	Never –	No	No exceptions

This drive to achieve a minimum of free energy may be interpreted as the driving force of a chemical reaction.

Chapter Summary

The following terms summarize all the concepts and ideas that were introduced in this chapter. You should be able to explain their meaning and how you would use them in chemistry. They appear in boldface type in this chapter to draw your attention to them. The boldface type also makes it easier for you to look them up if you need to. You could also use the search or "Google" action on your computer to get a quick and expanded explanation of these terms, laws, and formulas.

acid ionization constant
common ion effect
enthalpy (ΔH)
entropy (ΔS)
equilibrium
equilibrium constant
free energy
Gibbs free-energy equation
solubility product constant

Le Châtelier's Principle
Second Law of Thermodynamics

Practice Exercises

1. For the reaction $A + B \rightleftharpoons C + D$, the equilibrium constant can be expressed as:

   (A) $K_{eq} = \dfrac{[A][B]}{[C][D]}$

   (B) $K_{eq} = \dfrac{[C][B]}{[A][D]}$

   (C) $K_{eq} = \dfrac{[C][D]}{[A][B]}$

   (D) $K_{eq} = \dfrac{C \cdot D}{A \cdot B}$

2. The concentrations in an expression of the equilibrium constant are given in

   (A) mol/mL
   (B) g/L
   (C) gram-equivalents/L
   (D) mol/L

3. In the equilibrium expression for the reaction

   $$BaSO_4(s) \rightleftharpoons Ba^{2+}(aq) + SO_4^{2-}(aq)$$

   K_{sp} is equal to

   (A) $[Ba^{2+}][SO_4^{2-}]$

   (B) $\dfrac{[Ba^{2+}][SO_4^{2-}]}{BaSO_4}$

   (C) $\dfrac{[Ba^{2+}][SO_4^{2-}]}{[BaSO_4]}$

   (D) $\dfrac{[BaSO_4]}{[Ba^{2+}][SO_4^{2-}]}$

4. The K_w of water at 298 K is equal to

(A) 1×10^{-7}
(B) 1×10^{-17}
(C) 1×10^{-14}
(D) 1×10^{-1}

5. The pH of a solution that has a hydrogen ion concentration of 1×10^{-4} mole/liter is

(A) 4
(B) −4
(C) 10
(D) −10

6. The pH of a solution that has a hydroxide ion concentration of 1×10^{-4} mole/liter is

(A) 4
(B) −4
(C) 10
(D) −10

7. A small value for K_{eq}, the equilibrium constant, indicates that

(A) the concentration of the un-ionized molecules must be relatively small compared with the ion concentrations
(B) the concentration of the ionized molecules must be larger than the ion concentrations
(C) the substance ionizes to a large degree
(D) the concentration of the un-ionized molecules must be relatively large compared with the ion concentrations

8. In the Haber process for making ammonia, an increase in pressure favors

(A) the forward reaction
(B) the reverse reaction
(C) neither reaction

9. A change in which of these conditions will change the *K* of an equilibrium?

(A) Temperature
(B) Pressure
(C) Concentration of reactants
(D) Concentration of products

10. If $Ca(OH)_2$ is dissolved in a solution of NaOH, its solubility, compared with that in pure water, is

(A) increased
(B) decreased
(C) unaffected

The following questions are in the format that is used on the **SAT Subject Area Test in Chemistry**. If you are not familiar with these types of questions, study pages xii–xv before doing the remainder of the review questions.

Directions: The following set of lettered choices refers to the numbered questions immediately below it. For each numbered item, choose the one lettered choice that fits it best. Every choice in the set may be used once, more than once, or not at all.

Questions 11–15

(A) the free energy is always negative
(B) the free energy is negative at lower temperatures
(C) the free energy is negative at high temperatures
(D) the free energy is never negative
(E) the system is at equilibrium, and there is no net reaction

Complete the sentence with the appropriate choice.

11. When enthalpy is negative and entropy is positive,
12. When enthalpy is positive and entropy is positive,
13. When enthalpy is negative and entropy is negative,
14. When enthalpy is positive and entropy is negative,
15. When ΔG, free energy, is zero,

Answers and Explanations

1. **(C)** The correct setup of K_{eq} is the product of the concentration of the products over the products of the reactants.

2. **(D)** The concentrations in an expression of the equilibrium constant are given in moles/liter (mol/L).

3. **(A)** This reaction is the equilibrium between a precipitate and its ions in solution. Because $BaSO_4$ is a solid, it does not appear in the solubility product expression. Therefore, $K_{sp} = [Ba^{+2}][SO_4^{2-}]$.

4. **(C)** This is the K_w expression of water.

5. **(A)** Because the pH is defined as the negative of the log of the H^+ concentration, it is –(log of 10^{-4}), which is –(–4) or 4.

6. **(C)** pH + pOH = 14. In this problem, the pOH = –log [OH] = –log [10^{-4}] = –(–4) = 4. Placing this value in the equation, you have pH + 4 = 14. Solving for pH, you get 10.

7. **(D)** For K_{eq} to be a small value, the numerator of the expression must be small compared with the denominator. Because the numerator is the product of the concentrations of the ion concentrations and the denominator is the product of the un-ionized molecules, the concentration of the un-ionized molecules must be relatively large compared with the ion concentrations.

8. **(A)** In the reaction for the formation of ammonia, $N_2 + 3H_2 \leftrightarrow 2NH_3$ + heat (at equilibrium) An increase in pressure will cause the forward reaction to be favored, because the equation shows that four molecules of reactants are forming two molecules of products. This effect tends to reduce the increase in pressure by the formation of more ammonia.

9. **(A)** Only the changing of the temperature of the equilibrium reaction will change the *K* of the equilibrium.

10. **(B)** Because there already is a concentration of $(OH)^-$ ions from the NaOH in solution, this common ion effect will decrease the solubility of the $Ca(OH)_2$.

11. **(A)** When enthalpy is negative and entropy is positive, the free energy is always negative.

12. **(C)** When enthalpy is positive and entropy is positive, the free energy is negative at high temperatures.

13 **(B)** When enthalpy is negative and entropy is negative, the free energy is negative at lower temperatures.

14. **(D)** When enthalpy is positive and entropy is negative, the free energy is never negative.

15. **(E)** When ΔG, free energy, is zero, the system is at equilibrium and there is no net reaction.

Acids, Bases, and Salts CHAPTER 11

- Describe the properties of an Arrhenius acid and base, and know the name, formula, and degree of ionization of common acids and bases.
- Explain the Brønsted-Lowry Theory of acids and conjugate bases.
- Draw examples and explain the Lewis Theory of acids–bases.
- Determine pH and pOH of solutions.
- Solve titration problems and the use of indicators in the process.
- Describe how a buffer works.
- Explain the formation and naming of salts.
- Explain amphoteric substances in relation to acid–base theory.

What defines an acid? A base? A salt? You must know the different ways that these interact. You must also know how to determine the pH of a substance and how to neutralize that same substance.

DEFINITIONS AND PROPERTIES

Acids

There are some characteristic properties by which an acid may be defined. The most important are:

1. ***Water (aqueous) solutions of acids conduct electricity.*** The degree of conduction depends on the acid's degree of ionization. A few acids ionize almost completely, while others ionize to only a slight degree. The table below indicates some common acids and their degrees of ionization.

2. ***Acids will react with metals that are more active than hydrogen ions (see Table 8, page 210) to liberate hydrogen.*** (Some acids are also strong oxidizing agents and will not release hydrogen. Somewhat concentrated nitric acid is such an acid.)

3. ***Acids have the ability to change the color of indicators.*** Some common indicators are litmus and phenolphthalein. **Litmus** is a dyestuff obtained from plant life. When litmus is added to an acidic solution, or paper impregnated with litmus is dipped into an acid, the neutral purple color changes to pink-red. **Phenolphthalein** is pink in a basic solution and becomes colorless in a neutral or acid solution.

Degrees of Ionization of Common Acids

Completely or Nearly Completely Ionized	Moderately Ionized	Slightly Ionized
Nitric	Oxalic	Hydrofluoric
Hydrochloric	Phosphoric	Acetic
Sulfuric	Sulfurous	Carbonic
Hydriodic		Hydrosulfuric
Hydrobromic		(Most others)

4. ***Acids react with bases so that the properties of both are lost to form water and a salt.*** This is called **neutralization**. The general equation is:

$$\text{Acid} + \text{Base} \rightarrow \text{Salt} + \text{Water}$$

An example is:

$$Mg(OH)_2 + H_2SO_4 \rightarrow MgSO_4 + 2H_2O$$

5. ***Acids react with carbonates to release carbon dioxide.*** An example:

$$CaCO_3 + 2HCl \rightarrow CaCl_2 + H_2CO_3 \text{ (unstable and decomposes)}$$
$$\rightarrow H_2O + CO_2(g)$$

The most common theory used in first-year chemistry is the **Arrhenius Theory**, which states that an acid is a substance that yields hydrogen ions in an aqueous solution. Although we speak of these hydrogen ions in the solution, they are really not separate ions but become attached to the oxygen of the polar water molecule to form the H_3O^+ ion (the hydronium ion). Thus, it is really this hydronium ion we are concerned with in an acid solution.

The general reaction for the dissociation of an acid, HX, is commonly written as

$$HX \rightleftharpoons H^+ + X^-$$

To show the formation of the hydronium ion, H_3O^+, the complete equation is:

$$HX + H_2O \rightleftharpoons H_3O^+ + X^-$$

A list of common acids and their formulas is given in Chapter 4, Table 7; an explanation of the naming procedures for acids precedes Table 7.

Bases

Bases may also be defined by some operational definitions that are based on experimental observations. Some of the important ones are as follows:

1. ***Bases are conductors of electricity in an aqueous solution.*** Their degrees of conduction depend on their degrees of ionization. The degrees of ionization of some common bases are shown in the table below.

Degrees of Ionization of Common Bases

Completely or Nearly Completely Ionized	Slightly Ionized
Potassium hydroxide	Ammonium hydroxide
Sodium hydroxide	(All others)
Barium hydroxide	
Strontium hydroxide	
Calcium hydroxide	

2. ***Bases cause a color change in indicators.*** Litmus changes from red to blue in a basic solution, and phenolphthalein turns pink from its colorless state.
3. ***Bases react with acids to neutralize each other and form a salt and water.***
4. ***Bases react with fats to form a class of compounds called soaps.*** Earlier generations used this method to make their own soap.
5. ***Aqueous solutions of bases feel slippery, and the stronger bases are very caustic to the skin.***

The Arrhenius Theory defines a base as a substance that yields hydroxide ions (OH^-) in an aqueous solution.

Some common bases have familiar names, for example:

Sodium hydroxide	= lye, caustic soda
Potassium hydroxide	= caustic potash
Calcium hydroxide	= slaked lime, hydrated lime, limewater
Ammonium hydroxide	= ammonia water, household ammonia

Much of the sodium hydroxide produced today comes from the Hooker cell electrolysis apparatus. The electrolysis process for the decomposition of water was discussed in Chapter 7. When an electric current is passed through a saltwater solution, hydrogen, chlorine, and sodium hydroxide are the products. The formula for this equation is:

$$2NaCl + 2HOH \xrightarrow[\text{energy}]{\text{electrical}} H_2(g) + Cl_2(g) + 2NaOH$$

Broader Acid-Base Theories

Besides the common Arrhenius Theory of acids and bases discussed for aqueous solutions, two other theories, the Brønsted-Lowry Theory and the Lewis Theory, are widely used.

The Brønsted-Lowry Theory (1923) defines acids as proton donors and bases as proton acceptors. This definition agrees with the aqueous solution definition of an acid giving up hydrogen ions in solution, but goes beyond to other cases as well.

An example is the reaction of dry HCl gas with ammonia gas to form the white solid NH_4Cl.

$$HCl(g) + NH_3(g) \rightarrow NH_4^+(s) + Cl^-(s)$$

The HCl is the proton donor or acid, and the ammonia is a Brønsted-Lowry base that accepts the proton.

Conjugate Acids and Bases

In an acid–base reaction, the original acid gives up its proton to become a **conjugate base.** In other words, after losing its proton, the remaining ion is capable of gaining a proton, thus qualifying as a base. The original base accepts a proton, so it now is classified as a **conjugate acid** since it can release this newly acquired proton and thus behave like an acid.

Some examples are given below:

$$\underset{\text{Acid}}{HCl} + \underset{\text{Base}}{NH_3} \longrightarrow \underset{\text{Conjugate acid}}{NH_4^+} + \underset{\text{Conjugate base}}{Cl^-}$$

* * * *

$$\underset{\text{Acid}}{HC_2H_3O_2} + \underset{\text{Base}}{H_2O} \rightleftharpoons \underset{\text{Conjugate acid}}{H_3O^+} + \underset{\text{Conjugate base}}{C_2H_3O_2^-}$$

Strength of Conjugate Acids and Bases

The extent of the reaction between a Brønsted-Lowry acid and base depends on the relative strengths of the acids and bases involved. Consider the following example. Hydrochloric is a strong acid. It gives up protons readily. It follows that the Cl–ion has little tendency to attract and retain a proton. Consequently, the Cl–ion is an extremely weak base.

$$\underset{\textbf{strong acid}}{HCl(g)} + \underset{\textbf{base}}{H_2O(l)} \rightarrow \underset{\textbf{acid}}{H_3O^+(aq)} + \underset{\textbf{weak base}}{Cl^-(aq)}$$

This observation leads to an important conclusion: the stronger an acid is, the weaker its conjugate base; the stronger a base is, the weaker its conjugate acid. This concept allows strengths of different acids and bases to be compared to predict the outcome of a reaction. As an example, consider the reaction of perchloric acid, $HClO_4$, and water.

$$\underset{\textbf{stronger acid}}{HClO_4\,(aq)} + \underset{\textbf{stronger base}}{H_2O(l)} \rightarrow \underset{\textbf{weaker acid}}{H_3O^+(aq)} + \underset{\textbf{weaker base}}{ClO_4^-\,(aq)}$$

Another important general conclusion is that proton-transfer reactions favor the production of the weaker acid and the weaker base. For a reaction to approach completion, the reactants must be much stronger as an acid and as a base than the products.

The Lewis Theory (1916) defines acids and bases in terms of the electron-pair concept, which is probably the most generally useful concept of acids and bases. According to the Lewis definition, an acid is an electron-pair acceptor; and a base is an electron-pair donor. An example is the formation of ammonium ions from ammonia gas and hydrogen ions.

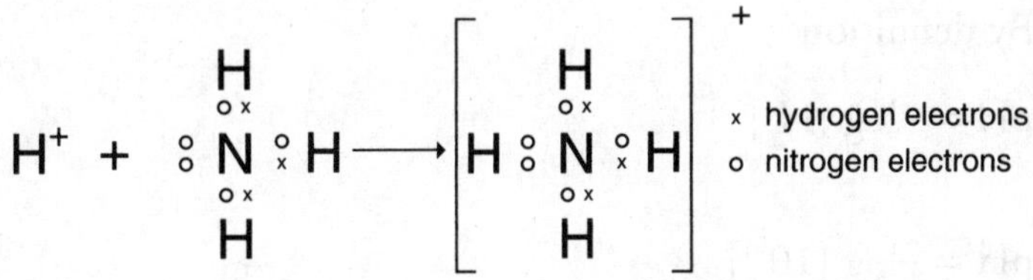

Notice that the hydrogen ion is in fact accepting the electron pair of the ammonia, so it is a Lewis acid. The ammonia is donating its electron pair, so it is a Lewis base.

Another example is boron trifluoride. It is an excellent Lewis acid. It forms a fourth covalent bond with many molecules and ions. Its reaction with a fluoride ion is shown below.

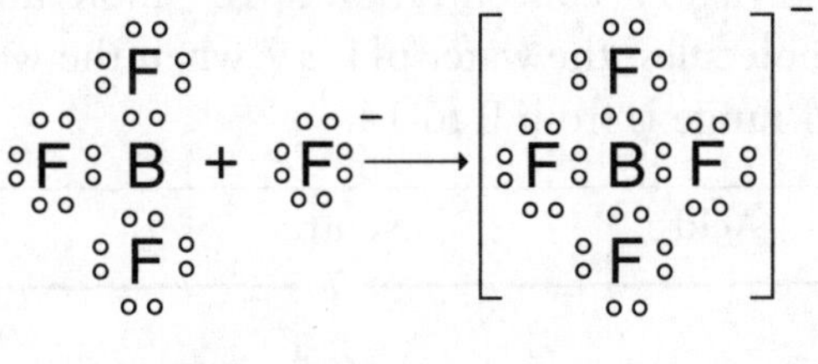

$$BF_3(aq) + F^-(aq) \rightarrow BF_4^-(aq)$$

The acid–base systems are summarized below.

Type	Acid	Base
Arrhenius	H^+ or H_3O^+ producer	OH– producer
Brønsted-Lowry	proton (H^+) donor	proton (H^+) acceptor
Lewis	electron-pair acceptor	electron-pair donor

Acid Concentration Expressed as pH

Frequently, acid and base concentrations are expressed by means of the **pH** system. The pH can be defined as –log $[H^+]$, where $[H^+]$ is the concentration of hydrogen ions expressed in moles per liter. The logarithm is the exponent of 10 when the number is written in the base 10. For example:

> **TIP**
>
> Remember:
> $pH = -\log[H_3O^+]$

$$100 = 10^2 \text{ so logarithm of 100, base } 10 = 2$$
$$10{,}000 = 10^4 \text{ so logarithm of 10,000, base } 10 = 4$$
$$0.01 = 10^{-2} \text{ so logarithm of 0.01, base } 10 = -2$$

The logarithms of more complex numbers can be found in a logarithm table. An example of a pH problem is:

Find the pH of a 0.1 molar solution of HCl.

1st step. Because HCl ionizes almost completely into H^+ and Cl^-, $[H^+] = 0.1$ mole/liter.

2nd step. By definition

$$pH = -\log [H^+]$$

so

$$pH = -\log [10^{-1}]$$

3rd step. The logarithm of 10^{-1} is –1

so

$$pH = -(-1)$$

4th step. The pH then is = 1.

Because water has a normal H^+ concentration of 10^{-7} mole/liter because of the slight ionization of water molecules, the water pH is 7 when the water is neither acid nor base. The normal pH range is from 0 to 14.

	Acid	Neutral	Base	
pH = 0 ←	⟷	7	⟷	→ 14
$[H_3O]^+ = 10^0$ ←	⟷	10^{-7}	⟷	→ 10^{-14}

The pOH is the negative logarithm of the hydroxide ion concentration:

$$pOH = -\log\ [OH^-]$$

If the concentration of the hydroxide ion is 10^{-9} M, then the pOH of the solution is +9.

From the equation

$$[H^+][OH^-] = 1.0 \times 10^{-14} \text{ at } 298 \text{ K}$$

the following relationship can be derived:

$$pH + pOH = 14.00$$

In other words, the sum of the pH and pOH of an aqueous solution at 298 K must always equal 14.00. For example, if the pOH of a solution is 9.00, then its pH must be 5.00.

SAMPLE PROBLEM: What is the pOH of a solution whose pH is 3.0?

Substituting 3.0 for pH in the expression

$$pH + pOH = 14.0$$

gives

$$3.0 + pOH = 14.0$$
$$pOH = 14.0 - 3.0$$
$$pOH = 11.0$$

INDICATORS

Some indicators can be used to determine pH because of their color changes somewhere along this pH scale. Some common indicators and their respective color changes are given below.

Indicator	pH Range of Color Change	Color Below Lower pH	Color Above Higher pH
Methyl orange	3.1–4.4	Red	Yellow
Bromthymol blue	6.0–7.6	Yellow	Blue
Litmus	4.5–8.3	Red	Blue
Phenolphthalein	8.3–10.0	Colorless	Red/Pink

TIP

Notice that each indicator has its own range of color change.

Here is an example of how to read this chart: At pH values below 4.5, litmus is red; above 8.3, it is blue. Between these values, it is a mixture of the two colors.

In choosing an indicator for a titration, we need to consider if the solution formed when the end point is reached has a pH of 7. Depending on the type of acid and base used, the resulting hydrolysis of the salt formed may cause it to be slightly acidic, slightly basic, or neutral. If the titration is of a strong acid and a strong base, the end point will be at pH 7 and practically any indicator can be used. This is because the addition of 1 drop of either reagent will change the pH at the end point by about 6 units. For titrations of strong acids and weak bases, we need an indicator, such as methyl orange, that changes color between 3.1 and 4.4 in the acid region. In the titration of a weak acid and a strong base, we should use an indicator that changes in the basic range. Phenolphthalein is a suitable choice for this type of titration because it changes color in the pH 8.3 to 10.0 range.

TITRATION—VOLUMETRIC ANALYSIS

Knowledge of the concentrations of solutions and the reactions they take part in can be used to determine the concentrations of "unknown" solutions or solids. The use of volume measurement in solving these problems is called **titration.**

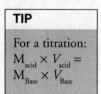

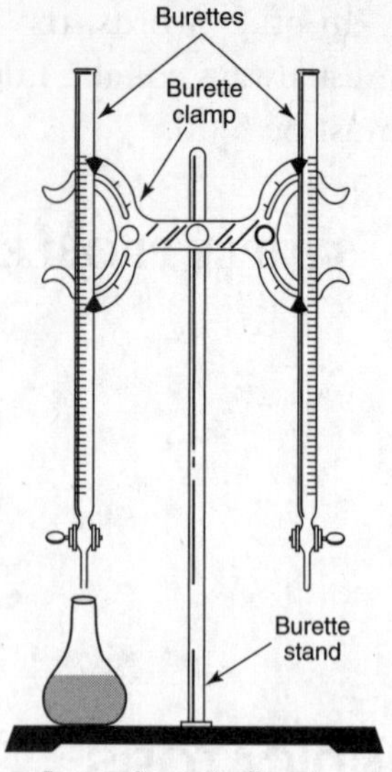

Burette Setup for Titration

A common example of a titration uses acid-base reactions. If you are given a base of known concentration, that is, a standard solution, let us say 0.10 M NaOH, and you want to determine the concentration of an HCl solution, you could **titrate** the solutions in the following manner.

First, introduce a measured quantity, 25.0 milliliters, of the NaOH into a flask by using a pipet or burette in a setup like the one in the accompanying diagram. Next, introduce 2 drops of a suitable indicator. Because NaOH and HCl are considered a strong base and a strong acid, respectively, an indicator that changes color in the middle pH range would be appropriate. Litmus solution would be one choice. It is blue in a basic solution but changes to red when the solution becomes acidic. Slowly introduce the HCl until the color change occurs. This point is called the **end point**. The point at which enough acid is added to neutralize all the standard solution in the flask is called the **equivalence point**.

Suppose 21.5 milliliters of HCl was needed to produce the color change. The reaction that occurred was

$$H^+(aq) + OH^-(aq) \rightarrow H_2O$$

until all the OH^- was neutralized; then the excess H^+ caused the litmus paper to change color.

TIP

For a titration: $M_{acid} \times V_{acid} = M_{Base} \times V_{Base}$

To solve the question of the concentration of NaOH, this equation is used:

$$M_{acid} \times V_{acid} = M_{base} \times V_{base}$$

Substituting the known amounts in this equation gives

$$x\,M_{acid} \times 21.5\text{ mL} = 0.1\text{ M} \times 25.0\text{ mL}$$
$$x = 0.116\text{ M}$$

In choosing an indicator for a titration, we need to consider whether the solution formed when the end point is reached has a pH of 7. Depending on the types of acid and base used, the resulting hydrolysis of the salt formed may cause the solution to be slightly acidic, slightly basic, or neutral. If a strong acid and a strong base are titrated, the end point will be at pH 7, and practically any indicator can be used because adding 1 drop of either reagent will change the pH at the end point by about 6 units. For titrations of strong acids and weak bases, we need an indicator, such as methyl orange, that changes color between 3.1 and 4.4 in the acid region. When titrating a weak acid and a strong base, we should use an indicator that changes in the basic range. Phenolphthalein is the suitable choice for this type of titration because it changes color in the pH 8.3 to 10.0 range.

The process of the neutralization reaction can be represented by a titration curve like the one below, which shows the titration of a strong acid with a strong base.

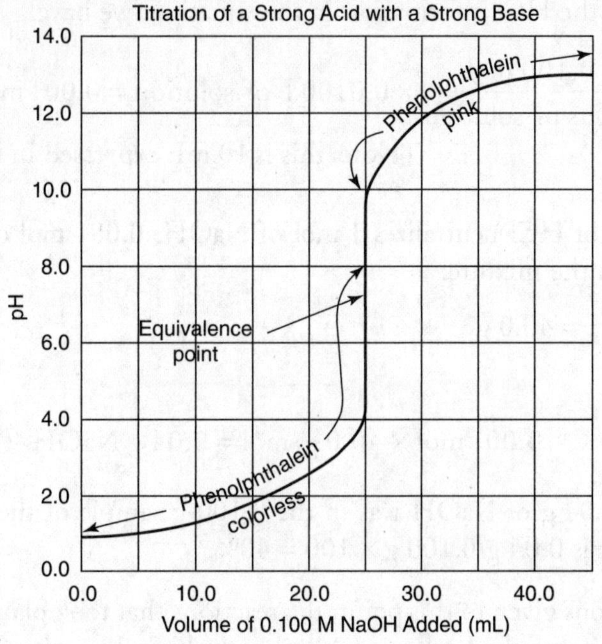

SAMPLE PROBLEMS:

1. Find the concentration of acetic acid in vinegar if 21.6 milliliters of 0.20 M NaOH is needed to titrate a 25-milliliter sample of the vinegar.

   Solution

   Using the equation $M_{acid} \times V_{acid} = M_{base} \times V_{base}$, we have

   $x\,M \times 25\text{ mL} = 0.2 \times 21.6\text{ mL}$

   $x = 0.17\ M_{acid}$ or $.17\ M_{acid}$

   Another type of titration problem involves a solid and a titrated solution.

2. A solid mixture contains NaOH and NaCl. If 10.0 milliliters of 0.100 M HCl is required to titrate a 0.100-gram sample of this mixture to its end point, what is the percent of NaOH in the sample?

   Solution

   Since

   $$\text{Molarity} = \frac{\text{No. of moles}}{\text{Liters of solution}}$$

   then

   $$M \times V = \frac{\text{No. of moles}}{\text{Liters of solution}} \times \text{Liters of solution} = \text{No. of moles}$$

Substituting the HCl information in the equation, we have

$$\frac{0.100 \text{ mol}}{\text{Liters of solution}} \times 0.0100 \text{ L of solution} = 0.001 \text{ mol}$$

(Note: this is 10 mL expressed in liters)

Since 1 mol of HCl neutralizes 1 mol of NaOH, 0.001 mol of NaOH must be present in the mixture.

1 mol NaOH = 40.0 g

so

$$0.001 \text{ mol} \times 40.0 \text{ g/mol} = 0.04 \text{ g NaOH}$$

Therefore 0.04 g of NaOH was in the 0.100-g sample of the solid mixture. The percent is 0.04 g/0.100 g × 100 = 40%.

In the explanations given to this point, the reactions that took place were between monoprotic acids (single hydrogen ions) and monobasic bases (one hydroxide ion per base). This means that each mole of acid had 1 mole of hydrogen ions available, and each mole of base had 1 mole of hydroxide ions available, to interact in the following reaction until the end point was reached:

$$H^+(aq) + OH^-(aq) \rightarrow 2H_2O$$

This is not always the case, however, and it is important to know how to deal with acids and bases that have more than one hydrogen ion and more than one hydroxide ion per formula. The following is an example of such a problem.

If 20.0 milliliters of an aqueous solution of calcium hydroxide, $Ca(OH)_2$, is used in a titration, and an appropriate indicator is added to show the neutralization point (end point), the few drops of indicator that are added can be ignored in the volume considerations. Therefore, if 25.0 milliliters of standard 0.050 M HCl is required to reach the end point, what was the original concentration of the $Ca(OH)_2$ solution?

The balanced equation for the reaction gives the relationship between the number of moles of acid reacting and the number of moles of base:

$$\underbrace{2HCL}_{\textbf{2 mol}} + \underbrace{Ca(OH)_2}_{\textbf{1 mol}} \rightarrow CaCl_2 + 2H_2O$$

The mole relationship here is that the number of moles of acid is twice the number of moles of base:

$$\text{No. of moles of acid} = \underset{\uparrow \text{ mole factor}}{2} \times \text{No. of moles of base}$$

Since the molar concentration of the acid times the volume of the acid gives the number of moles of acid:

$$M_a \times V_a = \text{moles of acid}$$

and the molar concentration of the base times the volume of the base gives the number of moles of base:

$$M_b \times V_b = \text{moles of base}$$

then, substituting these products into the mole relationship, we get

$$M_a V_a = 2M_b V_b$$

Solving for M_b gives

$$M_b = \frac{M_a V_a}{2V_b}$$

Substituting values, we get

$$M_b = \frac{0.050 \text{ mol/L} \times 0.0250 \not{L}}{2 \times 0.0200 \not{L}}$$

$$M_b = 0.0312 \text{ mol/L or } 0.0312 \text{ M}$$

BUFFER SOLUTIONS

Buffer solutions are equilibrium systems that resist changes in acidity and maintain constant pH when acids or bases are added to them. A typical laboratory buffer can be prepared by mixing equal molar quantities of a weak acid such as $HC_2H_3O_2$ and its salt, $NaC_2H_3O_2$. When a small amount of a strong base such as NaOH is added to the buffer, the acetic acid reacts (and consumes) most of the excess OH^- ion. The OH^- ion reacts with the H^+ ion from the acetic acid, thus reducing the H^+ ion concentration in this equilibrium:

$$HC_2H_3O_2 \rightleftharpoons H^+ + C_2H_3O_2^-$$

This reduction of H^+ causes a shift to the right, forming additional $C_2H_3O_2^-$ ions and H^+ ions. For practical purposes, each mole of OH^- added consumes 1 mole of $HC_2H_3O_2$ and produces 1 mole of $C_2H_3O_2^-$ ions.

When a strong acid such as HCl is added to the buffer, the H^+ ions react with the $C_2H_3O_2^-$ ions of the salt and form more undissociated $HC_2H_3O_2$. This does not alter the H^+ ion concentration. Proportional increases and decreases in the concentrations of $C_2H_3O_2^-$ and $HC_2H_3O_2$ do not significantly affect the acidity of the solution.

SALTS

A salt is an ionic compound containing positive ions other than hydrogen ions and negative ions other than hydroxide ions. The usual method of preparing a particular salt is by neutralizing the appropriate acid and base to form the salt and water.

TIP

There are five methods for preparing salts.

Five methods for preparing salts are as follows:

1. ***Neutralization reaction.*** An acid and a base neutralize each other to form the appropriate salt and water. For example:

$$\underset{\text{acid}}{2HCl} + \underset{\text{base}}{Ca(OH)_2} \rightarrow \underset{\text{salt}}{CaCl_2} + \underset{\text{water}}{2H_2O}$$

2. ***Single replacement reaction.*** An active metal replaces hydrogen in an acid. For example:

$$Mg(s) + H_2SO_4(aq) \qquad MgSO_4(aq) + H_2(g)$$

3. ***Direct combination of elements.*** An example of this method is the combination of iron and sulfur. In this reaction small pieces of iron are heated with powdered sulfur:

$$Fe + S \rightarrow \underset{\text{iron (II) sulfide}}{FeS}$$

4. ***Double replacement.*** When solutions of two soluble salts are mixed, they form an insoluble salt compound. For example:

$$AgNO_3(aq) + NaCl(aq) \qquad NaNO_3(aq) + AgCl(s)$$

5. ***Reaction of a metallic oxide with a nonmetallic oxide.*** For example:

$$MgO + SiO_2 \rightarrow MgSiO_3$$

The naming of salts is discussed on page 117.

AMPHOTERIC SUBSTANCES

Some substances, such as the HCO_3^- ion, the HSO_4^- ion, the H_2O molecule, and the NH_3 molecule, can act as either proton donors (acids) or proton receivers (bases), depending upon which other substances they come into contact with. These substances are said to be **amphoteric**. Amphoteric substances donate protons in the presence of strong bases and accept protons in the presence of strong acids.

Examples are the reactions of the bisulfate ion, HSO_4^-:

With a strong acid, HSO_4^- accepts a proton:

$$HSO_4^- + H^+ \rightarrow H_2SO_4$$

With a strong base, HSO_4^- donates a proton:

$$HSO_4^- + OH^- \rightarrow H_2O + SO_4^{2-}$$

ACID RAIN—AN ENVIRONMENTAL CONCERN

Acid rain is currently a subject of great concern in many countries around the world because of the widespread environmental damage it reportedly causes. It forms when the oxides of sulfur and nitrogen combine with atmospheric moisture to yield sulfuric and nitric acids—both known to be highly corrosive, especially to metals.

Once formed in the atmosphere, these acids can be carried long distances from their source before being deposited by rain. The pollution may also take the form of snow or fog or be precipitated in dry form. This dry form is just as damaging to the environment as the liquid form.

The problem of acid rain can be traced back to the beginning of the industrial revolution, and it has been growing ever since. The term "acid rain" has been in use for more than a century and is derived from atmospheric studies made in the region of Manchester, England. Although the severity of its effects has long been recognized in local settings, as indicated by spells of acid smog in heavily industrialized areas, its widespread destructiveness has been realized only in recent decades. One large area that has been studied extensively is northern Europe, where acid rain has eroded structures, injured crops and forests, and threatened or depleted life in freshwater lakes. In 1983 published reports claimed that 34% of the forested areas of West Germany had been damaged by acid rain. Many of the older monuments and buildings in Europe that have exposed metal have had to be refurbished. In 1989 the famous bronze horses on St. Mark Cathedral in Venice had to be removed for repair for this very reason.

> **TIP**
>
> **Acid rain** is the result of the formation of sulfuric acid from sulfur oxides reacting with water. Nitrogen oxides are also invovled.

Industrial emissions have been blamed as the major cause of acid rain. Industries have challenged this notion, because the chemical reactions involved in the production of acid rain in the atmosphere are complex and as yet little understood, and have stressed the need for further studies. Because of the high cost of pollution reduction, governments have tended to support this viewpoint. American studies released in the early 1980s, however, strongly implicated industries as the main source of acid rain in the eastern United States and Canada. In 1988, as part of the Long-Range Transboundary Air Pollution Agreement sponsored by the United Nations and the United States, along with 24 other countries, ratified a protocol freezing the rate of nitrogen oxide emissions at 1987 levels. The 1990 amendments to the Clean Air Act of 1967 put in place regulations to reduce the release of sulfur dioxide from power plants to 10 million tons per year by 2000. That achieved a 20 percent decrease in sulfur dioxide. And the attempts continue to further clean the air.

Chapter Summary

The following terms summarize all the concepts and ideas that were introduced in this chapter. You should be able to explain their meaning and how you would use them in chemistry. They appear in boldface type in this chapter to draw your attention to them. The boldface type also makes it easier for you to look them up if you need to. You could also use the search or "Google" action on your computer to get a quick and expanded explanation of these terms, laws, and formulas.

acid	conjugate base	neutralization
amphoteric	end point	pH
base	equivalence point	salt
buffer solution	indicator	titration
conjugate acid	litmus	volumetric analysis
Arrhenius Theory	Brønsted-Lowry Theory	Lewis Theory

Practice Exercises

1. The difference between HCl and $HC_2H_3O_2$ as acids is
   (A) the first has less hydrogen in solution
   (B) the second has more ionized hydrogen
   (C) the first is highly ionized
   (D) the second is highly ionized

2. The hydronium ion is represented as
   (A) H_2O^+
   (B) H_3O^+
   (C) HOH^+
   (D) H^-

3. H_2SO_4 is a strong acid because it is
   (A) slightly ionized
   (B) unstable
   (C) an organic compound
   (D) highly ionized

4. The common ionic reaction of an acid with a base involves ions of
   (A) hydrogen and hydroxide
   (B) sodium and chloride
   (C) hydrogen and hydronium
   (D) hydroxide and nitrate

5. Which pH is an acid solution?
   (A) 3
   (B) 7
   (C) 9
   (D) 10

6. The pH of a solution with a hydrogen ion concentration of 1×10^{-3} is
   (A) +3
   (B) −3
   (C) ±3
   (D) 1 + 3

7. According to the Brønsted-Lowry Theory, an acid is
   (A) a proton donor
   (B) a proton acceptor
   (C) an electron donor
   (D) an electron acceptor

8. A buffer solution
   (A) changes pH rapidly with the addition of an acid
   (B) does not change pH at all
   (C) resists changes in pH
   (D) changes pH only with the addition of a strong base

9. The point at which a titration is complete is called the
   (A) end point
   (B) equilibrium point
   (C) calibrated point
   (D) chemical point

10. If 10 mL of 1 M HCl was required to titrate a 20-mL NaOH solution of unknown concentration to its end point, what was the concentration of the NaOH?
    (A) 0.5 M
    (B) 1.5 M
    (C) 2 M
    (D) 2.5 M

Answers and Explanations

1. **(C)** The strength of an Arrhenius acid is determined by the degree of ionization of the hydrogens in the formula. The HCl ionizes to a great degree and is considered a strong acid, whereas $HC_2H_3O_2$, acetic acid, which is found in vinegar, ionizes only to a small degree and is considered a weak acid.

2. **(B)** The hydronium ion is written as H_3O^+.

3. **(D)** H_2SO_4 is a strong acid because it is highly ionized.

4 **(A)** The basic reaction between an acid and a base is$H^+ + OH- \rightarrow H_2O$ or $H_3O^+ + OH- \rightarrow 2H_2O$

5. **(A)** The pH scale is 0 to 14; the numbers below 7 are acid and those above 7 are increasingly basic.

6. **(A)** Because the pH is defined as the negative of the log of the H^+ concentration, it is –(log of 10^{-3}), which is –(–3) or +3.

7. **(A)** By definition, a Brønsted-Lowry acid is a proton donor.

8. **(C)** A buffer solution resists the changes in pH.

9. **(A)** The point in a titration when the "unknown" solution has been neutralized—in the case of an acid/base titration—by the "standard" solution of known concentration is called the end point or equivalence point.

10. **(A)** Use the equation $M_{acid} \times V_2 = M_{base} \times V_{base}$ and change mL to L by dividing by 1,000 mL/L to change 10 mL into .01 L and 20 mL into .02 L. Then substituting, you get $1\ M_{acid} \times .01\ L_{acid} = xM_{base} \times .02\ L_{base}$. Solving for *x*, you get $x = 0.5M_{base}$.

Oxidation–Reduction and Electrochemistry

CHAPTER 12

- Explain the reactions in a voltaic cell.
- Use electrode potentials to determine if a reaction will occur.
- Describe electroplating and how to determine the quantity of product plated.
- Explain the differences between voltaic and electrolytic cells.
- Explain how ordinary batteries work.
- Balance redox equations using oxidation numbers.

In the early 1830s, Michael Faraday discovered that water solutions of certain substances conduct an electric current. He called these substances **electrolytes**. Our definition today of an electrolyte is much the same. It is a substance that dissolves in water to form a solution that will conduct an electric current.

The usual apparatus to test for this conductivity is a lightbulb placed in series with two prongs that are immersed in the solution tested, as shown in Figure 35.

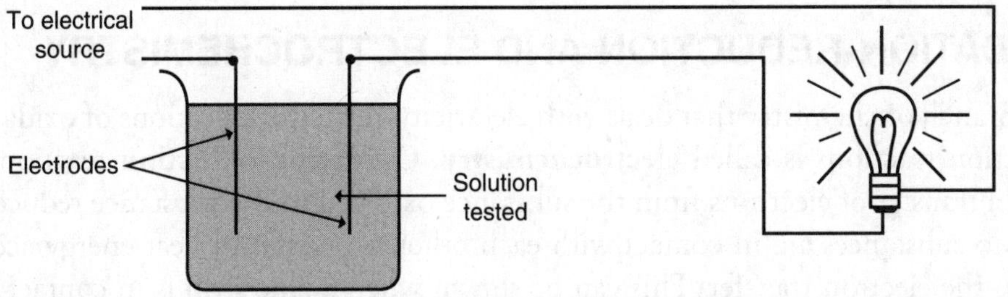

Figure 35. Test for Conductivity of Electrolytes

Using this type of apparatus, we can classify solutions as good, moderate, or poor electrolytes. If they do not conduct at all, they are called **nonelectrolytes**. The table below gives the classifications of some common substances.

Water Solutions Tested for Conductivity

Good	Poor	Nonelectrolyte
Sodium chloride	Acetic acid	Sugar
Hydrochloric acid	Ammonium hydroxide	Benzene

The reason that these substances conduct with varying degrees of efficiency is related to the number of ions in solution.

The ionic lattice substances like sodium chloride are dissociated by the water molecules so that the individual positive and negative ions are dispersed throughout the solution. In the case of a covalent bonded substance the degree of polarity determines the extent to which it will be ionized. The water molecules, which are polar themselves, can help weaken and finally break the polar covalent bonds by clustering around the substance. When the ions are formed in this manner, the process is called **ionization**; when the ionic lattice comes apart, the process is **dissociation**. Substances that are nonelectrolytes are usually bonded so that the molecule is a nonpolar molecule. The polar water molecule cannot orient itself around the molecule and cause its ionization.

Let us see how the current was carried through the apparatus in Figure 35. The electricity causes one electrode to become positively charged and one negatively charged. If the solution contains ions, they will be attracted to the electrode with the charge opposite to their own. This means that the positive ions migrate to the negative pole, and they are referred to as **cations**. The negative ions migrate toward the positive pole, and they are called **anions**. When these ions arrive at the respective electrodes, the negative ions give up electrons and the positive ions accept electrons, and we have a completed path for the electric current. The more highly ionized the substance, the more current flows and the brighter the lightbulb glows. It was the Swedish chemist Arrhenius who proposed a theory to explain the behavior of electrolytes in aqueous solutions.

REMEMBER

"**Leo** the lion says **Ger**" stands for **L**oss of **E**lectrons is **O**xidation and **G**ain of **E**lectrons is **R**eduction

OXIDATION-REDUCTION AND ELECTROCHEMISTRY

The branch of chemistry that deals with electricity-related applications of oxidation-reduction reactions is called **electrochemistry**. **Oxidation-reduction** reactions involve a transfer of electrons from the substance oxidized to the substance reduced. If the two substances are in contact with each other, a transfer of heat energy accompanies the electron transfer. This can be shown when a zinc strip is in contact with a copper(II) sulfate solution in a beaker, as shown in the accompanying diagram. The zinc strip loses electrons to the copper(II) ions in solution. The copper(II) ions accept the electrons and fall out of solution as copper atoms. As electrons are transferred between zinc atoms and copper(II) ions, energy is released in the form of heat, as indicated by a rise in temperature.

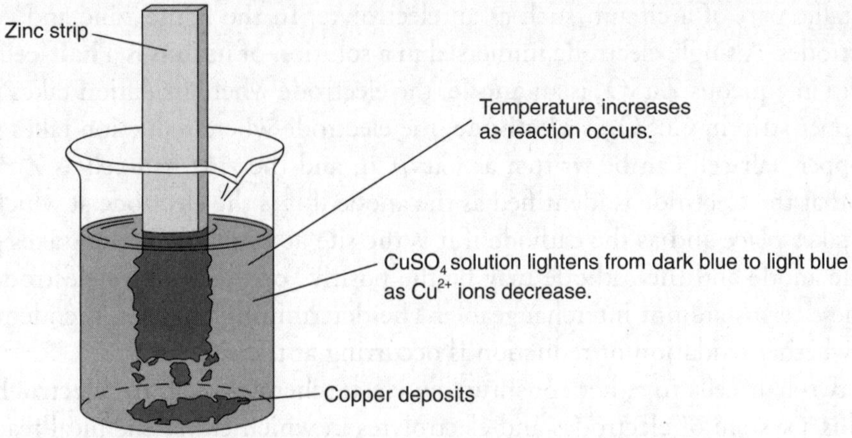

Voltaic Cells

In another example of an oxidation-reduction reaction, the substance that is oxidized during the reaction is separated from the substance that is reduced during the reaction. The electron transfer is accompanied by a transfer of electrical energy instead of heat. One means of separating oxidation and reduction half-reactions is with a porous barrier, which prevents the metal atoms of one half-reaction from mixing with the ions of the other half-reaction. Ions in the two solutions, however, can move through the porous barrier. Electrons can be transferred from one side to the other through an external connecting wire. Electric current moves in a closed loop path, or circuit, so this movement of electrons through the wire is balanced by the movement of ions in the solution, as shown in the figure below.

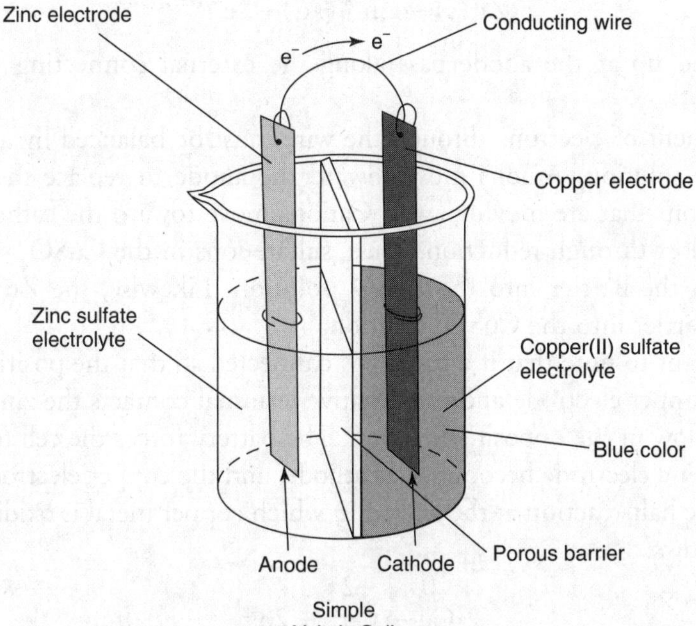

Simple Voltaic Cell

REMEMBER:

"An Ox (anode is oxidation) and a Red Cat" (reduction at cathode)

The zinc strip is in an aqueous solution of $ZnSO_4$; the copper strip, in an aqueous solution of $CuSO_4$. Both solutions conduct electricity, so they are classified as electrolytes. An **electrode** is a conductor used to establish electrical contact with a nonmetallic part of a circuit, such as an electrolyte. In the figure, zinc and copper are electrodes. A single electrode immersed in a solution of its ions is a **half-cell**. The zinc strip in aqueous $ZnSO_4$ is an **anode**, the electrode where oxidation takes place. The copper strip in $CuSO_4$ is a **cathode**, the electrode where reduction takes place. The copper half-cell can be written as Cu^{2+}/Cu, and the zinc half-cell as Zn^{2+}/Zn. Notice that the electrode is identified as the anode if it is the electrode at which oxidation takes place and as the cathode if it is the site at which reduction takes place. Both the anode and the cathode may be the positive or the negative electrode of a cell. These terms are not interchangeable. The determining factor in the identification is whether oxidation or reduction is occurring at the site.

The two half-cells together constitute an electrochemical cell. An **electrochemical cell** is a system of electrodes and electrolytes in which either chemical reactions produce electrical energy or an electric current produces chemical change. An electrochemical cell may be represented by the following notation: cathode | anode. For example, the cell made up of zinc and copper can be written as Cu | Zn. There are two types of electrochemical cells: voltaic (also called galvanic) and electrolytic.

If the redox reaction in an electrochemical cell occurs naturally and produces electrical energy as is shown in the drawing immediately above, the cell is a **voltaic cell**. Cations in the solution are reduced when they gain electrons at the surface of the cathode to become metal atoms. This half-reaction for the voltaic cell shown in the drawing is as follows:

$$Cu^{2+}(aq) + 2e^- \rightarrow Cu(s)$$

The half-reaction occurring at the negative electrode or anode is:

$$Zn(s) \rightarrow Zn^{2+}(aq) + 2e^-$$

Electrons given up at the anode pass along the external connecting wire to the cathode.

The movement of electrons through the wire must be balanced by a movement of ions in the solution. Anions move toward the anode to replace the negatively charged electrons that are moving away. Cations move toward the cathode as positive charge is lost through reduction. Thus, sulfate ions in the $CuSO_4$ solution can move through the barrier into the $ZnSO_4$ solution. Likewise, the Zn^{2+} ions pass through the barrier into the $CuSO_4$ solution.

It is important to note that if a battery is connected so that the positive terminal contacts the copper electrode and the negative terminal contacts the zinc electrode, the electrons flow in the opposite direction. The battery forces the cell to reverse its reaction; the zinc electrode becomes the cathode, and the copper electrode becomes the anode. The half-reaction at the anode, in which copper metal is oxidized, can be written as follows:

$$\overset{0}{Cu} \rightarrow \overset{+2}{Cu^{2+}} + 2e^{-2}$$

The reduction half-reaction of zinc at the cathode is written as follows:

$$\overset{+2}{Zn^{2+}} + 2e^- \rightarrow \overset{0}{Zn}$$

Electrode Potentials

In the discussion of acids and metals reacting to produce hydrogen (page 210), the activities of metals compared with the activity of hydrogen are shown. A more inclusive presentation of this activity of metals, called the **electromotive series**, is shown in Table 10.

From this chart you notice that zinc is above copper and, therefore, more active. This means zinc can displace copper ions in a solution of copper sulfate:

$$Zn^0 + Cu^{2+} + SO_4^{2-} \rightarrow Cu^0 + Zn^{2+} + SO_4^{2-}$$

The zero indicates the zinc atom is neutral. → The zinc ion carries a +2 charge.

The zinc atom must have lost 2 electrons to become Zn^{2+} ions:

$$Zn^0 \rightarrow Zn^{2+} + 2e^- \text{ (electrons)}$$

At the same time, the Cu^{2+} must have gained 2 electrons to become the Cu^0 atom:

$$Cu^{2+} + 2e^- \text{ (electrons)} \rightarrow Cu^0$$

These two equations are called **half-reactions**. The loss of electrons by the zinc is called oxidation; the gain of electrons by the copper ion, reduction. It is important to remember that the gain of electrons is **reduction** and the loss of electrons is **oxidation**. A way to remember this is the statement "LEO the lion says GER." LEO translates into "Loss of Electrons is Oxidation," and GER into "Gain of Electrons is Reduction." The fact that the zinc in this reaction is oxidized by giving up electrons makes it possible for electrons to be gained by the copper, which is acting as a **reducing agent**. Similarly, because the copper in this reaction is being reduced by gaining electrons, electrons can be lost by the zinc, which is acting as an **oxidizing agent**. Remember: The substance oxidized is called the reducing agent, and the substance reduced is the oxidizing agent.

REMEMBER

The reducing agent is oxidized.

The oxidizing agent is reduced.

The metal elements that lose electrons easily and become positive ions are placed high in the electromotive series. The metal elements that lose electrons with more difficulty are placed lower in Table 10.

The energy required to remove electrons from metallic atoms can be assigned numerical values called **electrode potentials**. The energies, E^0_r, required for the reduction of common elements are shown in Tables 10 and 11.

Table 10. Standard Reduction Potentials

Element			Standard Half-Cell Voltages Reduction Electron Reaction	Electrode Potential,*E^0
Potassium			K^+ +e^- → K	−2.93 V
Calcium			Ca^{2+} +$2e^-$ → Ca^-	−2.87 V
Sodium	Increasing	Increasing	Na^+ +e^- → Na	−2.71 V
Magnesium	Tendency	Tendency	Mg^{2+} +$2e^-$ → Mg^-	−2.37 V
Aluminum	for	for	Al^{3+} +$3e^-$ → Al^-	−1.67 V
Zinc	Atoms	Ions	Zn^{2+} +$2e^-$ → Zn	−0.76 V
Iron	to	to Gain	Fe^{2+} +$2e^-$ → Fe	−0.44 V
Tin	Lose	Electrons	Sn^{2+} +$2e^-$ → Sn	−0.14 V
Lead	Electrons	and	Pb^{2+} +$2e^-$ → Pb	−0.13 V
Hydrogen	and	Form	$2H^+$ +$2e^-$ → H_2	0.00 V
Copper	Form	Atoms	Cu^{2+} +$2e^-$ → Cu	+0.34 V
Mercury	Positive	of the	Hg_2^{2+} +$2e^-$ → 2Hg	+0.79 V
Silver	Ions	Metal	Ag^+ +e^- → Ag	+0.80 V
Mercury			Hg^{2+} +$2e^-$ → Hg	+0.85 V
Gold			Au^{3+} +$3e^-$ → Au	+1.50 V

A measure in volts of the tendency of atoms to gain or lose electrons.

These voltages depend on the nature of the reaction, the concentrations of reactants and products, and the temperature. Throughout this discussion, we use standard concentrations, that is, all ions or molecules in aqueous solution are at a concentration of 1 molar. Furthermore, all gases taking part in the reactions are at 1 atmosphere pressure, and the temperature is 25°C. The voltage measured under these conditions is called **standard voltage.**

The currently accepted convention is to give the potentials of half-reactions as reduction processes. For example:

> **TIP**
>
> If the E^0_{cell} is positive, the reaction will occur.

$$2H^+ + 2e^- \rightarrow H_2$$

The E^0 values corresponding to these half-reactions are called **standard reduction potentials.**

Some reduction electrode potentials are shown in the last column of Tables 10 and 11. Notice that hydrogen is used as the standard with an electrode potential of

TABLE 11. Activity of Nonmetals

Element			Reduction Electron Reaction	E^0
Fluorine	Increasing	Increasing	$F_2 + 2e^- \rightarrow 2F^-$	+2.87 V
	Tendency	Tendency		
Chlorine	for Atoms	for Ions	$Cl_2 + 2e^- \rightarrow 2Cl^-$	+1.36 V
	to Gain	to Lose		
Bromine	Electrons	Electrons	$Br_2 + 2e^- \rightarrow 2Br^-$	+1.09 V
	and Form	and Form		
Iodine	Negative	Atoms	$I_2 + 2e^- \rightarrow 2I^-$	+0.54 V
	Ions			

zero. These values help you to predict what reactions will occur and how readily they will occur.

The following examples will clarify the use of electrode potentials.

If magnesium reacts with chlorine, we can write the equation

$$Mg + Cl_2 \rightarrow MgCl_2$$

The two half-reactions with the electrode potentials are:

Oxidation half-reaction:	$Mg \rightarrow Mg^{2+} + 2e^-$	$E^0 = +2.37$ V
Reduction half-reaction:	$Cl_2 + 2e^- \rightarrow 2Cl^-$	$E^0 = +1.36$ V
Net reaction:	$Mg + Cl_2 \rightarrow Mg^{2+} + 2Cl^-$	$E^0_{cell} = +3.73$ V

In the net reaction, E^0 is a positive number. This indicates that the reaction occurs spontaneously.

You should also note that the total number of electrons lost in oxidation is equal to the total number of electrons gained so that the net reaction (arrived at by adding the two reactions) does not contain any electrons.

For sodium reacting with chlorine the equation is:

$$2Na + Cl_2 \rightarrow 2NaCl$$

The two half-reactions are:

Oxidation half-reaction:	$2Na \rightarrow 2Na^+ + 2e^-$	$E^0 = +2.71$ V
Reduction half-reaction:	$Cl_2 + 2e^- \rightarrow 2Cl^-$	$E^0 = +1.36$ V
Net reaction:	$2Na + Cl_2 \rightarrow 2Na^+ + 2Cl^-$	$E^0_{cell} = +4.07$ V

TIP

This note is important.

Note: The standard potentials (E^0) are *not* multiplied by the coefficients in calculating the E^0 for the reaction.

Again, the E^0 for this reaction is positive and the reaction is spontaneous.

The next example shows a negative E^0.

Copper metal placed in an acid solution is shown as follows:

Oxidation:	$Cu \rightarrow Cu^{2+} + 2e^-$	$E^0 = -0.34$ V
Reduction:	$2H^+ + 2e^- \rightarrow H_2$	$E^0 = \ \ 0.00$ V
Net reaction:	$Cu + 2H^+ \rightarrow Cu^{2+} + H_2$	$E^0_{cell} = -0.34$ V

Since E^0 is negative, we know the reaction will not take place.

Many times the reduction reactions with their E^0 values are shown for metals. If you must use the oxidation reactions of these metals, the equations must be reversed, and the sign of E^0 changed. For example, when a piece of copper is placed into a solution of silver ions:

Oxidation (reversed):	$Cu \rightarrow Cu^{2+} + 2e^-$	$E^0 = -0.34$ V
Reduction:	$2Ag^+ + 2e^- \rightarrow 2Ag^0$	$E^0 = +0.80$ V
	$Cu + 2Ag^+ \rightarrow Cu^{2+} + 2Ag^0$	$E^0 = +0.46$ V

This reaction will occur spontaneously since E^0 is positive.

- **SAMPLE PROBLEM:**

Calculate the cell voltage of the following reaction:

$$Zn(s)|\ Zn^{2+}(0.001\,M)\|\ Ag^+(0.1\,M)|\ Ag(s)$$

(Each side of the ‖ represents the half-reaction characters, with the concentration when appropriate.)

The reactions are:

Cathode reduction reaction:	$2Ag^+ + 2e^- \rightarrow 2Ag(s)$	$E^0 = +0.80$ V
Anode oxidation reaction:	$Zn(s) \rightarrow Zn^{2+} + 2e^-$	$E^0 = +0.76$ V
Cell reaction:	$2Ag^+ + Zn(s) \rightleftharpoons Zn^{2+} + 2Ag$	$E^0_{cell} = +1.56$ V

Since the cell voltage is positive, the reaction will occur.

Electrolytic Cells

In the second type of electrochemical reactions, the redox reactions do not occur spontaneously but can be forced to take place by supplying energy with an external current. These reactions are called **electrolytic reactions.** Some examples of this type of reaction are electroplating, electrolysis of salt solution, electrolysis of water, and electrolysis of molten salts. An example of the electrolytic setup is shown in Figure 36.

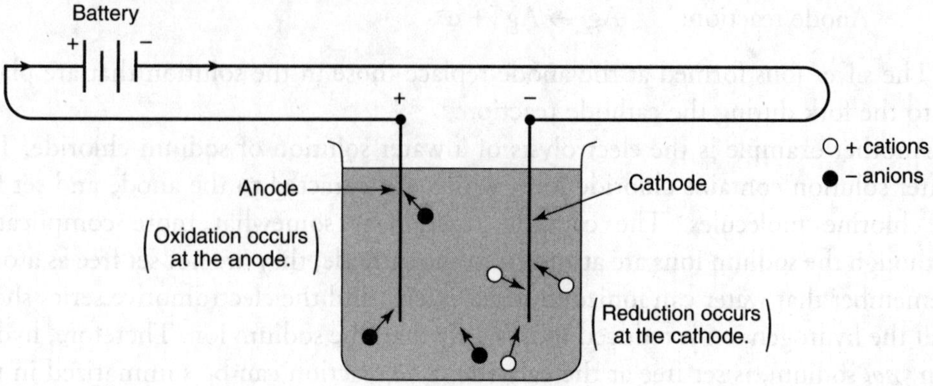

Figure 36. Electrochemical Reactions

If this solution contained Cu^{2+} ions and Cl^- ions, the half-reactions would be:

Anode half-reaction:		
Oxidation:	$2Cl^- \rightarrow Cl_2(g) + 2e^-$	$E^0 = -1.36$ V
Cathode half-reaction:		
Reduction:	$Cu^{2+} + 2e^- \rightarrow Cu^0$	$E^0 = +0.34$ V
	$Cu^2 + 2Cl^- \rightarrow Cu^0 + Cl_2(g)$	$E^0 = -1.02$ V

Notice that the E^0 for this reaction is negative, so an outside source of energy must be used to make it occur.

In **electroplating**, where electrolysis is used to coat a material with a layer of metal, the object to be plated is made the cathode in the reaction. A bar of the plating metal is made the anode, and the solution contains ions of the plating metal. See Figure 37 below.

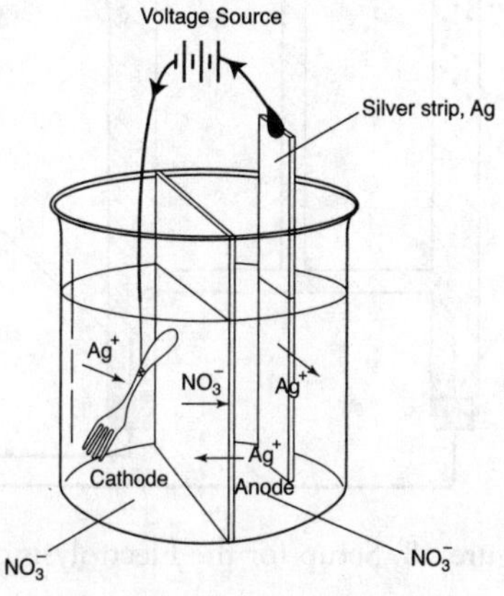

Figure 37. Electroplating a Metal Fork

In this example, silver nitrate, $AgNO_3$, can be used to silver plate. When dissolved in H_2O, it forms a solution containing silver ions, Ag^+. Assume a metal fork is made the cathode and a bar of silver is made the anode. When the current is switched on, the positive silver ions in the solution are attracted to the fork. When the silver ions make contact, they are reduced and change from ions to atoms of silver. These atoms gradually form a metallic coating on the fork. At the anode, oxidation occurs, and the anode itself is oxidized. These two half reactions are:

Cathode reaction: $Ag^+ + e^- \rightarrow Ag$
Anode reaction: $Ag \rightarrow Ag^+ + e^-$

The silver ions formed at the anode replace those in the solution that are plated onto the fork during the cathode reaction.

Another example is the electrolysis of a water solution of sodium chloride. This water solution contains chloride ions, which are attracted to the anode and set free as chlorine molecules. The cathode reaction is somewhat more complicated. Although the sodium ions are attracted to the cathode, they are not set free as atoms. Remember that water can ionize to some extent, and the electromotive series shows that the hydrogen ion is reduced more easily than the sodium ion. Therefore, hydrogen, *not* sodium, is set free at the cathode. The reaction can be summarized in this way:

Cathode reaction: $2H_2O + 2e^- \rightarrow H_2(g) + 2OH^-$
Reduction: $2Cl^- \rightarrow 2e^- + Cl_2$

Net reaction: $2H_2O + 2Cl^- \rightarrow H_2(g) + Cl_2(g) + 2OH^-$

Another example of electrolysis is the decomposition of water by using an apparatus like the one shown in Figure 38.

The solution in this apparatus contains distilled water and a small amount of H_2SO_4. The reason for adding H_2SO_4 is to make the solution an electrolyte because distilled water alone will not conduct an electric current. The solution, therefore, contains ions of H^+, HSO_4^-, and SO_4^{2-}.

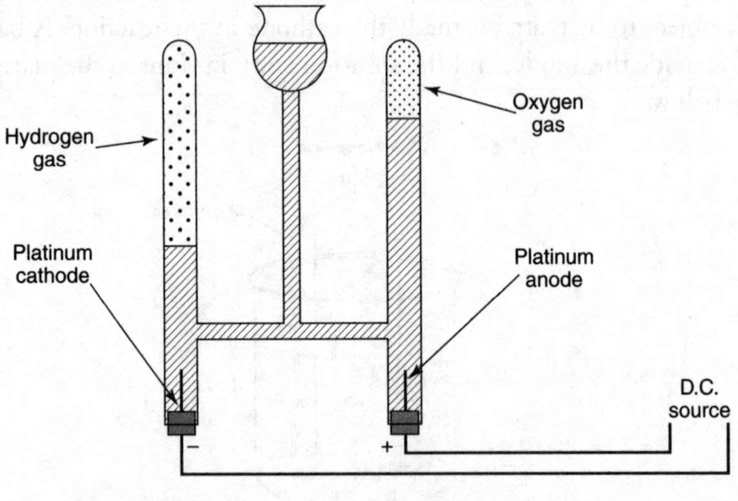

Figure 38. Setup for the Electrolysis of Water

The hydrogen ions (H^+) migrate to the cathode where they are reduced to hydrogen atoms and form hydrogen molecules (H_2) in the form of a gas. The SO_4^{2-} and HSO_4^- migrate to the anode but are not oxidized since the oxidation of water occurs more readily. These ions then are merely spectator ions. The oxidized water reacts as shown in the following half-reaction.

Cathode reaction:
Reduction: $4H_2O + 4e^- \rightarrow 2H_2(g) + 4OH^-$
Anode reaction:
Oxidation: $2H_2O \rightarrow O_2(g) + 4H^+ + 4e^-$

Net reaction: $2HOH \rightarrow 2H_2(g) + O_2(g)$

Notice that the equation shows 2 volumes of hydrogen gas are released while only 1 volume of oxygen gas is liberated. This agrees with the discussion of the composition of water in Chapter 7.

There are two important differences between the voltaic cell and the electrolytic cell:

1. The anode and cathode of an electrolytic cell are connected to a battery or other direct-current source, whereas a voltaic cell serves as a source of electrical energy.
2. In electrolytic cells, electrical energy from an external source causes nonspontaneous redox reactions to occur. In voltaic cells, however, spontaneous redox reactions produce electricity. Thus, in an electrolytic cell, electrical energy is converted into chemical energy; in a voltaic cell, chemical energy is converted into electrical energy.

Applications of Electrochemical Cells (Commercial Voltaic Cells)

One of the most common voltaic cells is the ordinary "dry cell" used in flashlights. Its makeup is shown in the drawing below, along with anode, cathode, and paste reactions.

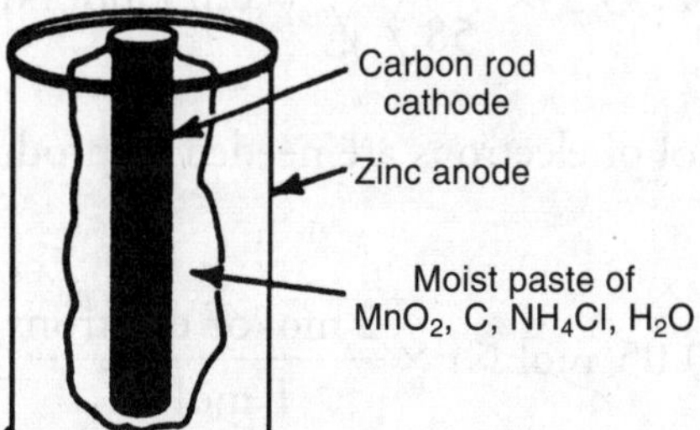

Anode reaction:
$Zn(s) \rightarrow Zn^{2+}(aq) + 2e^-$

Cathode reaction:
$MnO_2(s) + 4NH_4^+(aq) + e^- \rightarrow$

$Mn^{3+}(aq) + 4NH_3(aq) + 2\ H_2O$

The automobile lead storage battery is also a voltaic cell. When it discharges, the reactions are as shown in the drawing below.

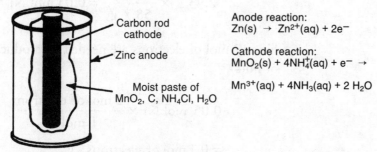

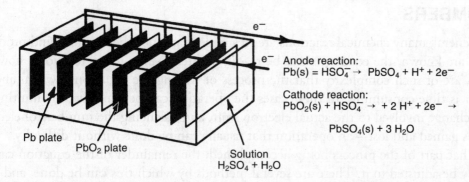

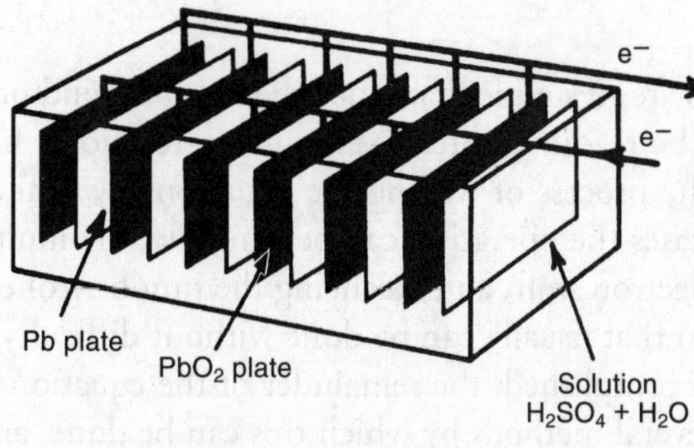

Anode reaction:
$Pb(s) = HSO_4^- \rightarrow PbSO_4 + H^+ + 2e^-$

Cathode reaction:
$PbO_2(s) + HSO_4^- \rightarrow + 2\ H^+ + 2e^- \rightarrow$

$PbSO_4(s) + 3\ H_2O$

QUANTITATIVE ASPECTS OF ELECTROLYSIS

Relationship Between Quantity of Electricity and Amount of Products

The amounts of products liberated at the electrodes of an electrolytic cell are related to the quantity of electricity passed through the cell and the electrode reactions.

In electrolysis, 1 mole of electrons is called 1 *faraday* of electric charge. In the electrolysis of molten NaCl, 1 faraday will liberate 1 mole of sodium atoms.

$$\text{Cathode reaction: } Na^+ + e^- \rightarrow Na(s)$$

At the same time, 1 mole of Cl^- ions at the anode will form 1 mole of chlorine atoms and thus 0.5 mole of chlorine molecules.

If the metallic ion had been Ca^{2+}, 1 faraday would release only 0.5 mole of calcium atoms since each calcium ion requires 2 electrons, as shown here:

$$Ca^{2+} + 2e^- \rightarrow Ca(s)$$

TIP

Use the equation to determine the quantity of electrons needed.

SAMPLE PROBLEM: How many moles of electrons are required to reduce 2.93 grams of nickel ions from melted NiCl2?

The reaction is

$$Ni^{2+} + 2e^- \rightarrow Ni(s)$$

The 2.93 g represents

$$2.93\ \cancel{g} \times \frac{1\text{ mol Ni}}{58.7\ \cancel{g}} = 0.05\text{ mol Ni}$$

Since 2 mol of electrons are needed to produce 1 mol of Ni(s),

$$0.05\ \cancel{\text{mol Ni}} \times \frac{2\text{ mol of electrons}}{1\ \cancel{\text{mol Ni}}}$$

$$= 0.1\text{ mol of electrons}$$

BALANCING REDOX EQUATIONS USING OXIDATION NUMBERS

In general, many chemical reactions are so simple that, once the reactants and products are known, the equations can be readily written. Many redox reactions, however, are of such complexity that the process of writing the equations by trial and error is time consuming. In these cases the operation can be simplified by limiting the change involved to the actual electron shift, and balancing the numbers of electrons gained and lost—an operation that usually can be done without difficulty.

That part of the process being accomplished, the remainder of the equation can easily be adjusted to it. There are several methods by which this can be done, and a number of techniques that can be employed for each method. We will show two methods: the electron shift method and the ion-electron method.

The Electron Shift Method

FIRST EXAMPLE:

$$\overset{+1\,-1}{HCl} + \overset{+4\,-2}{MnO_2} \rightarrow \overset{+1\,-2}{H_2O} + \overset{+2\,-1}{MnCl_2} + \overset{0}{Cl_2}$$

1. The molecular formulas are written in a statement of the reaction, and the oxidation states are assigned to the elements. These appear in the above equation.
2. Inspection shows that the oxidation state of the manganese atoms has changed from +4 to +2, and that the oxidation state of the chlorine atoms that emerge from the action as free elements has changed from −1 to 0:

   $1\,(Mn^{4+} + 2e^{+} \rightarrow Mn^{2+})$ Reduction

   $2\,(Cl^{1-} \rightarrow Cl^{0} + 1e^{-})$ Oxidation
3. The reason for multiplying the two half-reactions by 1 and 2 is to balance the electrons gained with the electrons lost.
4. It appears that $1Mn^{4+} + 2Cl^{1-} \rightarrow 1Mn^{2+} + 2Cl^{0}$.
5. Placing these coefficients in the equation gives

   $$2HCl + 1MnO_2 \rightarrow *?\ H_2O + 1MnCl_2 + Cl_2$$
6. From the numbers thus established, the remaining coefficients can easily be deduced. Two *more* molecules of HCl are required to furnish the chlorine for $MnCl_2$, and the two atoms of oxygen in MnO_2 form $2H_2O$.
7. *The final equation is:* $4HCl + MnO_2 \rightarrow 2H_2O + MnCl_2 + Cl_2$.

TIP

Notice that electrons gained are on the left side and electrons lost are on the right side.

Now consider a more complicated reaction. Only the elements that show a change in oxidation state are indicated.

SECOND EXAMPLE:

$$\overset{-1}{HCl} + \overset{+7}{KMnO_4} \rightarrow H_2O + KCl + \overset{+2}{MnCl_2} + \overset{0}{Cl_2}$$

The electron change is: $5(Cl^{-1} = Cl^{0} + 1e^{-})$

$1(Mn^{7+} + 5e^{-} = Mn^{2+})$

Therefore, $5HCl + 1KMnO_4 = \frac{5}{2}Cl_2 + 1MnCl_2$

↑ **5 atoms of chlorine equal $\frac{5}{2}\ Cl_2$ molecules**

Multiplying by 2 to avoid fractions gives:

$$10HCl + 2KMnO_4 \rightarrow 5Cl_2 + 2MnCl_2$$

Substituting in the original equation:

$$10HCl + 2KMnO_4 \rightarrow *?\ H_2O + *?\ KCl + 2MnCl_2 + 5Cl_2$$

**? indicates that this coefficient is not yet known.*

For the remainder of the coefficients, $2KMnO_4$ produces 2KCl and $8H_2O$, while 2KCl + $2MnCl_2$ call for six additional molecules of HCl. Finally the expression becomes

$$16HCl + 2KMnO_4 \rightarrow 8H_2O + 2KCl + 2MnCl_2 + 5Cl_2$$

The Ion-Electron Method

The second method is more complex but seems to represent the true mechanism of such reactions more closely.

In this method, only units that actually have individual existence (atoms, molecules, or ions) in the particular reaction being studied are taken into consideration. The principal oxidizing agent and the principal reducing agent are chosen from these (by a method to be indicated later). Then the electron loss or gain of each of these two principal actors is determined by taking into consideration the fact that, since electrons can neither be created nor destroyed, electrons lost by one of these actors must be gained by the other. This is accomplished by using two separate partial equations representing the changes undergone by each of the two principal actors.

Probably the best way to show how the method operates will be to follow in detail the steps taken in balancing two actual oxidation-reduction reactions.

FIRST EXAMPLE: Assume that the equation $K_2CrO_4 + HCl \rightarrow KCl + CrCl_3 + H_2O + Cl_2$ is to be balanced.

1. Determine which of the substances present are involved in the oxidation-reduction.

   This is done by listing all substances present on each side of the arrow and then crossing out those that appear on both sides of the arrow without being changed in any way.

   $$\cancel{K^+}, CrO_4^{2-}, H^+, Cl^- \rightarrow \cancel{K^+}, Cl^-, Cr^{3+}, H_2O, Cl_2$$

   Note that Cl^- is not crossed out although it appears on both sides because some of the Cl^- from the left side appears in a changed form, namely, Cl_2, on the right.

   The two substances on the left side that are *not* crossed out are the ones involved in the oxidation-reduction.

   If, as *in this case*, there are *more* than two, disregard H^+, OH^-, or H_2O; this will leave the two principal actors.

2. Indicate in two as yet unbalanced partial equations the fate of each of the two active agents thus:

   $$CrO_4^{2-} \rightarrow Cr^{3+}$$
   $$Cl^- \rightarrow Cl_2$$

3. Balance these equations chemically, inserting any substance necessary.

$$
\begin{aligned}
CrO_4^{2-} + 8H^+ &\rightarrow Cr^{3+} + 4H_2O \\
2Cl^- &\rightarrow Cl_2
\end{aligned}
$$

In the upper partial equation, H^+ had to be added in order to remove the oxygen from the CrO_4^{2-} ion.

H+ always is used for this purpose in acid solutions. If the solution is basic, H_2O must be used; thus:

$$CrO_4^{2-} + 4H_2O \rightarrow Cr^{3+} + 8OH^-$$

4. Balance these equations electrically by adding electrons *on either side* so that the total electric charge is the same on the left and right sides, thus:

$$
\begin{aligned}
CrO_4^{2-} + 8H^+ + 3e^- &\rightarrow Cr^{3+} + 4H_2O \\
2Cl^- &\rightarrow Cl_2 + 2e^-
\end{aligned}
$$

5. Add these partial equations.

But before we add we must realize that electrons can neither be created nor destroyed. Electrons are gained in one of these equations and lost in the other. THOSE GAINED IN THE ONE MUST COME FROM THE OTHER. Therefore we must multiply each of these equations through by numbers so chosen that the num-ber of electrons gained in one equation will be the same as the number lost in the other, thus:

$$
\begin{aligned}
2[CrO_4^{2-} + 8H^+ 3e^- &\rightarrow Cr^{3+} + 4H_2O] \\
3[2Cl^- &\rightarrow Cl_2 + 2e^-]
\end{aligned}
$$

THIS IS THE KEY TO THE METHOD!

Adding the multiplied equations, we get:

$$2CrO_4^{2-} + 16H^+ + 6Cl^- \rightarrow 2Cr^{3+} + 8H_2O + 3Cl_2$$

This sum tells us all that really happens in the oxidation-reduction, but if a conventionally balanced equation is desired for problem purposes or for some other reason, it can be obtained by carrying the coefficient into the original skeleton equation, thus:

$$2K_2CrO_4 + 16HCl \rightarrow 4KCl + 2CrCl_3 + 8H_2O + 3Cl_2$$

$\uparrow$

(4 inserted by inspection)

SECOND EXAMPLE:

$$2KMnO_4 + H_2O + KI \rightarrow 2MnO_2 + 2KOH + KIO_3$$

$$2[MnO_4^- + 2H_2O + 3e^- \rightarrow MnO_2 + 4OH^-]$$

Adding: $$I^- + 6OH^- \rightarrow IO_3^- + 3H_2O + 6e^-$$

$$2MnO_4^- + I^- + H_2O \rightarrow 2MnO_2 + IO_3^- + 2OH^-$$

Note that, because this reaction takes place in a basic solution, H_2O was used to remove oxygen in the upper partial.

Note also that in the second partial it was necessary to ADD oxygen and that this was done by means of OH^- ion.

If the solution had been *acid*, then H_2O would have been used for this purpose, thus:

$$I^- + 3H_2O \rightarrow IO_3^- + 6H^+ + 6e^-$$

In general, you will meet three types of partial equations:

1. Where electrons only are needed to balance, as in the second partial of the first example above.
2. Where oxygen must be removed from an ion.

   In acid solution, H^+ is used for this purpose as in the first partial of the first example above.

   In basic solution, H_2O is used for this as in the first partial of the second example above.
3. Where oxygen must be added to an ion.

   In basic solution, OH^- is used for this purpose as in the second partial of the second example above.

   In acid solution, H_2O is used for this as in:

   $$SO_3^{2-} + H_2O \rightarrow SO_4^{2-} + 2H^+ + 2e^-$$

TRY THESE PROBLEMS:

1. $Zn + HNO_3 \rightarrow Zn(NO_3)_2 + NH_4NO_3 + H_2O$ Ans. 4, 10, 4, 1, 3
2. $Cu + HNO_3 \rightarrow Cu(NO_3)_2 + H_2O + NO$ Ans. 3, 8, 3, 4, 2
3. $KMnO_4 + HCl \rightarrow KCl + MnCl_2 + H_2O + Cl_2$ Ans. 2, 16, 2, 2, 8, 5

TIP

This one is a challenge!

ANSWERS

1. The half-reactions balanced chemically:

$$Zn \rightarrow Zn^{2+}$$
$$NO_3^- + 10H^+ \rightarrow NH_4^+ + 3H_2O$$

Balancing electrically and adding:

$$4(Zn \rightarrow Zn^{2+} + 2e^-)$$
$$NO_3^- + 10H^+ + 8e^- \rightarrow NH_4^+ + 3H_2O$$
$$\overline{4Zn + NO_3^- + 10H^+ \rightarrow 4Zn^{2+} + NH_4^+ + 3H_2O}$$

In the equation:

$$\underline{4}Zn + \underline{10}HNO_3 \rightarrow \underline{4}Zn(NO_3)_2 + NH_4NO_3 + \underline{3}H_2O$$

2. The half-reactions balanced chemically:

$$Cu^0 \rightarrow CU^{2+}$$
$$NO_3^{1-} + 4H^+ \rightarrow NO + 2H_2O$$

Balancing electrically and adding:

$$3(Cu^0 \rightarrow Cu^{2+} + 2e^-$$
$$2(NO_3^{1-} + 4H^+ + 3e^- \rightarrow NO + 2H_2O)$$
$$\overline{3Cu + 8H^+ + 2NO_3^{3-} \rightarrow 3Cu^{2+} + 2NO + 4H_2O}$$

In the equation:

$$\underline{3}Cu + \underline{8}HNO_3 \rightarrow \underline{3}Cu(NO_3)_2 + \underline{4}H_2O + \underline{2}NO$$

3. The half-reactions balanced chemically:

$$MnO_4^- + 8H^+ \rightarrow Mn^{2+} + 4H_2O$$
$$2Cl^- \rightarrow Cl_2$$

Balancing electrically and adding:

$$2(MnO_4^- + 8H^+ + 5e^- \rightarrow Mn^{2+} + 4H_2O)$$
$$5(2\,Cl^- \rightarrow Cl_2^{\,0} + 2e^-)$$
$$\overline{2MnO_4^- + 16H^+ + 10Cl^- \rightarrow 2Mn^{2+} + 8H_2O + 5Cl_2}$$

In the equation:

$$\underline{2}KMnO_4 + \underline{16}HCl \rightarrow \underline{2}KCl + \underline{2}MnCl_2 + \underline{8}H_2O + \underline{5}Cl_2$$

sets this as 2

by inspection

Chapter Summary

The following terms summarize all the concepts and ideas that were introduced in this chapter. You should be able to explain their meaning and how you would use them in chemistry. They appear in boldface type in this chapter to draw your attention to them. The boldface type also makes it easier for you to look them up if you need to. You could also use the search or "Google" action on your computer to get a quick and expanded explanation of these terms, laws, and formulas.

anions
anode
cathode
cations
dissociation
electrochemical cell
electrode potential
electrolyte
electrolytic reaction
electromotive series
electroplating
half-cell electrolysis
ionization
oxidation
oxidizing agent
reducing agent
reduction
standard voltage
voltaic cell

Practice Exercises

1. Which of the following when dissolved in water and placed in the conductivity apparatus would cause the light to glow?
   (A) table salt
   (B) ethyl alcohol
   (C) sugar
   (D) glycerine

2. In question 1, the reason that a current could flow was that
   (A) ions combine to form molecules
   (B) molecules migrate to the charge plates
   (C) ions migrate to the charge plates
   (D) sparks cross the gap

3. The extent of ionization depends on the
   (A) nature of the solvent
   (B) nature of the solute
   (C) concentration of the solution
   (D) temperature of the solution
   (E) all the above

4. Which of the following is TRUE?
   (A) The number of positive ions in solution equals the number of negative ions.
   (B) The positive ions are called anions.
   (C) The positive ions are called cathodes.
   (D) The total positive charge equals the total negative charge in solution.
   (E) None of the above

5. The hydronium ion is represented as
   (A) H_2O^+
   (B) H_3O^+
   (C) HOH^+
   (D) H^-

6. In the electrolysis of copper chloride, the substance liberated at the anode is
   (A) copper
   (B) chlorine
   (C) hydrogen
   (D) copper chloride

7. Ions are particles that exist
   (A) only in water solutions
   (B) in some crystals
   (C) in polar covalent compounds
   (D) in covalent compounds that are not polar

8. Ionic compounds will conduct an electric current when they are
   (A) solidified
   (B) melted
   (C) frozen
   (D) dehydrated

9. The cathode in an electrochemical cell is the electrode that is
   (A) always negative
   (B) always positive
   (C) always neutral
   (D) the electrode at which reduction takes place

10. Electrolysis of a dilute solution of sodium chloride results in the cathode product
   (A) sodium
   (B) chlorine
   (C) hydrogen
   (D) oxygen

11. Electrode potentials are:

   $Zn^0 \rightarrow Zn^{2+} + 2e^-$ $E^0 = +0.76$ V

   $Au^0 \rightarrow Au^{3+} + 3e^-$ $E^0 = -1.42$ V

   If a gold foil were placed in a solution containing zinc ions, the reaction potential would be
   (A) −1.34 V
   (B) −2.18 V
   (C) −0.66 V
   (D) +2.18 V
   (E) +1.34 V

12. A positive reaction potential indicates that
   (A) the reaction will not occur
   (B) the reaction will occur and give off energy
   (C) the reaction will occur if heat or energy is added
   (D) the reaction will power an outside alternating electric current

Questions 13–15

The following elements are listed in order of decreasing reactivity as they appear in the electrochemical series.

Ca, Na, Mg, Zn, Fe, H, Cu, Hg, Ag, Au

13. The element that is the best reducing agent is
   (A) Ca
   (B) Au
   (C) H
   (D) Fe
   (E) Cu

14. Of the following, the element that does NOT react with hydrochloric acid to produce hydrogen gas is
   (A) Zn
   (B) Fe
   (C) Hg
   (D) Ca
   (E) Mg

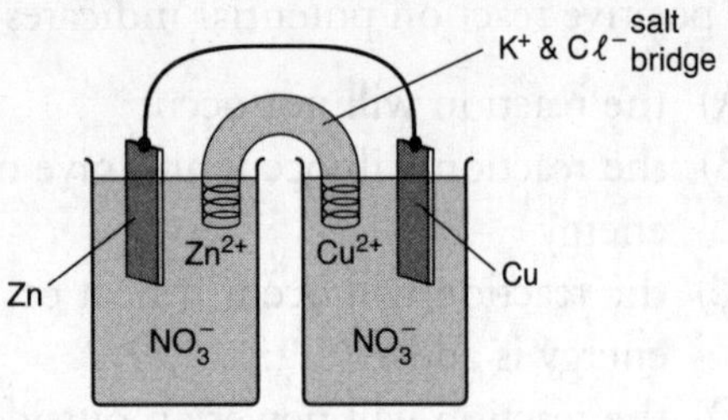

15. In the electrochemical cell shown above, which of the following half-reaction occurs at the anode?

(A) $Cu^{2+} + e^- \rightarrow Cu+$
(B) $Zn(s) \rightarrow Zn^{2+} + 2e^-$
(C) $Zn^{2+} + 2e^- \rightarrow Zn(s)$
(D) $Cu(s) \rightarrow Cu^{2+} + 2e^-$
(E) $Cu^{2+} + 2\,e^- \rightarrow Cu(s)$

Answers and Explanations

1. **(A)** Table salt is the only substance listed that has ions in its water solution and will show a current is moving through the system by lighting the conductivity apparatus.

2. **(C)** The current is carried through the solution by ions migrating to the charged plates that act as the anode and the cathode.

3. **(E)** The extent of ionization of a substance in solvent depends on all four of the listed factors.

4. **(D)** The only statement that is true is that the total positive charge in solution equals the total negative charge. The positive ions migrate to the negative pole called the cathode so they are called cations.

5. **(B)** The hydronium ion is written as H_3O^+.

6. **(B)** Because the anode is positively charged, it attracts the negatively charged Cl^- ions that are released from copper chloride. When it is oxidized by losing its electon, it forms a Cl_2 molecule with another atom of chlorine to be released as chlorine gas. The anode reaction would be:

$$2Cl^- \rightarrow Cl_2\uparrow + 2e^-$$

7. **(B)** Ions not only exist in some water solutions but also in ionic crystal lattices like sodium chloride.

8. **(B)** When an ionic compound is melted, the ions become free to migrate in the liquid and thereby carry an electric current.

9. **(D)** The cathode in an electrochemical cell is the electrode at which reduction occurs. The other pole is called the anode.

10. **(C)** The electrolysis of a dilute water solution of sodium chloride contains chloride ions, which are attracted to the anode and are set free as chlorine molecules. The cathode reaction is somewhat more complicated. Although the sodium ions are attracted to the cathode, they are not set free as atoms. Because water can ionize to some extent, and the standard electrode chart shows that the hydrogen ion is reduced more easily than the sodium ion, it is hydrogen, not sodium, that is set free at the cathode. The reactions are:

Cathode reaction; $2H_2O + 2e^- \rightarrow H_2\uparrow + 2OH^-$

Reduction: $2Cl^- \rightarrow Cl_2\uparrow + 2e^-$

Net reaction: $2H_2O + 2Cl^- \rightarrow H_2\uparrow + Cl_2\uparrow + 2OH^-$

11. **(B)** Because zinc is higher than gold in the standard reduction potentials listing, it will remain in the ionic form, whereas gold will favor the nonionized form. The two half-reactions that would have to occur with the electrode potentials for each are:

Oxidation half-reaction: $Zn^{2+} + 2e^- \rightarrow Zn\ E^0 = -0.76\ V$

Reduction half-reaction: $Au \rightarrow Au^{3+} + 3e^-\ E^0 = -1.42\ V$

Balancing the electrons: $3Zn^{2+} + 6e^- \rightarrow 3Zn\ ^*\ E^0 = -0.76\ V$
$2Au^{3+} + 6e^- \rightarrow 2Au\ ^*E^0 = -1.42\ V$

Net reaction: $3Zn + 2Au^{3+} \rightarrow 3Zn^{2+} + 2Au\ E^0 = -2.18\ V$

This negative cell potential indicates that these reactions would not occur.

Note:** The standard potentials (E^0) are ***not multiplied by the coefficients in calculating the E^0 for the reaction.

12. **(B)** A positive reaction potential indicates that the reaction will occur and that energy is given off.

13. **(A)** Ca is the most active, therefore, it loses electrons so that reduction can occur. Reducing agents are oxidized.

14. **(C)** Because Hg is below H in activity, it cannot displace it.

15. **(B)** Zn acts as the anode in this setup. If a power source were in the wiring, the zinc would be the cathode.

Some Representative Groups and Families

CHAPTER 13

- Describe the properties, both physical and chemical, of the major members of each group and family, and the common compounds formed by each.
- Write equations for major processes involving these elements.

In the following section, a brief description is given of some of the important and representative groups of elements usually discussed in most first-year chemistry courses.

SULFUR FAMILY

Since we discussed oxygen in Chapter 5, the most important element in this family left to discuss is sulfur.

Sulfur is found free in the volcanic regions of Japan, Mexico, and Sicily. It is removed from the rock mixtures by heating in retorts or furnaces.

Sulfuric Acid

Important properties of sulfuric acid. Sulfuric acid ionizes in two steps:

$$H_2SO_4 + H_2O \rightleftharpoons H_3O^+ + HSO_4^- \qquad K_{a_1} \text{ is very large.}$$
$$HSO_4^- + H_2O \rightleftharpoons H_3O^+ + SO_4^{2-} \qquad K_{a_2} \text{ is very small.}$$

to form a strong acid solution. The ionization is more extensive in a dilute solution. Most hydronium ions are formed in the first step. Salts formed with the HSO_4^- (bisulfate ion) are called **acid salts**; the SO_4^{2-} (sulfate ion) forms **normal salts**. Sulfuric acid reacts like other acids, as shown below:

With active metals:	$Zn + H_2SO_4 \rightarrow ZnSO_4 + H_2(g)$ (for dilute H_2SO_4)
With bases:	$2NaOH + H_2SO_4 \rightarrow Na_2SO_4 + 2H_2O$
With metal oxides:	$MgO + H_2SO_4 \rightarrow MgSO_4 + H_2O$
With carbonates:	$CaCO_3 + H_2SO_4 \rightarrow CaSO_4 + H_2O + CO_2(g)$

Table 12. Allotropic Forms of Sulfur

Characteristic	Allotropic Form Rhombic	Monoclinic	Amorphous
Shape	Rhombic or octahedral crystals 	Needle-shaped monoclinic crystals 	Noncrystalline
Color	Pale-yellow, opaque, brittle	Yellow, waxy, translucent, brittle	Dark, tough, elastic
Preparation	Crystallizes from CS_2 solution of sulfur	Crystallizes from molten sulfur as it cools	Obtained by quick-cooling very hot liquid sulfur
Stability	Stable below 96°C	Stable between 96° and 119°C	Unstable; slowly changes to rhombic

Sulfuric acid has other particular characteristics.

As an oxidizing agent:

$$Cu + 2H_2SO_4\,(concentrated) \rightarrow CuSO_4 + SO_2(g) + 2H_2O$$

As a dehydrating agent with carbohydrates:

$$C_{12}H_{22}O_{11}\,(sugar) \xrightarrow[H_2SO_4]{Conc.} 12C + 11H_2O$$

Important uses of H_2SO_4

1. Making other acids (because of its high boiling point, 338°C):

$$NaCl + H_2SO_4 \xrightarrow{heat} NaHSO_4 + HCl(g)$$

(With more heat and an excess of salt, the second hydrogen ion can also be removed to form the salt Na_2SO_4.)

2. Washing objectionable colors from gasoline made by the "cracking" process.
3. Dehydrating agent in making explosives, dyes, and drugs.
4. Freeing metals of iron or steel of scale and rust. This is called "pickling."
5. Electrolyte in ordinary lead storage batteries.
6. Precipitating bath in making rayon.

Other Important Compounds of Sulfur

Hydrogen sulfide is a colorless gas having an odor of rotten eggs. It is fairly soluble in water and is poisonous in rather small concentrations. It can be prepared by reacting ferrous sulfide with an acid, such as dilute HCl:

$$FeS + 2HCl \rightarrow FeCl_2 + H_2S(g)$$

Hydrogen sulfide burns in excess oxygen to form compounds of water and sulfur dioxide. If insufficient oxygen is available, some free sulfur will form. It is only a weak acid in a water solution. Hydrogen sulfide is used widely in qualitative laboratory tests since many metallic sulfides precipitate with recognizable colors. These sulfides are sometimes used as paint pigments. Some common sulfides and their colors are:

ZnS—White
CdS—Bright yellow
As_2S_3—Lemon yellow
Sb_2S_3—Orange
CuS—Black
HgS—Black
PbS—Brown-black

Another important compound of sulfur is **sulfur dioxide**. It is a colorless gas with a suffocating odor.

The structure of sulfur dioxide is a good example of a **resonance structure**. Its molecule is depicted in Figure 39.

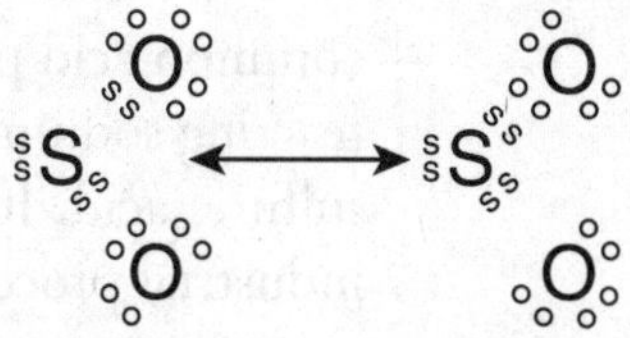

Figure 39. Sulfur Dioxide Molecule

TIP

Resonance is a hybrid of the two structures shown.

You will notice in Figure 39 that the covalent bonds between sulfur and oxygen are shown in one drawing as single bonds and in the other as double bonds. This signifies that the bonds between the sulfur and oxygens have been shown by experimentation to be neither single nor double bonds, but "hybrids" of the two. Sulfur trioxide, shown below, also is a resonance structure.

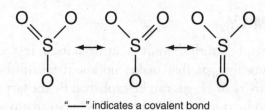

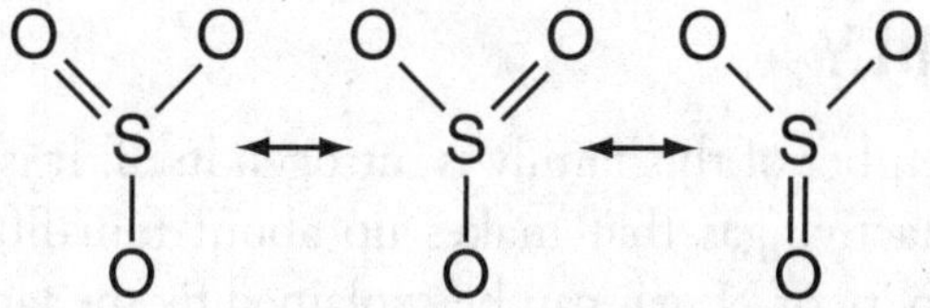

"—" indicates a covalent bond

HALOGEN FAMILY

The common members of the halogen family are shown in Table 13 with some important facts concerning them.

Table 13. Halogen Family

Item	Fluorine	Chlorine	Bromine	Iodine
Molecular formula	F_2	Cl_2	Br_2	I_2
Atomic number	9	17	35	53
Activity	Most active	←	to →	Least active
Outer energy level structure	7 electrons	7 electrons	7 electrons	7 electrons
State and color at room temperature	Gas—pale yellow	Gas—green	Liquid—dark red	Solid—purplish black crystals

Because each halogen lacks one electron in its outer principal energy level, these elements usually are acceptors of electrons (oxidizing agents). Fluorine is the most active nonmetal in the periodic chart.

Some Important Halides and Their Uses

Hydrochloric acid	—common acid prepared in the laboratory by reacting sodium chloride with concentrated sulfuric acid. It is used in many important industrial processes.
Silver bromide and silver iodide	—halides used on photographic film. Light intensity is recorded by developing as black metallic silver those portions of the film upon which the light fell during exposure.
Hydrofluoric acid	—acid used to etch glass by reacting with SiO_2 to release silicon fluoride gas. Also used to frost lightbulbs.

NITROGEN FAMILY

The most common member of this family is nitrogen itself. It is a colorless, odorless, tasteless, and rather inactive gas that makes up about four-fifths of the air in our atmosphere. The inactivity of N_2 gas can be explained by the fact that the two atoms of nitrogen are bonded by three covalent bonds (:N⋮⋮N:) that require a great deal of energy to break. Since nitrogen must be "pushed" into combining with other elements, many of its compounds tend to decompose violently with a release of the energy that went into forming them.

Nature "fixes" nitrogen, or makes nitrogen combine, by means of a nitrogen-fixing bacteria found in the roots of beans, peas, clover, and other leguminous plants. Discharges of lightning also cause some nitrogen fixation with oxygen to form nitrogen oxides.

Nitric Acid

An important compound of nitrogen is nitric acid. This acid is useful in making dyes, celluloid film, and many of the lacquers on cars.

Its physical properties are: it is a colorless liquid (when pure), it is one and one-half times as dense as water, it has a boiling point of 86°C, the commercial form is about 68% pure, and it is miscible with water in all proportions.

Its outstanding chemical properties are: the dilute acid shows the usual properties of an acid except that it rarely produces hydrogen when it reacts with metals, and it is quite unstable and decomposes as follows:

$$4HNO_3 \rightarrow 2H_2O + 4NO_2(g) + O_2(g)$$

Because of this ease of decomposition, nitric acid is a good oxidizing agent. When it reacts with metals, the nitrogen product formed will depend on the conditions of the reaction, especially the concentration of the acid, the activity of the metal, and the temperature. If the nitric acid is concentrated and the metal is copper, the principal reduction product will be nitrogen dioxide (NO_2), a heavy, red-brown gas with a pungent odor.

$$Cu + 4HNO_3 \rightarrow Cu(NO_3)_2 + 2NO_2(g) + 2H_2O$$

With dilute nitric acid, this reaction is:

$$3Cu + 8HNO_3 \rightarrow 3Cu(NO_3)_2 + 4H_2O + 2NO(g)$$

The product NO, called nitric oxide, is colorless and is immediately oxidized in air to NO_2 gas.

With still more dilute nitric acid, considerable quantities of nitrous oxide (N_2O) are formed; with an active metal like zinc, the product may be the ammonium ion (NH_4^+).

When nitric acid is mixed with hydrochloric acid, the mixture is called **aqua regia** because of its ability to dissolve gold. The great activity of aqua regia is due to the formation of nitrosyl chloride (NOCl) and chlorine (Cl_2). The Cl_2 reacts with gold to form the soluble compound of gold.

Other Important Compounds of Nitrogen

Ammonia is one of the oldest known compounds of nitrogen. In times past, it was prepared by distilling leather scraps, hoofs, and horns. It can be prepared in the laboratory by heating an ammonium salt with a strong base. An example is:

$$Ca(OH)_2 + (NH_4)_2SO_4 \rightarrow CaSO_4 + 2H_2O + 2NH_3(g)$$

Industrial methods include the **Haber process** (discussed on page 235), which combines nitrogen and hydrogen:

$$3H_2 + N_2 \rightarrow 2NH_3$$

and the destructive distillation of coal, which gives off NH_3 as a by-product.

The physical properties of ammonia are: it is a colorless, pungent gas that is extremely soluble in water; it is lighter than air; and it has a high critical temperature (the critical temperature of a gas is the temperature above which it cannot be liquefied by pressure alone).

Some other nitrogen compounds are summarized in Table 14.

Table 14. Some Nitrogen Compounds

Name	Formula	Notes of Importance
Nitrate ion	NO_3^-	1. Nitrates of metals can be decomposed with heat to release oxygen.
		2. Nitrate test—add freshly prepared ferrous sulfate solution; hold test tube at an angle and slowly add concentrated sulfuric acid down the side; if a brown ring forms just above the sulfuric acid level, it indicates a positive test for nitrate ion. (Could repeat using acetic instead of sulfuric acid; if brown ring again forms, it indicates a nitrite.)
Nitrous oxide	N_2O	1. Colorless; $1\frac{1}{2}$ times heavier than air; slight sweet smell and odor; supports combustion much like oxygen.
		2. Sometimes used as an anesthetic called "laughing gas."
Nitric oxide	NO	1. Colorless, slightly heavier than air, does not support combustion, is easily oxidized to NO_2 when exposed to air.
Nitrogen dioxide	NO_2	1. Reddish brown gas, 1.6 times heavier than air, very soluble in water, and extremely poisonous; has unpleasant odor.
Nitrogen tetroxide	N_2O_4	1. When NO_2 is cooled, its brown color fades to a pale yellow (N_2O_4). Two molecules of NO_2 combine to form this new compound.

Other Members of the Nitrogen Family

As you proceed down Table 15, the elements in the nitrogen family change from gases to volatile solids, and then to less volatile solids. At the same time acid-forming properties are decreasing while base-forming properties are increasing. These changes can be attributed to the increased atomic radii and the decreased ability to attract and hold the outer energy level electrons.

Table 15. The Nitrogen Family

Name	Symbol	Atomic Number	Properties and Uses
Nitrogen	N	7	(Discusssed)
Phosphorus	P	15	Two allotropic forms: white (kindling temperature near room temperature, thus kept under water) and red. When burned in oxygen, forms P_2O_5, the acid anhydride of phosphoric acid. When P_2O_5 reacts with water, it first forms metaphosphoric acid, HPO_3, then more water reacts to form the orthophosphoric acid. Phosphorus is used in making matches.
Arsenic	As	33	Slightly volatile gray solid. Most abundant compound is As_2O_3, which is used as a preservative, in medicines, and in glass making (highly poisonous).
Antimony	Sb	51	Lustrous gray solid. Forms very weak acid, mostly basic solutions.
Bismuth	Bi	83	Almost wholly metallic properties. Shiny, dense metal with slight reddish tinge. Low melting point; therefore used in alloys for electric fuses and fire-sprinkler systems. "Woods metal" is the most outstanding of these (melting point 70°C).

METALS

Properties of Metals

Some physical properties of metals are: they have metallic luster, they can conduct heat and electricity, they can be pounded into sheets (are malleable), they can be drawn into wires (are ductile), most have a silvery color, they have densities usually between 7 and 14 grams per cubic centimeter, and none is soluble in any ordinary solvent without a chemical change.

The general chemical properties of metals are: they are electropositive, and the more active metallic oxides form bases, although some metals form amphoteric hydroxides that can react as both acids and bases.

Metals can be removed from their ores by various means depending on their degrees of activity (see Table 16).

Table 16. Means of Extracting Metals

Occurrence	Activity Decreasing	Metal	Ease of Reduction	Reduced By
In combined form only		Potassium Calcium Sodium Magnesium Aluminum	Very difficult	Electrolysis
In combined form only		Manganese Zinc Chromium	Difficult	Carbon with diminishing difficulty
Usually combined		Iron Cadmium Cobalt Nickel	Difficult	Carbon with diminishing difficulty
Usually combined		Tin Lead	Easy	Carbon
Free and combined		HYDROGEN Antimony Bismuth Arsenic Copper	Easy	Carbon
Free and combined		Mercury Silver	Very easy	By heating oxides
Free		Platinum Gold	Easiest	All compounds decomposed by heat

In some cases the ore is concentrated before it is reduced. This might be done by a merely physical means, such as flotation, in which the ore is crushed, mixed with a water solution containing a suitable oil (e.g., pine oil), and then vigorously agitated so that a frothy mass of oil rises to the top with the adhering metal-bearing portion of the ore. The froth is removed from the top of the machine and filtered to recover the metal ore.

A chemical process is used to convert sulfide or carbonate ores to oxides. It is called **roasting** and consists of heating the compounds in the air. For example:

$$2ZnS + 3O_2 \rightarrow 2ZnO + 2SO_2(g)$$

and

$$PbCO_3 \rightarrow PbO + CO_2(g)$$

These oxides can then be treated with appropriate reducing agents to free the metal.

The following table summarizes the properties of the first two families of metals found in Group IA or 1 and Group IIA or 2 of the periodic chart. The table shows the distinct similarities in the properties and bonding of these metals. These similarities are directly related to the similarities that exist in the outer orbital electron configuration.

Element	Outer Orbital	Bonding and Charge	Physical Properties
Alkali Metals			
Lithium, Li	$2s^1$	Ionic, 1+	Silvery white
Sodium, Na	$3s^1$	Ionic, 1+	Silvery white
Potassium, K	$4s^1$	Ionic, 1+	Silvery white
Rubidium, Rb	$5s^1$	Ionic, 1+	Silvery white
Cesium, Cs	$6s^1$	Ionic, 1+	Silvery white
Francium, Fr	$7s^1$	Ionic, 1+	Silvery white
Alkaline-Earth Metals			
Beryllium, Be	$2s^2$	Ionic, 2+	Gray-white
Magnesium, Mg	$3s^2$	Ionic, 2+	Silvery white
Calcium, Ca	$4s^2$	Ionic, 2+	Silvery white
Strontium, Sr	$5s^2$	Ionic, 2+	Silvery white
Barium, Ba	$6s^2$	Ionic, 2+	Yellow-white; lumpy
Radium, Ra	$7s^2$	Ionic, 2+	White; radioactive

Some Important Reduction Methods

The **Hall process** is the electrolytic method of preparing **aluminum**. It uses molten cryolite (Na_3AlF_6) as the solvent for bauxite ore ($Al_2O_3 \cdot 2H_2O$). This mixture is electrolyzed in a rectangular iron tank lined with carbon, which acts as the cathode. The anodes are heavy carbon rods suspended in the tank. The aluminum released at the cathode is tapped periodically and allowed to flow out of the tank so that the process can continue.

The outstanding properties of aluminum are its light weight with high strength, its ability to form an aluminum oxide coating that protects this metal from further oxidation, and its ability to conduct an electric current.

The Dow process is an electrolytic method of preparing **magnesium** from seawater. The magnesium salts are precipitated with $Ca(OH)_2$ solution to form $Mg(OH)_2$. The hydroxide is then converted to a chloride by adding hydrochloric acid to the hydroxide. This chloride is mixed with potassium chloride, melted, and electrolyzed to release the magnesium at the cathode and chlorine at the anode.

Magnesium has the important property of being a light, rigid, and inexpensive metal. It, too, can form a protective adherent oxide coating.

Another major metal prepared by electrolysis is **copper**. After the ore is concentrated by flotation and roasted to change sulfides to oxides, it is heated in a converter to get impure copper, called "blister" copper. This copper is then used as the anodes in the electrolysis apparatus, in which pure copper is used to make the cathode plates. As the process progresses, the copper in the anodes becomes ions in the solution, which are reduced at the cathodes as pure copper.

Copper has wide usage in electrical appliances and wires because of its high conductivity. It is widely used in alloys of bronze (copper and tin) and brass (copper and zinc).

Iron ore is refined by reduction in a blast furnace, that is, a large, cylinder-shaped furnace charged with iron ore (usually hematite, Fe_2O_3), limestone, and coke. A hot air blast, often enriched with oxygen, is blown into the lower part of the furnace through a series of pipes called tuyeres. The chemical reactions that occur can be summarized as follows:

Burning coke: $2C + O_2 \rightarrow 2CO(g)$
$C + O_2 \rightarrow CO_2(g)$

Reduction of CO_2: $CO_2 + C \rightarrow 2CO(g)$

Reduction of ore: $Fe_2O_3 + 3CO \rightarrow 2Fe + 3CO_2(g)$
$Fe_2O_3 + 3C \rightarrow 2Fe + 3CO(g)$

Formation of slag: $CaCO_3 \rightarrow CaO + CO_2(g)$
$CaO + SiO_2 \rightarrow CaSiO_3$

The molten iron from the blast furnace is called **pig iron**.

From pig iron, the molten metal may undergo one of three steel-making processes that burn out impurities and set the contents of carbon, manganese, sulfur, phosphorus, and silicon. Often nickel and chromium are alloyed in steel to give the particular properties of hardness needed for tool parts. The three most important means of making steel involve the basic oxygen, the open-hearth, and the electric furnaces. The first two methods are the most common.

The basic oxygen furnace uses a lined "pot" into which the molten pig iron is poured. Then a high-speed jet of oxygen is blown from a water-cooled lance into the top of the pot. This "burns out" impurities to make a batch of steel rapidly and cheaply.

The open-hearth furnace is a large oven containing a dish-shaped area to hold the molten iron, scrap steel, and other additives with which it is charged. Alternating blasts of flame are directed across the surface of the melted metal until the proper proportions of additives are established for that "heat" so that the steel will have the particular properties needed by the customer. The tapping of one of these furnaces holding 50 to 400 tons of steel is a truly beautiful sight.

The final method of making steel involves the **electric furnace**. This method uses enormous amounts of electricity through graphite cathodes that are lowered into the molten iron to purify it and produce a high grade of steel.

Alloys

An **alloy** is a mixture of two or more metals. In a mixture certain properties of the metals involved are affected. Three of these are:

1. Melting point — The melting point of an alloy is lower than that of its components.
2. Hardness — An alloy is usually harder than the metals that compose it.
3. Crystal structure — The size of the crystalline particles in the alloy determines many of the physical properties. The size of these particles can be controlled by heat treatment. If the alloy cools slowly, the crystalline particles tend to be larger. Thus, by heating and cooling an alloy, its properties can be altered considerably.

Common alloys are:

1. Brass, which is made up of copper and zinc.
2. Bronze, which is made up of copper and tin.
3. Steel, which has controlled amounts of carbon, manganese, sulfur, phosphorus, and silicon, is alloyed with nickel and chromium.
4. Sterling silver, which is alloyed with copper.

Metalloids

In the preceding sections, representative metals and nonmetals have been reviewed, along with the properties of each. Some elements, however, are difficult to classify as one or the other. One example is carbon. The diamond form of carbon is a poor conductor, yet the graphite form conducts fairly well. Neither form looks metallic, so carbon is classified as a nonmetal.

Silicon looks like a metal. However, its conductivity properties are closer to those of carbon.

Since some elements are neither distinctly metallic nor clearly nonmetallic, a third class, called the **metalloids**, is recognized.

The properties of metalloids are intermediate between those of metals and those of nonmetals. Although most metals form ionic compounds, metalloids as a group may form ionic or covalent bonds. Under certain conditions pure metalloids conduct electricity, but do so poorly, and are thus termed **semiconductors**. This property makes the metalloids important in microcircuitry.

The metalloids are located in the periodic table along the heavy dark line that starts alongside boron and drops down in steplike fashion between the elements found lower in the table (see Figure 40).

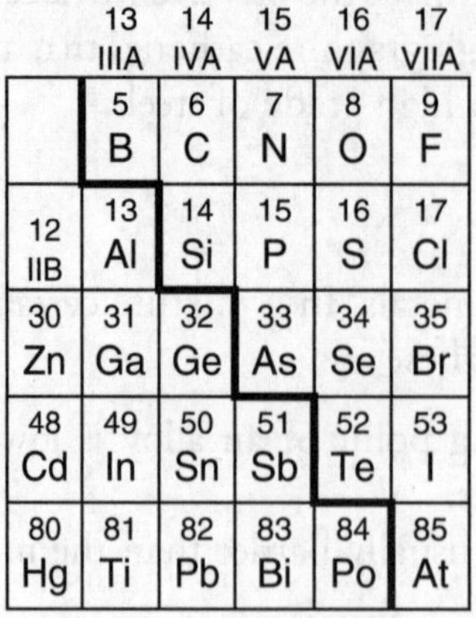
13 IIIA
14 IVA
15 VA
16 VIA
17 VIIA
5 B
6 C
7 N
8 O
9 F
12 IIB
13 Al
14 Si
15 P
16 S
17 Cl
30 Zn
31 Ga
32 Ge
33 As
34 Se
35 Br
48 Cd
49 In
50 Sn
51 Sb
52 Te
53 I
80 Hg
81 Ti
82 Pb
83 Bi
84 Po
85 At

Figure 40. Location of Metalloids

Chapter Summary

The following terms summarize all the concepts and ideas that were introduced in this chapter. You should be able to explain their meaning and how you would use them in chemistry. They appear in boldface type in this chapter to draw your attention to them. The boldface type also makes it easier for you to look them up if you need to. You could also use the search or "Google" action on your computer to get a quick and expanded explanation of these terms, laws, and formulas.

acid salt
allotropic form
alloy
electric furnace
Haber process
metalloid
normal salt
pig iron
resonance structure
rhombic, monoclinic, amorphous
semiconductor

Practice Exercises

1. The most active nonmetallic element is
   (A) chlorine
   (B) fluorine
   (C) oxygen
   (D) sulfur

2. The order of decreasing activity of the halogens is
   (A) Fl, Cl, I, Br
   (B) F, Cl, Br, I
   (C) Cl, F, Br, I
   (D) Cl, Br, I, F

3. A light-sensitive substance used on photographic films has the formula
   (A) AgBr
   (B) CaF_2
   (C) CuCl
   (D) $MgBr_2$

4. Sulfur dioxide is the anhydride of
   (A) hydrosulfuric acid
   (B) sulfurous acid
   (C) sulfuric acid
   (D) hyposulfurous acid

5. The charring action of sulfuric acid is due to its being
   (A) a strong acid
   (B) an oxidizing agent
   (C) a reducing agent
   (D) a dehydrating agent

6. Ammonia is prepared commercially by the
   (A) Ostwald process
   (B) arc process
   (C) Haber process
   (D) contact process

7. A nitrogen compound that has a color is
   (A) nitric oxide
   (B) nitrous oxide
   (C) nitrogen dioxide
   (D) ammonia

8. If a student heats a mixture of ammonium chloride and calcium hydroxide in a test tube, he will detect
   (A) no reaction
   (B) the odor of ammonia
   (C) the odor of rotten eggs
   (D) nitric acid fumes

9. The difference between ammonia and ammonium is
   (A) an electron
   (B) a neutron
   (C) a proton
   (D) radioactivity

10. An important ore of iron is
    (A) bauxite
    (B) galena
    (C) hematite
    (D) smithsonite

11. A reducing agent used in the blast furnace is
    (A) $CaCO_3$
    (B) CO
    (C) O_2
    (D) SiO_2

12. Aluminum is extracted from its oxide by
    (A) roasting
    (B) reduction with carbon
    (C) smelting
    (D) electrolysis

13. The placement of the halogen family in the Periodic Table explains which of the following statements?
    I. The most active nonmetallic element in the periodic table is fluorine.
    II. The normal physical state of the halogens goes from a solid to a gaseous state as you go down the family.
    III. The halogen elements become ions by filling the outermost *d* orbital.
    (A) I only
    (B) II only
    (C) I and II
    (D) II and III
    (E) I and III

14. Which of the following statements concerning metals are true?
    I. Aluminum is produced by the electrolytic Hall process from bauxite ore, $Al_2O_3 \cdot 2\ H_2O$.
    II. Brass is an alloy of copper and zinc.
    III. Silicon looks like a metal, but its conductivity is closer to that of carbon.
    (A) I only
    (B) II only
    (C) I and II
    (D) II and III
    (E) I, II, and III

Answers and Explanations

1. **(B)** Fluorine is the most active nonmetallic element because of its atomic structure. It needs only one electron to complete its outer shell and has the highest electronegativity.

2. **(B)** The order of activity of the halogens is from the smallest atomic radii to the largest straight down the family on the Periodic Table.

3. **(A)** Silver bromide, like many of the halogen salts, is light sensitive and is used in photographic films.

4. **(B)** The reaction of SO_2 with water forms sulfurous acid. The equation is: $SO_2 + H_2O \rightarrow H_2SO_3$.

5. **(D)** Sulfuric acid is a strong dehydrating agent and draws water to itself so strongly that is can char sucrose by withdrawing the hydrogen and oxygen from $C_{12}H_{22}O_{11}$.

6. **(C)** The Haber process is used to prepare ammonia commercially.

7. **(C)** The only nitrogen compound in the list that has a reddish brown color is nitrogen dioxide.

8. **(B)** The reaction of these two chemicals results in the production of ammonium hydroxide, which is unstable and forms ammonia gas and H_2O. The reaction is: $2NH_4Cl + Ca(OH)_2 \rightarrow 2NH_4OH + CaCl_2$

   Then, because the ammonium hydroxide is unstable at room temperatures, this reaction occurs: $NH_4OH \rightarrow NH_3\uparrow + H_2O$.

9. **(C)** The ammonia molecule is trigonal pyramidal with an unshared pair of electrons in one corner of the pyramid. This negative charge attracts a H^+ to form the ammonium ion, NH_4^+.

10. **(C)** An important ore of iron is hematite.

11. **(B)** CO, carbon monoxide, is used in the blast furnace as a reducing agent to react with oxide impurities.

12. **(D)** The famous Hall process is used to electrolytically produce aluminum.

13. **(A)** Because the halogen family's physical state goes from a gas to a solid and they form ions by completing the outer *p* orbital, the only true statement is I, that fluorine is the most active.

14 **(E)** All three statements are true.

Carbon and Organic Chemistry

CHAPTER 14

- Describe the bonding patterns of carbon and its allotropic forms.
- Explain the structural pattern and naming of the alkanes, alkenes, and alkynes, and their isomers.
- Show graphically how hydrocarbons can be changed and the development of these functional groups, their structures, and their names: alcohols, aldehydes, ketones, esters, and amines.

Carbon is unique. It forms inorganic substances such as carbon dioxide, graphite, and diamonds. It also forms organic substances without which life could not exist. It forms planar substances, tetrahedrons, and rings.

CARBON

Forms of Carbon

The element carbon occurs mainly in three allotropic forms: diamond, graphite, and amorphous (although some evidence shows the amorphous forms have some crystalline structure). In the mid-1980s, **fullerenes** were identified as a new allotropic form of carbon. They are found in soot that forms when carbon-containing materials are burned with limited oxygen. Their structure consists of near-spherical cages of carbon atoms resembling geodesic domes.

TIP

Allotropic forms:

diamond

graphite

amorphous

fullerenes

The diamond form has a close-packed crystal structure that gives it its property of extreme hardness. In it each carbon is bonded to four other carbons in a tetrahedron arrangement like this: These covalent solids form crystals that can be viewed as a single giant molecule made up of an almost endless number of covalent bonds. Because all of the bonds in this structure are equally strong, covalent solids are often very hard, and they are notoriously difficult to melt. Diamond is the hardest natural substance, and it melts at 3,550°C.

TIP

Diamond uses the sp^3 hybrid orbitals to explain its tetrahedron structure.

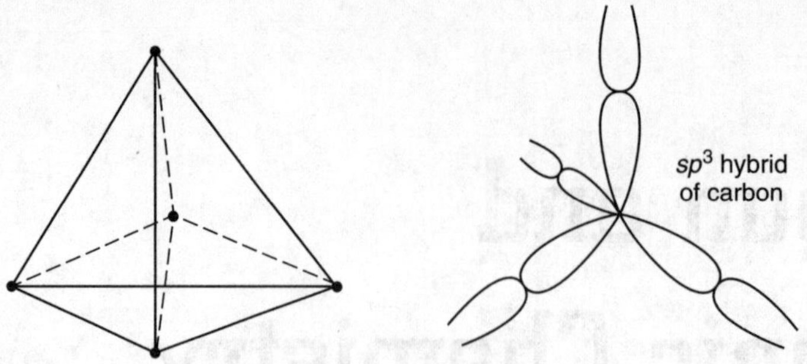

It has been possible to make synthetic diamonds in machines that subject carbon to extremely high pressures and temperatures. Most of these diamonds are used for industrial purposes, such as dies.

The graphite form is made up of planes of hexagonal structures that are weakly bonded to the planes above and below. This explains graphite's slippery feeling and makes it useful as a dry lubricant. Graphite is also mixed with clay to make "lead" for lead pencils. Its structure can be seen below. Graphite also has the property of being an electrical conductor.

TIP

Graphite uses the trigonal sp^2 hybrids.

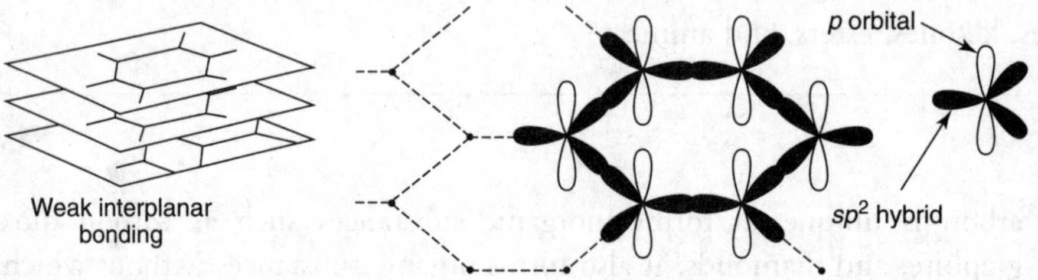

Some common amorphous forms of carbon are charcoal, coke, bone black, and lampblack.

The process of destructive distillation is used in the preparation of both charcoal and coke. This process involves the heating of a substance in an enclosed container so that volatile materials, which would cause wood and coal to burn with a flame, can be forced out. After these substances are removed by this process, the charcoal and coke merely glow when burned. Many of these volatile substances have proved to be useful by-products of destructive distillation. From the destructive distillation of coal we get coal gas, coal tar, and ammonia. Coal gas can be used as a fuel; coal tars are the primary source of most synthetic dyes and many other useful organic compounds; and ammonia is also a useful chemical.

Carbon Dioxide

Carbon dioxide (CO_2) is a widely distributed gas that makes up 0.04 percent of the air. There is a cycle that keeps this figure relatively stable. It is shown in Figure 41.

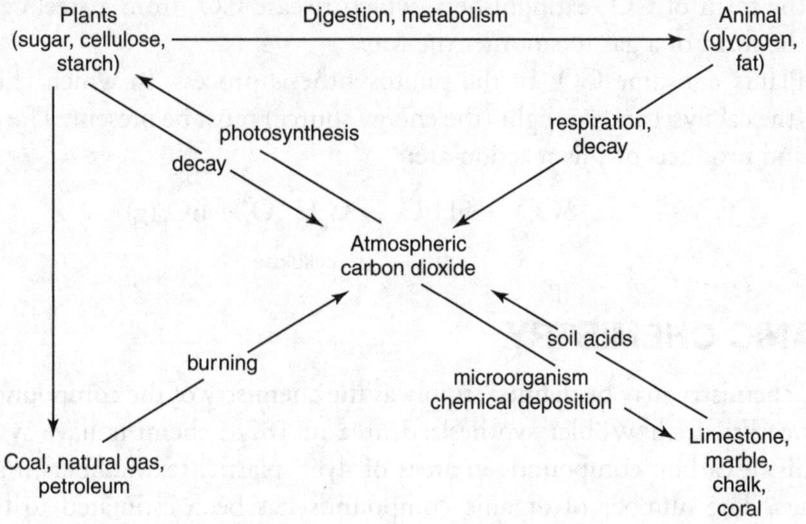

Figure 41. Carbon Dioxide Cycle

LABORATORY PREPARATION OF CO_2. The usual laboratory preparation consists of reacting calcium carbonate (marble chips) with hydrochloric acid, although any carbonate or bicarbonate and any common acid could be used. The gas is collected by water displacement or air displacement.

The test for carbon dioxide consists of passing it through limewater ($Ca(OH)_2$). If CO_2 is present the limewater turns cloudy because of the formation of a white precipitate of finely divided $CaCO_3$:

$$Ca(OH)_2 + CO_2 \rightarrow CaCO_3(s) + H_2O$$

Continued passing of CO_2 into the solution will eliminate the cloudy condition because the insoluble $CaCO_3$ becomes soluble calcium bicarbonate ($Ca(HCO_3)_2$):

$$CaCO_3(s) + H_2O + CO_2 \rightarrow Ca^{2+}(HCO_3^-)^2$$

This reaction can easily be reversed with increased temperature or decreased pressure. This is the way stalagmites and stalactites form on the floors and roofs of caves, respectively. The ground water containing calcium bicarbonate is deposited on the roof and floor of the cave and decomposes into solid calcium carbonate formations.

IMPORTANT USES OF CO_2

1. Because CO_2 is the acid anhydride of carbonic acid, it forms the acid when reacted with soft drinks, thus making them "carbonated" beverages.

$$CO_2 + H_2O \rightarrow H_2CO_3$$

2. Solid carbon dioxide (–78°C), or "dry ice," is used as a refrigerant because it has the advantage of not melting into a liquid; instead, it sublimes and in the process absorbs 3 times as much heat per gram as ice.
3. Fire extinguishers make use of CO_2 because of its properties of being $1\frac{1}{2}$ times heavier than air and not supporting ordinary combustion. It is used in

the form of CO_2 extinguishers, which release CO_2 from a steel cylinder in the form of a gas to smother the fire.

4. Plants consume CO_2 in the photosynthesis process, in which chlorophyll (the catalyst) and sunlight (the energy source) must be present. The reactants and products of this reaction are:

$$6CO_2 + 5H_2O \rightarrow C_6H_{10}O_5 + 6O_2(g)$$
cellulose

ORGANIC CHEMISTRY

Organic chemistry may be defined simply as the chemistry of the compounds of carbon. Since Friedrich Wöhler synthesized urea in 1828, chemists have synthesized thousands of carbon compounds in areas of dyes, plastic, textile fibers, medicines, and drugs. The number of organic compounds has been estimated to be in the neighborhood of a million and constantly increasing.

The carbon atom (atomic number 6) has four electrons in its outermost energy level, which show a tendency to be shared (electronegativity of 2.5) in covalent bonds. By this means, carbon bonds to other carbons, hydrogens, halogens, oxygen, and other elements to form the many compounds of organic chemistry.

HYDROCARBONS

Hydrocarbons, as the name implies, are compounds containing only carbon and hydrogen in their structures. The simplest hydrocarbon is methane, CH_4. As previously mentioned, this type of formula, which shows the kinds of atoms and their respective numbers, is called an **empirical** formula. In organic chemistry this is not sufficient to identify the compound it is used to represent. For example, the empirical formula C_2H_6O could denote either an ether or an ethyl alcohol. For this reason, a **structural** formula is used to indicate how the atoms are arranged in the molecule. The ether of C_2H_6O looks like this:

TIP
Organic chemistry makes use of structural formulas to show atomic arrangements.

$$\begin{array}{ccc} & H & & H \\ & | & & | \\ H- & C & -O- & C & -H \\ & | & & | \\ & H & & H \end{array}$$

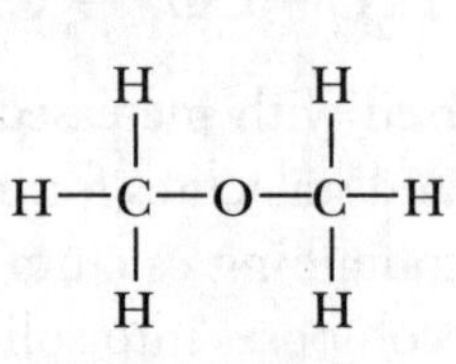

whereas the ethyl alcohol is represented by this structural formula:

$$\begin{array}{cc} H & H \\ | & | \\ H-C-C-OH \\ | & | \\ H & H \end{array}$$

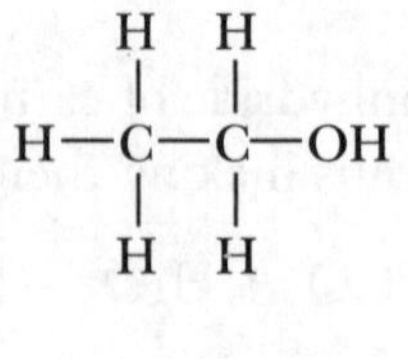

To avoid ambiguity, structural formulas are more often used than empirical formulas in organic chemistry. The structural formula of methane is

$$
\begin{array}{c}
\text{H} \\
| \\
\text{H} - \text{C} - \text{H} \\
| \\
\text{H}
\end{array}
$$

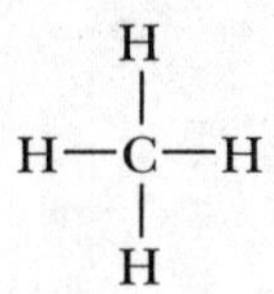

Alkane Series (Saturated)

Methane is the first member of a hydrocarbon series called the **alkanes** (or paraffin series). The general formula for this series is C_nH_{2n+2}, where n is the number of carbons in the molecule. Table 17 provides some essential information about this series. Since many other organic structures use the stem of the alkane names, you should learn these names and structures well. Notice that, as the number of carbons in the chain increases, the boiling point also increases. The first four alkanes are gases at room temperature; the subsequent compounds are liquid, then become more viscous with increasing length of the chain.

Since the chain is increased by a carbon and two hydrogens in each subsequent molecule, the alkanes are referred to as a **homologous** series.

The alkanes are found in petroleum and natural gas. They are usually extracted by fractional distillation, which separates the compounds by varying the temperature so that each vaporizes at its respective boiling point. Methane, which forms 90% of natural gas, can also be prepared in the laboratory by heating soda lime (containing NaOH) with sodium acetate:

$$NaC_2H_3O_2 + NaOH \rightarrow CH_4(g) + Na_2CO_3$$

When the alkanes are burned with sufficient air, the compounds formed are CO_2 and H_2O. An example is:

$$2C_2H_6 + 7O_2 \rightarrow 4CO_2(g) + 6H_2O(g)$$

The alkanes can be reacted with halogens so that hydrogens are replaced by a halogen atom: These are called **alkyl halides**.

$$
\begin{array}{c}
\text{H} \\
| \\
\text{H} - \text{C} - \text{H} \\
| \\
\text{H}
\end{array}
+ Br_2 \rightarrow
\begin{array}{c}
\text{H} \\
| \\
\text{H} - \text{C} - \text{Br} \\
| \\
\text{H}
\end{array}
+ HBr
$$

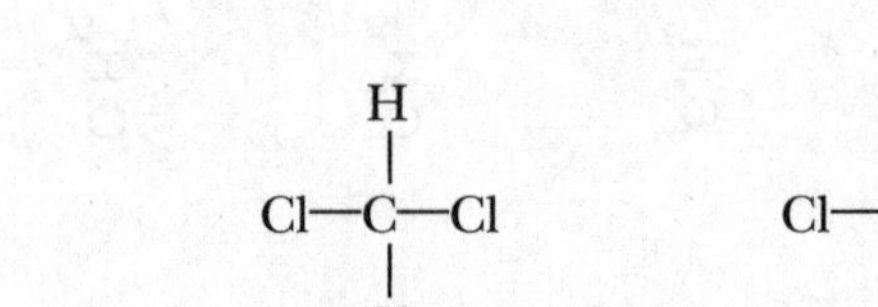

Some common substitution compounds of methane are:

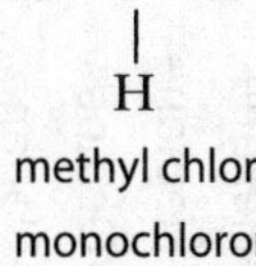

$\begin{array}{c} \text{H} \\ \| \\ \text{H} - \text{C} - \text{Cl} \\ \| \\ \text{H} \end{array}$	$\begin{array}{c} \text{H} \\ \| \\ \text{Cl} - \text{C} - \text{Cl} \\ \| \\ \text{Cl} \end{array}$	$\begin{array}{c} \text{Cl} \\ \| \\ \text{Cl} - \text{C} - \text{Cl} \\ \| \\ \text{Cl} \end{array}$
methyl chloride	chloroform	carbon tertrachloride
monochloromethane	trichloromethane	tetrachloromethane

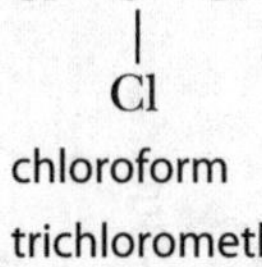

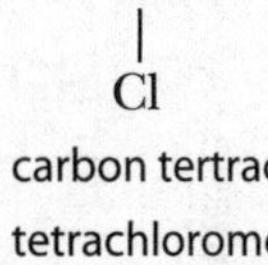

Table 17. The Alkanes

IUPAC Name	Molecular Formula	Number of Structural Isomers	Structure	State at Room Temperature	Boiling Point (°C)
Methane	CH_4	1	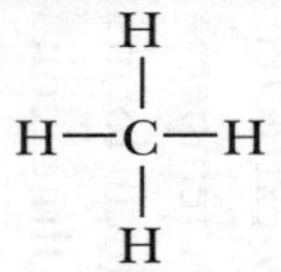	Gas	–162
Ethane	C_2H_6	1	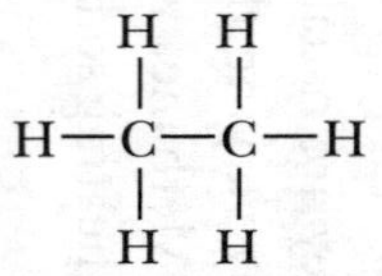		–89
Propane	C_3H_8	1	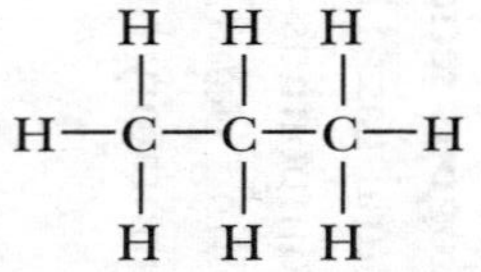		–42
n-Butane	C_4H_{10}	2	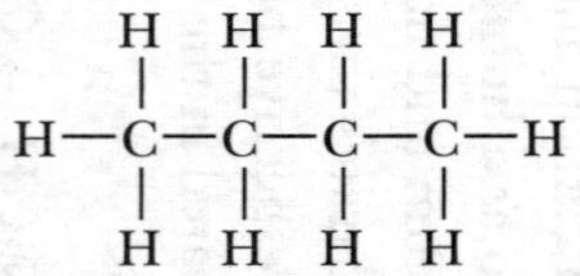		0
n-Pentane	C_5H_{12}	3	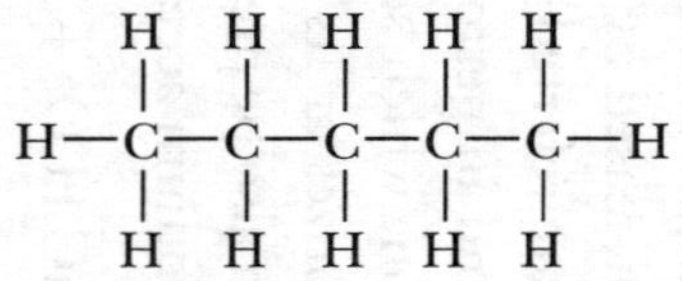	Liquid (*Note:* Solids at 17 carbons in the chain.)	36
n-Hexane	C_6H_{14}	5	$CH_3-CH_2-CH_2-CH_2-CH_2-CH_3$		69
n-Heptane	C_7H_{16}	7	$CH_3-CH_2-CH_2-CH_2-CH_2-CH_2-CH_3$		98
n-Octane	C_8H_{18}	18	$CH_3-CH_2-CH_2-CH_2-CH_2-CH_2-CH_2-CH_3$		126
n-Nonane	C_9H_{20}	35	$CH_3-CH_2-CH_2-CH_2-CH_2-CH_2-CH_2-CH_2-CH_3$		151
n-Decane	$C_{10}H_{22}$	75	$CH_3-CH_2-CH_2-CH_2-CH_2-CH_2-CH_2-CH_2-CH_2-CH_3$		174

NAMING ALKANE SUBSTITUTIONS. When an alkane hydrocarbon has an end hydrogen removed, it is referred to as an alkyl substituent or group. The respective name of each is the alkane name with -ane replaced by *-yl.* These are called **alkyl groups.**

TIP

Replace *-ane* with *-yl* to form alkyl groups.

Alkane	Alkyl Group	Compounds
methane 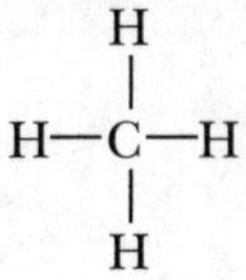	methyl 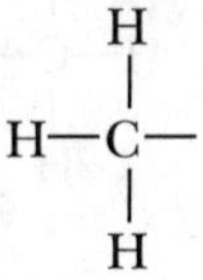	methyl bromide 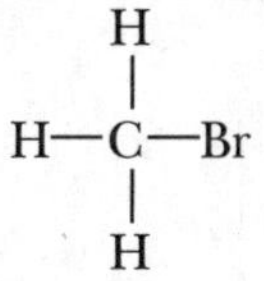
butane 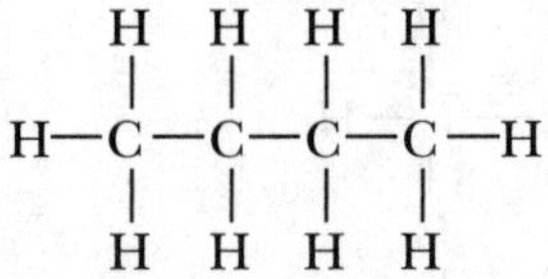	butyl 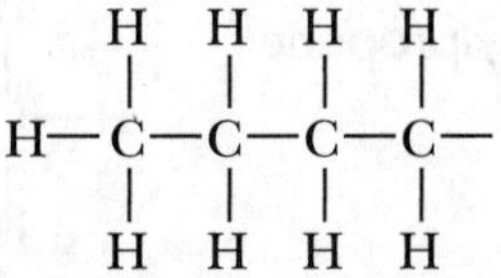	butyl chloride 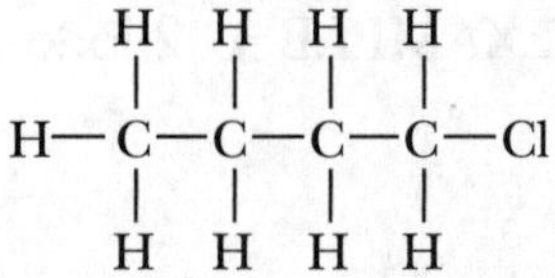

One method of naming a substitution product is to use the alkyl name for the respective chain and the halide as shown above.

The IUPAC system uses the name of the longest carbon chain as the parent chain. The carbon atoms are numbered in the parent chain to indicate where branching or substitution takes place. The direction of numbering is chosen so that the lowest numbers possible are given to the side chains. The complete name of the compound is arrived at by first naming the attached group, each of these being prefixed by the number of the carbon to which it is attached, and then the parent alkane. If a particular group appears more than once, the appropriate prefix (di, tri, and so on) is used to indicate how many times the group appears. A carbon atom number must be used to indicate the position of each such group. If two or more of the same group are attached to the same carbon atom, the number of the carbon atom is repeated. If two or more different substituted groups are in a name, they are arranged alphabetically.

EXAMPLE 1 2,2-dimethylbutane

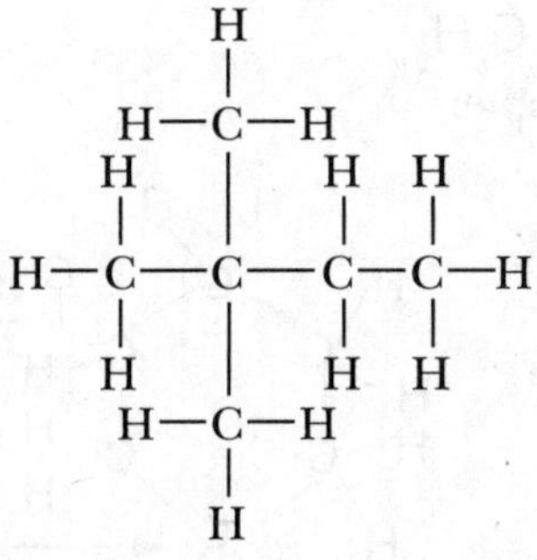

NOTE

Numbers have been added to the longest chain for identification only.

EXAMPLE 2 1,1-dichloro-3-ethyl-2,4-dimethylpentane

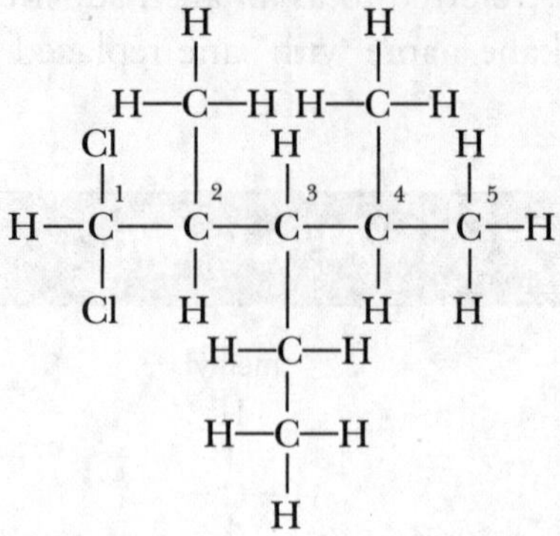

EXAMPLE 3 2-iodo-2-methylpropane

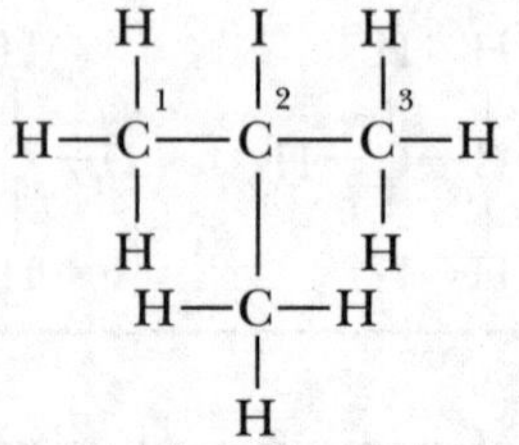

TIP

Cycloalkanes form single-bonded ring compounds.

CYCLOALKANES. Starting with propane in the alkane series, it is possible to get a ring form by attaching the two chain ends. This reduces the number of hydrogens by two.

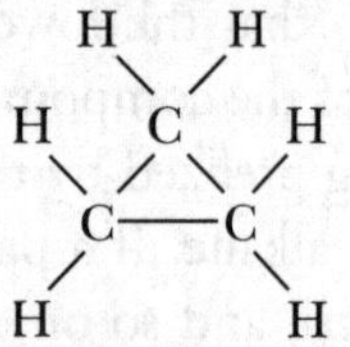

This hydrocarbon is called **cyclopropane.**

Cycloalkanes are named by adding the prefix cyclo- to the name of the straight-chain alkane with the same number of hydrocarbons, as shown above.

When there is only one alkyl group attached to the ring, no position number is necessary. When there is more than one alkyl group attached to the ring, the carbon atoms in the ring are numbered to give the lowest numbers possible to the alkyl groups. This means that one of the alkyl groups will always be in position 1. The general formula is C_nH_{2n}.

Here is an example:

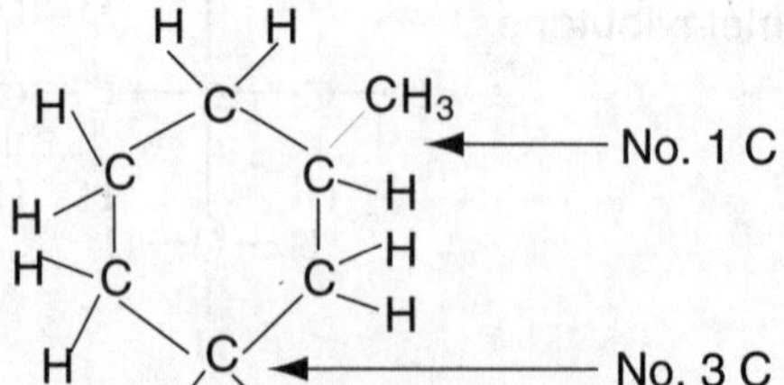

1,3-dimethylcyclohexane

If there are two or more alkyl groups attached to the ring, number the carbon atoms in the ring. Assign position number one to the alkyl group that comes first in alphabetical order, then number in the direction that gives the rest of the alkyl groups the lowest numbers possible.

Because all the members of the alkane series have single covalent bonds, this series and all such structures are said to be **saturated**.

If the hydrocarbon molecule contains double or triple covalent bonds, it is referred to as **unsaturated**.

PROPERTIES AND USES OF ALKANES. Properties for some straight-chain alkanes are indicated in Table 17. The trends in these properties can be explained by examining the structures of alkanes. The carbon-hydrogen bonds are nonpolar. The only forces of attraction between nonpolar molecules are weak intermolecular, or London dispersion, forces. These forces increase as the mass of a molecule increases.

The table also shows the physical states of alkanes. Smaller alkanes exist as gases at room temperature, while larger ones exist as liquids. Gasoline and kerosene consist mostly of liquid alkanes. Seventeen carbons are needed in the chain for the solid form to occur. Paraffin wax contains solid alkanes.

The differences in the boiling points of mixtures of the liquid alkanes found in petroleum make it possible to separate the various components by **fractional distillation**. This is the major industrial process used in refining petroleum into gasoline, kerosene, lubricating oils, and several other minor components.

Alkene Series (Unsaturated)

The **alkene** series has a double covalent bond between two adjacent carbon atoms. The general formula of this series is C_nH_{2n}. In naming these compounds, the suffix of the alkane is replaced by *-ene*. Two examples:

> **TIP**
>
> **Alkenes** have the form C_nH_{2n}

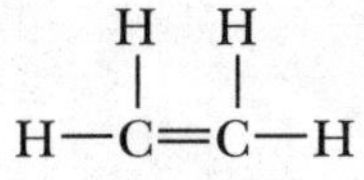

$$
\begin{array}{c}
\text{H} \quad\;\; \text{H} \\
|\qquad | \\
\text{H}-\text{C}=\text{C}-\text{H}
\end{array}
\qquad \text{ethene (common name: ethylene)}
$$

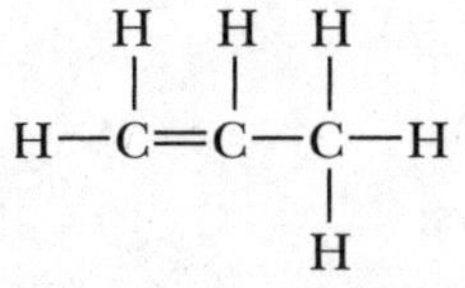

$$
\begin{array}{c}
\text{H} \quad\;\; \text{H} \quad\;\; \text{H} \\
|\qquad |\qquad | \\
\text{H}-\text{C}=\text{C}-\text{C}-\text{H} \\
|\\
\text{H}
\end{array}
\qquad \text{1-propene (common name: propylene)}
$$

Naming a more complex example is:

$$
\begin{array}{c}
\text{CH}_2-\text{CH}_3 \\
|\\
\text{CH}_2=\text{C}-\text{CH}_2-\text{CH}_2-\text{CH}_3
\end{array}
$$

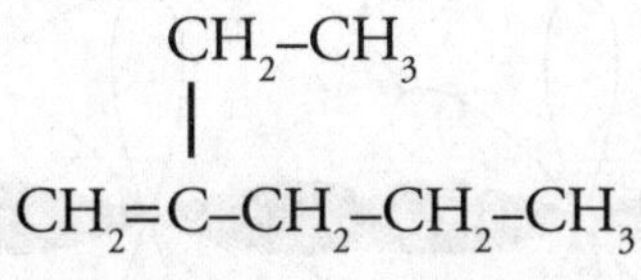

The position number and name of the alkyl group are in front of the double-bond position number. The alkyl group above is an ethyl group. It is on the second carbon atom of the parent hydrocarbon.

The name is: 2-ethyl-1-pentene

If the double bond occurs on an interior carbon, the chain is numbered so that the position of the double bond is designated by the lowest possible number assigned to the first doubly bonded carbon. For example:

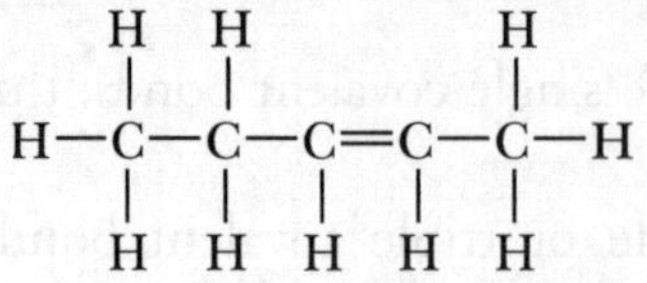

2-pentene

The bonding is more complex in the double covalent bond than in the single bonds in the molecule. Using the orbital pictures of the atom, we can show this as follows:

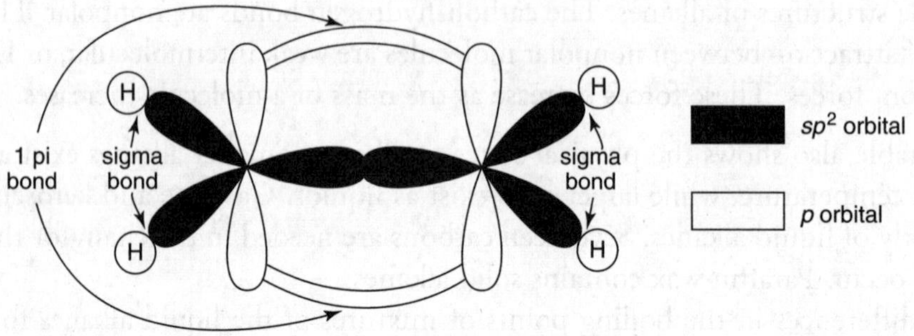

The two p lobes attached above and below constitute *one* bond called a pi (π) bond. The sp^2 orbital bonds between the carbons and with each hydrogen are referred to as sigma (σ) bonds.

TIP

Alkynes have the form C_nH_{2n-2}.

Alkyne Series (Unsaturated)

The **alkyne** series has a triple covalent bond between two adjacent carbons. The general formula of this series is C_nH_{2n-2}. In naming these compounds, the alkane suffix is replaced by *-yne*. Two examples:

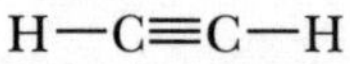

ethyne (common name: acetylene)

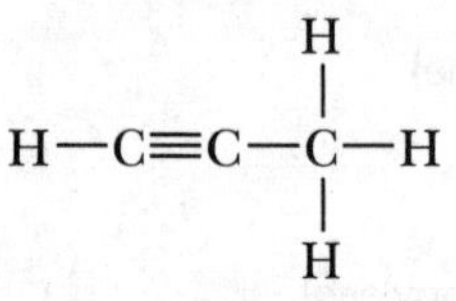

propyne

The orbital structure of ethyne can be shown as follows:

TIP

Notice that **pi bonds** are between p orbitals and that **sigma bonds** are between s and p orbitals.

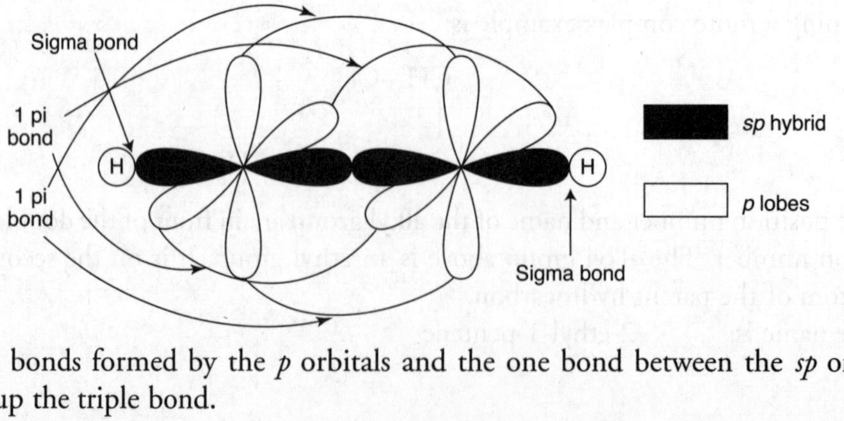

The bonds formed by the p orbitals and the one bond between the sp orbitals make up the triple bond.

The preceding examples show only one triple bond. If there is more than one triple bond, modify the suffix to indicate the number of triple bonds. For example, 2 would be a diyne, 3 would be a triyne, and so on. Next add the names of the alkyl groups if they are attached. Number the carbon atoms in the chain so that the first carbon atom in the triple bond nearest the end of the chain has the lowest number. If numbering from both ends gives the same positions for two triple bonds, then number from the end nearest the first alkyl group. Then, place the position numbers of the triple bonds immediately before the name of the parent hydrocarbon alkyne and place the alkyl group position numbers immediately before the name of the corresponding alkyl group.

Two more examples of alkynes are:

$CH_3–CH_2–CH_2–C≡CH$

1-pentyne

$CH≡C–CH–CH_3$
$\quad\quad\quad\ |$
$\quad\quad\ CH_3$

3-methyl-1-butyne

Naming a more complex example is:

$\quad\quad CH_2–CH_3$
$\quad\quad |$
$CH ≡ C–CH_2–CH_2–CH_3$

The position number and the name of the alkyl group are placed in front of the double-bond position number. The alkyl group above is an ethyl group. It is on the second carbon atom of the parent hydrocarbon.

The name is: 2-ethyl-1-pentyne

Aromatics

The aromatic compounds are unsaturated ring structures. The basic formula of this series is C_nH_{2n-6}, and the simplest compound is benzene (C_6H_6). The benzene structure is a resonance structure that is represented like this:

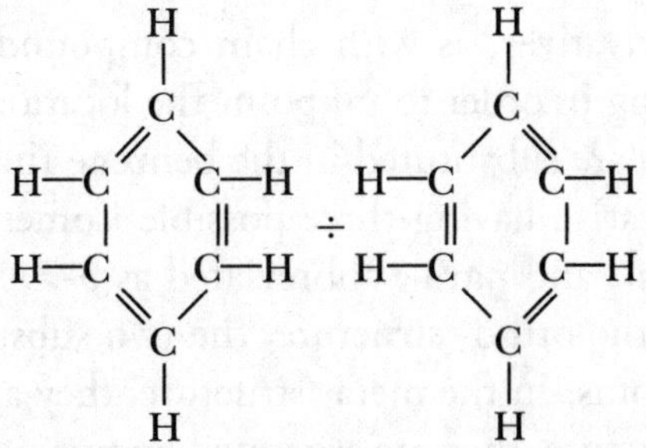

Note: The carbon-to-carbon bonds are neither single nor double bonds but hybrid bonds. This structural representation is called **resonance**.

TIP

C_6H_6, benzene is the simplest aromatic compound.

The orbital structure can be represented like this:

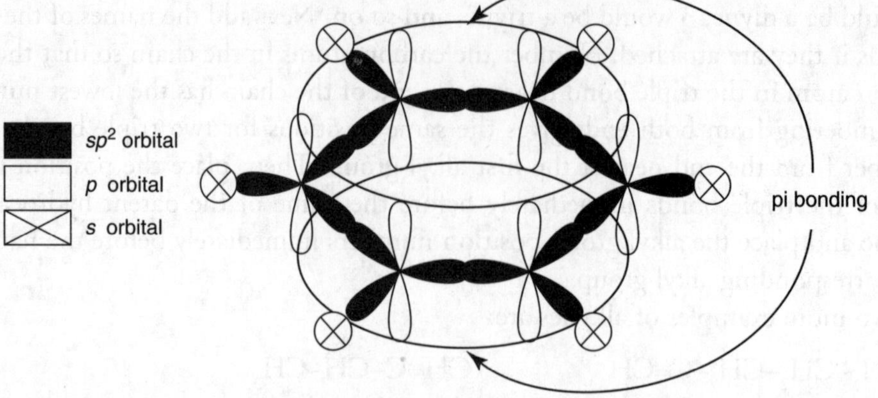

Most of the aromatics have an aroma, thus the name "aromatic."

The C_6H_5 group is a substituent called phenyl. This is the benzene structure with one hydrogen missing. If the phenyl substituent adds a methyl group, the compound is called toluene or methyl benzene.

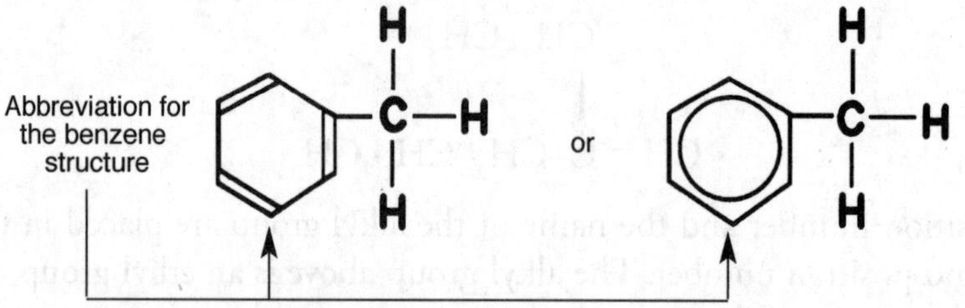

Two other members of the benzene series and their structures:

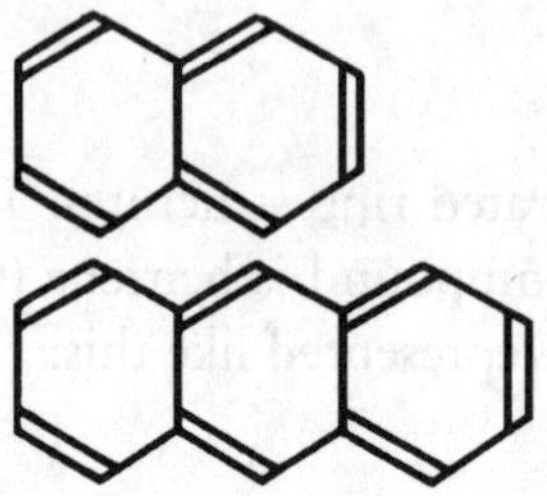

$C_{10}H_8$ naphthalene

$C_{14}H_{10}$ anthracene

The IUPAC system of naming benzene derivatives, as with chain compounds, involves numbering the carbon atoms in the ring in order to pinpoint the locations of the side chains. However, if only two groups are substituted in the benzene ring, the compound formed will be a benzene derivative having three possible isomeric forms. In such cases, the prefixes **ortho**-, **meta**-, and **para**-, abbreviated as *o*-, *m*-, and *p*-, may be used to name the isomers. In the ortho- structure, the two substituted groups are located on adjacent carbon atoms. In the meta- structure, they are separated by one carbon atom. In the para- structure, they are separated by two carbon atoms.

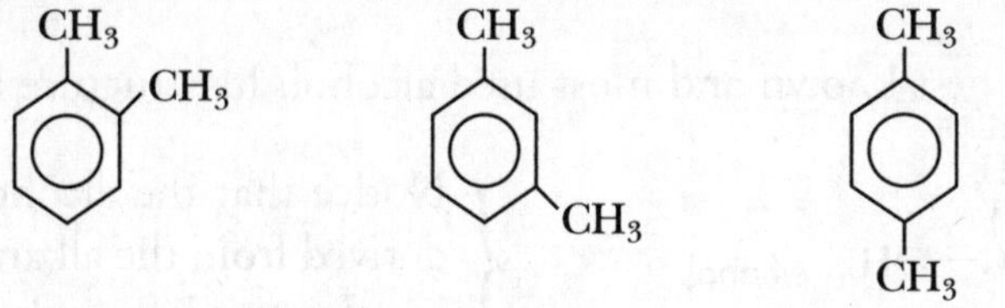

ortho- structure	meta- structure	para- structure
1,2-dimethylbenzene	1,3-dimethylbenzene	1,4-dimethylbenzene
or	or	or
o-xylene	*m*-xylene	*p*-xylene

Isomers

Many of the chain hydrocarbons can have the same formula, but their structures may differ. For example, butane is the first compound that can have two different structures or **isomers** for the same formula.

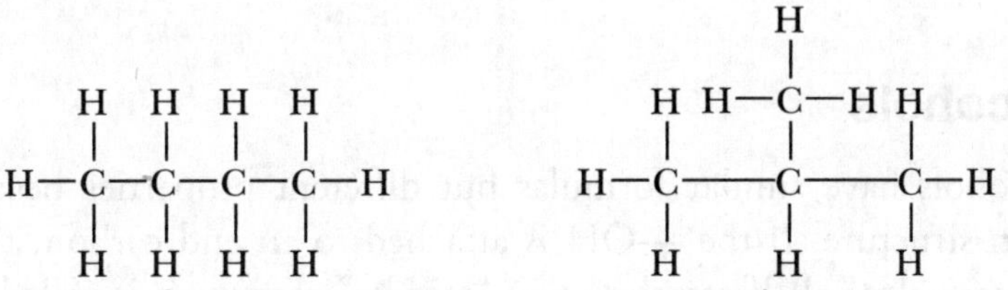

n-butane   isobutane

This isomerization can be shown by the following equation:

$$CH_3{-}CH_2{-}CH_2{-}CH_3 \xrightarrow[70\text{–}100^\circ C]{AlCl_3} CH_3{-}\underset{}{\overset{CH_3}{\overset{|}{C}H}}{-}CH_3$$

butane   isobutane

The isomers have different properties, both physical and chemical, from those of hydrocarbons with the normal structure.

HYDROCARBON DERIVATIVES

Alcohols—Methanol and Ethanol

The simplest alcohols are alkanes that have one or more hydrogen atoms replaced by the hydroxyl group, —OH. This is called its **functional group.**

> **TIP**
> Alcohol's functional group is R—OH. A hydroxyl group is attached to an alkyl stem.

- Methanol

Methanol is the simplest alcohol. Its structure is

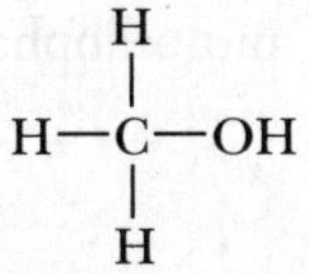

methanol or wood alcohol

PROPERTIES AND USES. Methanol is a colorless, flammable liquid with a boiling point of 65°C. It is miscible with water, is exceedingly poisonous, and can cause blindness if taken internally. It can be used as a fuel, as a solvent, and as a denaturant to make ethyl alcohol, unsuitable for drinking.

• Ethanol

Ethanol is the best known and most used alcohol. Its structure is

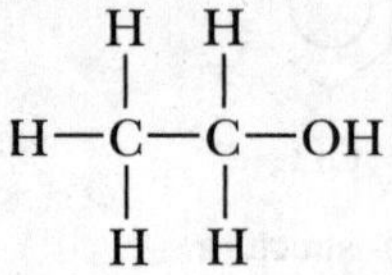

$$\begin{array}{c} \text{H} \quad \text{H} \\ | \qquad | \\ \text{H}-\text{C}-\text{C}-\text{OH} \\ | \qquad | \\ \text{H} \quad \text{H} \end{array} \quad \text{ethanol}$$

{ Notice that the alcohol names are derived from the alkanes by replacing the *e* ending with *-ol*. }

Its common names are ethyl alcohol and grain alcohol.

PROPERTIES AND USES. Ethanol is a colorless, flammable liquid with a boiling point of 78°C. It is miscible with water and is a good solvent for a wide variety of substances (these solutions are often referred to as "tinctures"). It is used as an antifreeze because of its low freezing point, -115°C, and for making acetaldehyde and ether.

Other Alcohols

Isomeric alcohols have similar formulas but different properties because of their differences in structure. If the —OH is attached to an end carbon, the alcohol is called a primary alcohol. If attached to a "middle" carbon, it is called a secondary alcohol. Some examples:

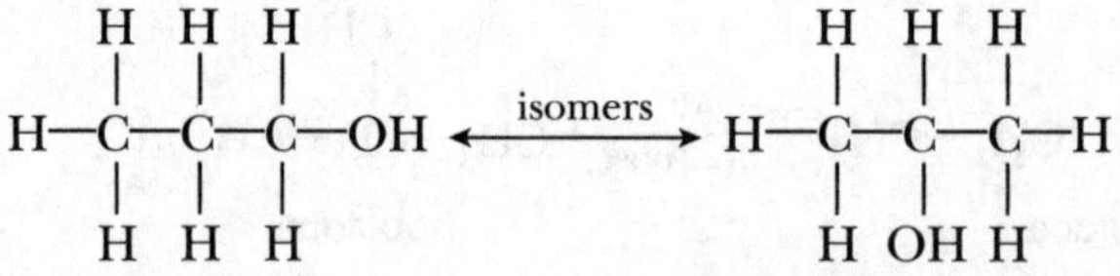

$$\begin{array}{c} \text{H} \quad \text{H} \quad \text{H} \\ | \quad | \quad | \\ \text{H}-\text{C}-\text{C}-\text{C}-\text{OH} \\ | \quad | \quad | \\ \text{H} \quad \text{H} \quad \text{H} \end{array} \xleftrightarrow{\text{isomers}} \begin{array}{c} \text{H} \quad \text{H} \quad \text{H} \\ | \quad | \quad | \\ \text{H}-\text{C}-\text{C}-\text{C}-\text{H} \\ | \quad | \quad | \\ \text{H} \quad \text{OH} \quad \text{H} \end{array}$$

1-propanol 2-propanol
or or
propyl alcohol isopropyl alcohol

The number in front of the name indicates to which carbon the —OH ion is attached.

Other alcohols with more than one —OH group:

$$\begin{array}{c} \text{H} \quad \text{H} \\ | \qquad | \\ \text{H}-\text{C}-\text{C}-\text{H} \\ | \qquad | \\ \text{OH} \quad \text{OH} \end{array}$$

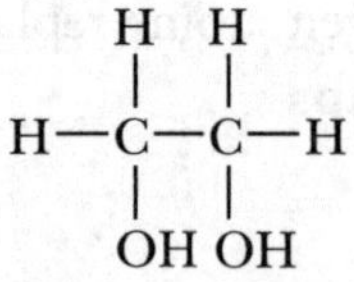

ethylene glycol
or
1,2-ethanediol

A colorless liquid, high boiling point, low freezing point. Used as permanent antifreeze in automobiles.

$$\begin{array}{c} \text{H} \quad \text{H} \quad \text{H} \\ | \quad | \quad | \\ \text{H}-\text{C}-\text{C}-\text{C}-\text{H} \\ | \quad | \quad | \\ \text{OH} \quad \text{OH} \quad \text{OH} \end{array}$$

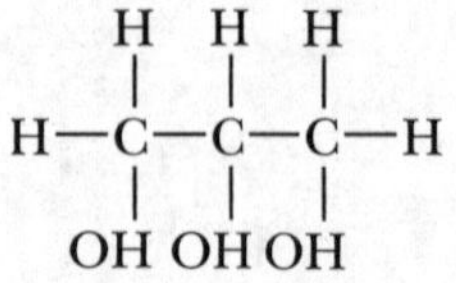

glycerine
or
glycerol
or
1,2,3-propanetriol

Colorless liquid, odorless, viscous, sweet taste. Used to make nitroglycerine, resins for paint, and cellophane.

Aldehydes

The functional group of an aldehyde is the $-C\!\!\begin{smallmatrix}\nearrow O\\ \searrow H\end{smallmatrix}$, formyl group. The general

formula is RCHO, where R represents a hydrocarbon radical.

PREPARATION FROM AN ALCOHOL. Aldehydes can be prepared by the oxidation of an alcohol. This can be done by inserting a hot copper wire into the alcohol. A typical reaction is:

$$\underset{\substack{\text{methanol}\\ \text{or}\\ \text{methyl alcohol}}}{\overset{\displaystyle H}{\underset{\displaystyle H}{H-\overset{\displaystyle |}{\underset{\displaystyle |}{C}}-OH}}} + [O] \xrightarrow{\substack{\text{mild oxidizing}\\ \text{agent}}} \left[\overset{\displaystyle H}{\underset{\displaystyle OH}{H-\overset{\displaystyle |}{\underset{\displaystyle |}{C}}-OH}}\right] \rightarrow \underset{\substack{\text{methanal}\\ \text{or}\\ \text{formaldehyde}}}{H-C\!\!\begin{smallmatrix}\nearrow O\\ \searrow H\end{smallmatrix}} + H_2O$$

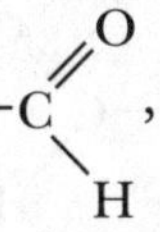

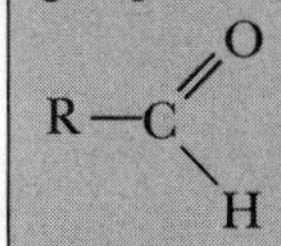

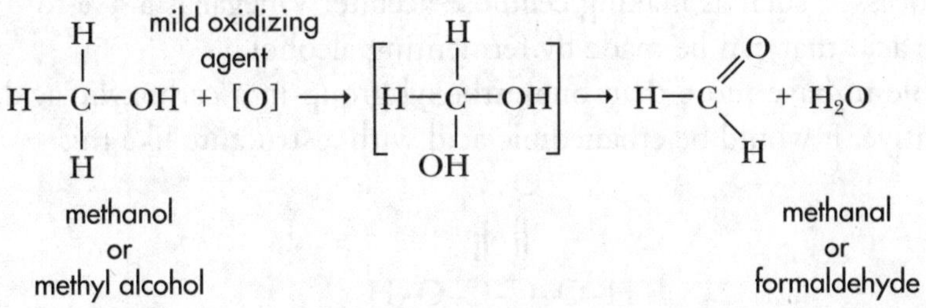

The middle structure is an intermediate structure; since two hydroxyl groups do not stay attached to the same carbon, it changes to the aldehyde by a water molecule "breaking away."

The aldehyde name is derived from the alcohol name by dropping the *-ol* and adding *-al*.

Ethanol forms ethanal (acetaldehyde) in the same manner.

Organic Acids or Carboxylic Acids

The functional group of an organic acid is the $-C\!\!\begin{smallmatrix}\nearrow O\\ \searrow OH\end{smallmatrix}$, carboxyl group. The general formula is R—COOH.

PREPARATION FROM AN ALDEHYDE. Organic acids can be prepared by the mild oxidation of an aldehyde. The simplest acid is methanoic acid, which is present in ants, bees, and other insects. A typical reaction is:

$$\underset{\substack{\text{methanal}\\ \text{or}\\ \text{formaldehyde}}}{H-C\!\!\begin{smallmatrix}\nearrow O\\ \searrow H\end{smallmatrix}} + [O] \rightarrow \underset{\substack{\text{methanoic acid}\\ \text{or}\\ \text{formic acid}}}{H-C\!\!\begin{smallmatrix}\nearrow O\\ \searrow OH\end{smallmatrix}}$$

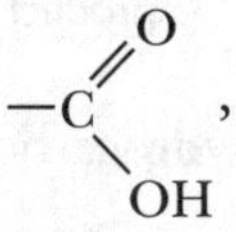

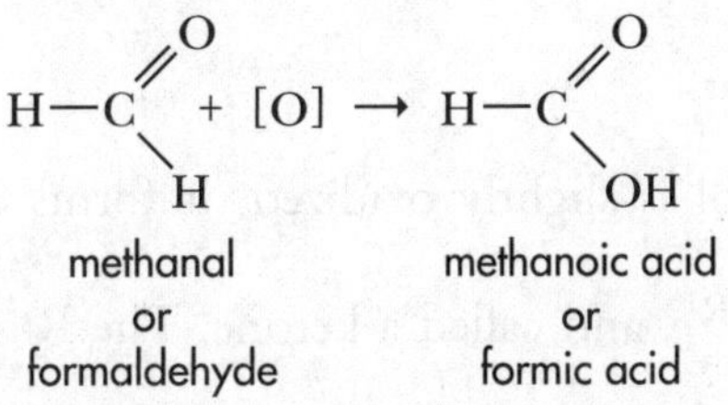

Notice that in the IUPAC system the name is derived from the alkane stem by adding *-oic*.

TIP

The aldehyde functional group is

$R-C\!\!\begin{smallmatrix}\nearrow O\\ \searrow H\end{smallmatrix}$

A formyl group is attached to an alkyl stem.

TIP

The carboxyl group is the functional group of an organic acid.

Ethanal can be oxidized to ethanoic acid:

$$H-\underset{\underset{H}{|}}{\overset{\overset{H}{|}}{C}}-C\overset{O}{\underset{H}{\diagdown}} + [O] \rightarrow H-\underset{\underset{H}{|}}{\overset{\overset{H}{|}}{C}}-C\overset{O}{\underset{OH}{\diagdown}}$$

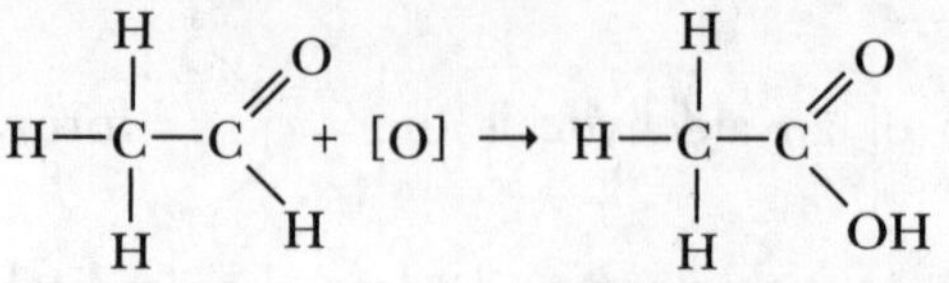

ethanal ethanoic acid
or
acetic acid

Acetic acid, as ethanoic acid is commonly called, is a mild acid that, in the concentrated form, is called glacial acetic acid. Glacial acetic acid is used in many industrial processes, such as making cellulose acetate. Vinegar is a 4% to 8% solution of acetic acid that can be made by fermenting alcohol.

It is possible to have more than one carboxyl group in a carboxylic acid. In the ethane derivative, it would be ethanedioic acid with a structure like this:

$$H-O-\overset{\overset{O}{\parallel}}{C}-\overset{\overset{O}{\parallel}}{C}-O-H$$

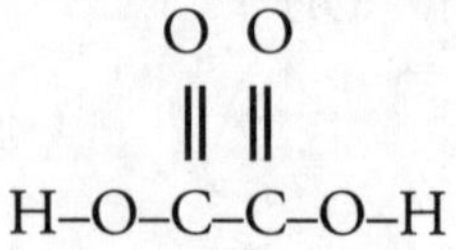

ethanedioic acid

Summary of Oxygen Derivatives

Functional Group

$$R-H \rightarrow R-Cl \rightarrow R-OH \rightarrow R^1CHO \rightarrow R^1-COOH$$

hydrocarbon chlorine alcohol aldehyde acid
 substitution
 product

(ending -*ol*) (ending -*al*) (ending -*oic*)

Note: R^1 indicates a hydrocarbon chain different from R by having one less carbon in the chain.

An actual example using ethane:

$$C_2H_6 \xrightarrow{Cl_2} C_2H_5Cl \xrightarrow{NaOH} C_2H_5OH \xrightarrow{[O]} CH_3CHO \xrightarrow{[O]} CH_3COOH$$

ethane monochloro- ethanol ethanal ethanoic
 ethane acid

TIP

Note that the functional group of ketones is

$$R-\overset{\overset{}{\underset{\underset{O}{\parallel}}{C}}}{}-R^1$$

Ketones

When a secondary alcohol is slightly oxidized, it forms a compound having the functional group $R-\overset{\underset{\underset{O}{\parallel}}{}}{C}-R^1$, and called a ketone. The R^1 indicates that this group need not be the same as R. An example is:

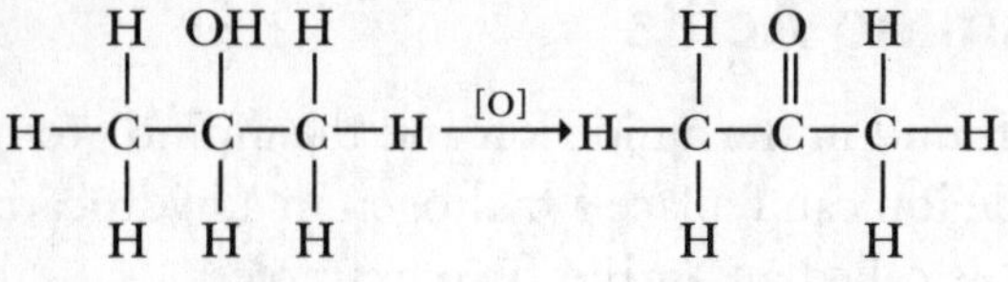

$$H-\underset{\underset{\displaystyle H}{|}}{\overset{\overset{\displaystyle H}{|}}{C}}-\underset{\underset{\displaystyle H}{|}}{\overset{\overset{\displaystyle OH}{|}}{C}}-\underset{\underset{\displaystyle H}{|}}{\overset{\overset{\displaystyle H}{|}}{C}}-H \xrightarrow{[O]} H-\underset{\underset{\displaystyle H}{|}}{\overset{\overset{\displaystyle H}{|}}{C}}-\overset{\overset{\displaystyle O}{\|}}{C}-\underset{\underset{\displaystyle H}{|}}{\overset{\overset{\displaystyle H}{|}}{C}}-H$$

2-propanol 2-propanone

or or

isopropyl alcohol acetone

In the IUPAC method the name of the ketone has the ending -*one* with a digit indicating the carbon that has the double-bonded oxygen. Another method of designating a ketone is to name the radicals on either side of the ketone structure and use the word **ketone**. In the preceding reaction, the product would be dimethyl ketone.

Note that both aldehydes and ketones contain the carbonyl group in their structures. In the aldehydes, it is at the end of the chain, and, in acids, it is the interior of the chain.

Ethers

When a primary alcohol, such as ethanol, is dehydrated with sulfuric acid, an ether forms. The functional group is R—O—R^1, in which R^1 may be the same hydrocarbon group, as shown in example 1 below, or a different hydrocarbon group, as shown in example 2.

> **TIP**
>
> The functional group of ethers is R—O—R^1

1.

$$H-\underset{\underset{\displaystyle H}{|}}{\overset{\overset{\displaystyle H}{|}}{C}}-\underset{\underset{\displaystyle H}{|}}{\overset{\overset{\displaystyle H}{|}}{C}}-OH + HO-\underset{\underset{\displaystyle H}{|}}{\overset{\overset{\displaystyle H}{|}}{C}}-\underset{\underset{\displaystyle H}{|}}{\overset{\overset{\displaystyle H}{|}}{C}}-H \xrightarrow{H_2SO_4} H-\underset{\underset{\displaystyle H}{|}}{\overset{\overset{\displaystyle H}{|}}{C}}-\underset{\underset{\displaystyle H}{|}}{\overset{\overset{\displaystyle H}{|}}{C}}-O-\underset{\underset{\displaystyle H}{|}}{\overset{\overset{\displaystyle H}{|}}{C}}-\underset{\underset{\displaystyle H}{|}}{\overset{\overset{\displaystyle H}{|}}{C}}-H + H_2O$$

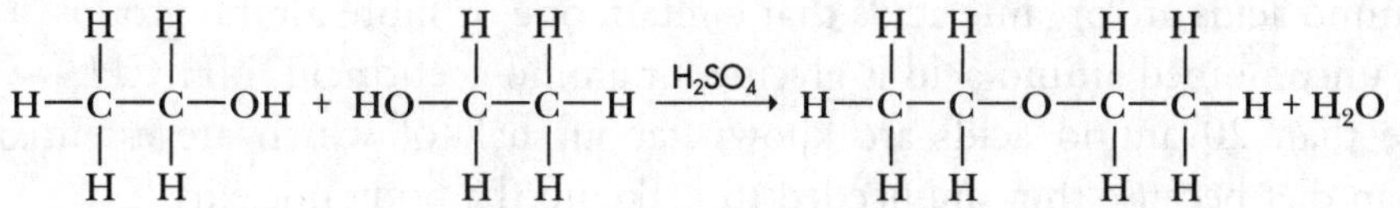

ethanol ethanol ethyl ether or diethyl ether

2. Another ether with unlike groups, R—O—R^1:

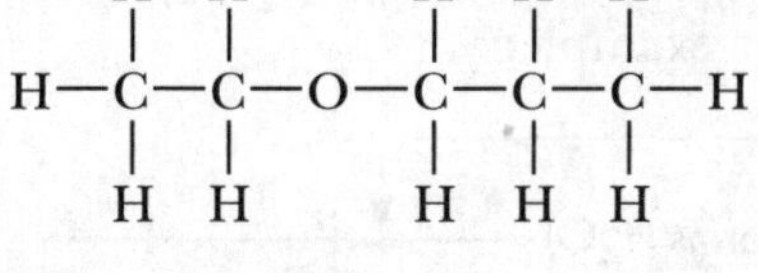

$$H-\underset{\underset{\displaystyle H}{|}}{\overset{\overset{\displaystyle H}{|}}{C}}-\underset{\underset{\displaystyle H}{|}}{\overset{\overset{\displaystyle H}{|}}{C}}-O-\underset{\underset{\displaystyle H}{|}}{\overset{\overset{\displaystyle H}{|}}{C}}-\underset{\underset{\displaystyle H}{|}}{\overset{\overset{\displaystyle H}{|}}{C}}-\underset{\underset{\displaystyle H}{|}}{\overset{\overset{\displaystyle H}{|}}{C}}-H \qquad \text{ethyl propyl ether}$$

In the IUPAC method, the ether name, as shown in the example, is made up of the alkyl attachments to the oxygen followed by the name ether.

Diethyl ether is commonly referred to as ether and is used as an anesthetic.

Amines and Amino Acids

The group NH_2^- is found in the amide ion and the amino group. Under the proper conditions, the amide ion can replace a hydrogen in a hydrocarbon compound. The resulting compound is called an **amine**. Two examples:

TIP

The amide functional group is

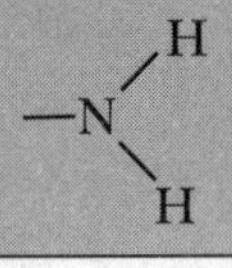

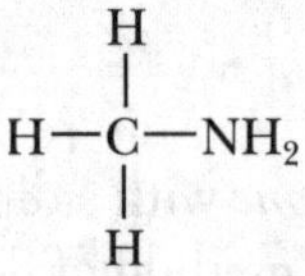

methylamine

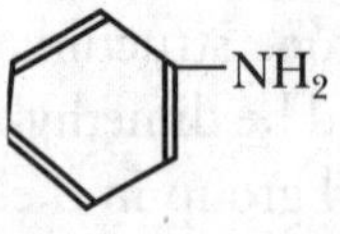

aniline

In *amides*, the NH_2^- group replaces a hydrogen in the carboxyl group. When naming amides, the *-ic* of the common name or the *-oic* of the IUPAC name of the parent acid is replaced by *-amide*. For example:

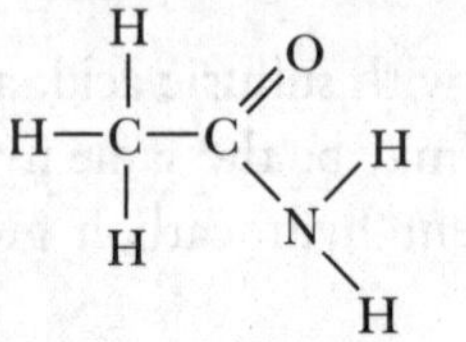

acetamide
or
ethanamide

TIP

Esters can be compared to inorganic salts.

Amino acids are organic acids that contain one or more amino groups. The simplest uncombined amino acid is glycine, or amino acetic acid, NH_2CH_2—COOH. More than 20 amino acids are known, about half of which are essential in the human diet because they are needed to make up the body proteins.

TIP

Note the functional group

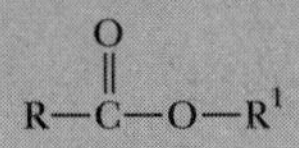

Esters

Esters are often compared to inorganic salts because their preparations are similar. To make a salt, you react the appropriate acid and base. To make an ester, you react the appropriate organic acid and alcohol. For example:

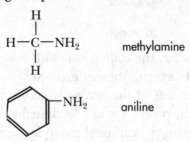

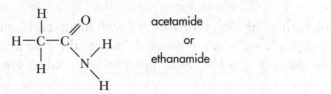

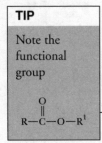

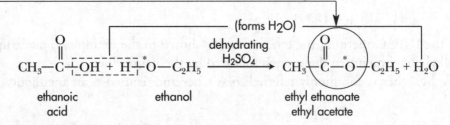

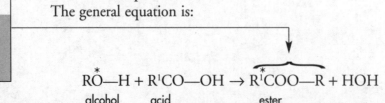

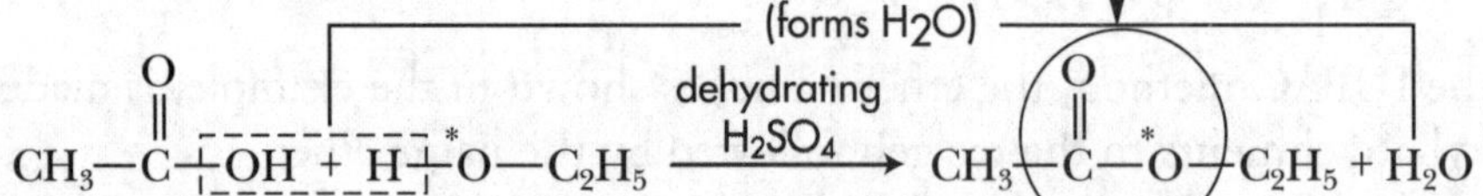

The name is made up of the alkyl substituent of the alcohol and the acid name, in which *-ic* is replaced with *-ate*.

The general equation is:

TIP

Note the functional group.

$$\overset{*}{R}O—H + R^1CO—OH \rightarrow \overbrace{R^1CO\overset{*}{O}}—R + HOH$$

alcohol acid ester

Esters usually have sweet smells and are used in perfumes and flavor extracts.

Chapter Summary

The following terms summarize all the concepts and ideas that were introduced in this chapter. You should be able to explain their meaning and how you would use them in chemistry. They appear in boldface type in this chapter to draw your attention to them. The boldface type also makes it easier for you to look them up if you need to. You could also use the search or "Google" action on your computer to get a quick and expanded explanation of these terms, laws, and formulas.

alcohol	amine	fullerene
aldehyde	amino acid	hydrocarbon
alkane	aromatics	hydrogenation
alkene	ester	isomer
alkyne	ether	ketone

Practice Exercises

1. Carbon atoms usually
   (A) lose 4 electrons
   (B) gain 4 electrons
   (C) form 4 covalent bonds
   (D) share the 2 electrons in the first principal energy level

2. Coke is produced from bituminous coal by
   (A) cracking
   (B) synthesis
   (C) substitution
   (D) destructive distillation

3. The usual method for preparing carbon dioxide in the laboratory is
   (A) heating a carbonate
   (B) fermentation
   (C) reacting an acid and a carbonate
   (D) burning carbonaceous materials

4. The precipitate formed when carbon dioxide is bubbled into limewater is
   (A) $CaCl_2$
   (B) H_2CO_3
   (C) CaO
   (D) $CaCO_3$

5. The "lead" in a lead pencil is
   (A) bone black
   (B) graphite and clay
   (C) lead oxide
   (D) lead peroxide

6. The first and simplest alkane is
   (A) ethane
   (B) methane
   (C) C_2H_2
   (D) methene
   (E) CCl_4

7. Slight oxidation of a primary alcohol gives
   (A) a ketone
   (B) an organic acid
   (C) an ether
   (D) an aldehyde
   (E) an ester

8. The characteristic group of the organic ester is
   (A) —CO—
   (B) —COOH
   (C) —CHO
   (D) —O—
   (E) —COO—

9. The organic acid that can be made from ethanol is
   (A) acetic acid
   (B) formic acid
   (C) C_3H_7OH
   (D) found in bees and ants
   (E) butanoic acid

10. An ester can be prepared by the reaction of
   (A) two alcohols
   (B) an alcohol and an aldehyde
   (C) an alcohol and an organic acid
   (D) an organic acid and an aldehyde
   (E) an acid and a ketone

11. Compounds that have the same composition but differ in their structural formulas
   (A) are used for substitution products
   (B) are called isomers
   (C) are called polymers
   (D) have the same properties
   (E) are usually alkanes

12. Ethene is the first member of the
   (A) alkane series
   (B) saturated hydrocarbons
   (C) alkyne series
   (D) unsaturated hydrocarbons
   (E) aromatic hydrocarbons

The following questions are in the format that is used on the **SAT Subject Test in Chemistry**. If you are not familiar with these types of questions, study pages xii–xv before doing the remainder of the review questions.

Directions: Each of the following sets of lettered choices refers to the numbered questions immediately below it. For each numbered item, choose the one lettered choice that fits it best. Every choice in a set may be used once, more than once, or not at all.

Questions 13–20

(A) $CH_3–CH_2–CH_3$

(B) $CH_3–C(=O)–OH$

(C) $CH_3–O–C_3H_7$

(D) $CH_3–C(=O)–CH_3$

(E) $CH_3–CH_2–N(H)(H)$

13. Which organic structure is ethylamine?

14. Which organic structure is methyl propyl ether?

15. Which organic structure is propane?

16. Which organic structure is ethanoic acid ?

17. Which organic structure is 2-propanone?

Questions 18–20

Using the same choices, match the functional groups named to the structure that contains it.

18. Which structure contains an organic acid functional group?

19. Which structure contains a ketone grouping?

20. Which structure contains an amine group?

Answers and Explanations

1. **(C)** Because carbon has 4 electrons in its outer energy level, it usually forms four covalent bonds to fill each of four sp^3 orbitals.

2. **(D)** Bituminous coal has too much of gaseous impurities to burn at a high temperature needed to refine iron ore. It is heated in coke ovens to form the hotter and cleaner burning coke.

3. **(C)** The reaction of an acid and a carbonate is the usual way to prepare CO_2.

4. **(D)** The reaction is: $CO_2 + Ca(OH)_2 \rightarrow CaCO_3\downarrow + H_2O$.

5. **(B)** The lead in a lead pencil is a mixture of graphite and clay.

6. **(B)** The first alkane is methane, CH_4.

7. **(D)** The slight oxidation of a primary alcohol produced an aldehyde.

   An example is :

$$CH_3\text{–}OH + [O] \rightarrow CH_2(OH)\text{–}OH \rightarrow H\text{–}C(=O)\text{–}H + H_2O$$

methanol (intermediate structure) methanal

8. **(E)** The functional group for an ester is shown as –COO–.

9. **(A)** Ethanol can be oxidized into the organic acid enthanoic acid, which has the common name of acetic acid.

10. **(C)** The formation of an ester is from the reaction of an organic acid and an alcohol. The general equation is:

$$RO\text{–}H + R^1CO\text{–}OH \rightarrow R^1COO\text{–}R + HOH$$

alcohol acid ester

11. **(B)** Isomers are compounds that have the same composition but differ in their structural formulas.

12. **(D)** Ethene is the first member of the alkene series that has one double bond. Because it has a double bond, it is said to be unsaturated. The alkane series, which has all single bonds between the carbon atoms in the chain, is called a saturated series.

13. **(E)** The -NH2 group, called an amine group, makes this structure ethylamine. This type of organic structure is called an amide.

14. **(C)** The methyl (CH_3–) group and the propyl (–C_3H_7) group attached to a center oxygen (-O-) makes this methyl propyl ether.

15. **(A)** Propane is the third member of the alkane series, which is made up of a chain of single-bonded carbons and hydrogens with the general formula of C_nH_{2n+2}.

16. **(B)** Ethanoic acid is composed of a methyl group attached to the carboxyl group (-COOH). The latter is the functional group for an organic acid.

17. **(D)** The propan- part of the name tells you its basic structure is from propane, which is a three-carbon alkane. The –one part tells you it is a ketone that has a double-bonded oxygen attached to the (2- in the name) second carbon in the chain.

18. **(B)** The ethanoic acid contains the carboxyl group (-COOH) that is shown in (B). This is the identifying functional group for organic acids.

19. **(D)** The functional group for ketones is a carbon in the chain double bonded to an oxygen atom ($-\overset{\overset{\displaystyle O}{||}}{C}-$). The number in the front tells you which carbon has the double-bonded oxygen attached to it.

20. **(E)** The amine group is the nitrogen with two hydrogens, (-NH_2), attached to a chain carbon. These are basic to the amino acid structures in the body.

Nucleonics

CHAPTER 15

- Explain the nature of radioactivity, including types and characteristics of each and the inherent dangers.
- Describe methods of detection of alpha, beta, and gamma rays.
- Explain the changes that occur in a decay series.
- Calculate the age of a substance using half-life.

The discovery of radioactivity came in 1896 (less than two months after Wilhelm Röntgen had announced the discovery of X-rays). Röntgen had pointed out that the X-rays came from a spot on a glass tube where a beam of electrons, in his experiments, was hitting, and that this spot simultaneously showed strong fluorescence. It occurred to Antoine-Henri Becquerel and others that X-rays might in some way be related to fluorescence (emission of light when exposed to some exciting agency) and to **phosphorescence** (emission of light *after* the exciting agency is removed).

Becquerel accordingly tested a number of phosphorescent substances to determine whether they emitted X-rays while phosphorescing. He had no success until he tried a compound of uranium. Then he found that the uranium compound, whether or not it was allowed to phosphoresce by exposure to light, continuously produced an emission that could penetrate lightproof paper and even thicker materials.

Becquerel determined that the compounds of uranium and the element itself produced **ionization** in the surrounding air. Thus either the ionizing effect, as indicated by the rate of discharge of a charged electroscope, or the degree of darkening of a photographic plate, could be used to measure the intensity of the invisible emission. Moreover, the emission from the uranium was continuous and perhaps even permanent, and required no energy from any external source. Yet, probably because of the current interest and excitement over X-rays, Becquerel's work received little attention until early in 1898, when Marie and Pierre Curie entered the picture (Pierre Curie was one of Becquerel's colleagues in Paris).

Searching for the source of the intense radiation in uranium ore, Marie and Pierre Curie used tons of it to isolate very small quantities of two new elements, radium and polonium, both radioactive. Along with Becquerel, the Curies shared the Nobel Prize in physics in 1903. Unfortunately, Pierre Curie died soon afterward from a tragic accident with a horse-drawn street cart. Marie Curie carried on their work and was chosen to succeed him at the Sorbonne as the first female professor at that university. In 1911, she received a second Nobel Prize for her work in chemistry with the newly discovered elements radium and polonium. She founded the Radium Institute in Paris and devoted much of her time to the applications of radioactivity.

THE NATURE OF RADIOACTIVE EMISSIONS

While the early separation experiments were in progress, an understanding was slowly being gained of the nature of the spontaneous emission from the various radioactive elements. Becquerel thought at first that there were simply X-rays, but THREE different kinds of radioactive emission, now called **alpha particles**, **beta particles**, and **gamma rays**, were soon found. We now know that alpha particles are positively charged particles of helium nuclei, beta particles are streams of high-speed electrons, and gamma rays are high-energy radiations similar to X-rays. The emission of these three types of radiation is depicted below.

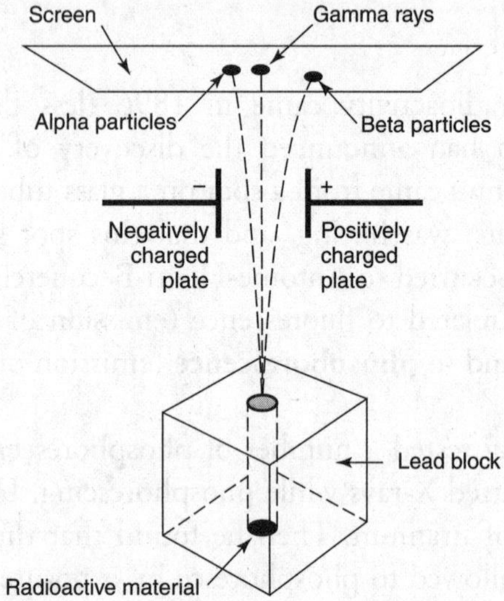

Deflection of Radioactive Emissions

The important characteristics of each type of radiation can be summarized as follows:

Alpha Particle (helium nucleus ${}^{4}_{2}He$) Positively charged, 2+

1. Ejection reduces the atomic number by 2, the atomic weight by 4 amu.
2. High energy, relative velocity.
3. Range: about 5 cm in air.
4. Shielding needed: stopped by the thickness of a sheet of paper, skin.
5. Interactions: produces about 100,000 ionizations per centimeter; repelled by the positively charged nucleus; attracts electrons, but does not capture them until its speed is much reduced.
6. An example: Thorium-230 has an unstable nucleus and undergoes radioactive decay through alpha emission. The nuclear equation that describes this reaction is:

$$ {}^{230}_{90}Th \rightarrow {}^{4}_{2}He + {}^{226}_{88}Ra $$

In a decay reaction like this, the initial element (thorium-230) is called the parent nuclide and the resulting element (radium-226) is called the daughter nuclide.

Beta Particle (fast electron) Negatively charged, 1-

1. Ejected when a *neutron* decays into a proton and an electron.
2. High velocity, low energy.
3. Range: about 12 m.
4. Shielding needed: stopped by 1 cm of aluminum or thickness of average book.
5. Interactions: weak because of high velocity, but produces about 100 ionizations per centimeter.
6. An example: Protactinium-234 is a radioactive nuclide that undergoes beta emission. The nuclear equation is:

$$ {}^{234}_{91}Pa \rightarrow {}^{234}_{92}U + {}^{0}_{-1}e $$

Gamma Radiation (electromagnetic radiation identical with light; high energy) No charge

1. Beta particles and gamma rays are usually emitted together; after a beta is emitted, a gamma ray follows.
2. Arrangement in nucleus is unknown. Same velocity as visible light.
3. Range: no specific range.
4. Shielding needed: about 13 cm of lead.
5. Interactions: weak of itself; gives energy to electrons, which then perform the ionization.

METHODS OF DETECTION OF ALPHA, BETA, AND GAMMA RAYS

All methods of detection of these radiations rely on their ability to ionize. Three methods are in common use.

1. ***Photographic plate.*** The fogging of a photographic emulsion led to the discovery of radioactivity. If this emulsion is viewed under a high-power microscope, it is seen that beta and gamma rays cause the silver bromide grains to develop in a scattered fashion. However, alpha particles, owing to the dense ionization they produce, leave a definite track of exposed grains in the emulsion. Hence not only is the alpha particle detected, but also its range (in the emulsion) can be measured. Special emulsions are capable of showing the beta-particle tracks. Film badges use exposure of film to measure radiation exposure of people working with radioactive materials.
2. ***Scintillation counter.*** A fluorescent screen (e.g., ZnS) will show the presence of electrons and X-rays, as already mentioned. If the screen is viewed with a magnifying eyepiece, small flashes of light, called scintillations, will be observed. By observing the scintillations, one not only can detect the presence of alpha particles, but also can actually count them.
3. ***Geiger counter.*** This instrument is perhaps the most widely used at the present time for determining individual radiation. It consists of a fine wire of tungsten mounted along the axis of a metal tube that contains a gas at reduced pressure. A difference of potential of about 1,000 volts is applied in such a way as to make the tube negative with respect to the wire. The voltage is high enough so that the electrons produced are accelerated by the electric field. Near the wire, where the field is strongest, the accelerated particles produce more ions, positive ones going to the walls, negative ones being collected by the wire. Any particle that will produce an ion gives rise to the same avalanche of ions, so the type of particle cannot be identified. However, each individual particle can be detected.

TIP

Know how to use half-life to determine the age of a substance.

See "Radioactive Dating" for more info.

DECAY SERIES, TRANSMUTATIONS, AND HALF-LIFE

The nuclei of uranium, radium, and other radioactive elements are continually disintegrating. It should be emphasized that spontaneous disintegration produces the gas known as radon. The time required for half of the atoms of a radioactive nuclide to decay is called its **half-life**.

For example, for radium, we know that, on the average, half of all the radium nuclei present will have disintegrated to radon in 1,590 years. In another 1,590 years, half of this remainder will decay, and so on. When a radium atom disintegrates, it loses an alpha particle, which eventually, upon gaining two electrons, becomes a neutral helium atom. The remainder of the atom becomes **radon**.

Such a conversion of an element to a new element (because of a change in the number of protons) is called a **transmutation**. This transmutation can be produced artificially by bombarding the nuclei of a substance with various particles from a particle accelerator, such as the cyclotron.

The following uranium-radium disintegration series shows how a radioactive atom may change when it loses each kind of particle. Note that an atomic number is shown by a subscript ($_{92}U$), and the isotopic mass by a superscript (^{238}U). The alpha particle is represented by the Greek symbol α, and the beta particle by β.

$$^{238}_{92}U \xrightarrow{-\alpha} {}^{234}_{90}Th \xrightarrow{-\beta} {}^{234}_{91}Pa \xrightarrow{-\beta} {}^{234}_{92}U \xrightarrow{-\alpha} {}^{230}_{90}Th \xrightarrow{-\alpha}$$

$$^{226}_{88}Ra \xrightarrow{-\alpha} {}^{222}_{86}Rn \xrightarrow{-\alpha} {}^{218}_{84}Po \xrightarrow{-\alpha} {}^{214}_{82}Pb \xrightarrow{-\beta} {}^{214}_{83}Bi \xrightarrow{-\beta}$$

$$^{214}_{84}Po \xrightarrow{-\alpha} {}^{210}_{82}Pb \xrightarrow{-\beta} {}^{210}_{83}Bi \xrightarrow{-\beta} {}^{210}_{84}Po \xrightarrow{-\alpha} {}^{206}_{82}Pb \text{ (stable)}$$

The changes that occur in radioactive reactions and the subatomic particles involved are summarized in the following charts.

Radioactive Decay and Nuclear Change

Type of Decay	Decay Particle	Particle Mass	Particle Charge	Change in Nucleon Number	Change in Atomic Number
Alpha decay	α	4	2+	Decreases by 4	Decreases by 2
Beta decay	β	0	1–	No change	Increases by 1
Gamma radiation	γ	0	0	No change	No change
Positron emission	β^+	0	1+	No change	Decreases by 1
Electron capture	e^-	0	1–	No change	Decreases by 1

Nuclear Symbols for Subatomic Particles

Particle	Symbols	Nuclear Symbols
Proton	p	^1_1p or ^1_1H
Neutron	n	^1_0n
Electron	e^- or β^-	$^0_{-1}e$ or $^0_{-1}\beta$
Positron	e^+ or β^+	$^0_{+1}e$ or $^0_{+1}\beta$
Alpha particle	α	^4_2He or $^4_{+2}\alpha$
Beta particle	β or β^-	$^0_{-1}e$ or $^0_{-1}\beta$
Gamma ray	γ	$^0_0\gamma$

This is shown graphically in the radioactive decay series below.

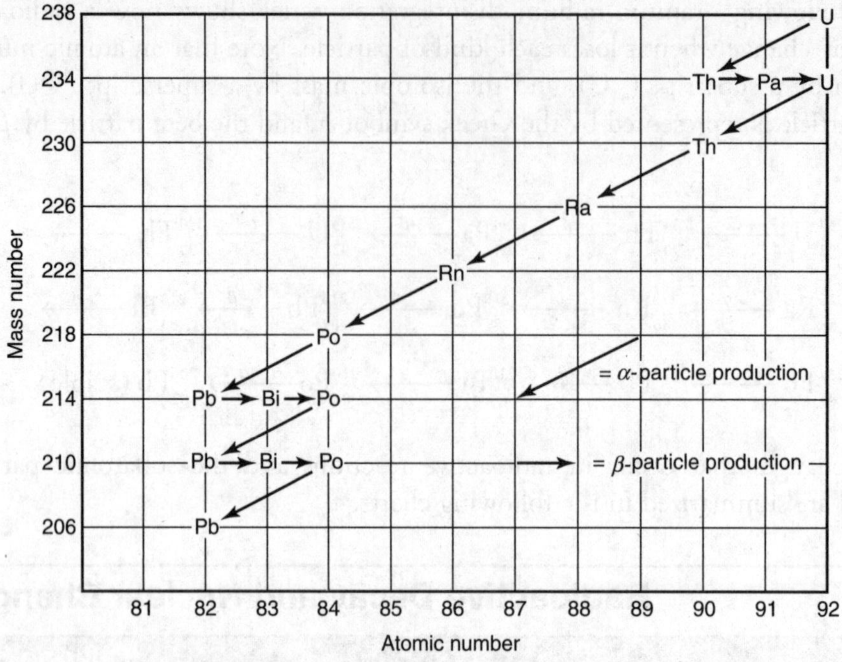

Uranium-238 Decay Series

The stability of an atom seems to be related to binding energy. The **binding energy** is the amount of energy released when a nucleus is formed from its component particles. If you add the masses of the components and compare this sum with the actual mass of the nucleus formed, there will be a small difference in these figures. This difference in mass can be converted to its energy equivalent using Einstein's equation, $E = mc^2$, where E is energy, m is mass, and c is the velocity of light. It is this energy that is called the binding energy.

It has been found that the lightest and heaviest elements have the smallest binding energy per nuclear particle and thus are less stable than the elements with intermediate atomic masses, which have the greatest binding energy.

The relationship of even-odd number of protons to the number of neutrons affects the stability of a nucleus. Many stable nuclei have even numbers of protons and neutrons; stability is less frequent in nuclei that have an even number of protons and an odd number of neutrons, or vice versa. Only a few stable nuclei that have odd numbers of protons and neutrons are known.

RADIOACTIVE DATING

A helpful application of radioactive decay is in the determination of the ages of substances such as rocks and relics that have bits of organic material trapped in them. Because carbon-14 has a half-life of about 5,700 years and occurs in the remains of organic materials, it has been useful in dating these materials. A small percentage of CO_2 in the atmosphere contains carbon-14. The stable isotope of carbon is carbon-12. Carbon-14 is a beta emitter and decays to form nitrogen-14:

$$^{14}_{6}C \rightarrow\ ^{14}_{7}N + ^{0}_{-1}e$$

In any living organism, the ratio of carbon-14 to carbon-12 is the same as in the atmosphere because of the constant interchange of materials between organism and surroundings. When an organism dies, this interaction stops, and the carbon-14 gradually decays to nitrogen. By comparing the relative amounts of carbon-14 and carbon-12 in the remains, the age of the organism can be established. Carbon-14 has a half-life of 5,700 years. If a sample of wood had originally contained 5 grams of carbon-14 and now had only half or 2.5 grams of carbon-14, its age would be 5,700 years. In other words, the old wood emits half as much beta radiation per gram of carbon as that emitted by living plant tissues. This method was used to determine the age of the Dead Sea Scrolls (about 1,900 years) and has been found to be in agreement with several other dating techniques.

Chapter Summary

The following terms summarize all the concepts and ideas that were introduced in this chapter. You should be able to explain their meaning and how you would use them in chemistry. They appear in boldface type in this chapter to draw your attention to them. The boldface type also makes it easier for you to look them up if you need to. You could also use the search or "Google" action on your computer to get a quick and expanded explanation of these terms, laws, and formulas.

alpha particle
beta particle
binding energy
gamma ray
Geiger counter
half-life
phosphorescence
radon
transmutation

Practice Exercises

1. Radioactive changes differ from ordinary chemical changes because radioactive changes
   (A) involve changes in the nucleus
   (B) are explosive
   (C) absorb energy
   (D) release energy
2. Isotopes of uranium have different
   (A) atomic numbers
   (B) atomic masses
   (C) numbers of planetary electrons
   (D) numbers of protons
3. Pierre and Marie Curie discovered
   (A) oxygen
   (B) hydrogen
   (C) chlorine
   (D) radium
4. The number of protons in the nucleus of an atom of atomic number 32 is
   (A) 4
   (B) 32
   (C) 42
   (D) 73

5. Atoms of ^{235}U and ^{238}U differ in structure by three
   (A) electrons
   (B) isotopes
   (C) neutrons
   (D) protons

6. The use of radioactive isotopes has produced promising results in the treatment of certain types of
   (A) cancer
   (B) heart disease
   (C) pneumonia
   (D) diabetes

7. The emission of a beta particle results in a new element with the atomic number
   (A) increased by 1
   (B) increased by 2
   (C) decreased by 1
   (D) decreased by 2

8. Einstein's formula for the Law of Conservation of Mass and Energy states that $E =$ ______.

9. The energy of the sun is thought to be produced by the fusion of hydrogen atoms into ______.

10. Can a mass spectrograph (or calutron) be used to separate ^{54}Mn from ^{54}Cr? ______.

Choose from this list the answer that best matches each numbered statement below.

A. Crooke	H. gamma ray
B. alpha particle	I. gaseous diffusion
C. beta particle	J. neptunium
D. Becquerel	K. plutonium
E. Einstein	L. ^{235}U
F. chain reaction	M. ^{238}U
G. deuteron	

11. Helium nucleus
12. Stable isotope of uranium
13. One method of separating isotopes
14. Most penetrating ray from radioactive decay
15. Mass-energy conversion

Before attempting to answer the following questions, (16–19), review the directions for this type of question, found on pages xiv–xv.

Record your answers here:

	I	II	CE*
16.	(T) (F)	(T) (F)	◯
17.	(T) (F)	(T) (F)	◯
18.	(T) (F)	(T) (F)	◯
19.	(T) (F)	(T) (F)	◯

	I		II
16.	Alpha particles are the heaviest type of radiation decay	BECAUSE	alpha emission particles consist of 2 protons and 2 neutrons.
17.	An emission of a β particle results in a decrease of the atomic number	BECAUSE	a β particle is an electron emitted from the nucleus.
18.	Isotopes can be separated by gaseous diffusion	BECAUSE	isotopes differ only in atomic mass.
19.	The most penetrating nuclear radiation is γ rays	BECAUSE	γ rays are the highest mass radioactive particles.

Answers and Explanations

1. **(A)** Radioactive changes differ because they involve changes in the nucleus.

2. **(B)** Isotopes differ in their atomic mass because of differences in the number of neutrons.

3. **(D)** The Curies discovered radium.

4. **(B)** The atomic number 32 gives the number of protons in the nucleus.

5. **(C)** These two isotopes, ^{235}U and ^{238}U, differ in their atomic mass by 3 neutrons.

6. **(A)** Radioactive isotopes have been successful in the treatment of certain cancers.

7. **(A)** A beta particle emission causes an increase of 1 in the atomic number.

8. mc^2 This is the famous Einstein equation that states that mass and energy are convertible into each other.

9. He Hydrogen atoms, through fusion, change to helium atoms and release huge amounts of energy.

10. No They both have an atomic mass of 54 and therefore cannot be separated by a mass spectrograph, because it depends on a difference in mass to achieve a separation of particles.

11. **(B)** A helium nucleus, two protons and two neutrons, is the makeup of an alpha particle.

12. **(M)** The stable isotope of uranium is ^{235}U.

13. **(I)** One method of separating isotopes is gaseous diffusion. Because the isotopes differ in mass, they pass through the diffusion mechanism at different rates and therefore separate.

14. **(H)** The gamma ray is the most penetrating ray of radioactivity since it has no mass.

15. **(E)** Einstein's famous equation that $E = mc^2$ is used to calculate the mass converted to energy in a nuclear reaction.

16. **(T, T, CE)** Alpha particles consist of 2 neutrons and 2 protons. This is essentially the nucleus of a helium atom. Because of this, it is the most massive of the radiation types.

17. **(F, T)** A beta discharge is the release of an electron from the nucleus with no change in the atomic mass and an increase of 1 in the atomic number.

18. **(T, T, CE)** Because isotopes differ in their atomic masses, they can be separated by taking advantage of this difference in mass and the consequent difference in their speed in gaseous diffusion.

19. **(T, F)** It is true that gamma rays are the most penetrating type of radiation, but it is because they consist of high-energy electromagnetic radiation, which can pass more easily through materials.

The Laboratory

CHAPTER **16**

- Follow laboratory safety rules.
- Understand basic chemistry setups.
- Know basic qualitative tests.

Laboratory setups vary from school to school depending on whether the lab is equipped with macro- or microscale equipment. Microlabs use specialized equipment that allows lab work to be done on a much smaller scale. The basic principles are the same as when using full-sized equipment, but microscale equipment lowers the cost of materials, results in less waste, and poses less danger. The examples in this book are of macroscale experiments.

Along with learning to use microscale equipment, most labs require a student to learn how to use technological tools to assist in experiments. The most common are:

Gravimetric balance with direct readings to thousandths of a gram instead of a triple-beam balance

pH meters that give pH readings directly instead of using indicators

Spectrophotometer, which measures the percentage of light transmitted at specific frequencies so that the molarity of a sample can be determined without doing a titration

Computer-assisted labs that use probes to take readings, e.g., temperature and pressure, so that programs available for computers can print out a graph of the relationship of readings taken over time

LABORATORY SAFETY RULES

The Ten Commandments of Lab Safety

The following is a summary of rules you should be well aware of in your own chemistry lab.

1. Dress appropriately for the lab. Wear safety goggles and a lab apron or coat. Tie back long hair. Do not wear open-toed shoes.
2. Know what safety equipment is available and how to use it. This includes the eyewash fountain, fire blanket, fire extinguisher, and emergency shower.
3. Know the dangers of the chemicals in use, and read labels carefully. Do not taste or sniff chemicals.

4. Dispose of chemicals according to instructions. Uses designated disposal sites, and follow the rules. Never return unneeded chemicals to the original containers.
5. Always add acid to water slowly to avoid splattering. This is especially important when using strong acids that can generate significant heat, form steam, and splash out of the container.
6. Never point heating test tubes at yourself or others. Be aware of reactions that are occurring so that you can remove them from the heat if necessary before they "shoot" out of the test tube.
7. Do not pipette anything by mouth! Never use your mouth as a suction pump, not even at home with toxic or flammable liquids.
8. Use the fume hood when dealing with toxic fumes! If you can smell them, you are exposing yourself to a dose that can harm you.
9. Do not eat or drink in the lab! It is too easy to take in some dangerous substance accidentally.
10. Follow all directions. Never haphazardly mix chemicals. Pay attention to the order in which chemicals are to be added to each other, and do not deviate!

SOME BASIC SETUPS

Throughout this book, drawings of laboratory setups that serve specific needs have been presented. You should be familiar with the assembly and use of each of these setups. The following list, with page references, will enable you to review them in context with their uses:

• Preparation of a gaseous product, nonsoluble in water, by water displacement from solid reactants. . . . . page 131

• Preparation of a gaseous product, nonsoluble in water, by water displacement from at least one reactant in solution. . . . . page 134

• Distillation of a liquid . . . . page 181

• Titration . . . . page 254

The following are additional laboratory setups with which you should be familiar:

1. PREPARATION OF A GASEOUS PRODUCT, SOLUBLE IN WATER AND LIGHTER THAN AIR, BY THE DOWNWARD DISPLACEMENT OF AIR. SEE FIGURE 44.

EXAMPLE: Preparation of ammonia (NH_3).

$$2NH_4Cl(s) + Ca(OH)_2(s) \rightarrow CaCl_2 + 2H_2O + 2NH_3(g)$$

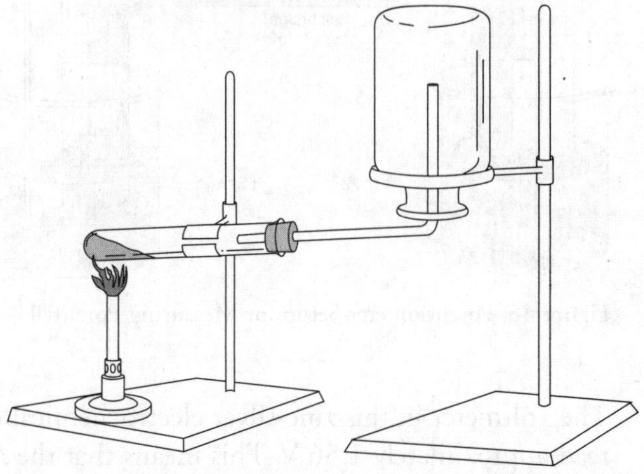

Figure 44. Preparation of Ammonia

2. SEPARATION OF A MIXTURE BY CHROMATOGRAPHY. SEE FIGURE 45.

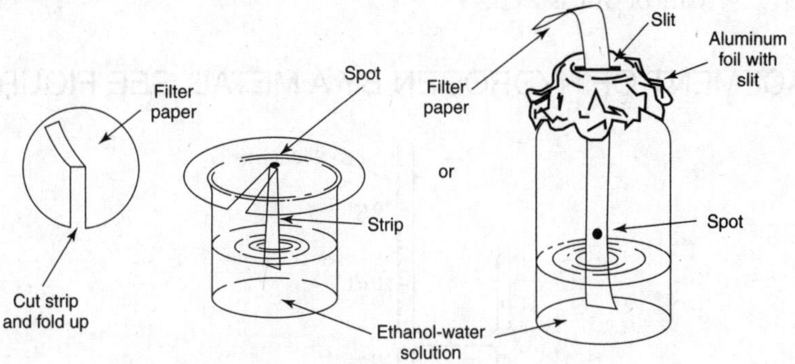

Figure 45. Chromatography Setup

EXAMPLE: Chromatography is a process used to separate parts of a mixture. The component parts separate as the solvent carrier moves past the spot of material to be separated by capillary action. Because of variations in solubility, attraction to the filter paper, and density, each fraction moves at a different rate. Once separation occurs, the fractions are either identified by color or removed for other tests. A usual example is the use of Shaeffer Skrip Ink No. 32, which separates into yellow, red, and blue streaks of dyes.

3. MEASURING POTENTIALS IN ELECTROCHEMICAL CELLS. SEE FIGURE 46.

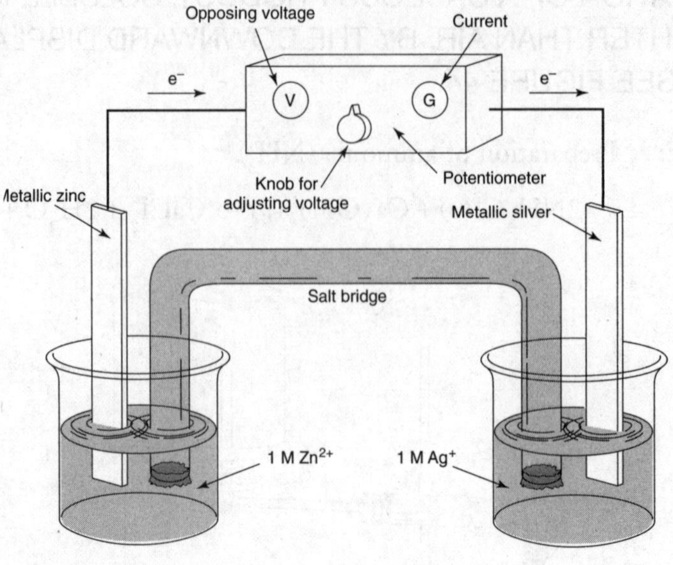

Figure 46. Potentiometer Setup for Measuring Potential

EXAMPLE: The voltmeter in this zinc-silver electrochemical cell would read approximately 1.56 V. This means that the Ag to Ag^+ half-cell has 1.56 V more electron-attracting ability than the Zn to Zn^{2+} half-cell. If the potential of the zinc half-cell were known, the potential of the silver half-cell could be determined by adding 1.56 V to the potential of the zinc half-cell. In a setup like this, only the difference in potential between two half-cells can be measured.

4. REPLACEMENT OF HYDROGEN BY A METAL. SEE FIGURE 47.

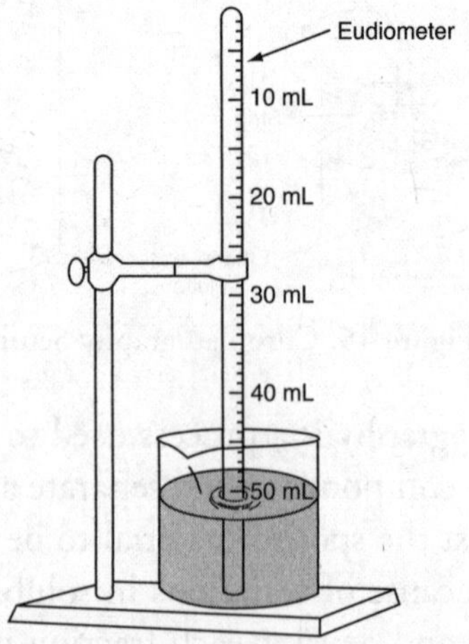

Figure 47. Eudiometer Apparatus

EXAMPLE: Measure the mass of a strip of magnesium with an analytical balance to the nearest 0.001 g. Using a coiled strip with a mass of about 0.040 g produces about 40 mL of H_2. Pour 5 mL of concentrated HCl into a eudiometer, and slowly fill the remainder with water. Try to minimize mixing. Lower the coil of Mg strip into the tube, invert it, and lower it to the bottom of the beaker. After the reaction is complete, you can measure the volume of the gas released and calculate the mass of hydrogen replaced by the magnesium. (Refer to Chapter 5 for a discussion of gas laws.)

5. USING DISPLACED WATER TO MEASURE THE VOLUME OF A GAS THAT IS PRODUCED. SEE FIGURE 48.

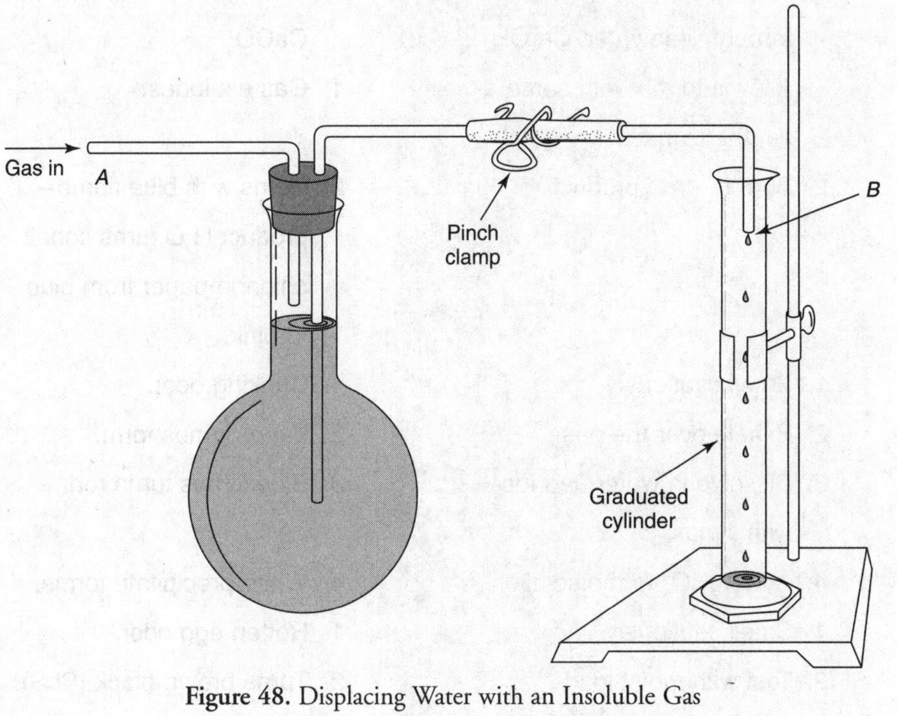

Figure 48. Displacing Water with an Insoluble Gas

EXAMPLE: In this setup, before the gas begins to be generated, you should release the pinch clamp and gently blow in the tube at *A* to get water up and into the remainder of the system. Close the pinch clamp once water starts to flow at *B*, discard any water in the cylinder, then connect the system to the generator. Now you are ready to read the quantity of gas evolved into the flask by reading the volume of water displaced into the graduated cylinder.

SUMMARY OF QUALITATIVE TESTS

I. Identification of Some Common Gases

Gas	Test	Result
Ammonia NH_3	1. Smell cautiously. 2. Test with litmus. 3. Expose to HCl fumes.	1. Sharp odor. 2. Red litmus turns blue. 3. White fumes form, NH_4Cl.
Carbon dioxide CO_2	1. Pass through limewater, $Ca(OH)_2$.	1. White precipitate forms, $CaCO_3$.
Carbon monoxide CO	1. Burn it and pass product through limewater, $Ca(OH)_2$.	1. White precipitate forms, $CaCO_3$.
Hydrogen H_2	1. Allow it to mix with some air, then ignite. 2. Burn it—trap product.	1. Gas explodes. 2. Burns with blue flame—product H_2O turns cobalt chloride paper from blue to pink.
Hydrogen chloride HCl	1. Smell cautiously. 2. Exhale over the gas. 3. Dissolve in water and test with litmus. 4. Add $AgNO_3$ to the solution.	1. Choking odor. 2. Vapor fumes form. 3. Blue litmus turns red. 4. White precipitate forms.
Hydrogen sulfide H_2S	1. Smell cautiously. 2. Test with moist lead acetate paper.	1. Rotten egg odor. 2. Turns brown-black (PbS).
Oxygen O_2	1. Insert glowing splint. 2. Add nitric oxide gas.	1. Bursts into flame. 2. Turns reddish brown.

II. Identification of Some Negative Ions

Ion	Test	Result
Acetate $C_2H_3O_2^-$	Add concentrated H_2SO_4 and warm gently.	Odor of vinegar released.
Carbonate CO_3^-	Add HCl acid; pass released gas through limewater.	White, cloudy precipitate forms.
Chloride Cl^-	1. Add silver nitrate solution. 2. Then add nitric acid, later followed by ammonium hydroxide.	1. White precipitate forms. 2. Precipitate insoluble in HNO_3 but dissolves in NH_4OH.
Hydroxide OH^-	Test with red litmus paper.	Turns blue.
Sulfate SO_4^-	Add solution of $BaCl_2$, then HCl.	White precipitate forms; insoluble in HCl.
Sulfide S^{2-}	Add HCl and test gas released with lead acetate paper.	Gas, with rotten egg odor, turns paper brown-black.

III. Identification of Some Positive Ions

Ion	Test	Result
Ammonium NH_4^+	Add strong base (NaOH); heat gently.	Odor of ammonia.
Ferrous Fe^{2+}	Add solution of potassium ferri-cyanide, $K_3Fe(CN)_6$.	Dark blue precipitate forms (Turnball's blue).
Ferric Fe^{3+}	Add solution of potassium ferro-cyanide, $K_4Fe(CN)_6$.	Dark blue precipitate forms (Prussian blue).
Hydrogen H^+	Test with blue litmus paper.	Turns red.

IV. Qualitative Tests of Some Metals

FLAME TESTS. Carefully clean a platinum wire by dipping it into dilute HNO_3 and heating in the Bunsen flame. Repeat until the flame is colorless. Dip heated wire into the substance being tested (either solid or solution), and then hold it in the hot outer part of the Bunsen flame.

Compound of	Color of Flame
Sodium (Na)	Yellow
Potassium (K)	Violet (use cobalt-blue glass to screen out Na impurities)
Lithium (Li)	Crimson
Calcium (Ca)	Orange-red
Barium (Ba)	Green
Strontium (Sr)	Bright red

HYDROGEN SULFIDE TESTS. Bubble hydrogen sulfide gas through the solution of a salt of the metal being tested. Check color of the precipitate formed.

Compound of	Color of Sulfide Precipitate
Lead (Pb)	Brown-black (PbS)
Copper (Cu)	Black (CuS)
Silver (Ag)	Black (Ag_2S)
Mercury (Hg)	Black (HgS)
Nickel (Ni)	Black (NiS)
Iron (Fe)	Black (FeS)
Cadmium (Cd)	Yellow (CdS)
Arsenic (As)	Light yellow (As_2S_3)
Antimony (Sb)	Orange (Sb_2S_3)
Zinc (Zn)	White (ZnS)
Bismuth (Bi)	Brown (Bi_2S_3)

Chapter Summary

The following terms summarize all the concepts and ideas that were introduced in this chapter. You should be able to explain their meaning and how you would use them in chemistry. They appear in boldface type in this chapter to draw your attention to them. The boldface type also makes it easier for you to look them up if you need to. You could also use the search or "Google" action on your computer to get a quick and expanded explanation of these terms, laws, and formulas.

gravimetric balance with direct readings
pH meters
spectrophotometer
computer-assisted labs

Practice Exercises

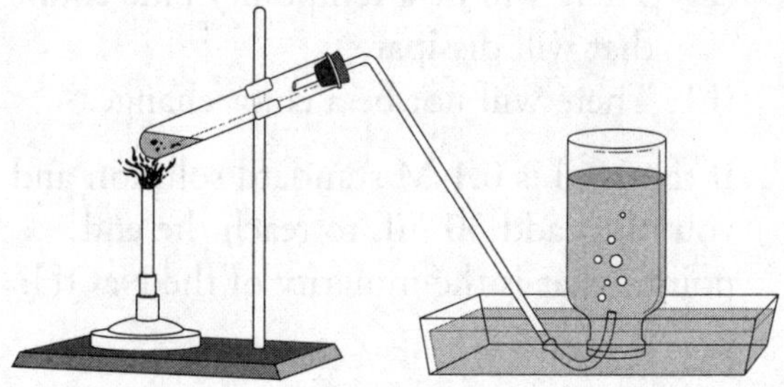

1. In the reaction setup shown above, which of the following are true?

   I. This setup can be used to prepare a soluble gas by water displacement.
   II. This setup involves a decomposition reaction if the substance heated is potassium chlorate.
   III. This setup can be used to prepare an insoluble gas by water displacement.

   (A) I only
   (B) II only
   (C) I and III
   (D) II and III
   (E) I, II, and III

<u>Questions 2–4</u> refer to the following diagram:

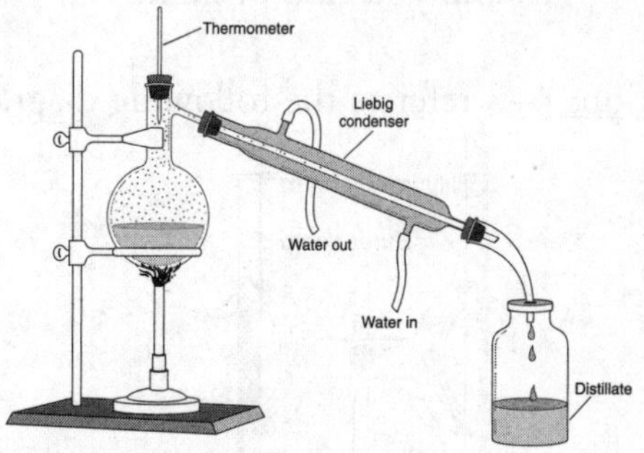

(A) Around the thermometer
(B) In the condenser
(C) In the circulating water
(D) In the heated flask
(E) In the distillate

2. In this laboratory setup for distillation, where does the vaporization take place?

3. If the liquid being distilled contains dissolved magnesium chloride, where will it be found after distillation is completed?

4. If the liquid being distilled contains dissolved ammonia gas, where will it be bound after distillation is completed?

5. If the flame used to heat a flask is an orange color and blackens the bottom of the flask, what correction should you make to solve this problem?

    (A) Move the flask farther from the flame.
    (B) Move the flask closer to the flame.
    (C) Allow less air into the collar of the burner.
    (D) Allow more air into the collar of the burner.
    (E) The problem is in the supply of the gas, and you cannot fix it.

<u>Questions 6–8</u> refer to the following diagram:

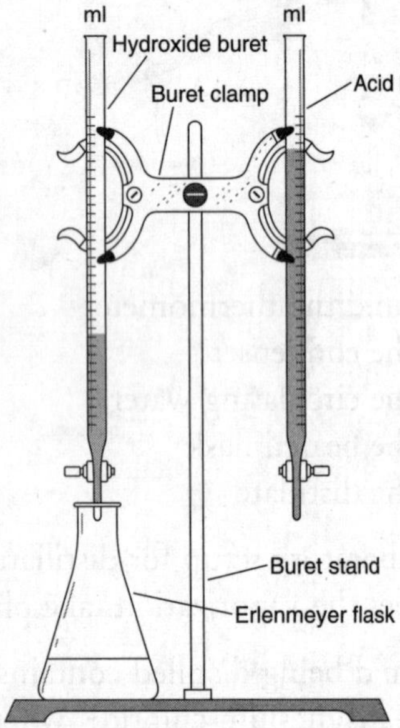

6. In the above titration setup, if you introduce 15 mL of the NaOH with an unknown molarity into the flask and then add 5 drops of phenolphthalein indicator, what will you observe?

    (A) A pinkish color will appear throughout the solution.
    (B) A blue color will appear throughout the solution.
    (C) There will be a temporary pinkish color that will dissipate.
    (D) There will be a temporary blue color that will dissipate.
    (E) There will not be a color change.

7. If the HCl is 0.1 M standard solution and you must add 30 mL to reach the end point, what is the molarity of the NaOH?

    (A) 0.1 M
    (B) 0.2 M
    (C) 0.3 M
    (D) 1 M
    (E) 2 M

8. When is the end point reached and the volume of the HCl recorded in this reaction?

    I. When the color first disappears and returns in the flask.
    II. When equal amounts of HCl and NaOH are in the flask.
    III. When the color disappears and does not return in the flask.

    (A) I only
    (B) III only
    (C) I and III
    (D) II and III
    (E) I, II, and III

<u>Questions 9–11</u> refer to the following diagram:

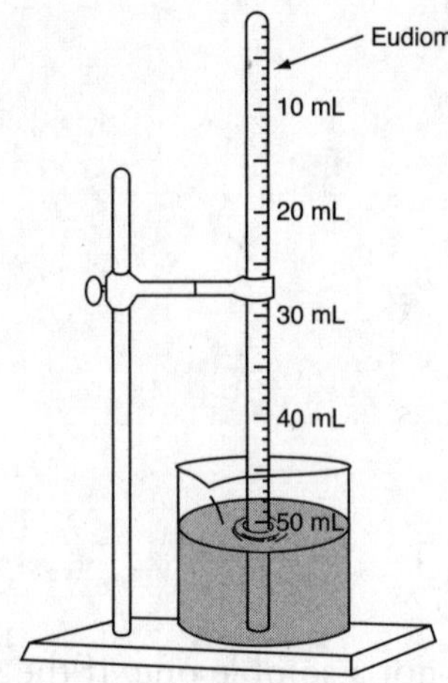

In this setup, a clean strip of magnesium with a mass of 0.040 g was introduced into the bottom of the tube, which contained a dilute solution of HCl, and allowed to react completely. The hydrogen gas formed was collected and the following data recorded:

Air pressure in the room = 730 mm Hg
Temperature of the water solution = 302 K
Vapor pressure of water at 302 K = 30.0 mm Hg

The gas collected did not fill the eudiometer. The height of the meniscus above the level of the water was 40.8 mm.

9. What is the theoretical yield (in mL) at STP of hydrogen gas produced when the 0.040 g of Mg reacted completely?

(A) 10 mL
(B) 25 mL
(C) 37 mL
(D) 46 mL
(E) 51 mL

10. What is the correction to the atmospheric pressure due to the 40.8 mm height of the solution in the tube and above the level in the beaker?

(A) 3.0 mm Hg
(B) 6.0 mm Hg
(C) 13.6 mm Hg
(D) 27.2 mm Hg
(E) 40.8 mm Hg

11. What is the pressure of the collected gas once you have also corrected for the vapor pressure of the water?

(A) 730 mm Hg
(B) 727 mm Hg
(C) 30.0 mm Hg
(D) 697 mm Hg
(E) 760 mm Hg

<u>Questions 12–14</u>

(A) The rule is to add concentrated acid to water slowly.
(B) The rule is to add water to the concentrated acid slowly.
(C) Carefully replace unused or excess chemicals into their properly labeled containers from which they came.
(D) Flush eyes with water at the eyewash fountain for at least 15 minutes, and then report the accident for further help.
(E). Dispose of chemicals in the proper places and following posted procedures. Do not return them to their original containers.

12. Which of the above choices is the proper way to dilute a concentrated acid?

13. How do you properly dispose of chemicals not needed in the experiment?

14. What should you do if a chemical splatters into your eye?

15. What instrument is used in chemistry labs to measure the molarity of a colored solution by measuring the light transmitted through it?

(A) Electronic gravimetric balance
(B) pH meter
(C) Spectrophotometer
(D) Computer assisted probes
(E) Galvanometer

Answers and Explanations

1. **(D)** This setup can be used to prepare an insoluble gas but not a soluble one. If the substance is potassium chlorate, if does decompose into potassium chloride and oxygen.

2. **(D)** Vaporization occurs in the heated flask.

3. **(D)** The magnesium chloride will be left behind in the heated flask since it is not volatile as the liquid boils off.

4. **(E)** Since the dissolved ammonia is volatile below the boiling point of water, it will be found in the distillate. Some will also escape as a gas.

5. **(D)** One of the basic adjustments to a burner is to assure enough air is mixing with the gas to form a blue cone-shaped flame. With insufficient air, carbon deposits will form on the flask and the flame will be orange.

6. **(A)** Phenolphthalein indicator is colorless below a pH of 8.3 but is red to pink in basic solutions above this pH. In this NaOH solution, it will be red to pink.

7. **(B)** Calculate the molarity by using the formula:

$$M_{acid} \times V_{acid} = M_{base} \times V_{base}$$
$$0.1\text{ M} \times 30\text{ mL} = M_{base} \times 15\text{ mL}$$
$$M_{base} = 0.2\text{ M}$$

Notice that as long as the units of volume are the same, they cancel out of the equation.

8. **(B)** The end point is reached when the color of the indicator disappears and does not return. The color will first disappear temporarily before the end point but finally will not return.

9. **(C)** The theoretical yield at STP can be found from the chemical equation of the reaction:

$$\begin{array}{ccccccccc} 0.040 \text{ g Mg} & & & & & & x\text{ L} \\ Mg & + & 2HCl & \rightarrow & MgCl_2 & + & H_2\ (g) \\ 24.0 \text{ g/mol} & & & & & & 22.4 \text{ L} \end{array}$$

Solving for x L = .037 L or 37 mL

10. **(A)** The 40.8 mm of water being held up in the tube by atmospheric pressure can be changed to its equivalent height of mercury by dividing by 13.6, since 1 mm of Hg = 13.6 mm of water.

$$40.8 \text{ mm } H_2O \times \frac{1 \text{ mm Hg}}{13.6 \text{ mm } H_2O} = 3 \text{ mm Hg}$$

By correcting the atmospheric pressure, we get: 730 mm Hg – 3 mm Hg = 727 mm Hg

11. **(D)** Correcting the previous pressure by subtracting the given amount for the vapor pressure of water at 302 K gives you 727 mm Hg – 30 mm Hg vapor pressure = 697 mm Hg as the final pressure.

12. **(A)** The correct and safe way to dilute concentrated acids is to add water down the side of the beaker slowly and be aware of any heat buildup.

13. **(E)** You never return chemicals or solutions to their original containers for fear of contaminating the original source.

14. **(D)** It is essential to get your eyes washed of any chemicals. Know where the eyewash fountains are, and know how to use them.

15. **(C)** One of the new technological additions to chemistry labs is the spectron 20 that uses the absorption of light waves to do qualitative and quantitative investigations in the lab.

PART 3

PRACTICE TESTS

Practice Subject Tests in Chemistry

The Subject Tests in Chemistry are planned to test the principles and concepts drawn from the factual material found largely in inorganic chemistry and, to a much lesser extent, in organic chemistry. Only a few questions are asked concerning industrial or analytical chemistry.

According to a description of the test by the College Entrance Examination Board, it includes the Kinetic-Molecular Theory and the three states of matter; atomic theory and structure and the Periodic Table of the Elements; nuclear reactions; quantitative relations as applied to chemical formulas and equations; chemical bonding and molecular structure, and their relations to properties; the nature of chemical reactions, including acid-base reactions, oxidation-reduction reactions, ionic reactions, and other chemical changes occurring in solution; energy changes accompanying chemical reactions; interpretation of chemical equilibria and reaction rates; solution phenomena; electrochemistry, nuclear chemistry, and radiochemistry; physical and chemical properties of the more familiar metals, transition elements, and nonmetals and of the more familiar compounds; understanding and interpretation of laboratory procedures and observations.

You will be provided with a Periodic Table to use during the test. All necessary information regarding atomic numbers and atomic masses is given on the chart. Before you attempt any of the practice tests, read the information in the Introduction and the material that precedes the Diagnostic Test in the front of this book. When you understand the information there, are aware of the types of questions and their respective directions, and know how the test will be scored, you are ready to take Practice Test 1.

Remember that you have 1 hour and that you **may not use a calculator**. Use the Periodic Table provided for the Diagnostic Test in the front of this book, and record your answers in the appropriate spaces on the answer sheet.

Answer Sheet

PRACTICE TEST 1

Determine the correct answer for each question. Then, using a No. 2 pencil, blacken completely the oval containing the letter of your choice.

1 Ⓐ Ⓑ Ⓒ Ⓓ Ⓔ
2 Ⓐ Ⓑ Ⓒ Ⓓ Ⓔ
3 Ⓐ Ⓑ Ⓒ Ⓓ Ⓔ
4 Ⓐ Ⓑ Ⓒ Ⓓ Ⓔ
5 Ⓐ Ⓑ Ⓒ Ⓓ Ⓔ
6 Ⓐ Ⓑ Ⓒ Ⓓ Ⓔ
7 Ⓐ Ⓑ Ⓒ Ⓓ Ⓔ
8 Ⓐ Ⓑ Ⓒ Ⓓ Ⓔ
9 Ⓐ Ⓑ Ⓒ Ⓓ Ⓔ
10 Ⓐ Ⓑ Ⓒ Ⓓ Ⓔ
11 Ⓐ Ⓑ Ⓒ Ⓓ Ⓔ
12 Ⓐ Ⓑ Ⓒ Ⓓ Ⓔ
13 Ⓐ Ⓑ Ⓒ Ⓓ Ⓔ
14 Ⓐ Ⓑ Ⓒ Ⓓ Ⓔ
15 Ⓐ Ⓑ Ⓒ Ⓓ Ⓔ
16 Ⓐ Ⓑ Ⓒ Ⓓ Ⓔ
17 Ⓐ Ⓑ Ⓒ Ⓓ Ⓔ
18 Ⓐ Ⓑ Ⓒ Ⓓ Ⓔ
19 Ⓐ Ⓑ Ⓒ Ⓓ Ⓔ
20 Ⓐ Ⓑ Ⓒ Ⓓ Ⓔ
21 Ⓐ Ⓑ Ⓒ Ⓓ Ⓔ
22 Ⓐ Ⓑ Ⓒ Ⓓ Ⓔ
23 Ⓐ Ⓑ Ⓒ Ⓓ Ⓔ

ON THE ACTUAL CHEMISTRY TEST, THE FOLLOWING TYPE OF QUESTION MUST BE ANSWERED ON A SPECIAL SECTION (LABELED "CHEMISTRY") AT THE LOWER LEFT-HAND CORNER OF PAGE 2 OF YOUR ANSWER SHEET. THESE QUESTIONS WILL BE NUMBERED BEGINNING WITH 101 AND MUST BE ANSWERED ACCORDING TO THE DIRECTIONS.

CHEMISTRY* Fill in oval CE only if II is a correct explanation of I.

	I	II	CE*
101	Ⓣ Ⓕ	Ⓣ Ⓕ	◯
102	Ⓣ Ⓕ	Ⓣ Ⓕ	◯
103	Ⓣ Ⓕ	Ⓣ Ⓕ	◯
104	Ⓣ Ⓕ	Ⓣ Ⓕ	◯
105	Ⓣ Ⓕ	Ⓣ Ⓕ	◯
106	Ⓣ Ⓕ	Ⓣ Ⓕ	◯
107	Ⓣ Ⓕ	Ⓣ Ⓕ	◯
108	Ⓣ Ⓕ	Ⓣ Ⓕ	◯
109	Ⓣ Ⓕ	Ⓣ Ⓕ	◯
110	Ⓣ Ⓕ	Ⓣ Ⓕ	◯
111	Ⓣ Ⓕ	Ⓣ Ⓕ	◯
112	Ⓣ Ⓕ	Ⓣ Ⓕ	◯
113	Ⓣ Ⓕ	Ⓣ Ⓕ	◯
114	Ⓣ Ⓕ	Ⓣ Ⓕ	◯
115	Ⓣ Ⓕ	Ⓣ Ⓕ	◯
116	Ⓣ Ⓕ	Ⓣ Ⓕ	◯

ON THE ACTUAL CHEMISTRY TEST, THE REMIANING QUESTIONS MUST BE ANSWERED BY RETURNING TO THE SECTION OF YOUR ANSWER SHEET YOU STARTED FOR CHEMISTRY.

24 Ⓐ Ⓑ Ⓒ Ⓓ Ⓔ
25 Ⓐ Ⓑ Ⓒ Ⓓ Ⓔ
26 Ⓐ Ⓑ Ⓒ Ⓓ Ⓔ
27 Ⓐ Ⓑ Ⓒ Ⓓ Ⓔ
28 Ⓐ Ⓑ Ⓒ Ⓓ Ⓔ
29 Ⓐ Ⓑ Ⓒ Ⓓ Ⓔ
30 Ⓐ Ⓑ Ⓒ Ⓓ Ⓔ
31 Ⓐ Ⓑ Ⓒ Ⓓ Ⓔ
32 Ⓐ Ⓑ Ⓒ Ⓓ Ⓔ
33 Ⓐ Ⓑ Ⓒ Ⓓ Ⓔ
34 Ⓐ Ⓑ Ⓒ Ⓓ Ⓔ
35 Ⓐ Ⓑ Ⓒ Ⓓ Ⓔ
36 Ⓐ Ⓑ Ⓒ Ⓓ Ⓔ
37 Ⓐ Ⓑ Ⓒ Ⓓ Ⓔ
38 Ⓐ Ⓑ Ⓒ Ⓓ Ⓔ
39 Ⓐ Ⓑ Ⓒ Ⓓ Ⓔ
40 Ⓐ Ⓑ Ⓒ Ⓓ Ⓔ
41 Ⓐ Ⓑ Ⓒ Ⓓ Ⓔ
42 Ⓐ Ⓑ Ⓒ Ⓓ Ⓔ
43 Ⓐ Ⓑ Ⓒ Ⓓ Ⓔ
44 Ⓐ Ⓑ Ⓒ Ⓓ Ⓔ
45 Ⓐ Ⓑ Ⓒ Ⓓ Ⓔ
46 Ⓐ Ⓑ Ⓒ Ⓓ Ⓔ
47 Ⓐ Ⓑ Ⓒ Ⓓ Ⓔ
48 Ⓐ Ⓑ Ⓒ Ⓓ Ⓔ
49 Ⓐ Ⓑ Ⓒ Ⓓ Ⓔ
50 Ⓐ Ⓑ Ⓒ Ⓓ Ⓔ
51 Ⓐ Ⓑ Ⓒ Ⓓ Ⓔ
52 Ⓐ Ⓑ Ⓒ Ⓓ Ⓔ
53 Ⓐ Ⓑ Ⓒ Ⓓ Ⓔ
54 Ⓐ Ⓑ Ⓒ Ⓓ Ⓔ
55 Ⓐ Ⓑ Ⓒ Ⓓ Ⓔ
56 Ⓐ Ⓑ Ⓒ Ⓓ Ⓔ
57 Ⓐ Ⓑ Ⓒ Ⓓ Ⓔ
58 Ⓐ Ⓑ Ⓒ Ⓓ Ⓔ
59 Ⓐ Ⓑ Ⓒ Ⓓ Ⓔ
60 Ⓐ Ⓑ Ⓒ Ⓓ Ⓔ
61 Ⓐ Ⓑ Ⓒ Ⓓ Ⓔ
62 Ⓐ Ⓑ Ⓒ Ⓓ Ⓔ
63 Ⓐ Ⓑ Ⓒ Ⓓ Ⓔ
64 Ⓐ Ⓑ Ⓒ Ⓓ Ⓔ
65 Ⓐ Ⓑ Ⓒ Ⓓ Ⓔ
66 Ⓐ Ⓑ Ⓒ Ⓓ Ⓔ
67 Ⓐ Ⓑ Ⓒ Ⓓ Ⓔ
68 Ⓐ Ⓑ Ⓒ Ⓓ Ⓔ
69 Ⓐ Ⓑ Ⓒ Ⓓ Ⓔ

Practice Test 1

Note: For all questions involving solutions and/or chemical equations, assume that the system is in water unless otherwise stated.

Part A

Directions: Every set of the given lettered choices below refers to the numbered statements or formulas immediately following it. Choose the one lettered choice that best fits each statement or formula and then fill in the corresponding oval on the answer sheet. Each choice may be used once, more than once, or not at all in each set.

Questions 1–9

Periodic Table (abbreviated)

^{3}Li		…				(D)	^{10}Ne
	(A)	…			(C)		
(B)	^{20}Ca		…				(E)

1. The most electronegative element
2. The element with a possible oxidation number of -2
3. The element that would react in a 1 : 1 ratio with (D)
4. The element with the smallest ionic radius
5. The element with the smallest first ionization potential
6. The element with a complete *p* orbital as its outermost energy level
7. A member of the alkali metals family
8. A noble gas
9. The element that would react most actively when placed in water to form a strong base

Questions 10–12 refer to the following heating curve for water:

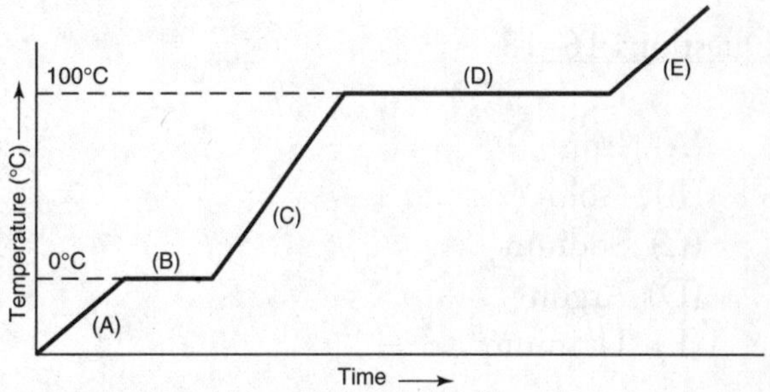

10. In which part of the curve is the state of H_2O only a solid?
11. Which part of the graph shows a phase change requiring the greatest amount of energy?
12. Where is the temperature of H_2O changing at 4.18 J/g°C or 1 cal/g°C?

GO ON TO THE NEXT PAGE.

Questions 13–15 refer to the following diagram:

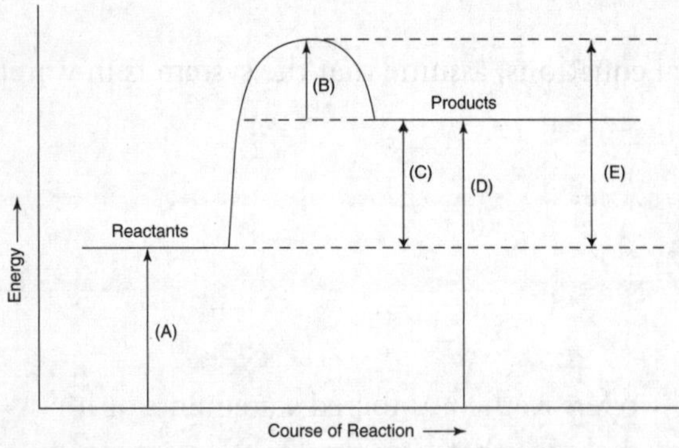

13. Indicates the activation energy of the forward reaction
14. Indicates the activation energy of the reverse reaction
15. Indicates the difference between the activation energies for the reverse and forward reactions and equals the energy change in the reaction

Questions 16–18

(A) Iron
(B) Gold
(C) Sodium
(D) Argon
(E) Uranium

16. A metal element that resists reaction with acids
17. A monoatomic element with filled *p* orbitals
18. A transition element that occurs when the inner 3*d* orbital is partially filled

Questions 19 and 20

(A) Radioactive isotope
(B) Monoclinic crystal
(C) Sulfur trioxide
(D) Phosphate
(E) Sulfite

19. A substance that exhibits a resonance structure
20. A product formed from a base reacting with sulfurous acid

Questions 21–23

(A) Dilute
(B) Concentrated
(C) Unsaturated
(D) Saturated
(E) Supersaturated

21. The condition, unrelated to quantities, that indicates that the rate going into solution is equal to the rate coming out of solution
22. The condition that exists when a water solution that has been at equilibrium and saturated is heated to a higher temperature with a higher solubility, but no additional solute is added
23. The descriptive term that indicates there is a large quantity of solute, compared with the amount of solvent, in a solution

Part B

ON THE ACTUAL CHEMISTRY TEST, THE FOLLOWING TYPE OF QUESTION MUST BE ANSWERED ON A SPECIAL SECTION (LABELED "CHEMISTRY") AT THE LOWER LEFT-HAND CORNER OF PAGE 2 OF YOUR ANSWER SHEET. THESE QUESTIONS WILL BE NUMBERED BEGINNING WITH 101 AND MUST BE ANSWERED ACCORDING TO THE FOLLOWING DIRECTIONS.

Directions: Every question below contains two statements, I in the left-hand column and II in the right-hand column. For each question, decide if statement I is true or false and if statement II is true or false and fill in the corresponding T or F ovals on your answer sheet. *Fill in oval CE only if statement II is a correct explanation of statement I.

Sample Answer Grid:

CHEMISTRY * Fill in oval CE only if II is a correct explanation of I.

	I		II		CE*
101.	T	F	T	F	◯

	I		II
101.	Nonmetallic oxides are usually acid anhydrides	BECAUSE	nonmetallic oxides form acids when placed in water.
102.	When HCl gas and NH_3 gas come into contact, a white smoke forms	BECAUSE	NH_3 and HCl react to form a white solid, ammonium chlorate.
103.	The reaction of barium chloride and sodium sulfate does not go to completion	BECAUSE	the compound barium sulfate is formed as an insoluble precipitate.
104.	When two elements react exothermically to form a compound, the compound should be relatively stable	BECAUSE	the release of energy from a combination reaction indicates that the compound formed is at a lower energy level than the reactants and thus relatively stable.
105.	The ion of a nonmetallic atom is larger in radius than the atom	BECAUSE	when a nonmetallic ion is formed, it gains electrons in the outer orbital and thus increases the size of the electron cloud around the nucleus.

	I		II
106.	Oxidation and reduction occur together	BECAUSE	in redox reactions, electrons must be gained and lost.
107.	Decreasing the atmospheric pressure on a pot of boiling water causes the water to stop boiling	BECAUSE	changes in pressure are directly related to the boiling point of water.
108.	The reaction of hydrogen with oxygen to form water is an exothermic reaction	BECAUSE	water molecules have polar covalent bonds.
109.	Atoms of different elements can have the same mass number	BECAUSE	the atoms of each element have a characteristic number of protons in the nucleus.
110.	The transmutation decay of $^{238}_{92}U$ can be shown as $^{238}_{92}U \rightarrow {}^{234}_{90}Th + {}^{4}_{2}He$	BECAUSE	the transmutation of $^{238}_{92}U$ is accompanied by the release of a beta particle.
111.	$^{13}_{6}C$ and $^{14}_{6}C$ are isotopes of the element carbon	BECAUSE	isotopes of an element have the same number of protons in the nucleus but have a different number of neutrons.
112.	The Cu^{2+} ion needs to be oxidized to form Cu metal	BECAUSE	oxidation is a gain of electrons.
113.	The volume of a gas at 373 K and a pressure of 600 millimeters of mercury will be decreased at STP	BECAUSE	decreasing the temperature and increasing the pressure will cause the volume to decrease because $V_2 = V_1 \times \frac{P_1}{P_2} \times \frac{T_2}{T_1}$.
114.	The pH of a 0.01 molar solution of HCl is 2	BECAUSE	dilute HCl dissociates into two essentially ionic particles.
115.	Nuclear fusion on the sun converts hydrogen to helium with a release of energy	BECAUSE	some mass is converted to energy in a solar fusion.
116.	The "bullet" usually used to initiate the fusion of ^{235}U is a neutron	BECAUSE	capture of the neutron by the ^{235}U nucleus causes an unstable condition that leads to its disintegration.

Part C

Directions: Every question or incomplete statement below is followed by five suggested answers or completions. Choose the one that is best in each case and then fill in the corresponding oval on the answer sheet.

24. What is the approximate formula mass of $Ca(NO_3)_2$?

(A) 70
(B) 82
(C) 102
(D) 150
(E) 164

25. In this reaction: $XClO_3 + A \rightarrow XCl + O_2(g) + A$, which substance is the catalyst?

(A) X
(B) $XClO_3$
(C) A
(D) XCl
(E) O_2

26. The normal electron configuration for ethyne (acetylene) is

(A) H:C::C:H
(B) H:C̈:C̈:H
(C) H·C:::C·H
(D) H:C⁝⁝C:H
(E) H:Ċ̣:Ċ̣:H

27. According to the Kinetic-Molecular Theory, molecules increase in kinetic energy when they

(A) are mixed with other molecules at lower temperature
(B) are frozen into a solid
(C) are condensed into a liquid
(D) are heated to a higher temperature
(E) collide with each other in a container at a lower temperature

28. How many atoms are represented in the formula $Ca_3(PO_4)_2$?

(A) 5
(B) 8
(C) 9
(D) 12
(E) 13

29. All of the following have covalent bonds EXCEPT

(A) HCl
(B) CCl_4
(C) H_2O
(D) CsF
(E) CO_2

30. Which of the following is (are) the WEAKEST attractive force?

(A) Dipole-dipole forces
(B) Coordinate covalent bonding
(C) Covalent bonding
(D) Polar covalent bonding
(E) Ionic bonding

GO ON TO THE NEXT PAGE.

31. Which of these resembles the molecular structure of the water molecule?

(A) 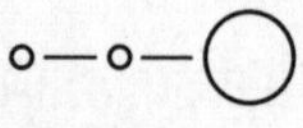

(B) 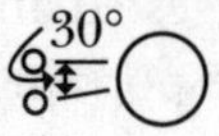

(C) 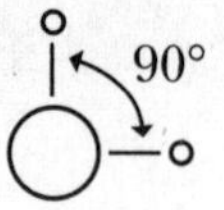

(D) 

(E) 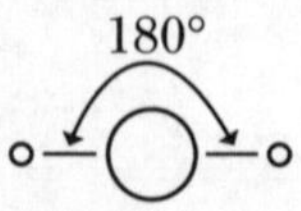

32. The two most important considerations in deciding whether a reaction will occur spontaneously are

(A) the stability and state of the reactants
(B) the energy gained and the heat evolved
(C) a negative value for ΔH and a positive value for ΔS
(D) a positive value for ΔH and a negative value for ΔS
(E) the endothermic energy and the structure of the products

33. The reaction of an acid such as HCl and a base such as NaOH always

(A) forms a precipitate
(B) forms a volatile product
(C) forms an insoluble salt and water
(D) forms a sulfate salt and water
(E) forms a salt and water

34. The oxidation number of sulfur in H_2SO_4 is

(A) +2
(B) +3
(C) +4
(D) +6
(E) +8

35. Balance this redox reaction by using the half-reaction method of electron exchange:

$$KMnO_4 + H_2SO_3 \rightarrow K_2SO_4 + MnSO_4 + H_2SO_4 + H_2O$$

Which of the following partial equations is the correct reduction half-reaction for the *balanced* equation?

(A) $5SO_3^{2-} + 5H_2O \rightarrow 5SO_4^{2-} + 10H^+ + 10e^-$
(B) $2MnO_4^- + 16H^+ + 10e^- \rightarrow 2Mn^{2+} + 8H_2O$
(C) $SO_3^{2-} \rightarrow SO_4^{2-} + 2e^-$
(D) $SO_3^{2-} + 2H^+ \rightarrow SO_4^{2-} + H_2O + 2e^-$
(E) $Mn^{7+} \rightarrow Mn^{2+} + 5e^-$

36. Which of the following when placed into water will test as an acid solution?

I. $HCl(g) + H_2O$
II. Excess $H_3O^+ + H_2O$
III. $CuSO_4(s) + H_2O$

(A) I only
(B) III only
(C) I and II only
(D) II and III only
(E) I, II, and III

37. The property of matter that is independent of its surrounding conditions and position is

(A) volume
(B) density
(C) mass
(D) weight
(E) state

38. Where are the highest ionization energies found in the Periodic Table?

(A) Upper left corner
(B) Lower left corner
(C) Upper right corner
(D) Lower right corner
(E) Middle of transition elements

39. Which of the following pairs of compounds can be used to illustrate the Law of Multiple Proportions?

(A) NO and NO_2
(B) CH_4 and CO_2
(C) ZnO_2 and $ZnCl_2$
(D) NH_3 and NH_4Cl
(E) H_2O and HCl

40. In this equilibrium reaction: $A + B \rightleftharpoons AB +$ heat (in a closed container), how could the forward reaction rate be increased?

I. By increasing the concentration of AB
II. By increasing the concentration of A
III. By removing some of product AB

(A) I only
(B) III only
(C) I and III only
(D) II and III only
(E) I, II, and III

41. For the reaction of sodium with water, the balanced equation using the smallest whole numbers has which of the following coefficients?

I. 1
II. 2
III. 3

(A) I only
(B) III only
(C) I and II only
(D) II and III only
(E) I, II, and III

42. If 10 liters of CO gas react with sufficient oxygen for a complete reaction, how many liters of CO_2 gas are formed?

(A) 5
(B) 10
(C) 15
(D) 20
(E) 40

43. If 49 grams of H_2SO_4 react with 80 grams of NaOH, how much reactant will be left over after the reaction is complete?

(A) 24.5 g H_2SO_4
(B) none of either compound
(C) 20 g NaOH
(D) 40 g NaOH
(E) 60 g NaOH

44. If the molar concentration of Ag^+ ions in 1 liter of a saturated water solution of silver chloride is 1.38×10^{-5} mole/liter, what is the K_{sp} of this solution?

(A) 0.34×10^{-10}
(B) 0.69×10^{-5}
(C) 1.90×10^{-10}
(D) 2.76×10^{-5}
(E) 2.76×10^{-10}

45. If the density of a diatomic gas at STP is 1.43 grams/liter, what is the gram-molecular mass of the gas?

(A) 14.3 g
(B) 32.0 g
(C) 48.0 g
(D) 64.3 g
(E) 224 g

46. From 2 moles of $KClO_3$ how many liters of O_2 can be produced at STP by decomposition of all the $KClO_3$?

(A) 11.2
(B) 22.4
(C) 33.6
(D) 44.8
(E) 67.2

47. Which value best determines whether a reaction is spontaneous?

(A) change in Gibbs free energy, ΔG
(B) change in entropy, ΔS
(C) change in kinetic energy, ΔKE
(D) change in enthalpy, ΔH
(E) change in heat of formation, ΔH^0_f

GO ON TO THE NEXT PAGE.

Questions 48–52 refer to the following experimental setup and data:

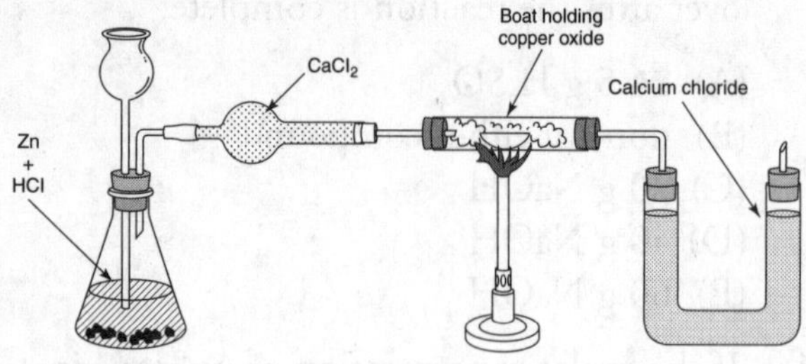

Recorded data:

Before: CuO + porcelain boat = 62.869 g
$CaCl_2$ + U-tube = 80.483 g
After: Porcelain boat + contents = 54.869 g
$CaCl_2$ + U-tube = 89.483 g

Reactions: $Zn + 2HCl \rightarrow ZnCl_2 + H_2$
then $H_2 + CuO \rightarrow H_2O + Cu$

48. What type of reaction occurred in the porcelain boat?

(A) electrolysis
(B) double displacement
(C) reduction and oxidation
(D) decomposition or analysis
(E) combination or synthesis

49. Why was the $CaCl_2$ tube placed between the generator and the tube containing the porcelain boat?

(A) To absorb evaporated HCl
(B) To absorb evaporated H_2O
(C) To slow down the gases released
(D) To absorb the evaporated Zn particles
(E) To remove the initial air that passes through the tube

50. How many grams of hydrogen were used in the formation of the water that was a product?

(A) 1
(B) 2
(C) 4
(D) 8
(E) 9

51. What conclusion can you draw from this experiment?

(A) Hydrogen diffuses faster than oxygen.
(B) Hydrogen is lighter than oxygen.
(C) The molar mass of oxygen is 32 g/mol.
(D) Water is a triatomic molecule with polar characteristics.
(E) Water is formed from hydrogen and oxygen in a ratio of 1 : 8, by mass.

52. Which of the following is an observation rather than a conclusion?

(A) A substance is an acid if it changes litmus paper from blue to red.
(B) A gas is lighter than air if it escapes from a bottle left mouth upward.
(C) The gas H_2 forms an explosive mixture with air.
(D) Air is mixed with hydrogen gas and ignited; it explodes.
(E) A liquid oil is immiscible with water because it separates into a layer above the water.

53. When HCl fumes and NH_3 fumes are introduced into opposite ends of a long, dry glass tube, a white ring forms in the tube. Which statement explains this phenomenon?

(A) NH_4Cl forms.
(B) HCl diffuses faster.
(C) NH_3 diffuses faster.
(D) The ring occurs closer to the end into which HCl was introduced.
(E) The ring occurs in the very middle of the tube.

54. The correct formula for calcium hydrogen sulfate is

(A) CaH_2SO_4
(B) $CaHSO_4$
(C) $Ca(HSO_4)_2$
(D) Ca_2HSO_4
(E) $Ca_2H_2SO_4$

55. Forty grams of sodium hydroxide is dissolved in enough water to make 1 liter of solution. What is the molarity of the solution?

    (A) 0.25 M
    (B) 0.5 M
    (C) 1 M
    (D) 1.5 M
    (E) 4 M

56. For a saturated solution of salt in water, which statement is true?

    (A) All dissolving has stopped.
    (B) Crystals begin to grow.
    (C) An equilibrium has been established.
    (D) Crystals of the solute will visibly continue to dissolve.
    (E) The solute is exceeding its solubility.

57. In which of the following series is the pi bond present in the bonding structure?

    I. Alkane
    II. Alkene
    III. Alkyne

    (A) I only
    (B) III only
    (C) I and III only
    (D) II and III only
    (E) I, II, and III

<u>Questions 58 and 59</u> refer to the following setup:

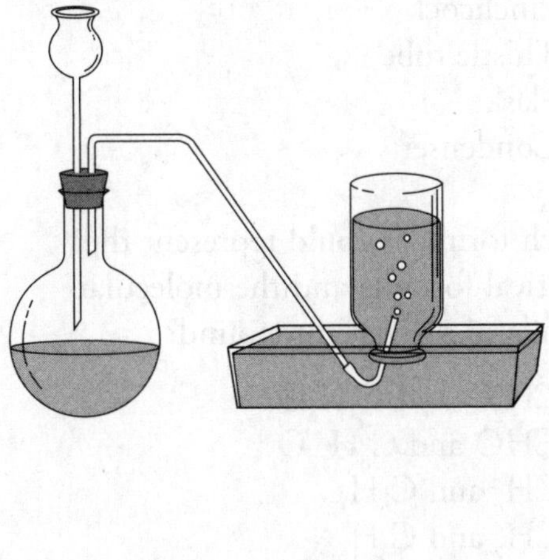

58. Why could you NOT use this setup for preparing H_2 if the generator contained Zn + vinegar?

    (A) Hydrogen would not be produced.
    (B) The setup of the generator is improper.
    (C) The generator must be heated with a burner.
    (D) The delivery tube setup is wrong.
    (E) The gas cannot be collected over water.

59. In a proper laboratory setup for collecting a gas by water displacement, which of these gases could NOT be collected over H_2O because of its solubility?

    (A) CO_2
    (B) NO
    (C) O_2
    (D) NH_3
    (E) CH_4

60. What is the approximate percentage of oxygen in the formula mass of $Ca(NO_3)_2$?

    (A) 28
    (B) 42
    (C) 58
    (D) 68
    (E) 84

GO ON TO THE NEXT PAGE.

61. For the following reaction:

$$N_2O_4(g) \rightleftharpoons 2NO_2(g),$$

the K_{eq} expression is

(A) $K_{eq} = \dfrac{[N_2O_4]}{[NO_2]}$

(B) $K_{eq} = \dfrac{[N_2O_4]}{[NO_2]}$

(C) $K_{eq} = \dfrac{[NO_2]}{[N_2O_4]}$

(D) $K_{eq} = \dfrac{[NO_2]^2}{[N_2O_4]}$

(E) $K_{eq} = \dfrac{[N_2O_4]^2}{[NO_2]}$

62. What is the K_{eq} for the reaction in question 61 if at equilibrium the concentration of N_2O_4 is 4×10^{-2} mole/liter and that of NO_2 is 2×10^{-2} mole/liter?

(A) 1×10^{-2}
(B) 2×10^{-2}
(C) 4×10^{-2}
(D) 4×10^{-4}
(E) 8×10^{-2}

63. How much water, in liters, must be added to 0.5 liter of 6 M HCl to make the solution 2 M?

(A) 0.33
(B) 0.5
(C) 1
(D) 1.5
(E) 2

64. Four grams of hydrogen are ignited with 4 grams of oxygen. How many grams of water can be formed?

(A) 0.5
(B) 2.5
(C) 4.5
(D) 8
(E) 36

65. Which structure is an ester?

(A) 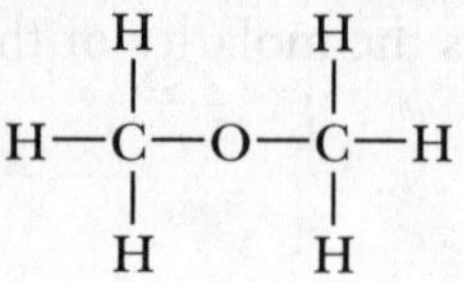

(B) 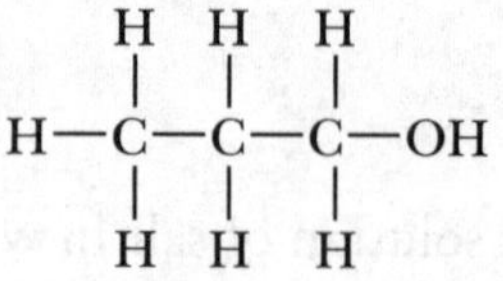

(C) 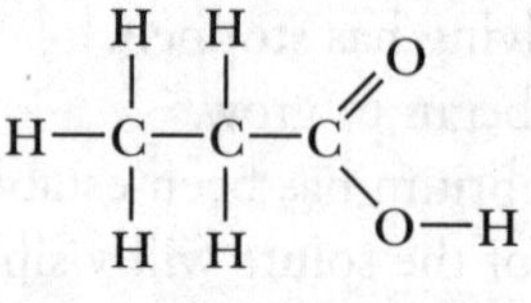

(D) 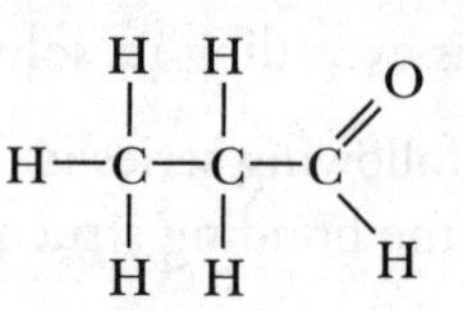

(E) 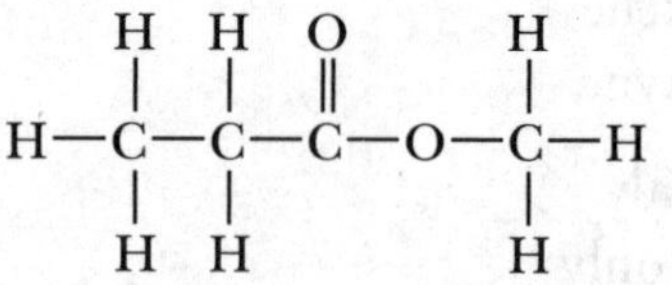

66. What piece of apparatus can be used to introduce more liquid into a reaction and also serve as a pressure valve?

(A) Stopcock
(B) Pinchcock
(C) Thistle tube
(D) Flask
(E) Condenser

67. Which formulas could represent the empirical formula and the molecular formula of a given compound?

(A) CH_2O and $C_4H_6O_4$
(B) CHO and $C_6H_{12}O_6$
(C) CH_4 and C_5H_{12}
(D) CH_2 and C_3H_6
(E) CO and CO_2

68. The reaction $Fe \rightarrow Fe^{2+} + 2e^-$ (+0.44 volt) would occur spontaneously with which of the following?

I. $Pb \rightarrow Pb^{2+} + 2e^-$ (+0.13 volt)
II. $Cu \rightarrow Cu^{2+} + 2e^-$ (−0.34 volt)
III. $2Ag^+ + 2e^- \rightarrow 2Ag^0$ (+0.80 volt)

(A) I only
(B) III only
(C) I and III only
(D) II and III only
(E) I, II, and III

69. $2Na(s) + Cl_2(g) \rightarrow 2NaCl(s) + 822$ kJ

How much heat is released by the above reaction if 0.5 mole of sodium reacts completely with chlorine?

(A) 205 kJ
(B) 411 kJ
(C) 822 kJ
(D) 1,644 kJ
(E) 3,288 kJ

STOP

If you finish before one hour is up, you may go back to check your work or complete unanswered questions.

Answers and Explanations for Test 1

1. **(D)** The most electronegative element, F, would be found in the upper right corner of the table; the noble gases are exceptions at the far right.

2. **(C)** Elements in the group with (C) have a possible oxidation number of −2.

3. **(B)** Elements in the group with (B) react in a 1 : 1 ratio with elements in the group with (D), since one has an electron to lose and the other one needs an electron to complete its outer energy level.

4. **(A)** (A) loses 2 electrons to form an ion whose remaining electrons, being close to the nucleus, are pulled in closer because of the unbalanced 2+ charge.

5. **(B)** Since (B) has only one electron in the outer 4*s* orbital, it can more easily be removed than can an electron from the 3*s* orbital of (A), which is closer to the positive nucleus.

6. **(E)** All Group VIII elements have a complete *p* orbital as the outer energy level. This explains why these elements are "inert."

7. **(B)** The Group IA elements comprise the alkali-metals family.

8. **(E)** The Group VIII elements are the noble gases.

9. **(B)** The alkali metals react with water to form a strong base. Activity increases down this family.

10. **(A)** H_2O is ice in Part A.

11. **(D)** H_2O changes state at Parts B and D. The heat of vaporization at D (540 cal/g) is greater than the heat of fusion (80 cal/g) at B.

12. **(C)** Water is heating at 1°C/cal/g in Part C.

13. **(E)** This is the energy needed to start the forward reaction.

14. **(B)** The part indicated by B represents the activation energy for the reverse reaction.

15. **(C)** The net energy released is the endothermic quantity indicated by C.

16. **(B)** Gold is known as a noble metal because of its resistance to acids. Aqua regia, a mixture of HNO_3 and HCl, will react with gold.

17. **(D)** Argon is the only element among the choices that is monoatomic in the molecular form.

18. **(A)** Iron has 5 electrons in the *d* orbitals, which are partially filled.

19. **(C)** Only sulfur trioxide has a resonance structure (as shown here):

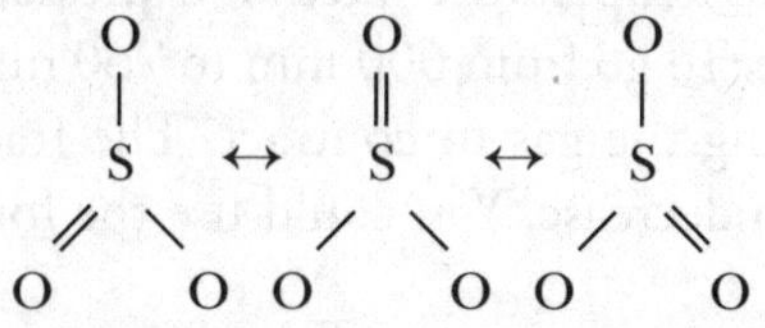

20. **(E)** Sulfurous acid reacts with a base to form a sulfite salt.

21. **(D)** The condition described is the equilibrium that exists at saturation.

22. **(C)** With the increased temperature more solute may go into solution; therefore the solution is now unsaturated.

23. **(B)** The term "concentrated" means that there is a large amount of solute in the solvent.

101. **(T, T, CE)** Nonmetallic oxides are usually acid anhydrides, and they form acids in water.

102. **(T, F)** The white smoke formed is ammonium chloride, not ammonium chlorate.

103. **(F, T)** The reaction of barium chloride and sodium sulfate *does* go to completion since barium sulfate is a precipitate.

104. **(T, T, CE)** The product of an exothermic reaction is relatively stable because it is at a lower energy level than the reactants.

105. **(T, T, CE)** Both statements are true, and the reason explains the assertion.

106. **(T, T, CE)** Both statements are true, and the reason explains the assertion.

107. **(F, T)** The statement is false while the reason is true. Decreasing the pressure on a boiling pot will only cause the water to boil more vigorously.

108. **(T, T)** The statements are true, but the reason doesn't explain the assertion.

109. **(T, T)** The statements are true, but the reason doesn't explain the assertion.

110. **(T, F)** The transmutation equation is true and shows the release of an alpha particle, not a beta particle.

111. **(T, T, CE)** The two configurations of carbon are isotopes because they have the same atomic number but different mass numbers because ${}^{13}_{6}C$ has 6 protons and 7 neutrons while ${}^{14}_{6}C$ has 6 protons and 8 neutrons.

112. **(F, F)** Both the statement and the reason are false.

113. **(T, T, CE)** In going from 100°C to 0°C, the volume decreases as the gas gets colder; therefore, the temperature fraction expressed in kelvins must be $\frac{273}{373}$ to decrease the volume. To go from 600 mm to 760 mm of pressure increases the pressure, thus causing the gas to contract. The fraction must then be $\frac{600}{760}$ to cause the volume to decrease. You could use the formula

$$V_2 = V_1 \times \frac{T_2}{T_1} \times \frac{P_1}{P_2}$$

114. **(T, T)** Since HCl is a strong acid and ionizes completely in dilute solution of water (the $[H^+]$ and $[H_3O^+]$ are the same thing), the molar concentration of a 0.01 molar solution is 1×10^{-2} mol/L.

$$\begin{aligned} pH &= -\log[H^+] \\ pH &= -\log[1\times10^{-2}] \\ pH &= -(-2) = 2 \end{aligned}$$

The pH is 2, but the reason, although true, does not explain the statement.

115. **(T, T, CE)** Both statements are true, and the reason explains the assertion.

116. **(F, T)** The statement should refer to fission, not fusion. The reason is true.

24. **(E)** The total formula mass is:

Ca	= 40
2N	= 28
6O	= 96
Total	164

25. **(C)** The catalyst, by definition, is not consumed in the reaction and ends up in its original form as one of the products.

26. **(D)** The ethyne molecule is the first member of the acetylene series with a general formula of C_nH_{2n-2}. It contains a triple bond between the two C atoms: H:C⁝⁝C:H.

27. **(D)** Heating molecules increases their kinetic energy.

28. **(E)** 3Ca + 2P + 8O = 13 atoms.

29. **(D)** Cesium and fluorine are from the most electropositive and electronegative portions, respectively, of the periodic chart, and thus form an ionic bond by cesium giving an electron to fluorine to form the respective ions.

30. **(A)** Dipole-dipole forces are due to the weak attraction of the nuclear positive charge of one atom to the negative electron field of an adjacent atom. They are much weaker than the others named.

31. **(D)** The molecular structure of water is that of a polar covalent compound with the hydrogens 105° apart.

32. **(C)** The most important considerations for a spontaneous reaction are (1) that the reaction is exothermic with a negative ΔH, so that once started it tends to continue on its own because of the energy released, and (2) that the reaction tends to go to the highest state of randomness, shown by a positive value for ΔS.

33. **(E)** Normal H^+ acids and OH^- bases form water and a salt (not necessarily a soluble salt).

34. **(D)** Every compound has a charge of 0. H usually has +1, and O usually has −2, so

$$\begin{aligned} H_2 &= +2 \\ O_4 &= -8 \\ S &= x \\ \hline \text{Total} &= 0 \\ +2 - 8 + x &= 0 \\ x &= -2 + 8 = +6 \end{aligned}$$

35. **(B)** In the balanced equation the two half-reactions are:

$$5SO_3^{2-} + 5H_2O \rightarrow 5SO_4^{2-} + 10H^+ + 10e^-$$
$$2MnO_4^- + 16H^+ + 10e^- \rightarrow 2Mn^{2+} + 8H_2O$$
$$\overline{2KMnO_4 + 5H_2SO_3 \rightarrow K_2SO_4 + 2MnSO_4 + 2H_2SO_4 + 3H_2O}$$

36. **(E)** HCl and H_3O^+ give acid solutions, as does $CuSO_4$ (the salt of a weak base and a strong acid) when it hydrolyzes in water. I, II, and III are correct.

37. **(C)** Mass is a constant and is not dependent on position or surrounding conditions.

38. **(C)** The complete outer energy levels of electrons of the smallest noble gases have the highest ionization potential.

39. **(A)** Only NO and NO_2 fit the definition of the Law of Multiple Proportions, in which one substance stays the same and the other varies in units of whole integers.

40. **(D)** Increasing the concentration of one or both of the reactants and removing some of the product formed would cause the forward reaction to increase in rate to try to regain the equilibrium condition. II and III are correct.

41. **(C)** I and II are correct. The equation is $2Na + 2H_2O \rightarrow 2NaOH + H_2(g)$. The coefficients include a 1 and a 2.

42. **(B)** $2CO + O_2 \rightarrow 2CO_2$ indicates 2 volumes of CO react with 1 volume of O_2 to form 2 volumes of CO_2. Therefore, 10 L of CO form 10 L of CO_2.

43. **(D)** $H_2SO_4 + 2NaOH \rightarrow 2H_2O + Na_2SO_4$ is the equation for this reaction. 1 mol H_2SO_4 = 98 g. 1 mol NaOH = 40 g. Then 49 g of H_2SO_4 = $\frac{1}{2}$ mol H_2SO_4. The equation shows that 1 mol of H_2SO_4 reacts with 2 mol of NaOH for a ratio of 1 : 2. Therefore, $\frac{1}{2}$ mol of H_2SO_4 reacts with 1 mol of NaOH in this reaction. Since 1 mol of NaOH equals 40 g, and 80 g of NaOH is given, 40 g of it will remain after the reaction has gone to completion.

44. **(C)** $K_{sp} = [Ag^+][Cl^-] = 1.9 \times 10^{-10}$. The molar concentrations of Ag^+ and Cl^- are 1.38×10^{-5} mol/L, so

$$[Ag^+][Cl^-] = K_{sp}$$
$$[1.38 \times 10^{-5}][1.38 \times 10^{-5}] = 1.9 \times 10^{-10}$$

45. **(B)** If 1.43 g is the mass of 1 L, then the mass of 22.4 L, which is the gram-molecular volume of a gas at STP, will give the gram-molecular mass. Then, 22.4 L × 1.43 g/L = 32.0 g, the gram-molecular mass.

46. **(E)** The equation is:

$$2KClO_3 \rightarrow 2KCl + 3O_2$$

This shows that 2 mol of $KClO_3$ yields 3 mol of O_2. Three moles of O_2 = 3 × 22.4 L = 67.2 L of O_2.

47. **(A)** Gibbs free energy combines the overall energy changes and the entropy change. The formula is $\Delta G = \Delta H - T\Delta S$. Only if ΔG is negative will the reaction be spontaneous in the forward direction.

48. **(C)** The CuO was reduced while the H_2 was oxidized, forming H_2O + Cu. The reduction and oxidation reaction is:

$$CuO + H_2 \rightarrow H_2O + Cu$$

49. **(B)** The purpose of this $CaCl_2$ tube is to absorb any water evaporated from the generator. If it were not present, some water vapor would pass through to the final drying tube and cause the weight of water gained there to be larger than it should be from the reaction only.

50. **(A)** Since 8 g of O_2 was lost by the CuO, the weight of water gained by the U tube came from 8 g of O_2 and 1 g of H_2, to make 9 g of water that it absorbed.

51. **(E)** The weight ratio of water is 1 g of H_2 to 8 g of O_2 or 1 : 8. The other statements are true but are not conclusions from this experiment.

52. **(D)** This is the only observation of the group. All the others are conclusions. Remember that an observation is only what you see, smell, taste, or measure with a piece of equipment.

53. **(A)** The phenomenon is the formation of the white ring, which is NH_4Cl. Although the distance traveled by each gas could be measured to verify Graham's Law of Gaseous Diffusion, this was not asked. The relationship is that the diffusion rate is inversely proportional to the square root of the gas's molecular weight.

54. **(C)** Since Ca^{2+} and HSO_4^- combine, the formula is $Ca(HSO_4)_2$.

55. **(C)** NaOH is 40 g/mol. 40 g in 1 L makes a 1 M solution.

56. **(C)** A saturated solution represents a condition where the solute is going into solution as rapidly as some solute is coming out of the solution.

57. **(D)** II and III are correct since the double-bonded carbons in the alkene and the triple-bonded carbons in the alkyne series have pi bonds.

58. **(B)** The thistle tube is not below the level of the liquid in the generator and the gas would escape into the air. (Vinegar is an acid and would produce hydrogen.)

59. **(D)** NH_3 is very soluble and could not be collected in this manner. All others are not sufficiently soluble to hamper this method of collection.

60. **(C)** The formula (mass) is the total of (Ca =) 40 + (2N =) 28 + (6O =) 96, or 164. Since oxygen is 96 amu of 164 total, the percentage is 96/164 × 100 = 58.5%.

61. **(D)** The K_{eq} expression consists of the concentration(s) of the products over those of the reactants, with the coefficients becoming exponents:

$$K_{eq} = \frac{[NO_2]^2}{[N_2O_4]}$$

62. **(A)** $K_{eq} = \frac{[2 \times 10^{-2}]^2}{[4 \times 10^{-2}]} = \frac{4 \times 10^{-4}}{4 \times 10^{-2}}$
$= 1 \times 10^{-2}$

63. **(C)** In dilution problems, the expressions $M_1 \times V_1 = M_2 \times V_2$ can be used. Substituting (6 M) (0.5 L) = (2 M) (x L) gives $x = 1.5$ L total volume. Since there was 0.5 L to begin with, an additional 1 L must be added.

64. **(C)** The reaction equation and information given can be set up like this:

Given		Given		
4 g		4 g		x g
$2H_2$	+	O_2	→	$2H_2O$
4 g		32 g		36 g

Studying this shows that the limiting element will be the 4 g of oxygen since 4 g of H_2 would require 32 g of O_2. The solution setup is

$$\frac{4\ \text{g}\ O_2}{32\ \text{g}\ O_2} = \frac{x\ \text{g}\ H_2O}{36\ \text{g}\ H_2O}$$

$$x = 4.5\ \text{g}\ H_2O$$

65. **(E)** The functional group of an ester is

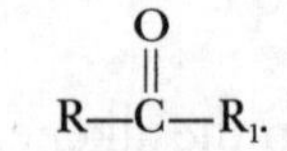

This appears only in (E).

66. **(C)** The thistle tube serves both these purposes.

67. **(D)** The empirical formula is a representation of the elements in their simplest ratio. Therefore, CH_2 is the simplest ratio of the molecular formula C_3H_6.

68. **(E)** I, II, and III would occur since their reduction reactions would be −0.13, +0.34, +0.80 V, respectively. These numbers added to +0.44 V separately give a positive E^0 for the reaction with Fe.

69. **(A)** In the equation, 2 mol of Na release 822 kJ of heat. If only 0.5 mol of Na is consumed, only one-fourth as much heat will be released: $\frac{1}{4} \times 822 = 205$ kJ.

CALCULATING YOUR SCORE

Your score on practice Test 1 can now be computed manually. The actual test is scored by machine, but the same method is used to arrive at the raw score. You get one point for each correct answer. For each wrong answer, you lose one-fourth of a point. Questions that you omit or that have more than one answer are not counted. On your answer sheet mark all correct answers with a "C" and all incorrect answers with as "X".

Determining Your Raw Test Score

Total the number of correct answers you have recorded on your answer sheet. It should be the same as the total of all the numbers you place in the block in the lower left corner of each area of the Subject Area summary in the next section.

A. Enter the total number of correct answers here: ________
Now count the number of wrong answers you recorded on your answer sheet.
B. Enter the total number of wrong answers here: ________
Multiply the number of wrong answers in B by 0.25.
C. Enter that product here: ________
Subtract the result in C from the total number of right answers in A.
D. Enter the result of your subtraction here: ________
E. Round the result in D to the nearest whole number: ________.
This is your raw test score.

Conversion of Raw Scores to Scaled Scores

Your raw score is converted by the College Board into a scaled score. The College Board scores range from 200 to 800. This conversion is done to ensure that a score earned on any edition of a particular SAT Subject Test in Chemistry is comparable to the same scaled score earned on any other edition of the same test. Because some editions of the tests may be slightly easier or more difficult than others, scaled scores are adjusted so that they indicate the same level of performance regardless of the edition of the test taken and the ability of the group that takes it. Consequently, a specific raw score on one edition of a particular test will not necessarily translate to the same scaled score on another edition of the same test.

Because the practice tests in this book have no large population of scores with which they can be scaled, scaled scores cannot be determined.

Results from previous SAT Chemistry tests appear to indicate that the conversion of raw scores to scaled scores GENERALLY follows this pattern:

Raw Score	Scaled Score	Raw Score	Scaled Score
85–75	800–800	30–25	560–540
75–70	800–780	25–20	540–520
70–65	780–750	20–15	520–490
65–60	750–720	15–10	490–460
60–55	720–700	10–5	460–430
55–50	700–680	5–0	430–410
50–45	680–650	0 to –5	410–380
45–40	650–620	–5 to –10	380–360
40–35	620–590	–10 to –15	360–330
35–30	590–560		

Note that this scale provides only a *general idea* of what a raw score may translate into on a scaled score range of 800–200. Scaling on every test is usually slightly different. Some students who had taken the SAT Subject Test in Chemistry after using this book had reported that they have scored slightly higher on the SAT test than on the practice tests in this book. They *all* reported that preparing well for the test paid off in a better score!

DIAGNOSING YOUR NEEDS

After taking Practice Test 1, check your answers against the correct ones. Then fill in the chart below.

In the space under each question number, place a check if you answered that question correctly.

EXAMPLE:

If your answer to question 5 was correct, place a check in the appropriate box.

Next, total the check marks for each section and insert the number in the designated block. Now do the arithmetic indicated, and insert your percent for each area.

Subject Area*	(✓) Questions Answered Correctly
I. Atomic Theory and Structure, including periodic relationships ☐ No. of checks ÷ 8 × 100 = ______%	6, 8, 105, 109, 111, 31, 38
II. Nucleonics ☐ No. of checks ÷ 2 × 100 = ______%	115, 116
III. Chemical Bonding and Molecular Structure ☐ No. of checks ÷ 9 × 100 = ______%	3, 17, 18, 19, 29, 30, 39, 54, 57
IV. States of Matter and Kinetic Molecular Theory of Gases ☐ No. of checks ÷ 7 × 100 = ______%	10, 11, 12, 107, 108, 113, 27
V. Solutions, including concentration units, solubility, and colligative properties ☐ No. of checks ÷ 5 × 100 = ______%	21, 103, 44, 55, 63
VI. Acids and Bases ☐ No. of checks ÷ 7 × 100 = ______%	20, 101, 102, 114, 33, 36, 53
VII. Oxidation-Reduction and Electrochemistry ☐ No. of checks ÷ 6 × 100 = ______%	106, 112, 34, 35, 48, 68
VIII. Stoichiometry ☐ No. of checks ÷ 10 × 100 = ______%	24, 28, 41, 42, 43, 45, 46, 50, 60, 64
IX. Reaction Rates ☐ No. of checks ÷ 2 × 100 = ______%	25, 32

**The subject areas have been expanded to identify specific areas in the text.*

Subject Area*	(✓) Questions Answered Correctly					
X. Equilibrium	22	40	56	61	62	
☐ No. of checks ÷ 5 × 100 = _____%						
XI. Thermodynamics: energy changes in chemical reactions, randomness, and criteria for spontaneity	13	14	15	104	47	64
☐ No. of checks ÷ 6 × 100 = _____%						
XII. Descriptive Chemistry: physical and chemical properties of elements and their familiar compounds; organic chemistry; periodic properties	1	2	4	5	7	9
	16	23	26	37	65	67
☐ No. of checks ÷ 12 × 100 = _____%						
XIII. Laboratory: equipment, procedures, observations, safety, calculations, and interpretation of results	49	51	52	58	59	66
☐ No. of checks ÷ 6 × 100 = _____%						

**The subject areas have been expanded to identify specific areas in the text.*

To develop a study plan, refer to pages 33–35.

Answer Sheet

PRACTICE TEST 2

Determine the correct answer for each question. Then, using a No. 2 pencil, blacken completely the oval containing the letter of your choice.

1 (A) (B) (C) (D) (E)
2 (A) (B) (C) (D) (E)
3 (A) (B) (C) (D) (E)
4 (A) (B) (C) (D) (E)
5 (A) (B) (C) (D) (E)
6 (A) (B) (C) (D) (E)
7 (A) (B) (C) (D) (E)
8 (A) (B) (C) (D) (E)
9 (A) (B) (C) (D) (E)
10 (A) (B) (C) (D) (E)
11 (A) (B) (C) (D) (E)
12 (A) (B) (C) (D) (E)
13 (A) (B) (C) (D) (E)
14 (A) (B) (C) (D) (E)
15 (A) (B) (C) (D) (E)
16 (A) (B) (C) (D) (E)
17 (A) (B) (C) (D) (E)
18 (A) (B) (C) (D) (E)
19 (A) (B) (C) (D) (E)
20 (A) (B) (C) (D) (E)
21 (A) (B) (C) (D) (E)
22 (A) (B) (C) (D) (E)
23 (A) (B) (C) (D) (E)

ON THE ACTUAL CHEMISTRY TEST, THE FOLLOWING TYPE OF QUESTION MUST BE ANSWERED ON A SPECIAL SECTION (LABELED "CHEMISTRY") AT THE LOWER LEFT-HAND CORNER OF PAGE 2 OF YOUR ANSWER SHEET. THESE QUESTIONS WILL BE NUMBERED BEGINNING WITH 101 AND MUST BE ANSWERED ACCORDING TO THE DIRECTIONS.

CHEMISTRY* Fill in oval CE only if II is a correct explanation of I.

	I		II		CE*
101	(T)	(F)	(T)	(F)	○
102	(T)	(F)	(T)	(F)	○
103	(T)	(F)	(T)	(F)	○
104	(T)	(F)	(T)	(F)	○
105	(T)	(F)	(T)	(F)	○
106	(T)	(F)	(T)	(F)	○
107	(T)	(F)	(T)	(F)	○
108	(T)	(F)	(T)	(F)	○
109	(T)	(F)	(T)	(F)	○
110	(T)	(F)	(T)	(F)	○
111	(T)	(F)	(T)	(F)	○
112	(T)	(F)	(T)	(F)	○
113	(T)	(F)	(T)	(F)	○
114	(T)	(F)	(T)	(F)	○
115	(T)	(F)	(T)	(F)	○
116	(T)	(F)	(T)	(F)	○

ON THE ACTUAL CHEMISTRY TEST, THE REMIANING QUESTIONS MUST BE ANSWERED BY RETURNING TO THE SECTION OF YOUR ANSWER SHEET YOU STARTED FOR CHEMISTRY.

24 (A) (B) (C) (D) (E)
25 (A) (B) (C) (D) (E)
26 (A) (B) (C) (D) (E)
27 (A) (B) (C) (D) (E)
28 (A) (B) (C) (D) (E)
29 (A) (B) (C) (D) (E)
30 (A) (B) (C) (D) (E)
31 (A) (B) (C) (D) (E)
32 (A) (B) (C) (D) (E)
33 (A) (B) (C) (D) (E)
34 (A) (B) (C) (D) (E)
35 (A) (B) (C) (D) (E)
36 (A) (B) (C) (D) (E)
37 (A) (B) (C) (D) (E)
38 (A) (B) (C) (D) (E)
39 (A) (B) (C) (D) (E)
40 (A) (B) (C) (D) (E)
41 (A) (B) (C) (D) (E)
42 (A) (B) (C) (D) (E)
43 (A) (B) (C) (D) (E)
44 (A) (B) (C) (D) (E)
45 (A) (B) (C) (D) (E)
46 (A) (B) (C) (D) (E)
47 (A) (B) (C) (D) (E)
48 (A) (B) (C) (D) (E)
49 (A) (B) (C) (D) (E)
50 (A) (B) (C) (D) (E)
51 (A) (B) (C) (D) (E)
52 (A) (B) (C) (D) (E)
53 (A) (B) (C) (D) (E)
54 (A) (B) (C) (D) (E)
55 (A) (B) (C) (D) (E)
56 (A) (B) (C) (D) (E)
57 (A) (B) (C) (D) (E)
58 (A) (B) (C) (D) (E)
59 (A) (B) (C) (D) (E)
60 (A) (B) (C) (D) (E)
61 (A) (B) (C) (D) (E)
62 (A) (B) (C) (D) (E)
63 (A) (B) (C) (D) (E)
64 (A) (B) (C) (D) (E)
65 (A) (B) (C) (D) (E)
66 (A) (B) (C) (D) (E)
67 (A) (B) (C) (D) (E)
68 (A) (B) (C) (D) (E)
69 (A) (B) (C) (D) (E)

Practice Test 2

Note: For all questions involving solutions and/or chemical equations, assume that the system is in water unless otherwise stated.

Part A

Directions: Every set of the given lettered choices below refers to the numbered statements or formulas immediately following it. Choose the one lettered choice that best fits each statement or formula and then fill in the corresponding oval on the answer sheet. Each choice may be used once, more than once, or not at all in each set.

Questions 1–4

(A) Law of Definite Composition
(B) Nuclear fusion
(C) van der Waals forces
(D) Graham's Law of Diffusion (Effusion)
(E) Triple point

1. At a particular temperature and pressure, three states of a substance may coexist.
2. The combining of nuclei to release energy.
3. The ratio of the rate of movement of hydrogen gas compared with the rate of oxygen gas is 4 : 1.
4. The molecules of nitrogen monoxide and nitrogen dioxide differ by a multiple of the mass of one oxygen.

Questions 5–7 refer to the following diagram:

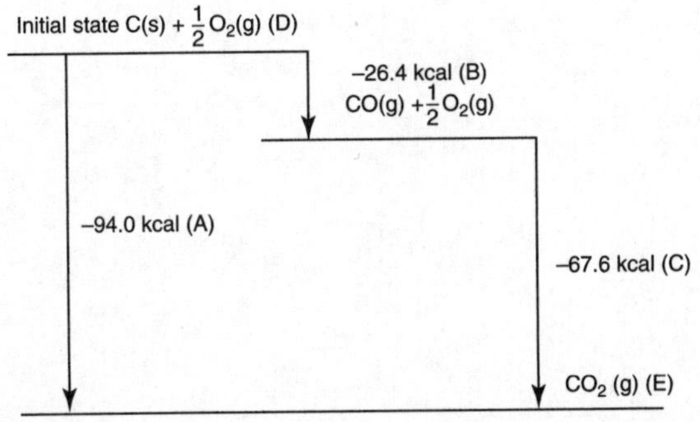

5. The ΔH of the reaction to form CO from $C + O_2$
6. The ΔH of the reaction to form CO_2 from $CO + O_2$
7. The ΔH of the reaction to form CO_2 from $C + O_2$

Questions 8–11

(A) Hydrogen bond
(B) Ionic bond
(C) Polar covalent bond
(D) Nonpolar covalent bond
(E) Metallic bond

8. The type of bond between atoms of potassium and chloride when they form a crystal of potassium chloride
9. The type of bond between the atoms in a nitrogen molecule
10. The type of bond between the atoms in a molecule of CO_2 (electronegativity difference = 1)
11. The type of bond between the atoms of calcium in a crystal of calcium

GO ON TO THE NEXT PAGE.

Questions 12–14 refer to the following phase diagram for CO_2:

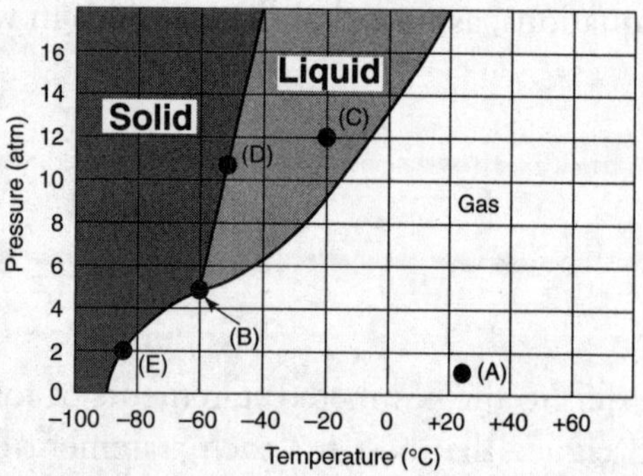

12. The point at which all three states of CO_2 can exist
13. The point at which CO_2 can exist only as a liquid
14. The point at which CO_2 can exist as a solid and a gas under 2 atmospheres of pressure

Questions 15–23 refer to the following graph, which shows the variation of the first ionization potential with respect to increasing atomic numbers:

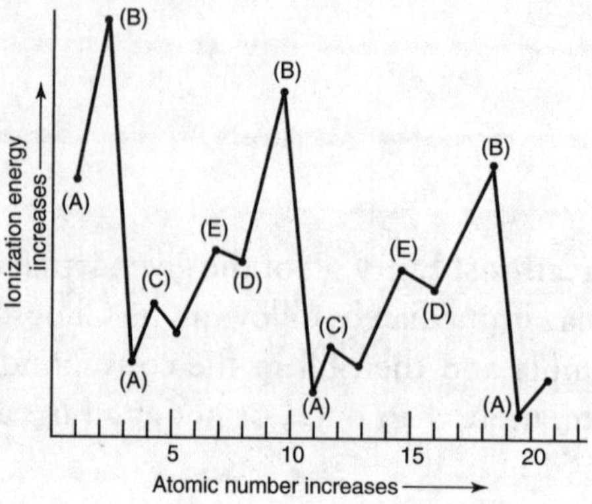

15. The atoms likely to react with water to release hydrogen
16. Nonmetals that are all found in the gaseous state at STP
17. The noble gases
18. The alkali metals
19. The half-filled condition of the *p* orbitals
20. The filled *s* orbitals with the exception of helium
21. The beginning of pairing in the *p* orbitals
22. The most active metals
23. The filled *p* orbitals

Part B

ON THE ACTUAL CHEMISTRY TEST, THE FOLLOWING TYPE OF QUESTION MUST BE ANSWERED ON A SPECIAL SECTION (LABELED "CHEMISTRY") AT THE LOWER LEFT-HAND CORNER OF PAGE 2 OF YOUR ANSWER SHEET. THESE QUESTIONS WILL BE NUMBERED BEGINNING WITH 101 AND MUST BE ANSWERED ACCORDING TO THE FOLLOWING DIRECTIONS.

Directions: Every question below contains two statements, I in the left-hand column and II in the right-hand column. For each question, decide if statement I is true or false and if statement II is true or false and fill in the corresponding T or F ovals on your answer sheet. *Fill in oval CE only if statement II is a correct explanation of statement I.

Sample Answer Grid:

CHEMISTRY * Fill in oval CE only if II is a correct explanation of I.

	I	II	CE*
101.	(T) (F)	(T) (F)	( )

	I		II
101.	The structure of SO_3 is shown by using more than one structural formula	BECAUSE	SO_3 is very unstable and resonates between these possible structures.
102.	When the ΔG of a reaction at a given temperature is negative, the reaction occurs spontaneously	BECAUSE	when ΔG is negative, ΔH is also negative.
103.	One mole of CO_2 has a greater mass than 1 mole of H_2O	BECAUSE	the molecular mass of CO_2 is greater than the molecular mass of H_2O.
104.	Hydrosulfuric acid is often used in qualitative tests	BECAUSE	$H_2S(aq)$ reacts with many metallic ions to give colored precipitates.
105.	Crystals of sodium chloride go into solution in water as ions	BECAUSE	the sodium ion has a 1+ charge and the chloride ion has a 1− charge and they are hydrated by the water molecules.
106.	In an equilibrium reaction, if the concentration of the reactants is increased, the reaction will increase its forward rate	BECAUSE	when a stress is applied to a reaction in equilibrium, the equilibrium shifts in the direction that opposes the stress.

GO ON TO THE NEXT PAGE.

	I		II
107.	The $\Delta H_{reaction}$ of a particular reaction can be arrived at by the summation of the $\Delta H_{reaction}$ values of two or more reactions that, added together, give the $\Delta H_{reaction}$ of the particular reaction	BECAUSE	Hess's Law conforms to the First Law of Thermodynamics, which states that the total energy of the universe is a constant.
108.	In a reaction that has both a forward and a reverse reaction, $A + B \rightleftharpoons AB$, when only A and B are introduced into a reacting vessel, the forward reaction rate is the highest at the beginning and begins to decrease from that point until equilibrium is reached	BECAUSE	the reverse reaction does not begin until equilibrium is reached.
109.	At equilibrium, the forward reaction and reverse reaction stop	BECAUSE	at equilibrium, the reactants and products have reached the equilibrium concentrations.
110.	The hydrid orbital form of carbon in acetylene is believed to be the *sp* form	BECAUSE	C_2H_2 is a linear compound with a triple bond between the carbons.
111.	The weakest of the bonds between molecules are coordinate covalent bonds	BECAUSE	coordinate covalent bonds represent the weak attractive force of the electrons of one molecule for the positively charged nucleus of another.
112.	A saturated solution is not necessarily concentrated	BECAUSE	*dilute* and *concentrated* are terms that relate only to the relative amount of solute dissolved in the solvent.
113.	Lithium is the most active metal in the first group of the Periodic Table	BECAUSE	lithium has only one electron in the outer energy level.
114.	The anions migrate to the cathode in an electrolytic cell	BECAUSE	positively charged ions are attracted to the negatively charged cathode.
115.	The atomic number of a neutral atom that has a mass of 39 and has 19 electrons is 19	BECAUSE	the number of protons in a neutral atom is equal to the number of electrons.
116.	For an element with an atomic number of 17, the most probable oxidation number is +1	BECAUSE	the outer energy level of the halogen family has a tendency to add one electron to itself.

Part C

Directions: Each of the questions or incomplete statements below is followed by five suggested answers or completions. Select the one that is best in each case and then fill in the corresponding oval on the answer sheet.

24. All of the following involve a chemical change EXCEPT

(A) the formation of HCl from H_2 and Cl_2
(B) the color change when NO is exposed to air
(C) the formation of steam from burning H_2 and O_2
(D) the solidification of vegetable oil at low temperatures
(E) the odor of NH_3 when NH_4Cl is rubbed together with $Ca(OH)_2$ powder

25. When most fuels burn, the products include carbon dioxide and

(A) hydrocarbons
(B) hydrogen
(C) water
(D) hydroxide
(E) hydrogen peroxide

26. In the metric system, the prefix *kilo-* means

(A) 10^0
(B) 10^{-1}
(C) 10^{-2}
(D) 10^2
(E) 10^3

27. How many atoms are in 1 mole of water?

(A) 3
(B) 54
(C) 6.02×10^{23}
(D) $2(6.02 \times 10^{23})$
(E) $3(6.02 \times 10^{23})$

28. Which of the following elements normally exist as monoatomic molecules?

(A) Cl
(B) H
(C) O
(D) N
(E) He

29. The shape of a PCl_3 molecule is described as

(A) bent
(B) trigonal planar
(C) linear
(D) trigonal pyramidal
(E) tetrahedral

30. The complete loss of an electron of one atom to another atom with the consequent formation of electrostatic charges is referred to as

(A) a covalent bond
(B) a polar covalent bond
(C) an ionic bond
(D) a coordinate covalent bond
(E) a pi bond between *p* orbitals

31. In the electrolysis of water, the cathode reaction is

(A) $2H_2O(\ell) + 2e^- \rightarrow H_2(g) + 2OH^- + O_2(g)$
(B) $2H_2O(\ell) \rightarrow \frac{1}{2}O_2(g) + 2H^+ + 2e^-$
(C) $2OH^- + 2e^- \rightarrow O_2(g) + H_2(g)$
(D) $2H^+ + 2e^- \rightarrow H_2(g)$
(E) $2H_2O(\ell) + 4e^- \rightarrow O_2(g) + 2H_2(g)$

Practice Test 2

GO ON TO THE NEXT PAGE.

32. Which of the following is a substance particle with the LEAST mass?

(A) Alpha particle
(B) Beta particle
(C) Proton
(D) Neutron
(E) Gamma ray

33. If a radioactive element with a half-life of 100 years is found to have transmutated so that only 25% of the original sample remains, what is the age, in years, of the sample?

(A) 25
(B) 50
(C) 100
(D) 200
(E) 400

34. What is the pH of an acetic acid solution if the $[H_3O^+] = 1 \times 10^{-4}$ mole/liter?

(A) 1
(B) 2
(C) 3
(D) 4
(E) 5

35. The polarity of water is useful in explaining which of the following?

I. The solution process
II. The ionization process
III. The high conductivity of distilled water

(A) I only
(B) II only
(C) I and II only
(D) II and III only
(E) I, II, and III

36. When sulfur dioxide is bubbled through water, the solution will contain

(A) sulfurous acid
(B) sulfuric acid
(C) hyposulfuric acid
(D) persulfuric acid
(E) anhydrous sulfuric acid

37. Four grams of hydrogen gas at STP contain

(A) 6.02×10^{23} atoms
(B) 12.04×10^{23} atoms
(C) 12.04×10^{46} atoms
(D) 1.2×10^{23} molecules
(E) 12.04×10^{23} molecules

38. Analysis of a gas gave: C = 85.7% and H = 14.3%. If the formula mass of this gas is 42 atomic mass units, what are the empirical formula and the true formula?

(A) CH; C_4H_4
(B) CH_2; C_3H_6
(C) CH_3; C_3H_9
(D) C_2H_2; C_3H_6
(E) C_2H_4; C_3H_6

39. Which fraction would be used to correct a given volume of gas at 30°C to its new volume when it is heated to 60°C and the pressure is kept constant?

(A) $\frac{30}{60}$

(B) $\frac{60}{30}$

(C) $\frac{273}{333}$

(D) $\frac{303}{333}$

(E) $\frac{333}{303}$

40. What would be the predicted freezing point of a solution that has 684 grams of sucrose (1 mol = 342 g) dissolved in 2,000 grams of water?

(A) −1.86°C or 271.14 K
(B) −0.93°C or 272.07 K
(C) −1.39°C or 271.61 K
(D) −2.48°C or 270.52 K
(E) −3.72°C or 269.28 K

41. What is the approximate pH of a 0.005 M solution of H_2SO_4?

(A) 1
(B) 2
(C) 5
(D) 9
(E) 13

42. How many grams of NaOH are needed to make 100 grams of a 5% solution?

(A) 2
(B) 5
(C) 20
(D) 40
(E) 95

43. For the Haber process: $N_2 + 3H_2 \rightleftharpoons 2NH_3$ + heat (at equilibrium), which of the following statements concerning the reaction rate is/are true?

I. The reaction to the right will increase when pressure is increased.
II. The reaction to the right will decrease when the temperature is increased.
III. The reaction to the right will decrease when NH_3 is removed from the chamber.

(A) I only
(B) II only
(C) I and II only
(D) II and III only
(E) I, II, and III

44. If you titrate 1 M H_2SO_4 solution against 50 milliliters of 1 M NaOH solution, what volume of H_2SO_4, in milliliters, will be needed for neutralization?

(A) 10
(B) 25
(C) 40
(D) 50
(E) 100

45. How many grams of CO_2 can be prepared from 150 grams of calcium carbonate reacting with an excess of hydrochloric acid solution?

(A) 11
(B) 22
(C) 33
(D) 44
(E) 66

Question 46 refers to the following diagram:

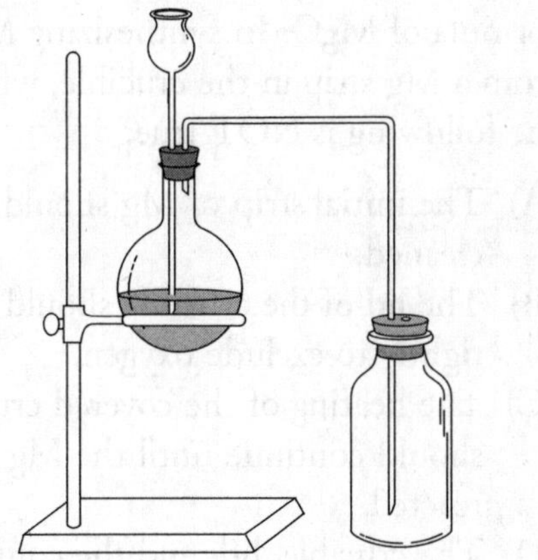

46. The diagram represents a setup that may be used to prepare and collect

(A) NH_3
(B) NO
(C) H_2
(D) SO_3
(E) CO_2

Practice Test 2

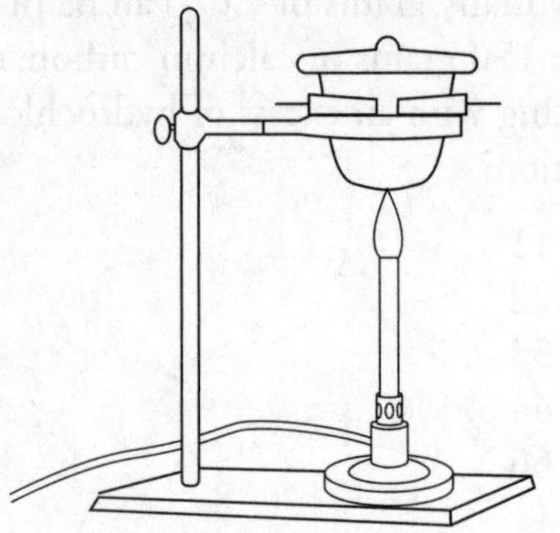

47. The lab setup shown above was used for the gravimetric analysis of the empirical formula of MgO. In synthesizing MgO from a Mg strip in the crucible, which of the following is NOT true?

    (A) The initial strip of Mg should be cleaned.
    (B) The lid of the crucible should fit tightly to exclude oxygen.
    (C) The heating of the covered crucible should continue until the Mg is fully reacted.
    (D) The crucible, lid, and the contents should be cooled to room temperature before measuring their mass.
    (E) When the Mg appears to be fully reacted, the crucible lid should be partially removed and heating continued.

<u>Questions 48–50</u> refer to the following experimental setup and data:

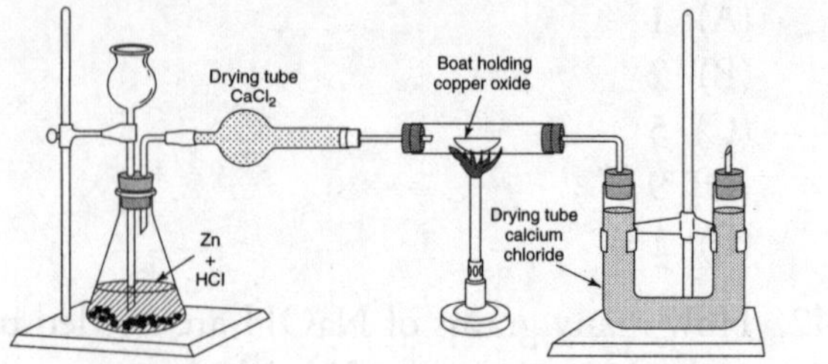

Recorded data:

Weight of U-tube .... 20.36 g
Weight of U-tube and calcium chloride before .... 39.32 g
Weight of U-tube and calcium chloride after .... 57.32 g
Weight of boat and contents before .... 30.23 g
Weight of boat and contents after .... 14.23 g
Weight of boat .... 5.00 g

48. What is the reason for the first $CaCl_2$ drying tube?

    (A) Generate water
    (B) Absorb hydrogen
    (C) Absorb water that evaporates from the flask
    (D) Decompose the water from the flask
    (E) Act as a catalyst for the combination of hydrogen and oxygen

49. What conclusion can be derived from the data collected?

    (A) Oxygen was lost from the $CaCl_2$.
    (B) Oxygen was generated in the U-tube.
    (C) Water was formed from the reaction.
    (D) Hydrogen was absorbed by the $CaCl_2$.
    (E) CuO was formed in the decomposition.

50. What is the ratio of the mass of water formed to the mass of hydrogen used in the formation of water?

(A) 1 : 8
(B) 1 : 9
(C) 8 : 1
(D) 9 : 1
(E) 8 : 9

51. What is the mass, in grams, of 1 mole of $KAl(SO_4)_2 \cdot 12H_2O$?

(A) 132
(B) 180
(C) 394
(D) 474
(E) 516

52. What mass of aluminum will be completely oxidized by 2 moles of oxygen at STP?

(A) 18 g
(B) 37.8 g
(C) 50.4 g
(D) 72.0 g
(E) 100.8 g

53. In general, when metal oxides react with water, they form solutions that are

(A) acidic
(B) basic
(C) neutral
(D) unstable
(E) colored

Questions 54–56 refer to the following diagram:

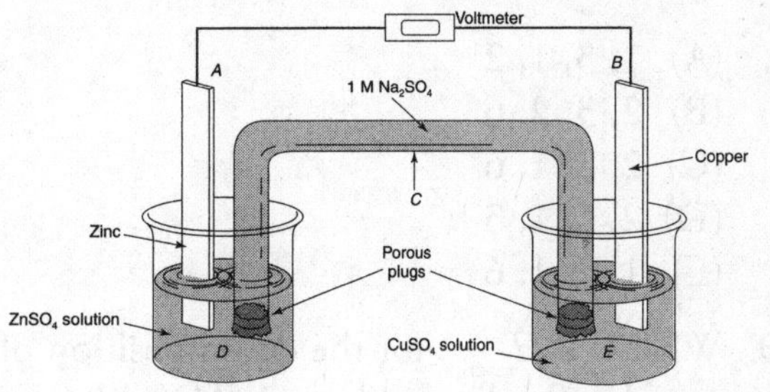

54. The oxidation reaction will occur at

(A) *A*
(B) *B*
(C) *C*
(D) *D*
(E) *E*

55. The apparatus at *C* is called the

(A) anode
(B) cathode
(C) salt bridge
(D) ion bridge
(E) osmotic bridge

56. The standard potentials of the metals are:

$Zn^{2+} + 2e^- \rightleftharpoons Zn^0$ $E^0 = -0.76$ volt

$Cu^0 \rightleftharpoons Cu^{2+} + 2e^-$ $E^0 = -0.34$ volt

What will be the voltmeter reading for this reaction?

(A) +1.10
(B) −1.10
(C) +0.42
(D) −0.42
(E) −1.52

57. How many liters of oxygen (STP) can be prepared from the decomposition of 212 grams of sodium chlorate (1 mol = 106 g)?

(A) 11.2
(B) 22.4
(C) 44.8
(D) 67.2
(E) 78.4

GO ON TO THE NEXT PAGE.

58. In this equation: $Al(OH)_3 + H_2SO_4 \rightarrow Al_2(SO_4)_3 + H_2O$, the whole-number coefficients of the balanced equation are

(A) 1, 3, 1, 2
(B) 2, 3, 2, 6
(C) 2, 3, 1, 6
(D) 2, 6, 1, 3
(E) 1, 3, 1, 6

59. What is $\Delta H_{reaction}$ for the decomposition of 1 mole of sodium chlorate? (ΔH_f^0 values: $NaClO_3(s) = -85.7$ kcal/mol, $NaCl(s) = -98.2$ kcal/mol, $O_2(g) = 0$ kcal/mol)

(A) −183.9 kcal
(B) −91.9 kcal
(C) +45.3 kcal
(D) +22.5 kcal
(E) −12.5 kcal

60. Isotopes of an element are related because which of the following is (are) the same in these isotopes?

I. Atomic mass
II. Atomic number
III. Arrangement of orbital electrons

(A) I only
(B) II only
(C) I and II only
(D) II and III only
(E) I, II, and III

61. In the reaction of zinc with dilute HCl to form H_2, which of the following will increase the reaction rate?

I. Increasing the temperature
II. Increasing the exposed surface of zinc
III. Using a more concentrated solution of HCl

(A) I only
(B) II only
(C) I and III only
(D) II and III only
(E) I, II, and III

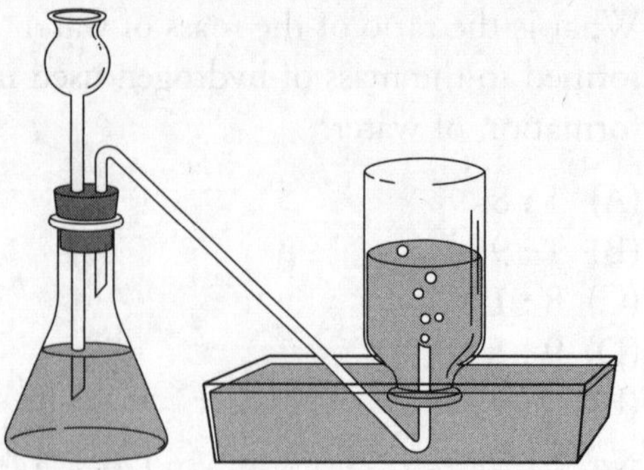

62. The laboratory setup shown above can be used to prepare a

(A) gas lighter than air and soluble in water
(B) gas heavier than air and soluble in water
(C) gas soluble in water that reacts with water
(D) gas insoluble in water
(E) gas that reacts with water

63. In this reaction: $CaCO_3 + 2HCl \rightarrow CaCl_2 + H_2O + CO_2$. If 4 moles of HCl are available to the reaction with an unlimited supply of $CaCO_3$, how many moles of CO_2 can be produced at STP?

(A) 1
(B) 1.5
(C) 2
(D) 2.5
(E) 3

64. A saturated solution of $BaSO_4$ at 25°C contains 3.9×10^{-5} mole/liter of Ba^{2+} ions. What is the K_{sp} of this salt?

(A) 3.9×10^{-5}
(B) 3.9×10^{-6}
(C) 2.1×10^{-7}
(D) 1.5×10^{-8}
(E) 1.5×10^{-9}

65. If 0.1 mole of K_2SO_4 was added to the solution in question 64, what would happen to the Ba^{2+} concentration?

(A) It would increase.
(B) It would decrease.
(C) It would remain the same.
(D) It would first increase, then decrease.
(E) It would first decrease, then increase.

66. Which of the following will definitely cause the volume of a gas to increase?

I. Decreasing the pressure with the temperature held constant.
II. Increasing the pressure with a temperature decrease.
III. Increasing the temperature with a pressure increase.

(A) I only
(B) II only
(C) I and III only
(D) II and III only
(E) I, II, and III

67. The number of oxygen atoms in 0.5 mole of $Al_2(CO_3)_3$ is

(A) 4.5×10^{23}
(B) 9.0×10^{23}
(C) 3.6×10^{24}
(D) 2.7×10^{24}
(E) 5.4×10^{24}

Question 68 refers to a solution of 1 M acid, HA, with $K_a = 1 \times 10^{-6}$.

68. What is the H_3O^+ concentration? (Assume $[HA] = 1$, $[H_3O^+] = x[A^-] = x$.)

(A) 1×10^{-5}
(B) 1×10^{-4}
(C) 1×10^{-2}
(D) 1×10^{-3}
(E) 0.9×10^{-3}

69. What is the percent dissociation of acetic acid in a 0.1 M solution if the $[H_3O^+]$ is 1×10^{-3} mole/liter?

(A) 0.01%
(B) 0.1%
(C) 1.0%
(D) 1.5%
(E) 2.0%

STOP

If you finish before one hour is up, you may go back to check your work or complete unanswered questions.

Answers and Explanations for Test 2

1. **(E)** A phase diagram shows that all three states can exist at the triple point.

2. **(B)** The combining of nuclei is called nuclear fusion.

3. **(D)** According to Graham's Law of Gaseous Diffusion (or Effusion), the rate of diffusion is inversely proportional to the square root of the molecular weight.

   Then

$$\frac{\text{rate}_H}{\text{rate}_O} = \frac{\sqrt{\text{mol. mass } O_2}}{\sqrt{\text{mol. mass } H_2}} = \frac{\sqrt{32}}{\sqrt{2}} = \sqrt{\frac{16}{1}} = \frac{4}{1}$$

4. **(A)** The Law of Definite Composition states that, when compounds form, they always form in the same ratio by mass. Water, for instance, always forms in a ratio of 1 : 8 of hydrogen to oxygen by mass. For nitrogen monoxide (NO) and nitrogen dioxide (NO_2) the difference in molecular mass is one atomic mass of oxygen.

5. **(B)** The first step is the ΔH for $C + \frac{1}{2}O_2 \rightarrow CO$. It releases 26.4 kcal of heat. This is written as –26.4 because it is exothermic.

6. **(C)** This is the second step on the diagram. It releases –67.6 kcal of heat.

7. **(A)** To arrive at the ΔH, take the total drop (–94.0 kcal) or add these reactions:

$$\begin{array}{lll} C + \frac{1}{2}O_2 & \rightarrow CO & -26.4 \text{ kcal} \\ CO + \frac{1}{2}O_2 & \rightarrow CO_2 & -67.6 \text{ kcal} \\ \hline C + O_2 & \rightarrow CO_2 & -94.0 \text{ kcal} \end{array}$$

8. **(B)** Potassium and chlorine have a large enough difference in their electronegativities to form ionic bonds. The respective positions of these two elements in the periodic chart also are indicative of the large difference in their electronegativity values.

9. **(D)** Two atoms of an element that forms a diatomic molecule always have a nonpolar covalent bond between them since the electron attraction or electronegativity of the two atoms is the same.

10. **(C)** Electronegativity differences between 0.5 and 1.7 are usually indicative of polar covalent bonding. CO_2 is an interesting example of a *nonpolar molecule* with polar covalent bonds since the bonds are symmetrical in the molecule.

11. **(E)** Calcium is a metal and forms a metallic bond between atoms.

12. **(B)** The point at which all three states can exist is called the triple point.
13. **(C)** The liquid state can exist only in (C).
14. **(E)** The only point at which CO_2 can exist as both a solid and a gas under 2 atmospheres of pressure is (E).
15. **(A)** These are the alkali metals.
16. **(B)** Because of complete outer energy levels these elements have very low boiling points.
17. **(B)** The complete outer energy level makes it difficult to remove an electron; thus the high peaks are the noble gases.
18. **(A)** With only one electron in the outer energy level, the alkali metals lose an electron relatively easily.
19. **(E)** There is a slight increase in the stability of the *p* orbitals when each has one electron—thus the little peaks at (E).
20. **(C)** The filled *s* orbital shows a slight increase in stability—thus the little peaks at (C).
21. **(D)** The dip at the (D) areas shows the first pairing of electrons in the *p* orbitals and indicates that this first paired electron has a lower ionization potential than in the atom, which has only one electron in each *p* orbital.
22. **(A)** The most active metals, the alkalis, have the lowest ionization potentials.
23. **(B)** The filled *p* orbitals occur in the noble gases.

101. **(T, F)** Sulfur trioxide is shown by three structural formulas because each bond is "hybrid" of a single and double bond. Resonance in chemistry does not mean that the bonds resonate between the structures shown in the structural drawing.
102. **(T, F)** When ΔG is negative in the Gibbs equation, the reaction is spontaneous. However, the total equation determines this, not just the ΔH. The Gibbs equation is:

$$\Delta G = \Delta H - T\Delta S.$$

103. **(T, T, CE)** One mole of each of the gases contains 6.02×10^{23} molecules, but their molecular masses are different. CO_2 is found by adding one C = 12 and two O = 32, or a total of 44 amu. The H_2O, however, adds up to two H = 2 plus one O = 16, or a total of 18 amu. Thus it is true that 1 mol of CO_2 at 44 g/mol is heavier than 1 mol of H_2O at 18 g/mol.

Practice Test 2

104. **(T, T, CE)** Hydrosulfuric acid is a weak acid but is used in qualitative tests because of the distinctly colored precipitates of sulfides that it forms with many metallic ions.

105. **(T, T, CE)** Sodium chloride is an ionic crystal, not a molecule, and its ions are hydrated by the polar water molecules.

106. **(T, T, CE)** This statement and its reason fit the rules of Le Châtelier's Principle.

107. **(T, T, CE)** The statement is true, and the reason is also true and explains the statement.

108. **(T, F)** The statement is true, but not the reason. In an equilibrium reaction, concentrations can be shown to progress like this:

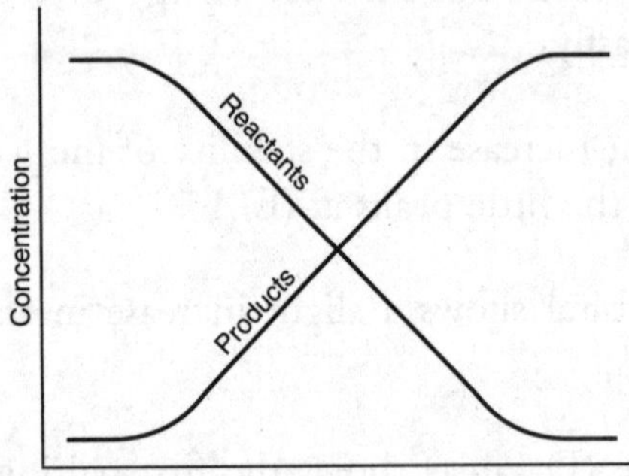

until equilibrium is reached. Then the concentrations stabilize.

109. **(F, T)** The forward and reverse reactions are occurring at equal rates when equilibrium is reached. The reactions do not stop. The concentrations remain the same at this point.

110. **(T, T, CE)** Since acetylene is known to be a linear molecule with a triple bond between the two carbons, the *sp* orbitals along the central axis with the hydrogens bonded on either end fit the experimental evidence.

111. **(F, F)** The weakest bonds between molecules are van der Waals forces, not coordinate covalent bonds.

112. **(T, T, CE)** The terms *dilute* and *concentrated* merely indicate a relatively large amount of solvent and a small amount of solvent, respectively. You can have a dilute saturated solution if the solute is only slightly soluble.

113. **(F, T)** Cs, not Li, is the most active Group I metal because Cs has (a) the largest atomic radius, thus making it easier to lose its outer energy level electron, and (b) the intermediate electrons help screen the positive attraction of the nucleus, also increasing the ease with which the outer electron is lost.

114. **(F, T)** The cations are positively charged ions and migrate to the cathode, while the anions are negatively charged and migrate to the anode.

115. **(T, T, CE)** There are as many electrons as there are protons in a neutral atom, and the atomic number represents the number of each.

116. **(F, T)** The first two principal energy levels fill up at 2 and 8 electrons, respectively. That leaves 7 electrons to fill the 3*s* and 3*p* orbitals like this: $3s^2$, $3p^5$. With only one electron missing in the 3*p* orbitals, the most likely oxidation number is −1.

24. **(D)** The solidification of vegetable oil is merely a physical change, like the formation of ice from liquid water at lower temperatures. All the other choices involve actual recombinations of atoms and thus are chemical changes.

25. **(C)** Water is formed because most common fuels contain hydrogen in their structures.

26. **(E)** The other choices, in order, represent 1, $\frac{1}{10}$ or deci-, $\frac{1}{100}$ or centi-, and 100 or hect-.

27. **(E)** One mole of any substance contains 6.02×10^{23} molecules. Since each water molecule is triatomic, there would be $3(6.02 \times 10^{23})$ atoms present.

28. **(E)** The noble gases are all monoatomic because of their complete outer energy levels. A rule to help you remember diatomic gases is: Gases ending in *-gen* or *-ine* usually form diatomic molecules.

29. **(D)** By both the VSEPR (valence shell electron pair repulsion) method and the orbital structure method, the PCl_3 molecule is trigonal pyramidal:

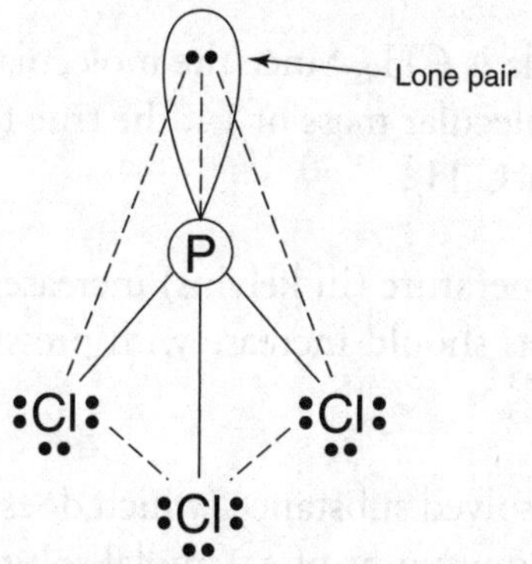

30. **(C)** The complete loss and gain of electrons is an ionic bond. All other bonds indicated are "sharing of electrons" type bonds or some form of covalent bonding.

31. **(D)** The cathode reaction releases only H_2 gas. This half-reaction is as given in (D).

32. **(B)** The beta particle is a high-speed electron and has the smallest mass of the first four choices. Gamma rays are electromagnetic waves.

Practice Test 2

33. **(D)** If 25% of the sample now remains, then 100 years ago 50% would be present. If you go back another 100 years, the sample would contain 100% of the radioactive element. Therefore, the sample is 100 + 100 = 200 years old.

34. **(D)** $pH = -\log[H_3O^+] = -\log[1 \times 10^{-4}] = -(-4) = 4$.

35. **(C)** Only I and II are true. Distilled water does not significantly conduct an electric current. The polarity of the water molecule is helpful in ionization and in causing substances to go into solution.

36. **(A)** SO_2 is the acid anhydride of H_2SO_3 or sulfurous acid. $H_2O + SO_2 \rightarrow H_2SO_3$.

37. **(E)** Four grams of hydrogen gas at STP represent 2 mol of hydrogen since 2 g is the gram-molecular mass of hydrogen. Each mole of a gas contains 6.02×10^{23} molecules, so 2 mol contains $2 \times 6.02 \times 10^{23}$ or 12.04×10^{23} molecules.

38. **(B)** To solve percent composition problems, first divide the % given by the atomic mass:

$$12\overline{)85.7} = 7.14 \qquad 1\overline{)14.3} = 14.3$$

Then divide by the smallest quotient to get small whole numbers:

$$7.14\overline{)7.14} = 1.0 \qquad 7.14\overline{)14.3} = 2.0$$

The empirical formula is CH_2. Since the molecular mass is 42 and the empirical formula has a molecular mass of 14, the true formula must be 3 times the empirical formula, or C_3H_6.

39. **(E)** Because the temperature (in kelvins) increases from 303 K to 333 K, the volume of the gas should increase with pressure held constant. The correct fraction is $\frac{333}{303}$.

40. **(A)** One mole of dissolved substance (which does not ionize) causes a 1.86°C drop in the freezing point of a 1 molal solution. Since 2,000 g of water were used, the solution has $\frac{684}{342}$ or 2 mol in 2,000 g of water. Then

$$\frac{2\text{ mol}}{2{,}000\text{ g}} = \frac{x\text{ mol}}{1{,}000\text{ g}}$$

$$x = \frac{2{,}000}{2{,}000} = 1.00\text{ mol}/1{,}000\text{ g}$$

The freezing point is depressed $1 \times -1.86° = -1.86°C$ or 271.14 K.

Practice Test 2

41. **(B)** The pH is $-\log[H^+]$. A 0.005 molar solution of H_2SO_4 ionizes in a dilute solution to release two H^+ ions per molecule of H_2SO_4. Therefore the molar concentration of H^+ ion is 2×0.005 mol/L or 0.010 mol/L. Substituting this in the formula, you have:

$$pH = -\log[0.01] = -\log[1 \times 10^{-2}]$$

The log of a number is the exponent to which the base 10 is raised to express that number in exponential form:

$$-\log[1 \times 10^{-2}] = -[-2] = 2$$

42. **(B)** If the solution is to be 5% sodium hydroxide, then 5% of 100 g is 5 g. Percent is always by mass unless otherwise specified.

43. **(C)** Because this equation is exothermic, higher temperatures will decrease the reaction to the right and increase the reaction to the left, so II is true.

Also, I is true because with an increase of pressure the reaction will try to relieve that pressure by going in the direction that has the least volume: in this reaction, to the right. Statement III is false because removing product in this reaction would increase the forward reaction. Statements I and II are true.

44. **(B)** The reaction is:

$H_2SO_4 + 2NaOH \rightarrow Na_2SO_4 + H_2O$
1 mol   2 mol
acid   base

1 mol acid = $\frac{1}{2}$ (mol) of base

Since

molarity × volume (L) = moles,

then

$$M_aV_a = \tfrac{1}{2}M_bV_b$$
$$(1\,M)(x\,L) = \tfrac{1}{2}(1\,M)(0.05\,L)$$
$$x\,L = \tfrac{1}{2}(0.05\,L)$$
$$x\,L = 0.025\,L \text{ or } 25\text{ mL}$$

45. **(E)** The reaction is:

150 g
$CaCO_3 + 2HCl \rightarrow CaCl_2 + H_2O + CO_2$

The gram-molecular mass of calcium is 100 g. Then 150 g = $\frac{150}{100}$ or 1.5 mol of calcium carbonate. According to the equation, 1 mol of $CaCO_3$ yields 1 mol of carbon dioxide, so 1.5 mol of $CaCO_3$ yields 1.5 mol of CO_2.

The gram-molecular mass of CO_2 = 44 g:

1.5 mol of CO_2 = 1.5 × 44 g = 66 g CO_2

46. **(E)** The other choices are wrong because:

   (A) is lighter than air
   (B) reacts with air
   (C) is lighter than air
   (D) needs heat to be evolved

47. **(B)** The Mg needs oxygen to form MgO; so the lid cannot be tightly sealed. Oxygen is needed for the Mg to oxidize to MgO. All other choices are true.

48. **(C)** To ensure that all the water vapor collected in the U-tube comes from the reaction, the first drying tube is placed in the path of the hydrogen to absorb any evaporated water.

49. **(C)** Calcium chloride is deliquescent, and its weight gain of water indicates that water was formed from the reaction.

50. **(D)**

$CaCl_2$ and U-tube after	= 57.32 g
$CaCl_2$ and U-tube before	= 39.32 g
Mass of water formed	= 18.00 g
CuO and boat before	= 30.23 g
CuO and boat after	= 14.23 g
Mass of oxygen reacted	= 16.00 g
So, mass of water	= 18 g
mass of oxygen	= 16 g
Mass of hydrogen	= 2 g

The ratio of mass of water to mass of hydrogen is 18 : 2 or 9 : 1.

51. **(D)** 1 K = 39, 1Al = 27, $2(SO_4) = 2(32 + 16 \times 4) = 192$, and $12H_2O = 12(2 + 16) = 216$. This totals 474 g.

52. **(D)** $\underset{108\text{ g}}{\overset{x\text{ g}}{4Al}} + \underset{3\text{ mol}}{\overset{2\text{ mol}}{3O_2}} \rightarrow 2Al_2O_3$

$$\frac{x}{108} = \frac{2}{3}, \text{ so } x = 72\text{ g}$$

Or, using the mole method: 44.8 L = 2 mol

$\underset{4\text{ mol}}{\overset{x\text{ mol}}{4Al}} + \underset{3\text{ mol}}{\overset{2\text{ mol}}{3O_2}} \rightarrow 2Al_2O_3$

Practice Test 2

This shows that:

$$\frac{x \text{ mol Al}}{4 \text{ mol Al}} = \frac{2 \text{ mol } O_2}{3 \text{ mol } O_2}$$

$$x = \frac{2 \text{ mol } O_2 \times 4 \text{ mol Al}}{3 \text{ mol } O_2} = \tfrac{8}{3} \text{ mol Al}$$

Since molar mass of Al = 27/mol

$x = \tfrac{8}{3}$ mol Al × 27g/mol = 72 g Al

53. **(B)** Metal oxides are generally basic anhydrides.

54. **(A)** Since A is the anode, the oxidation (or loss of electrons) will occur on this pole.

55. **(C)** "Salt bridge" is the correct terminology.

56. **(A)** $Zn \rightarrow Zn^{2+} + e^-$ $E^0 = +0.76$ V

$Cu^{2+} + 2e^- \rightarrow Cu^0 E^0 = +0.34$ V

Total is = +1.10 V

Voltmeter will read +1.10 V.

57. **(D)** $\overset{212\text{ g}}{2NaClO_3} \rightarrow 2NaCl + \overset{x\text{ L}}{3O_2}$

$$\frac{212 \text{ g}}{106 \text{ g/mol}} = 2 \text{ mol}$$

Equation shows

2 mol → 3 mol O_2

3 mol × 22.4 L/mol = 67.2 L

58. **(C)** The balanced equation has the coefficients 2, 3, 1, and 6: $2Al(OH)_3 + 3H_2SO_4 \rightarrow Al_2(SO_4)_3 + 6H_2O$.

59. **(E)** The reaction is $NaClO_3 \rightarrow NaCl + \tfrac{3}{2}O_2$.

$\Delta H_{reaction} = \Delta H_{f(products)} - \Delta H_{f(reactants)}$

$\Delta H_{reaction} = (-98.2 + 0) - (-85.7)$

$\Delta H_{reaction} = -12.5$ kcal

60. **(D)** II and III are identical; isotopes differ only in the number of neutrons in the nucleus and this affects the atomic mass only.

61. **(E)** I, II, and III will increase the rate of this reaction.

62. **(D)** This setup depends on water displacement of an insoluble gas.

63. **(C)** The coefficients give the molar relations, so 2 mol of HCl give off 1 mol of CO_2. Given 4 mol of HCl, you have

$$\frac{4}{2} = \frac{x}{1}, \text{ then } x = \frac{4}{2} = 2 \text{ mol}$$

64. **(E)** $K_{sp} = [Ba^{2+}]\,[SO_4^{\,2-}]$ since $[Ba^{2+}]$ is given as $[3.9 \times 10^{-5}]$, and the equation $BaSO_4 \rightarrow Ba^{2+} + SO_4^{\,2-}$ shows there will be as many $SO_4^{\,2-}$ ions as Ba^{2+} ions, then both $[Ba^{2+}]$ and $[SO_4^{\,2-}]$ will equal 3.9×10^{-5}. So,

$K_{sp} = [3.9 \times 10^{-5}][3.9 \times 10^{-5}]$

$K_{sp} = 1.5 \times 10^{-9}$

65. **(B)** The introduction of the "common ion" $SO_4^{\,2-}$ at 0.1 molar forces the equilibrium to shift to the left and reduce the Ba^{2+} concentration.

66. **(A)** According to the gas laws, only I will cause an increase in the volume of a confined gas.

67. **(D)** In 1 mol of $Al_2(CO_3)_3$, nine oxygens (three carbonates with three oxygen atoms each) are in each molecule, or 9 mol of O atoms are in 1 mol of $Al_2(CO_3)_3$. Because only 0.5 mol is given, there are $\frac{1}{2}$ (9) or 4.5 mol of O atoms. In 4.5 mol of oxygen, there are

$$4.5 \cancel{\text{ mol O}} \text{ atoms} \times \frac{6.0 \times 10^{23} \text{ atoms}}{1 \cancel{\text{ mol O}} \text{ atoms}}$$

27.0×10^{23} atoms or 2.7×10^{24} atoms.

68. **(D)** When HA ionizes, it forms equal amounts of H^+ and A^- ions, but these amounts are very small because the K_a is very small. K_a can be expressed as $[H^+]\,[A^-]/[HA]$. Because you are told to assume [HA] = 1, you have:

$$\frac{[x][x]}{[1]} = 1 \times 10^{-6}$$
$$x^2 = 1 \times 10^{-6}$$
$$x = 1 \times 10^{-3}$$

69. **(C)** Percent dissociation =

$$\frac{\text{moles/liter of } H_3O^+}{\text{original concentration}} \times 100$$

$$\% \text{ dissociation} = \frac{1 \times 10^{-3}}{1 \times 10^{-1}} \times 100 = 1\%$$

Practice Test 2

CALCULATING YOUR SCORE

Your score on Practice Test 2 can now be computed manually. The actual test is scored by machine, but the same method is used to arrive at the raw score. You get one point for each correct answer. For each wrong answer, you lose one-fourth of a point. Questions that you omit or that have more than one answer are not counted. On your answer sheet mark all correct answers with a "C" and all incorrect answers with an "X".

Determining Your Raw Test Score

Total the number of correct answers you have recorded on your answer sheet. It should be the same as the total of all the numbers you place in the block in the lower left corner of each area of the Subject Area summary in the next section.

A. Enter the total number of correct answers here: ________
   Now count the number of wrong answers you recorded on your answer sheet.
B. Enter the total number of wrong answers here: ________
   Multiply the number of wrong answers in B by 0.25.
C. Enter that product here: ________
   Subtract the result in C from the total number of right answers in A.
D. Enter the result of your subtraction here: ________
E. Round the result in D to the nearest whole number: ________.
   This is your raw test score.

Conversion of Raw Scores to Scaled Scores

Your raw score is converted by the College Board into a scaled score. The College Board scores range from 200 to 800. This conversion is done to ensure that a score earned on any edition of a particular SAT Subject Test in Chemistry is comparable to the same scaled score earned on any other edition of the same test. Because some editions of the tests may be slightly easier or more difficult than others, scaled scores are adjusted so that they indicate the same level of performance regardless of the edition of the test taken and the ability of the group that takes it. Consequently, a specific raw score on one edition of a particular test will not necessarily translate to the same scaled score on another edition of the same test.

Because the practice tests in this book have no large population of scores with which they can be scaled, scaled scores cannot be determined.

Results from previous SAT Chemistry tests appear to indicate that the conversion of raw scores to scaled scores GENERALLY follows this pattern:

Raw Score	Scaled Score	Raw Score	Scaled Score
85–75	800–800	30–25	560–540
75–70	800–780	25–20	540–520
70–65	780–750	20–15	520–490
65–60	750–720	15–10	490–460
60–55	720–700	10–5	460–430
55–50	700–680	5–0	430–410
50–45	680–650	0 to –5	410–380
45–40	650–620	–5 to –10	380–360
40–35	620–590	–10 to –15	360–330
35–30	590–560		

Note that this scale provides only a *general idea* of what a raw score may translate into on a scaled score range of 800–200. Scaling on every test is usually slightly different. Some students who had taken the SAT Subject Test in Chemistry after using this book had reported that they have scored slightly higher on the SAT test than on the practice tests in this book. They *all* reported that preparing well for the test paid off in a better score!

DIAGNOSING YOUR NEEDS

After taking Practice Test 2, check your answers against the correct ones. Then fill in the chart below.

In the space under each question number, place a check if you answered that question correctly.

EXAMPLE:

If your answer to question 5 was correct, place a check in the appropriate box.

Next, total the check marks for each section and insert the number in the designated block. Now do the arithmetic indicated and insert your percent for each area.

Subject Area*	(✓) Questions Answered Correctly
I. Atomic Theory and Structure, including periodic relationships ☐ No. of checks ÷ 8 × 100 = ______%	19, 20, 21, 23, 101, 115, 28, 60
II. Nucleonics ☐ No. of checks ÷ 3 × 100 = ______%	2, 32, 33
III. Chemical Bonding and Molecular Structure ☐ No. of checks ÷ 9 × 100 = ______%	4, 8, 9, 10, 11, 110, 111, 29, 30
IV. States of Matter and Kinetic Molecular Theory of Gases ☐ No. of checks ÷ 7 × 100 = ______%	3, 12, 13, 14, 27, 39, 65
V. Solutions, including concentration units, solubility, and colligative properties ☐ No. of checks ÷ 4 × 100 = ______%	105, 112, 40, 44
VI. Acids and Bases ☐ No. of checks ÷ 6 × 100 = ______%	34, 36, 41, 53, 68, 69
VII. Oxidation-Reduction and Electrochemistry ☐ No. of checks ÷ 6 × 100 = ______%	114, 116, 31, 35, 54, 55
VIII. Stoichiometry ☐ No. of checks ÷ 12 × 100 = ______%	103, 37, 38, 42, 45, 50, 51, 52, 57, 58, 63, 67
IX. Reaction Rates ☐ No. of checks ÷ 2 × 100 = ______%	43, 61

*The subject areas have been expanded to identify specific areas in the text.

Subject Area*	(✓) Questions Answered Correctly					
X. Equilibrium	106	108	109	64	66	
☐ No. of checks ÷ 5 × 100 = ______%						
XI. Thermodynamics: energy changes in chemical reactions, randomness, and criteria for spontaneity	5	6	7	102	107	59
☐ No. of checks ÷ 6 × 100 = ______%						
XII. Descriptive Chemistry: physical and chemical properties of elements and their familiar compounds; organic chemistry; periodic properties	1	15	16	17	18	22
	104	113	24	25	26	56
☐ No. of checks ÷ 12 × 100 = ______%						
XIII. Laboratory: equipment, procedures, observations, safety, calculations, and interpretation of results	46	47	48	49	62	
☐ No. of checks ÷ 6 × 100 = ______%						

**The subject areas have been expanded to identify specific areas in the text.*

To develop a study plan, refer to pages 33–35.

Answer Sheet

PRACTICE TEST 3

Determine the correct answer for each question. Then, using a No. 2 pencil, blacken completely the oval containing the letter of your choice.

1	Ⓐ	Ⓑ	Ⓒ	Ⓓ	Ⓔ
2	Ⓐ	Ⓑ	Ⓒ	Ⓓ	Ⓔ
3	Ⓐ	Ⓑ	Ⓒ	Ⓓ	Ⓔ
4	Ⓐ	Ⓑ	Ⓒ	Ⓓ	Ⓔ
5	Ⓐ	Ⓑ	Ⓒ	Ⓓ	Ⓔ
6	Ⓐ	Ⓑ	Ⓒ	Ⓓ	Ⓔ
7	Ⓐ	Ⓑ	Ⓒ	Ⓓ	Ⓔ
8	Ⓐ	Ⓑ	Ⓒ	Ⓓ	Ⓔ
9	Ⓐ	Ⓑ	Ⓒ	Ⓓ	Ⓔ
10	Ⓐ	Ⓑ	Ⓒ	Ⓓ	Ⓔ
11	Ⓐ	Ⓑ	Ⓒ	Ⓓ	Ⓔ
12	Ⓐ	Ⓑ	Ⓒ	Ⓓ	Ⓔ
13	Ⓐ	Ⓑ	Ⓒ	Ⓓ	Ⓔ
14	Ⓐ	Ⓑ	Ⓒ	Ⓓ	Ⓔ
15	Ⓐ	Ⓑ	Ⓒ	Ⓓ	Ⓔ
16	Ⓐ	Ⓑ	Ⓒ	Ⓓ	Ⓔ
17	Ⓐ	Ⓑ	Ⓒ	Ⓓ	Ⓔ
18	Ⓐ	Ⓑ	Ⓒ	Ⓓ	Ⓔ
19	Ⓐ	Ⓑ	Ⓒ	Ⓓ	Ⓔ
20	Ⓐ	Ⓑ	Ⓒ	Ⓓ	Ⓔ
21	Ⓐ	Ⓑ	Ⓒ	Ⓓ	Ⓔ
22	Ⓐ	Ⓑ	Ⓒ	Ⓓ	Ⓔ
23	Ⓐ	Ⓑ	Ⓒ	Ⓓ	Ⓔ

ON THE ACTUAL CHEMISTRY TEST, THE FOLLOWING TYPE OF QUESTION MUST BE ANSWERED ON A SPECIAL SECTION (LABELED "CHEMISTRY") AT THE LOWER LEFT-HAND CORNER OF PAGE 2 OF YOUR ANSWER SHEET. THESE QUESTIONS WILL BE NUMBERED BEGINNING WITH 101 AND MUST BE ANSWERED ACCORDING TO THE DIRECTIONS.

CHEMISTRY* Fill in oval CE only if II is a correct explanation of I.

	I		II		CE*
101	Ⓣ	Ⓕ	Ⓣ	Ⓕ	○
102	Ⓣ	Ⓕ	Ⓣ	Ⓕ	○
103	Ⓣ	Ⓕ	Ⓣ	Ⓕ	○
104	Ⓣ	Ⓕ	Ⓣ	Ⓕ	○
105	Ⓣ	Ⓕ	Ⓣ	Ⓕ	○
106	Ⓣ	Ⓕ	Ⓣ	Ⓕ	○
107	Ⓣ	Ⓕ	Ⓣ	Ⓕ	○
108	Ⓣ	Ⓕ	Ⓣ	Ⓕ	○
109	Ⓣ	Ⓕ	Ⓣ	Ⓕ	○
110	Ⓣ	Ⓕ	Ⓣ	Ⓕ	○
111	Ⓣ	Ⓕ	Ⓣ	Ⓕ	○
112	Ⓣ	Ⓕ	Ⓣ	Ⓕ	○
113	Ⓣ	Ⓕ	Ⓣ	Ⓕ	○
114	Ⓣ	Ⓕ	Ⓣ	Ⓕ	○
115	Ⓣ	Ⓕ	Ⓣ	Ⓕ	○
116	Ⓣ	Ⓕ	Ⓣ	Ⓕ	○

ON THE ACTUAL CHEMISTRY TEST, THE REMIANING QUESTIONS MUST BE ANSWERED BY RETURNING TO THE SECTION OF YOUR ANSWER SHEET YOU STARTED FOR CHEMISTRY.

24	Ⓐ	Ⓑ	Ⓒ	Ⓓ	Ⓔ
25	Ⓐ	Ⓑ	Ⓒ	Ⓓ	Ⓔ
26	Ⓐ	Ⓑ	Ⓒ	Ⓓ	Ⓔ
27	Ⓐ	Ⓑ	Ⓒ	Ⓓ	Ⓔ
28	Ⓐ	Ⓑ	Ⓒ	Ⓓ	Ⓔ
29	Ⓐ	Ⓑ	Ⓒ	Ⓓ	Ⓔ
30	Ⓐ	Ⓑ	Ⓒ	Ⓓ	Ⓔ
31	Ⓐ	Ⓑ	Ⓒ	Ⓓ	Ⓔ
32	Ⓐ	Ⓑ	Ⓒ	Ⓓ	Ⓔ
33	Ⓐ	Ⓑ	Ⓒ	Ⓓ	Ⓔ
34	Ⓐ	Ⓑ	Ⓒ	Ⓓ	Ⓔ
35	Ⓐ	Ⓑ	Ⓒ	Ⓓ	Ⓔ
36	Ⓐ	Ⓑ	Ⓒ	Ⓓ	Ⓔ
37	Ⓐ	Ⓑ	Ⓒ	Ⓓ	Ⓔ
38	Ⓐ	Ⓑ	Ⓒ	Ⓓ	Ⓔ
39	Ⓐ	Ⓑ	Ⓒ	Ⓓ	Ⓔ
40	Ⓐ	Ⓑ	Ⓒ	Ⓓ	Ⓔ
41	Ⓐ	Ⓑ	Ⓒ	Ⓓ	Ⓔ
42	Ⓐ	Ⓑ	Ⓒ	Ⓓ	Ⓔ
43	Ⓐ	Ⓑ	Ⓒ	Ⓓ	Ⓔ
44	Ⓐ	Ⓑ	Ⓒ	Ⓓ	Ⓔ
45	Ⓐ	Ⓑ	Ⓒ	Ⓓ	Ⓔ
46	Ⓐ	Ⓑ	Ⓒ	Ⓓ	Ⓔ
47	Ⓐ	Ⓑ	Ⓒ	Ⓓ	Ⓔ
48	Ⓐ	Ⓑ	Ⓒ	Ⓓ	Ⓔ
49	Ⓐ	Ⓑ	Ⓒ	Ⓓ	Ⓔ
50	Ⓐ	Ⓑ	Ⓒ	Ⓓ	Ⓔ
51	Ⓐ	Ⓑ	Ⓒ	Ⓓ	Ⓔ
52	Ⓐ	Ⓑ	Ⓒ	Ⓓ	Ⓔ
53	Ⓐ	Ⓑ	Ⓒ	Ⓓ	Ⓔ
54	Ⓐ	Ⓑ	Ⓒ	Ⓓ	Ⓔ
55	Ⓐ	Ⓑ	Ⓒ	Ⓓ	Ⓔ
56	Ⓐ	Ⓑ	Ⓒ	Ⓓ	Ⓔ
57	Ⓐ	Ⓑ	Ⓒ	Ⓓ	Ⓔ
58	Ⓐ	Ⓑ	Ⓒ	Ⓓ	Ⓔ
59	Ⓐ	Ⓑ	Ⓒ	Ⓓ	Ⓔ
60	Ⓐ	Ⓑ	Ⓒ	Ⓓ	Ⓔ
61	Ⓐ	Ⓑ	Ⓒ	Ⓓ	Ⓔ
62	Ⓐ	Ⓑ	Ⓒ	Ⓓ	Ⓔ
63	Ⓐ	Ⓑ	Ⓒ	Ⓓ	Ⓔ
64	Ⓐ	Ⓑ	Ⓒ	Ⓓ	Ⓔ
65	Ⓐ	Ⓑ	Ⓒ	Ⓓ	Ⓔ
66	Ⓐ	Ⓑ	Ⓒ	Ⓓ	Ⓔ
67	Ⓐ	Ⓑ	Ⓒ	Ⓓ	Ⓔ
68	Ⓐ	Ⓑ	Ⓒ	Ⓓ	Ⓔ
69	Ⓐ	Ⓑ	Ⓒ	Ⓓ	Ⓔ

Answer Sheet

[illegible]

Determine the correct answer for each question. Then, using a No. 2 pencil, blacken completely the oval containing the letter of your choice.

[illegible]

Practice Test 3

Note: For all questions involving solutions and/or chemical equations, assume that the system is in water unless otherwise stated.

Part A

Directions: Every set of the given lettered choices below refers to the numbered statements or formulas immediately following it. Choose the one lettered choice that best fits each statement or formula and then fill in the corresponding oval on the answer sheet. Each choice may be used once, more than once, or not at all in each set.

Questions 1–4 refer to the following diagram:

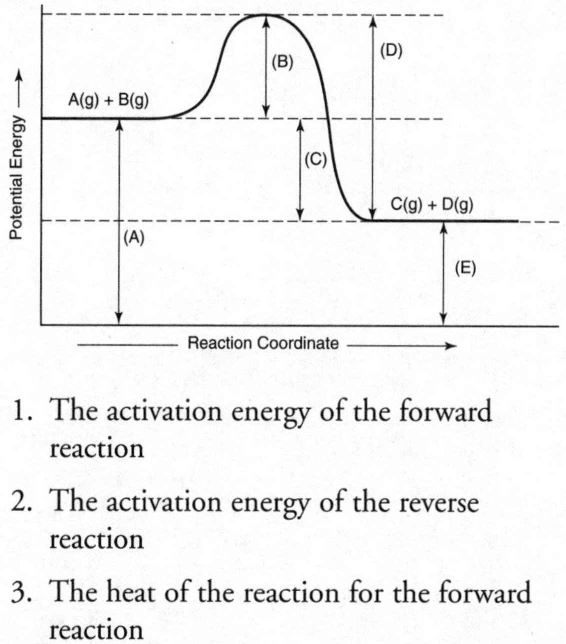

1. The activation energy of the forward reaction
2. The activation energy of the reverse reaction
3. The heat of the reaction for the forward reaction
4. The potential energy of the reactants

Questions 5–7 refer to the following diagram:

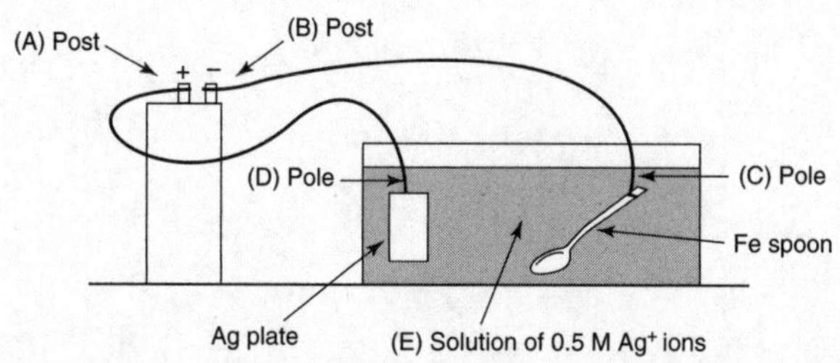

5. To plate silver on the spoon, the position to which the wire from the spoon must be connected
6. The position of the anode
7. The position from which silver that is plated out emerges

Questions 8–11 refer to the following phase diagram:

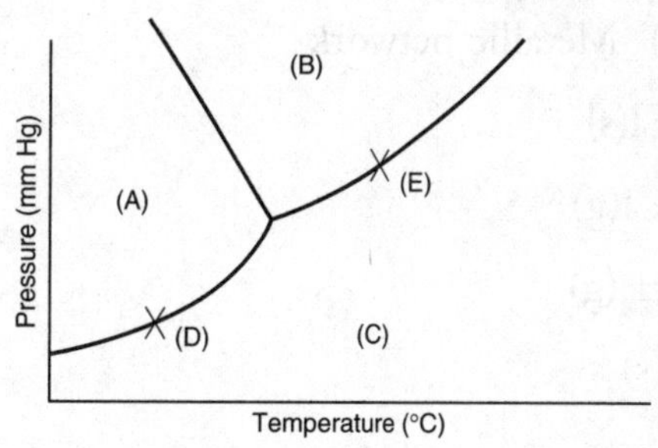

8. The solid area
9. The liquid area
10. The gaseous area
11. The area that can have both liquid and gas

GO ON TO THE NEXT PAGE.

Questions 12–14

(A) 1
(B) 2
(C) 3
(D) 4
(E) 5

12. When the following equation: $Cu + HNO_3 \rightarrow Cu(NO_3)_2 + H_2O + NO$ is balanced, what will be the coefficient, in the lowest whole number, of Cu?

13. If 6 moles of Cu reacted according to the above *balanced* equation, what will be the number of moles of NO that would be formed?

14. If $Cu(NO_3)_2$ goes into solution as ions, what will be the number of ions into which it will dissociate?

Questions 15–18

(A) Ionic substance
(B) Polar covalent substance
(C) Nonpolar covalent substance
(D) Amorphous substance
(E) Metallic network

15. KCl(s)

16. HCl(g)

17. CH_4(g)

18. Li(s)

Questions 19–23

(A) Brownian movement
(B) Litmus paper reaction
(C) Phenolphthalein reaction
(D) Dehydration
(E) Deliquescent

19. The reason why a blue crystal of $CuSO_4 \cdot 5H_2O$ turns white when heated

20. The zigzag path of colloidal particles in light

21. The pink color in a basic solution

22. The pink color in an acid solution

23. The adsorbtion of water to the surface of a crystal

Part B

ON THE ACTUAL CHEMISTRY TEST, THE FOLLOWING TYPE OF QUESTION MUST BE ANSWERED ON A SPECIAL SECTION (LABELED "CHEMISTRY") AT THE LOWER LEFT-HAND CORNER OF PAGE 2 OF YOUR ANSWER SHEET. THESE QUESTIONS WILL BE NUMBERED BEGINNING WITH 101 AND MUST BE ANSWERED ACCORDING TO THE FOLLOWING DIRECTIONS.

Directions: Every question below contains two statements, I in the left-hand column and II in the right-hand column. For each question, decide if statement I is true or false and if statement II is true or false and fill in the corresponding T or F ovals on your answer sheet. *Fill in oval CE only if statement II is a correct explanation of statement I.

Sample Answer Grid:

CHEMISTRY * Fill in oval CE only if II is a correct explanation of I.

	I		II		CE*
101.	(T)	(F)	(T)	(F)	( )

	I		II
101.	Elements in the upper right corner of the Periodic Table are active nonmetals	BECAUSE	nonmetals have larger ionic radii than their atomic radii.
102.	A synthesis reaction that is nonspontaneous and has a negative value for its heat of reaction will not occur until some heat is added	BECAUSE	nonspontaneous exothermic reactions need enough activation energy to get them started.
103.	Transition elements in a particular period may have the same oxidation number	BECAUSE	they have a complete outer energy level.
104.	When a crystal is added to a supersaturated solution of itself, the crystal does not appear to change	BECAUSE	the supersaturated solution is holding more solute than its normal solubility.
105.	Equilibrium is a static condition	BECAUSE	at equilibrium, the forward reaction rate equals the reverse reaction rate.
106.	The ionic bond is the strongest bond	BECAUSE	ionic bonds have electrostatic attraction due to the loss and gain of electron(s).

GO ON TO THE NEXT PAGE.

	I		II
107.	Pressure applied to a gaseous system in equilibrium favors the forward reaction	BECAUSE	the product side of an equilibrium would show an increase in gas pressure.
108.	If the forward reaction of an equilibrium is exothermic, adding heat to the system favors the reverse reaction	BECAUSE	additional heat causes a stress on the system, and the system moves in the direction that releases the stress.
109.	An element that has an electron configuration of $1s^2\ 2s^2\ 2p^6\ 3s^2\ 3p^6\ 3d^3\ 4s^2$ is a transition element	BECAUSE	the transition elements from scandium to zinc are filling the $3d$ orbitals.
110.	The most electronegative elements in the periodic chart are found among nonmetals	BECAUSE	electronegativity is a measure of the ability of an atom to draw valence electrons to itself.
111.	Basic anhydrides react in water to form bases	BECAUSE	metallic oxides react with water to form solutions that have an excess of hydroxide ions.
112.	There are 3 moles of atoms in 18 grams of water	BECAUSE	there are 6×10^{23} atoms in 1 mole.
113.	Benzene is a good electrolyte	BECAUSE	a good electrolyte has charged ions that carry the electric current.
114.	Normal butyl alcohol and 2-butanol are isomers	BECAUSE	isomers vary in the number of neutrons in the nucleus of the atom.
115.	The reaction of $CaCO_3$ and HCl goes to completion	BECAUSE	reactions that form a precipitate tend to go to completion.
116.	A large number of alpha particles were deflected in the Rutherford experiment	BECAUSE	alpha particles that came close to the nucleus of the gold atoms were deflected.

Part C

Directions: Every question or incomplete statement below is followed by five suggested answers or completions. Choose the one that is best in each case and then fill in the corresponding oval on the answer sheet.

24. What are the simplest whole-number coefficients that balance this equation?

$\ldots C_4H_{10} + \ldots O_2 \rightarrow \ldots CO_2 + \ldots H_2O$

(A) 1, 6, 4, 2
(B) 2, 13, 8, 10
(C) 1, 6, 1, 5
(D) 3, 10, 16, 20
(E) 4, 26, 16, 20

25. How many atoms are present in the formula $KAl(SO_4)_2$?

(A) 7
(B) 9
(C) 11
(D) 12
(E) 13

26. All of the following are compounds EXCEPT

(A) copper sulfate
(B) carbon dioxide
(C) washing soda
(D) air
(E) lime

27. What volume of gas, in liters, would 1.5 moles of hydrogen occupy at STP?

(A) 11.2
(B) 22.4
(C) 33.6
(D) 44.8
(E) 67.2

28. What is the maximum number of electrons held in the *d* orbitals?

(A) 2
(B) 6
(C) 8
(D) 10
(E) 14

29. If an element has an atomic number of 11, it will combine most readily with an element that has an electron configuration of

(A) $1s^2\ 2s^2\ 2p^6\ 3s^2\ 3p^1$
(B) $1s^2\ 2s^2\ 2p^6\ 3s^2\ 3p^2$
(C) $1s^2\ 2s^2\ 2p^6\ 3s^2\ 3p^3$
(D) $1s^2\ 2s^2\ 2p^6\ 3s^2\ 3p^4$
(E) $1s^2\ 2s^2\ 2p^6\ 3s^2\ 3p^5$

30. An example of a physical property is

(A) rusting
(B) decay
(C) souring
(D) low melting point
(E) high heat of formation

31. A gas at STP that contains 6.02×10^{23} atoms and forms diatomic molecules will occupy

(A) 11.2 L
(B) 22.4 L
(C) 33.6 L
(D) 67.2 L
(E) 1.06 qt

GO ON TO THE NEXT PAGE.

32. When excited electrons cascade to lower energy levels in an atom,
    (A) visible light is always emitted
    (B) the potential energy of the atom increases
    (C) the electrons always fall back to the first energy level
    (D) the electrons fall indiscriminately to all levels
    (E) the electrons fall back to a lower unfilled energy level

33. Mass spectroscopy uses the concept that
    (A) charged particles are evenly deflected in a magnetic field
    (B) charged particles are deflected in a magnetic field inversely to the mass of the particles
    (C) particles of heavier mass are deflected in a magnetic field to a greater degree than lighter particles
    (D) particles are evenly deflected in a magnetic field

34. The bond that includes an upper and a lower sharing of electron orbitals is called
    (A) a pi bond
    (B) a sigma bond
    (C) a hydrogen bond
    (D) a covalent bond
    (E) an ionic bond

35. What is the boiling point of water at the top of Pike's Peak?
    (A) It is 100°C.
    (B) It is > 100°C since the pressure is less than at ground level.
    (C) It is < 100°C since the pressure is less than at ground level.
    (D) It is > 100°C since the pressure is greater than at ground level.
    (E) It is < 100°C since the pressure is greater than at ground level.

36. The atomic structure of the alkane series contains the hybrid orbitals designated as
    (A) sp
    (B) sp^2
    (C) sp^3
    (D) sp^3d^2
    (E) sp^4d^3

37. Which of the following is (are) true for this reaction?

$$Cu + 4HNO_3 \rightarrow Cu(NO_3)_2 + 2H_2O + 2NO_2\uparrow$$

    I. It is an oxidation-reduction reaction.
    II. Copper is oxidized.
    III. The oxidation number of nitrogen goes from +5 to +4.

    (A) I only
    (B) III only
    (C) I and II only
    (D) II and III only
    (E) I, II, and III

38. Which of the following properties can be attributed to water?
    I. It has a permanent dipole moment attributed to its molecular structure.
    II. It is a very good conductor of electricity.
    III. It has its polar covalent bonds with hydrogen on opposite sides of the oxygen atom, so that the molecule is linear.

    (A) I only
    (B) III only
    (C) I and II only
    (D) II and III only
    (E) I, II, and III

39. Which of the following define(s) an acid according to conventional acid theories?

I. It is a good proton donor.
II. It is a good electron pair acceptor.
III. It has an excess of H_3O^+ in solution.

(A) I only
(B) III only
(C) I and II only
(D) II and III only
(E) I, II, and III

40. A nuclear reactor must include which of the following parts?

I. Electric generator
II. Fissionable fuel elements
III. Moderator

(A) I only
(B) III only
(C) I and II only
(D) II and III only
(E) I, II, and III

41. Which of the following salts will hydrolyze in water to form basic solutions?

I. NaCl
II. $CuSO_4$
III. K_3PO_4

(A) I only
(B) III only
(C) I and II only
(D) II and III only
(E) I, II, and III

42. When 1 mole of NaCl is dissolved in 1,000 grams of water, the boiling point of the water is changed to

(A) 100.51°C
(B) 101.02°C
(C) 101.53°C
(D) 101.86°C
(E) 103.62°C

43. What is the structure associated with the BF_3 molecule?

(A) Linear
(B) Trigonal planar
(C) Tetrahedron
(D) Trigonal pyramidal
(E) Bent or V-shaped

Questions 44 and 45 refer to the following setup:

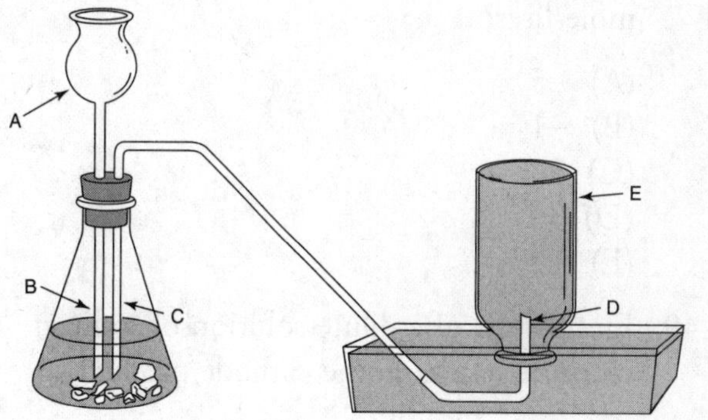

44. What letter designates an error in this laboratory setup?

(A) A
(B) B
(C) C
(D) D
(E) E

45. If the reaction in question 44 created a gas, where would the contents of the flask be expelled under these conditions?

(A) A
(B) B
(C) C
(D) D
(E) E

46. The most active nonmetal has

(A) a high electronegativity
(B) a low electronegativity
(C) a medium electronegativity
(D) large atomic radii
(E) a deliquescent property

Practice Test 3

GO ON TO THE NEXT PAGE.

47. In the reaction $Fe + S \rightarrow FeS$, which is true?

(A) $Fe + 2e^- \rightarrow Fe^{2+}$
(B) $Fe \rightarrow Fe^{2+} + 2e^-$
(C) $Fe^{2+} \rightarrow Fe + 2e^-$
(D) $S \rightarrow S^{2-} + 2e^-$
(E) $S^{-2} + 2e^- \rightarrow S$

48. What is the pH of a solution with a hydroxide ion concentration of 0.00001 mole/liter?

(A) −5
(B) −1
(C) 5
(D) 9
(E) 14

49. Electrolysis of a dilute solution of sodium chloride results in the cathode product

(A) sodium
(B) hydrogen
(C) chlorine
(D) oxygen
(E) peroxide

50. .....$C_2H_4(g)$ + ...$O_2(g)$ → .....$CO_2(g)$ + ...$H_2O(\ell)$

If the equation for the above reaction is balanced with whole-number coefficients, what is the coefficient for oxygen gas?

(A) 1
(B) 2
(C) 3
(D) 4
(E) 5

51. Five liters of gas at STP have a mass of 12.5 grams. What is the molar mass of the gas?

(A) 12.5 g/mol
(B) 25.0 g/mol
(C) 47.5 g/mol
(D) 56.0 g/mol
(E) 125 g/mol

52. A compound whose molecular mass is 90 grams contains 40.0% carbon, 6.67% hydrogen, and 53.33% oxygen. What is the true formula of the compound?

(A) $C_2H_2O_4$
(B) CH_2O_4
(C) C_3H_6O
(D) C_3HO_3
(E) $C_3H_6O_3$

53. How many moles of CaO are needed to react with an excess of water to form 370 grams of calcium hydroxide?

(A) 1
(B) 2
(C) 3
(D) 4
(E) 5

54. To what volume, in milliliters, must 50.0 milliliters of 3.50 M H_2SO_4 be diluted in order to make 2 M H_2SO_4?

(A) 25
(B) 60.1
(C) 87.5
(D) 93.2
(E) 101

55. A small value of K_{eq} indicates that equilibrium occurs

(A) at a low product concentration
(B) at a high product concentration
(C) after considerable time
(D) with the help of a catalyst
(E) with no forward reaction

56. A student measured 10.0 milliliters of an HCl solution into a beaker and titrated it with a standard NaOH solution that was 0.09 M. The initial NaOH burette reading was 34.7 milliliters while the final reading showed 49.2 milliliters.

What is the molarity of the HCl solution?

(A) 0.13
(B) 0.47
(C) 0.52
(D) 1.57
(E) 2.43

57. A student made the following observations in the laboratory:

(a) Sodium metal reacted vigorously with water while a strip of magnesium did not seem to react at all.
(b) The magnesium strip reacted with dilute hydrochloric acid faster than an iron strip.
(c) A copper rivet suspended in silver nitrate solution was covered with silver-colored stalactites in several days, and the resulting solution had a blue color.
(d) Iron filings dropped into the blue solution were coated with an orange color.

The order of *decreasing* strength as reducing agents is:

(A) Na, Mg, Fe, Ag, Cu
(B) Mg, Na, Fe, Cu, Ag
(C) Ag, Cu, Fe, Mg, Na
(D) Na, Fe, Mg, Cu, Ag
(E) Na, Mg, Fe, Cu, Ag

58. A student placed water, sodium chloride, potassium dichromate, sand, chalk, and hydrogen sulfide into a distilling flask and proceeded to distill. What ingredient besides water would be found in the distillate?

(A) Sodium chloride
(B) Chalk
(C) Sand
(D) Hydrogen sulfide
(E) Chrome sulfate

59. Which of these statements is NOT correct?

(A) In an exothermic reaction, ΔH is negative and the enthalpy decreases.
(B) In an endothermic reaction, ΔH is positive and the enthalpy increases.
(C) In a reaction where ΔG is negative, the forward reaction is spontaneous.
(D) In a reaction where ΔG is positive, ΔS may also be positive.
(E) In a reaction where ΔH is positive and ΔS is negative, the forward reaction is spontaneous.

60. A student filled a steam-jacketed eudiometer with 32 milliliters of oxygen and 4 milliliters of hydrogen over mercury. How much of which gas would be left uncombined after the mixture was sparked?

(A) None of either
(B) 3 mL H_2
(C) 24 mL O_2
(D) 28 mL O_2
(E) 30 mL O_2

61. What would be the *total* volume, in milliliters, of gases in question 60 after sparking?

(A) 16
(B) 24
(C) 34
(D) 36
(E) 40

GO ON TO THE NEXT PAGE.

62. How can the addition of a catalyst affect an exothermic reaction?

I. Speed up the reaction.
II. Slow down the reaction.
III. Increase the amount of product formed.

(A) I only
(B) II only
(C) I and II only
(D) II and III only
(E) I, II, and III

63. In which period of the periodic table is the most electronegative element found?

(A) 1
(B) 2
(C) 3
(D) 4
(E) 5

64. What could be the equilibrium constant for this reaction: $aA + bB \rightleftharpoons cC + dD$, if A and D are solids?

(A) $\frac{[C]^c[D]^d}{[A]^a[B]^b}$

(B) $\frac{[A]^a[B]^b}{[C]^c[D]^d}$

(C) $\frac{[C]^c}{[B]^b}$

(D) $\frac{[C]^c[D]^d}{[A]^a}$

(E) $[A]^a[B]^b[C]^c[D]^d$

65. Which of the following does NOT react with a dilute solution of sulfuric acid?

(A) $NaNO_3$
(B) Na_2S
(C) Na_3PO_4
(D) Na_2CO_3
(E) NaOH

66. Which of these statements is the best explanation for the sp^3 hybridization of carbon's electrons?

(A) The new orbitals are one *s* orbital and three *p* orbitals.
(B) The *s* electron is promoted to the *p* orbitals.
(C) The *s* orbital is deformed into a *p* orbital.
(D) Four new and equivalent orbitals are formed.
(E) The *s* orbital electron loses energy to fall back into a partially filled *p* orbital.

67. The intermolecular force that is most significant in explaining the variation of the boiling point of water from the boiling points of similarly structured molecules is

(A) hydrogen bonding
(B) van der Waals forces
(C) covalent bonding
(D) ionic bonding
(E) coordinate covalent bonding

68. If *K* for the reaction $H_2 + I_2 \rightleftharpoons 2HI$ is equal to 45.9 at 450°C, and 1 mole of H_2 and 1 mole of I_2 are introduced into a 1-liter box at that temperature, what will be the expression for *K* at equilibrium?

(A) $\frac{[x^2]^2}{[1-x][1-x]}$

(B) $\frac{[2x]^2}{[1-x][1-x]}$

(C) $\frac{[2x]^2}{[x][x]}$

(D) $\frac{[1-x][1-x]}{[2x]^2}$

(E) $\frac{[1-x][1-x]}{[x^2]^2}$

69. What is the molar mass of a nonionizing solid if 10 grams of this solid, dissolved in 200 grams of water, formed a solution that froze at −3.72°C?

(A) 25 g/mol
(B) 50 g/mol
(C) 100 g/mol
(D) 150 g/mol
(E) 1,000 g/mol

STOP

If you finish before one hour is up, you may go back to check your work or complete unanswered questions.

Answers and Explanations for Test 3

1. **(B)** The activation energy of the forward reaction is the energy needed to begin the reaction.

2. **(D)** For the reverse reaction to occur, activation energy equal to the sum of (B) + (C) is needed. This is shown by (D).

3. **(C)** The heat of the reaction is the heat liberated between the level of potential energy of the reactants and that of the products. This is quantity (C) on the diagram.

4. **(A)** The potential energy of the reactants is the total of the original potential energies of the reactants shown by (A).

5. **(B)** The spoon must be made the cathode to attract the Ag^+ ions.

6. **(D)** The silver plate is the anode.

7. **(E)** The solution of Ag^+ provides the silver for plating.

8. **(A)**
9. **(B)** } In a phase diagram, the zones are as shown below.
10. **(C)**

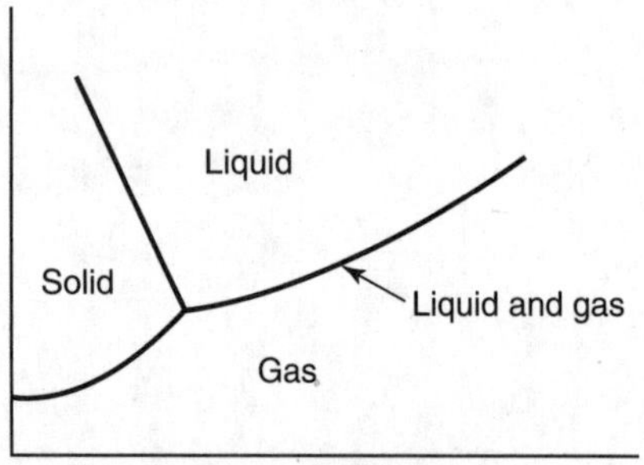

11. **(E)** The boundry of liquid and gas is shown on the diagram.

12–14. The balanced equations with half-reactions are as follows:

12. **(C)**

$$3Cu^0 \rightarrow 3Cu^{2+} + 3(2e^-)$$

$$2NO_3^- + 8H^+ + 2(3e^-) \rightarrow 2NO + 4H_2O$$

$$3Cu + 8HNO_3 \rightarrow 3Cu(NO_3)_2 + 4H_2O + 2NO$$

13. **(D)** The coefficients show that 3 mol of Cu produces 2 mol of NO, so 6 mol of Cu produces 4 mol of NO.

14. **(C)** The expression $Cu(NO_3)_2 \rightarrow Cu^{2+} + 2NO_3^-$ shows that the dissociation yields 3 ions.

15. **(A)** KCl is ionic because it is the product of a very active metal combining with a very active nonmetal.

16. **(B)** The electronegativity difference between H and Cl is between 0.5 and 1.7. This indicates an unequal sharing of electrons, which results in a polar covalent bond.

17. **(C)** Because the polar bonds are symmetrically arranged in the methane molecule, the molecule is nonpolar covalent.

18. **(E)** Lithium (Li) is a metal.

19. **(D)** When hydrated copper sulfate is heated, the crystal crumples as the water is forced out of the structure, and a white powder is the result.

20. **(A)** Brownian movement is due to molecular collisions with colloidal particles, which knock the particles about in a zigzag path noted by the reflected light from these particles.

21. **(C)** The indicator phenolphthalein turns pink in a basic solution.

22. **(B)** Litmus paper turns pink in an acid solution.

23. **(E)** A substance that is deliquescent, such as a crystal, draws water to its surface. At times it can draw enough water to form a water solution.

101. **(T, T)** The assertion is true because the nonmetals have a tendency to gain electrons readily. The reason is a true statement but does not explain the assertion.

102. **(T, T, CE)** The assertion is explained by the reason. The graphic display of this is:

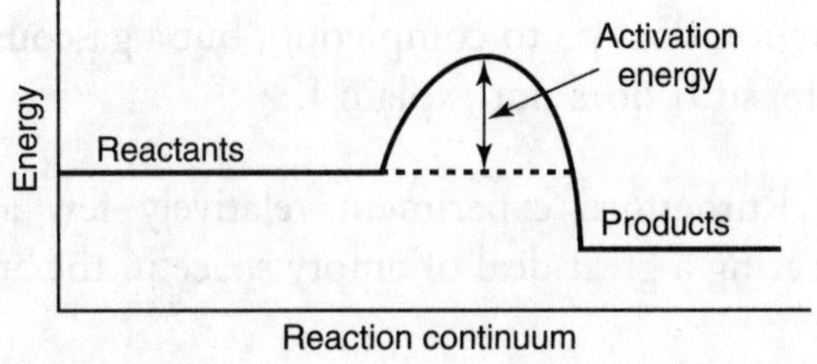

103. **(T, F)** The assertion is true but the reasoning is false. Transition elements have incomplete inner energy levels that are being filled with the additional

Practice Test 3

electrons, thus leaving the outer energy level the same in most cases. As a result, these elements have common oxidation numbers.

104. **(F, T)** The assertion is false and the reason is true. A supersaturated solution is holding more than its normal solubility, and the addition of a crystal causes crystallization to occur.

105. **(F, T)** Equilibrium is a dynamic condition because of the reason stated. The assertion is false; the reason, true.

106. **(T, T, CE)** Both statements are true, and the reason explains why ionic bonding is the strongest.

107. **(F, F)** An equilibrium system must have a gaseous reactant or product for pressure to affect the equilibrium. Then increased pressure will cause the reaction to go in the direction that reduces the concentration of gaseous substances.

108. **(T, T, CE)** The assertion is explained by the reason; both are true.

109. **(T, T, CE)** Both the assertion and the reason are true; they explain that the element's orbital designation places it in the first transition series of filling the 3*d* orbitals.

110. **(T, T, CE)** The assertion is true; the reason is true and explains why nonmetals have the highest electronegativity.

111. **(T, T, CE)** The assertion is explained by the reason; both are true.

112. **(T, T)** There are 3 moles of atoms in 18 g of water because 18 g is 1 mol of water molecules and each molecule has three atoms. The reason does not explain the assertion but is also true.

113. **(F, T)** Benzene is a nonionizing substance and therefore a nonelectrolyte. The reason is a true statement.

114. **(T, F)** Isomers have the same empirical formula but vary in their structural formulas.

115. **(T, T)** The reaction does go to completion, but a gaseous product is formed, not a precipitate, so II does not explain I.

116. **(F, T)** In the Rutherford experiment relatively few alpha particles were deflected, indicating a great deal of empty space in the atom. The reason is a true statement.

24. **(B)** The correct coefficients are 2, 13, 8, and 10.

25. **(D)** 1K + 1Al + 2S + 8O = 12 total

26. **(D)** Air is a mixture; all others are compounds. Washing soda (C) is sodium carbonate, and lime (E) is calcium oxide.

27. **(C)** One mole of a gas at STP occupies 22.4 L, so 1.5 mol × 22.4 L = 33.6 L.

28. **(D)** The maximum number of electrons in each kind of orbital is:

s = 2 in one orbital

p = 6 in three orbitals

d = 10 in five orbitals

f = 14 in seven orbitals

29. **(E)** The element with atomic number 11 is sodium with 1 electron in the 3s orbital. It would readily combine with the element that has $3p^5$ as the outer orbital since it needs only 1 more electron to fill it.

30. **(D)** The only physical property named in the list is low melting point.

31. **(A)** If the gas is diatomic, then 6.02×10^{23} atoms will form $6.02 \times 10^{23}/2$ molecules. At STP, 6.02×10^{23} molecules occupy 22.4 L, so half that number will occupy 11.2 L.

32. **(E)** Cascading excited electrons can fall only to lower energy levels that are unfilled.

33. **(B)** Mass spectroscopy uses a magnetic field to separate isotopes by bending their path. The lighter ones are bent farther than the heavier ones.

34. **(A)** The pi bond is a bond between two p orbitals, like this:

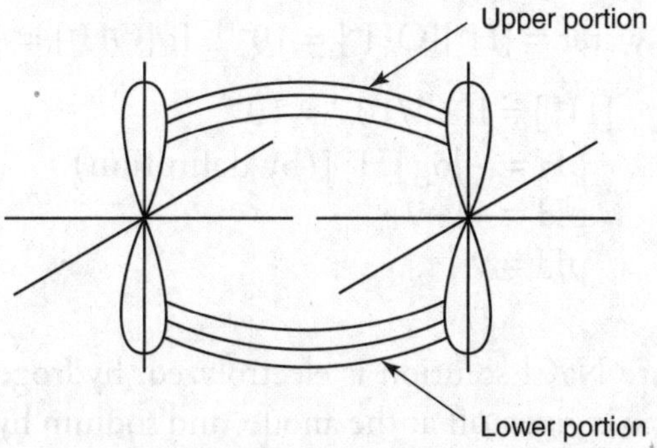

35. **(C)** At Pike's Peak (alt. approx. 14,000 ft) the pressure is lower than at ground level; therefore the vapor pressure at a lower temperature will equal the outside pressure and boiling will occur.

36. **(C)** The alkanes contain the sp^3 hybrid orbitals.

37. **(E)** I, II, and III are correct.

38. **(A)** Only I is correct.

39. **(E)** I, II, and III are acid definitions.

40. **(D)** I is not necessary for the reactor, but often nuclear energy is used to operate an electric generator. The others, II and III, are necessary for fuel and neutron-speed control, respectively.

41. **(B)** III is a salt from a strong base and a weak acid, which hydrolyzes to form a basic solution with water.

42. **(B)** Since the boiling point is increased by 0.51°C for each mole of particles, 1 mol of NaCl $\rightarrow Na^+ + Cl^-$ gives 2 mol of particles. Therefore the boiling point will be 1.02° higher or 101.02°C.

43. **(B)** The VSEPR model shows BF_3 is trigonal planar and so is related to the triangle shape on one plane.

44. **(C)** The delivery tube is below the fluid level in the flask and will cause liquid to be forced up the thistle tube when gas is evolved in the reaction.

45. **(A)** At first the fluid will be expelled up the thistle tube by the gas generated and exerting pressure in the reaction flask. When the level of the fluid falls below the end of the thistle tube, the gas will then be released through the thistle tube.

46. **(A)** The most active nonmetal has a high attraction for another electron—thus high electronegativity.

47. **(B)** Fe loses electrons to form the Fe^{2+} ion.

48. **(D)** The K_w of water = $[H^+][OH^-] = 10^{-14}$. If $[OH^-] = 10^{-5}$ mol/L, then

$$[H^+] = 10^{-14}/10^{-5} = 10^{-9}$$
$$pH = -\log[H^+]\,(\text{by definition})$$
$$pH = -[-9]$$
$$pH = 9$$

49. **(B)** When dilute NaCl solution is electrolyzed, hydrogen is given off at the cathode, chlorine is given off at the anode, and sodium hydroxide is left in the container.

50. **(C)** The correctly balanced equation is

$$C_2H_4(g) + 3O_2(g) \rightarrow 2CO_2(g) + 2H_2O(\ell)$$

51. **(D)** One mole of a gas at STP occupies 22.4 L.

$$\text{If } \frac{5\text{ L}}{12.5\text{ g}}, \text{ then}$$

$$\frac{22.4\text{ L}}{\text{gram-molecular mass}}$$

$$\frac{5\text{ L}}{12.5\text{ g}} = \frac{22.4\text{ L}}{x\text{ g}}$$

$$x = 56.0\text{ g/mol}$$

52. **(E)** To find the simple or empirical formula, divide each % by the element's atomic mass.

Carbon
$40 \div 12 = 3.333$

Hydrogen
$6.67 \div 1 = 6.67$

Oxygen
$53.33 \div 16 = 3.33$

Next, divide each quotient by the smallest quotient in an attempt to get small whole numbers.

$$3.33 \div 3.33 = 1\text{ C}$$
$$6.67 \div 3.33 = 2\text{ H}$$
$$3.33 \div 3.33 = 1\text{ O}$$

The simplest formula is CH_2O, which has a molecular mass of 30. The true molecular mass is given as 90, which is three times the simplest. Therefore the true formula is $C_3H_6O_3$.

53. **(E)** The reaction is

$$\begin{array}{c} 370\text{ g} \\ CaO + H_2O \rightarrow Ca(OH)_2 \end{array}$$

$Ca(OH)_2$ molecular mass = 74

$370\text{ g} \div 74 = 5$ mol of $Ca(OH)_2$ is wanted. The reaction shows 1 mol of CaO produces 1 mol of $Ca(OH)_2$, so the answer is 5 mol.

54. **(C)** In dilution problems, this formula can be used:

$$M_{before} \times V_{before} = M_{after} \times V_{after}$$

Substituting gives:

$$3.5 \times 50 = 2 \times (?x)$$

$x = 87.5$ mL, new volume after dilution

Practice Test 3

55. **(A)** For K_{eq} to be small, the numerator, which is made up of the concentration(s) of the product(s) at equilibrium, must be smaller than the denominator.

56. **(A)** The amount of NaOH used is

$49.2 - 34.7 = 14.5$ mL

Using $M_1 \times V_1 = M_2 \times V_2$ gives

$0.09\text{ M} \times 14.5\text{ mL} = M_2 \times 10\text{ mL}$

$M_2 = 0.13\text{ M}$

57. **(E)** The reactions recorded indicated that the ease of losing electrons is greater in sodium than magnesium, greater in magnesium than iron, greater in iron than copper, and finally greater in copper than silver.

58. **(D)** Distillation removes only dissolved solids from the distillate. The volatile gases, such as H_2S, will be carried into the distillate.

59. **(E)** All the first four statements are correct.

The Gibbs free-energy equation is:

$\Delta G = \Delta H - T\Delta S$

In choice (E), if ΔH is positive and ΔS is negative, then ΔG will definitely be positive, which means that the forward reaction will not occur spontaneously.

60. **(E)** H_2 to O_2 ratio by volume is 2 : 1 in the formation of water. Therefore, 4 mL H_2 will react with 2 mL of O_2 to make 4 mL of steam.

$$2H_2(g) + O_2(g) \rightarrow 2H_2O(g)$$

This leaves 30 mL of O_2 uncombined.

61. **(C)** There will be 30 mL of O_2 + 4 mL of steam = 34 mL total.

62. **(A)** By definition, a catalyst can be used to speed up a reaction, so I is correct.

63. **(B)** The most electronegative element is fluorine (F), found in period 2.

64. **(C)** Solids are incorporated into the *K* value and therefore do not appear on the right side of the equation.

65. **(A)** Only $NaNO_3$ will not react because it requires heat to react.

66. **(D)** When hybridization forms the sp^3 orbitals, four entirely new orbitals, different from but equivalent to the former *s* and *p* orbitals, result.

67. **(A)** Hydrogen bonding between water molecules causes the boiling point to be higher than would be expected.

68. **(B)** At the beginning of the reaction

$$[H_2] = 1 \text{ mol/L}$$
$$[I_2] = 1 \text{ mol/L}$$
$$[HI] = 0$$

At equilibrium

$$(H_2 + I_2 \rightleftharpoons 2HI)$$
(Let x = moles/liter of H_2 and I_2 in HI form)

$$[H_2] = (1 - x) \text{ mol/L}$$
$$[I_2] = (1 - x) \text{ mol/L}$$
$$[HI] = 2x \text{ mol/L}$$

Then, substituting the above values into the equation, you get:

$$K = \frac{[HI]^2}{[H_2][I_2]}$$
$$= \frac{[2x]^2}{[1-x][1-x]} = 45.9$$

69. **(A)** 10 g/200 g of water = 50 g/1,000 g of water (5 times as much)

The freezing point depression, 3.72°, is divided by 1.86°, which is the depression caused by 1 mol in 1,000 g of water, to find how many moles are dissolved.

3.72° ÷ 1.86° = 2 mol

If 50 g caused this depression and is equal to 2 mol, then 1 mol would be $\frac{1}{2}$ of 50 g, or 25 g. So, the molar mass of the solid is 25 g/mol.

CALCULATING YOUR SCORE

Your score on Practice Test 3 can now be computed manually. The actual test will be scored by machine, but the same method is used to arrive at the raw score. You get one point for each correct answer. For each wrong answer, you lose one-fourth of a point. Questions that you omit or that have more than one answer are not counted. On your answer sheet mark all correct answers with a "C" and all incorrect answers with an "X".

Determining Your Raw Test Score

Total the number of correct answers you have recorded on your answer sheet. It should be the same as the total of all the numbers you place in the block in the lower left corner of each area of the Subject Area summary in the next section.

A. Enter the total number of correct answers here: ________
   Now count the number of wrong answers you recorded on your answer sheet.
B. Enter the total number of wrong answers here: ________
   Multiply the number of wrong answers in B by 0.25.
C. Enter that product here: ________
   Subtract the result in C from the total number of right answers in A.
D. Enter the result of your subtraction here: ________
E. Round the result in D to the nearest whole number: ________.
   This is your raw test score.

Conversion of Raw Scores to Scaled Scores

Your raw score is converted by the College Board into a scaled score. The College Board scores range from 200 to 800. This conversion is done to ensure that a score earned on any edition of a particular SAT Subject Test in Chemistry is comparable to the same scaled score earned on any other edition of the same test. Because some editions of the tests may be slightly easier or more difficult than others, scaled scores are adjusted so that they indicate the same level of performance regardless of the edition of the test taken and the ability of the group that takes it. Consequently, a specific raw score on one edition of a particular test will not necessarily translate to the same scaled score on another edition of the same test.

Because the practice tests in this book have no large population of scores with which they can be scaled, scaled scores cannot be determined.

Results from previous SAT Chemistry tests appear to indicate that the conversion of raw scores to scaled scores GENERALLY follows this pattern:

Raw Score	Scaled Score	Raw Score	Scaled Score
85–75	800–800	30–25	560–540
75–70	800–780	25–20	540–520
70–65	780–750	20–15	520–490
65–60	750–720	15–10	490–460
60–55	720–700	10–5	460–430
55–50	700–680	5–0	430–410
50–45	680–650	0 to –5	410–380
45–40	650–620	–5 to –10	380–360
40–35	620–590	–10 to –15	360–330
35–30	590–560		

Note that this scale provides only a *general idea* of what a raw score may translate into on a scaled score range of 800–200. Scaling on every test is usually slightly different. Some students who had taken the SAT Subject Test in Chemistry after using this book had reported that they have scored slightly higher on the SAT test than on the practice tests in this book. They *all* reported that preparing well for the test paid off in a better score!

DIAGNOSING YOUR NEEDS

After taking Practice Test 3, check your answers against the correct ones. Then fill in the chart below.

In the space under each question number, place a check if you answered that question correctly.

EXAMPLE:

If your answer to question 5 was correct, place a check in the appropriate box.

Next, total the check marks for each section and insert the number in the designated block. Now do the arithmetic indicated and insert your percent for each area.

Subject Area*	(✓) Questions Answered Correctly											
I. Atomic Theory and Structure, including periodic relationships					110	112	116	28	32	34	66	
☐ No. of checks ÷ 7 × 100 = _____%												
II. Nucleonics										33	40	
☐ No. of checks ÷ 2 × 100 = _____%												
III. Chemical Bonding and Molecular Structure		15	16	17	18	106	25	29	38	43	67	
☐ No. of checks ÷ 10 × 100 = _____%												
IV. States of Matter and Kinetic Molecular Theory of Gases					8	9	10	11	27	31	35	
☐ No. of checks ÷ 7 × 100 = _____%												
V. Solutions, including concentration units, solubility, and colligative properties							19	20	104	42	54	
☐ No. of checks ÷ 5 × 100 = _____%												
VI. Acids and Bases						21	22	113	115	39	41	48
☐ No. of checks ÷ 7 × 100 = _____%												
VII. Oxidation-Reduction and Electrochemistry						5	6	7	37	49	47	
☐ No. of checks ÷ 6 × 100 = _____%												
VIII. Stoichiometry		12	13	14	24	50	51	52	53	60	61	69
☐ No. of checks ÷ 11 × 100 = _____%												
IX. Reaction Rates											1	108
☐ No. of checks ÷ 2 × 100 = _____%												

*The subject areas have been expanded to identify specific areas in the text.

Subject Area*	(✓) Questions Answered Correctly					
X. Equilibrium		105	107	55	64	68
☐ No. of checks ÷ 5 × 100 = _____%						
XI. Thermodynamics: energy changes in chemical reactions, randomness, and criteria for spontaneity	2	3	4	102	59	62
☐ No. of checks ÷ 6 × 100 = _____%						
XII. Descriptive Chemistry: physical and chemical properties of elements and their familiar compounds; organic chemistry; periodic properties	23	101	103	109	111	114
	26	30	36	46	63	65
☐ No. of checks ÷ 12 × 100 = _____%						
XIII. Laboratory: equipment, procedures, observations, safety, calculations, and interpretation of results		44	45	56	57	58
☐ No. of checks ÷ 5 × 100 = _____%						

**The subject areas have been expanded to identify specific areas in the text.*

To develop a study plan, refer to pages 33–35.

Answer Sheet

PRACTICE TEST 4

Determine the correct answer for each question. Then, using a No. 2 pencil, blacken completely the oval containing the letter of your choice.

1 (A) (B) (C) (D) (E)
2 (A) (B) (C) (D) (E)
3 (A) (B) (C) (D) (E)
4 (A) (B) (C) (D) (E)
5 (A) (B) (C) (D) (E)
6 (A) (B) (C) (D) (E)
7 (A) (B) (C) (D) (E)
8 (A) (B) (C) (D) (E)
9 (A) (B) (C) (D) (E)
10 (A) (B) (C) (D) (E)
11 (A) (B) (C) (D) (E)
12 (A) (B) (C) (D) (E)
13 (A) (B) (C) (D) (E)
14 (A) (B) (C) (D) (E)
15 (A) (B) (C) (D) (E)
16 (A) (B) (C) (D) (E)
17 (A) (B) (C) (D) (E)
18 (A) (B) (C) (D) (E)
19 (A) (B) (C) (D) (E)
20 (A) (B) (C) (D) (E)
21 (A) (B) (C) (D) (E)
22 (A) (B) (C) (D) (E)
23 (A) (B) (C) (D) (E)

ON THE ACTUAL CHEMISTRY TEST, THE FOLLOWING TYPE OF QUESTION MUST BE ANSWERED ON A SPECIAL SECTION (LABELED "CHEMISTRY") AT THE LOWER LEFT-HAND CORNER OF PAGE 2 OF YOUR ANSWER SHEET. THESE QUESTIONS WILL BE NUMBERED BEGINNING WITH 101 AND MUST BE ANSWERED ACCORDING TO THE DIRECTIONS.

CHEMISTRY* Fill in oval CE only if II is a correct explanation of I.

	I	II	CE*
101	(T) (F)	(T) (F)	○
102	(T) (F)	(T) (F)	○
103	(T) (F)	(T) (F)	○
104	(T) (F)	(T) (F)	○
105	(T) (F)	(T) (F)	○
106	(T) (F)	(T) (F)	○
107	(T) (F)	(T) (F)	○
108	(T) (F)	(T) (F)	○
109	(T) (F)	(T) (F)	○
110	(T) (F)	(T) (F)	○
111	(T) (F)	(T) (F)	○
112	(T) (F)	(T) (F)	○
113	(T) (F)	(T) (F)	○
114	(T) (F)	(T) (F)	○
115	(T) (F)	(T) (F)	○
116	(T) (F)	(T) (F)	○

ON THE ACTUAL CHEMISTRY TEST, THE REMIANING QUESTIONS MUST BE ANSWERED BY RETURNING TO THE SECTION OF YOUR ANSWER SHEET YOU STARTED FOR CHEMISTRY.

24 (A) (B) (C) (D) (E)
25 (A) (B) (C) (D) (E)
26 (A) (B) (C) (D) (E)
27 (A) (B) (C) (D) (E)
28 (A) (B) (C) (D) (E)
29 (A) (B) (C) (D) (E)
30 (A) (B) (C) (D) (E)
31 (A) (B) (C) (D) (E)
32 (A) (B) (C) (D) (E)
33 (A) (B) (C) (D) (E)
34 (A) (B) (C) (D) (E)
35 (A) (B) (C) (D) (E)
36 (A) (B) (C) (D) (E)
37 (A) (B) (C) (D) (E)
38 (A) (B) (C) (D) (E)
39 (A) (B) (C) (D) (E)
40 (A) (B) (C) (D) (E)
41 (A) (B) (C) (D) (E)
42 (A) (B) (C) (D) (E)
43 (A) (B) (C) (D) (E)
44 (A) (B) (C) (D) (E)
45 (A) (B) (C) (D) (E)
46 (A) (B) (C) (D) (E)
47 (A) (B) (C) (D) (E)
48 (A) (B) (C) (D) (E)
49 (A) (B) (C) (D) (E)
50 (A) (B) (C) (D) (E)
51 (A) (B) (C) (D) (E)
52 (A) (B) (C) (D) (E)
53 (A) (B) (C) (D) (E)
54 (A) (B) (C) (D) (E)
55 (A) (B) (C) (D) (E)
56 (A) (B) (C) (D) (E)
57 (A) (B) (C) (D) (E)
58 (A) (B) (C) (D) (E)
59 (A) (B) (C) (D) (E)
60 (A) (B) (C) (D) (E)
61 (A) (B) (C) (D) (E)
62 (A) (B) (C) (D) (E)
63 (A) (B) (C) (D) (E)
64 (A) (B) (C) (D) (E)
65 (A) (B) (C) (D) (E)
66 (A) (B) (C) (D) (E)
67 (A) (B) (C) (D) (E)
68 (A) (B) (C) (D) (E)
69 (A) (B) (C) (D) (E)

Practice Test 4

Note: For all questions involving solutions and/or chemical equations, assume that the system is in water unless otherwise stated.

Part A

Directions: Every set of the given lettered choices below refers to the numbered statements or formulas immediately following it. Choose the one lettered choice that best fits each statement or formula and then fill in the corresponding oval on the answer sheet. Each choice may be used once, more than once, or not at all in each set.

Questions 1–3 refer to the following graphs:

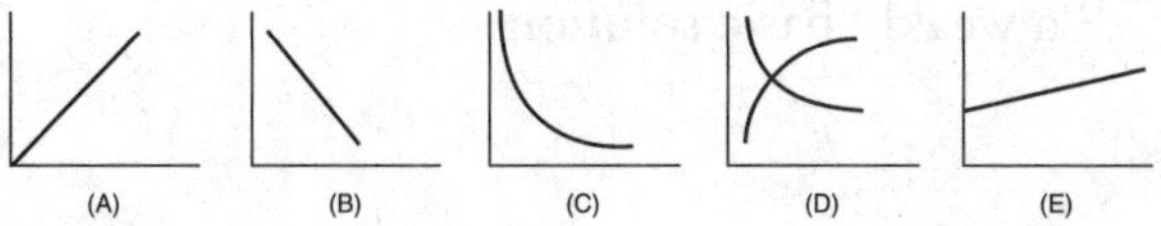

1. The graph that best shows the relationship of gas volume to temperature, with pressure held constant
2. The graph that best shows the relationship of gas volume to pressure, with temperature held constant
3. The graph that best shows the relationship of the number of grams of a solid that are soluble in 100 grams of H_2O at varying temperatures if the solubility begins at a small quantity and increases at a slow, steady pace as the temperature is increased

Questions 4–7

(A) A molecule
(B) A mixture of compounds
(C) An isotope
(D) An isomer
(E) An acid salt

4. The simplest unit of water that retains its properties
5. A commercial cake mix
6. An atom with the same number of protons as another atom of the same element but a different number of neutrons
7. Classification of $NaHCO_3$

Questions 8–10

(A) 1
(B) 7
(C) 9
(D) 10
(E) 14

8. The atomic number of an atom with an electron dot arrangement similiar to $:\overset{\cdot\cdot}{\underset{\cdot\cdot}{I}}:$
9. The number of atoms represented in the formula $Al(OH)_3$
10. The number that represents the most acid pH

Questions 11–14

(A) Density
(B) Equilibrium constant
(C) Molar mass
(D) Freezing point
(E) Molarity

11. Can be expressed in moles of solute per liter of solution

GO ON TO THE NEXT PAGE.

12. Can be expressed in grams per liter of a gas
13. Will NOT be affected by changes in temperature and pressure
14. At STP, can be used to determine the molecular mass of a pure gas

Questions 15–18

(A) Buffer
(B) Indicator
(C) Arrhenius acid
(D) Arrhenius base
(E) Neutral condition

15. Resists a rapid change of pH
16. Exhibits different colors in acidic and basic solutions
17. At 25°C, the aqueous solution has a pH < 7.
18. At 25°C, the aqueous solution has a pH > 7.

Questions 19–23

(A) H_2
(B) NH_3
(C) CO_2
(D) HCl
(E) O_2

19. A gas produced by the reaction of zinc with hydrochloric acid
20. A gas combustion product that is heavier than air
21. A gas produced by the heating of potassium chlorate
22. A gas that is slightly soluble in water and gives a weakly acid solution
23. A gas that is very soluble in water and gives a weakly basic solution

Part B

ON THE ACTUAL CHEMISTRY TEST, THE FOLLOWING TYPE OF QUESTION MUST BE ANSWERED ON A SPECIAL SECTION (LABELED "CHEMISTRY") AT THE LOWER LEFT-HAND CORNER OF PAGE 2 OF YOUR ANSWER SHEET. THESE QUESTIONS WILL BE NUMBERED BEGINNING WITH 101 AND MUST BE ANSWERED ACCORDING TO THE FOLLOWING DIRECTIONS.

Directions: Every question below contains two statements, I in the left-hand column and II in the right-hand column. For each question, decide if statement I is true or false and if statement II is true or false and fill in the corresponding T or F ovals on your answer sheet. *Fill in oval CE only if statement II is a correct explanation of statement I.

Sample Answer Grid:

CHEMISTRY * Fill in oval CE only if II is a correct explanation of I.

	I		II		CE*
101.	T	F	T	F	◯

	I		II
101.	According to the Kinetic Molecular Theory, the particles of a gas are in random motion above absolute zero	BECAUSE	the degree of random motion of gas molecules varies inversely with the temperature of the gas.
102.	An electron has wave properties as well as particle properties	BECAUSE	the design of a particular experiment determines which properties are verified.
103.	The alkanes are considered a homologous series	BECAUSE	homologous series have the same functional group but differ in formula by the addition of a fixed group of atoms.
104.	When an atom of an active metal becomes an ion, the radius of the ion is less than that of the atom	BECAUSE	the nucleus of an active metallic ion has less positive charge than the electron "cloud."
105.	When the heat of formation for a compound is negative, ΔH is negative	BECAUSE	a negative heat of formation indicates that a reaction is exothermic with a negative enthalpy change.

GO ON TO THE NEXT PAGE.

	I		II
106.	Water is a polar substance	BECAUSE	the sharing of the bonding electrons in water is equal.
107.	A catalyst accelerates a chemical reaction	BECAUSE	a catalyst lowers the activation energy of the reaction.
108.	Copper is an oxidizing agent in the reaction with silver nitrate solution	BECAUSE	copper loses electrons in a reaction with silver ions.
109.	The rate of diffusion (or effusion) of hydrogen gas compared with that of helium gas is 1 : 4	BECAUSE	the rate of diffusion (or effusion) of gases varies inversely as the square root of the molecular mass.
110.	A gas heated from 10°C to 100°C at constant pressure will increase in volume	BECAUSE	as Charles's Law states, if the pressure remains constant, the volume varies directly as the absolute temperature varies.
111.	The Gibbs free-energy equation can be used to predict the solubility of a solute	BECAUSE	the solubility of most salts increases as temperature increases.
112.	The complete electrolysis of 45 grams of water will yield 40 grams of H_2 and 5 grams of O_2	BECAUSE	water is composed of hydrogen and oxygen in a ratio of 8 : 1 by mass.
113.	320 calories or 1.34×10^3 joules of heat will melt 4 grams of ice at 0°C	BECAUSE	the heat of fusion of water is 80 calories per gram or 3.34×10^2 joules per gram.
114.	When 2 liters of oxygen gas react with 2 liters of hydrogen completely, the limiting factor is the volume of the oxygen	BECAUSE	the coefficients in balanced equations of gaseous reactions give the volume relationships of the envolved gases.
115.	Water is a good solvent for ionic and/or polar covalent substances	BECAUSE	water shows hydrogen bonding between oxygen atoms.
116.	Ammonia gas, NH_3, has a smaller density than argon gas, Ar, at STP	BECAUSE	the density of a gas at STP is found by dividing the molar mass by 22.4 liters.

Part C

Directions: Every question or incomplete statement below is followed by five suggested answers or completions. Choose the one that is best in each case and then fill in the corresponding oval on the answer sheet.

24. In this graphic representation of a chemical reaction, which arrow depicts the activation energy of the forward reaction?

(A) *A*
(B) *B*
(C) *C*
(D) *D*
(E) *E*

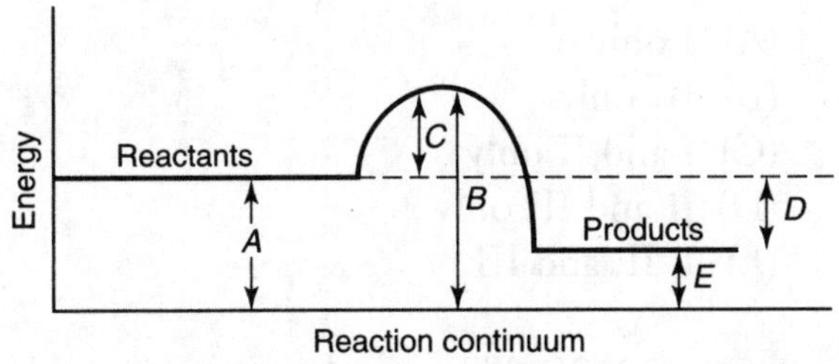

25. How many liters (STP) of O_2 can be produced by completely decomposing 2 moles of $KClO_3$?

(A) 11.2
(B) 22.4
(C) 33.6
(D) 44.8
(E) 67.2

26. Which of the following is the correct structural representation of sodium?

(A) 11 p 11 n Nucleus and electron configuration:

$1s^2\ 2s^2\ 2p^6\ 3s^2\ 3p^6\ 4s^2\ 4p^3$

(B) 11 p 12 n Nucleus and electron configuration:

$1s^2\ 2s^2\ 2p^6\ 3s^2\ 3p^6\ 4s^2\ 3d^1\ 4p^2$

(C) 23 p 23 n Nucleus and electron configuration:

$1s^2\ 2s^2\ 2p^6\ 3s^1$

(D) 23 p 23 n Nucleus and electron configuration:

$1s^2\ 2s^2\ 2p^6\ 3s^2\ 3p^6\ 3d^3\ 4s^2$

(E) 11 p 12 n Nucleus and electron configuration:

$1s^2\ 2s^2\ 2p^6\ 3s^1$

GO ON TO THE NEXT PAGE.

27. Which of the following statements is true?

(A) A catalyst cannot lower the activation energy.
(B) A catalyst can lower the activation energy.
(C) A catalyst affects only the activation energy of the forward reaction.
(D) A catalyst affects only the activation energy of the reverse reaction.
(E) A catalyst is permanently changed after the activation energy is reached.

28. If the molecular mass of NH_3 is 17, what is the density of this compound at STP?

(A) 0.25 g/L
(B) 0.76 g/L
(C) 1.52 g/L
(D) 3.04 g/L
(E) 9.11 g/L

29. Which bond(s) is (are) ionic?

I. H—Cl(g)
II. S—Cl(g)
III. Cs—F(s)

(A) I only
(B) III only
(C) I and II only
(D) II and III only
(E) I, II, and III

30. Aromatic hydrocarbons are represented by which of the following?

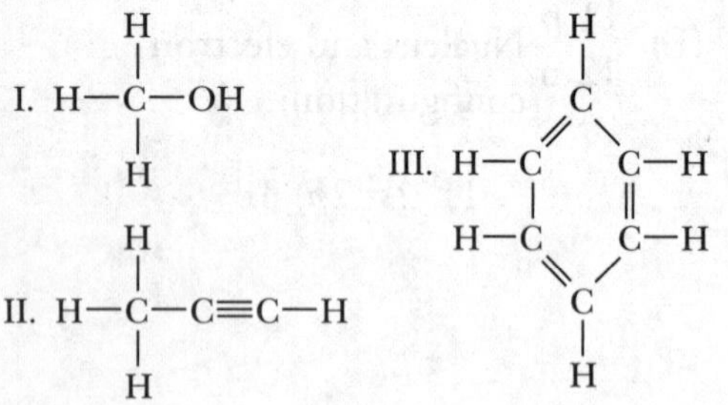

(A) I only
(B) III only
(C) I and II only
(D) II and III only
(E) I, II, and III

31. According to placement in the Periodic Table, which statement(s) regarding the first ionization energies of certain elements should be true?

I. Li has a higher value than Na.
II. K has a higher value than Cs.
III. Na has a higher value than Al.

(A) I only
(B) III only
(C) I and II only
(D) II and III only
(E) I, II, and III

32. Correctly expressed half-reactions include which of the following?

I. $CrO_4^{2-} + 8H^+ + 6e^- \rightarrow Cr^{3+} + 4H_2O$
II. $I^- + 6OH^- \rightarrow IO_3^- + 3H_2O + 6e^-$
III. $MnO_4^- + 2H_2O + 3e^- \rightarrow MnO_2 + 4OH^-$

(A) I only
(B) III only
(C) I and II only
(D) II and III only
(E) I, II, and III

Questions 33–35. What is the apparent oxidation number of the underlined element in the compound

33. Na<u>N</u>O_3?

(A) +1
(B) +2
(C) +3
(D) +4
(E) +5

34. <u>Ca</u>SO_4?

(A) +1
(B) −1
(C) +2
(D) −2
(E) +3

35. $\underline{N}H_3$?

(A) +2
(B) −2
(C) +3
(D) −3
(E) +5

36. An atom with an electron configuration of $1s^2\ 2s^2\ 2p^6\ 3s^2\ 3p^4$ will probably exhibit which oxidation state?

(A) +2
(B) −2
(C) +3
(D) −3
(E) +5

37. In the Lewis dot structure X:, what is the predictable oxidation number?

(A) +1
(B) −1
(C) +2
(D) −2
(E) +3

Questions 38–40 refer to the following pieces of lab equipment assembled by a student:

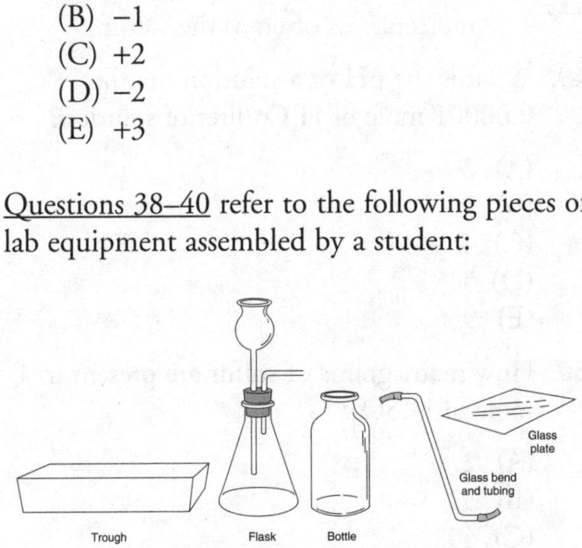

38. When the pieces are assembled properly, they can be used to prepare a gas by a reaction that takes place when a solid

(A) is heated
(B) is exposed
(C) reacts with a liquid
(D) is heated in a vacuum
(E) is decomposed by a catalyst

39. This apparatus suggests that the student plans to collect a gas that

(A) supports combustion
(B) is heavier than air
(C) is not flammable
(D) is insoluble in water
(E) reacts with water

40. If the student planned to prepare hydrogen, what would also be needed?

(A) mercuric oxide
(B) acid plus zinc
(C) potassium chlorate
(D) carbon disulfide
(E) benzene

41. If you collected hydrogen gas by the displacement of water and under the conditions shown:

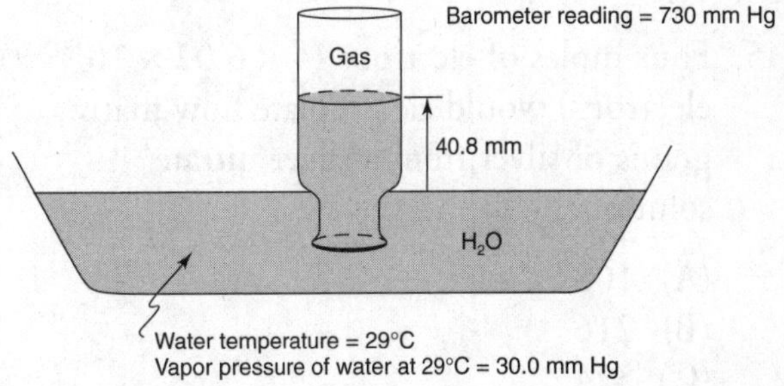

which of the following would give you the pressure of the hydrogen in the bottle?

(A) 730 mm − 40.8 mm
(B) 730 mm − 30.0 mm
(C) 730 mm − 30 mm/13.6 + 40.8 mm
(D) 730 mm − 30 mm/13.6 − 40.8 mm
(E) 730 mm − 40.8 mm/13.6 − 30.0 mm

42. What occurs when a reaction is at equilibrium and more reactant is added to the container?

(A) The equilibrium remains unchanged.
(B) The forward reaction rate increases.
(C) The reverse reaction rate increases.
(D) The forward reaction rate decreases.
(E) The reverse reaction rate decreases.

GO ON TO THE NEXT PAGE.

43. How much heat energy is released when 8 grams of hydrogen are burned? The thermal equation is $2H_2(g) + O_2(g) \rightarrow 2H_2O(g) +$ 483.6 kJ.

(A) 241.8 kJ
(B) 483.6 kJ
(C) 967.2 kJ
(D) 1,934 kJ
(E) 3,869 kJ

44. Would a spontaneous reaction occur between zinc *ions* and gold *atoms*?

$Zn^{2+} + 2e^- \rightleftharpoons Zn^0 \quad E^0 = -0.76$ volt
$Au^{3+} + 3e^- \rightleftharpoons Au^0 \quad E^0 = +1.42$ volts

(A) yes—Reaction potential 2.18 V
(B) no—Reaction potential −2.18 V
(C) yes—Reaction potential 0.66 V
(D) no—Reaction potential −0.66 V
(E) yes—Reaction potential 0.56 V

45. Four moles of electrons ($4 \times 6.02 \times 10^{23}$ electrons) would electroplate how many grams of silver from a silver nitrate solution?

(A) 108
(B) 216
(C) 324
(D) 432
(E) 540

46. A 5 M solution of HCl has how many moles of H^+ ion in 1 liter?

(A) 0.5
(B) 1.0
(C) 2.0
(D) 2.5
(E) 5.0

47. What is the K_{sp} for silver acetate if a saturated solution contains 2×10^{-3} moles of silver ion/liter of solution?

(A) 2×10^{-3}
(B) 2×10^{-6}
(C) 4×10^{-3}
(D) 4×10^{-6}
(E) 4×10^{6}

48. The following data were obtained for H_2O and H_2S:

	Formula Mass	*Freezing Point (°C)*	*Boiling Point (°C)*
H_2O	18	0	100
H_2S	34	−83	−60

What is the best explanation for the variation of physical properties between these two compounds?

(A) The H_2S has stronger bonds between molecules.
(B) The H_2O has a great deal of hydrogen bonding.
(C) The bond angles differ by about 15°.
(D) The formula mass is of prime importance.
(E) The oxygen atom has a smaller radius and thus cannot bump into other molecules as often as the sulfur.

49. What is the pH of a solution that has 0.00001 mole of H_3O^+/liter of solution?

(A) 2
(B) 3
(C) 4
(D) 5
(E) 9

50. How many grams of sulfur are present in 1 mole of H_2SO_4?

(A) 2
(B) 32
(C) 49
(D) 64
(E) 98

51. What is the approximate mass, in grams, of 1 liter of nitrous oxide, N_2O, at STP?

(A) 1
(B) 2
(C) 11.2
(D) 22
(E) 44

52. If the simplest formula of a substance is CH_2 and its molecular mass is 56, what is its true formula?

(A) CH_2
(B) C_2H_4
(C) C_3H_4
(D) C_4H_8
(E) C_5H_{10}

Questions 53 and 54 refer to the following diagrams of two methods of collecting gases:

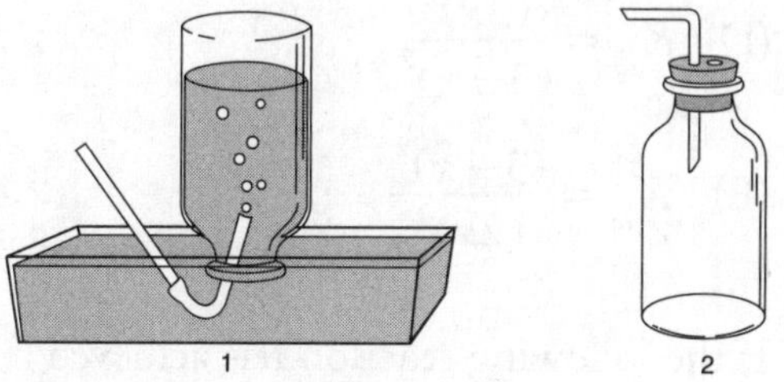

53. Method 1 is best suited to collect

(A) a gas denser than air
(B) a gas less dense than air
(C) a gas that is insoluble in water
(D) a gas that is soluble in water
(E) a gas that has a distinct color

54. Which of these gases, because of its density and solubility, should be collected by Method 2?

(A) NH_3
(B) H_2
(C) HCl
(D) CO_2
(E) He

55. What is the molar mass of $CaCO_3$?

(A) 68 g/mol
(B) 75 g/mol
(C) 82 g/mol
(D) 100 g/mol
(E) 116 g/mol

56. What volume, in liters, will be occupied at STP by 4 grams of H_2?

(A) 11.2
(B) 22.4
(C) 33.6
(D) 44.8
(E) 56.0

57. How many moles of KOH are needed to neutralize 196 grams of sulfuric acid? (H_2SO_4 = 98 amu)

(A) 1.0
(B) 1.5
(C) 2.0
(D) 4.0
(E) 6.0

58. What volume, in liters, of NH_3(g) is produced when 22.4 liters of N_2(g) are made to combine completely with a sufficient quantity of H_2(g) under appropriate conditions?

(A) 11.2
(B) 22.4
(C) 44.8
(D) 67.2
(E) 89.6

59. What volume, in liters, of SO_2 will result from the complete burning of 64 grams of sulfur?

(A) 2
(B) 11.2
(C) 44.8
(D) 126
(E) 158

Practice Test 4

60. The amount of energy required to melt 5 grams of ice at 0°C would also heat 1 gram of water at 4°C to what condition? (Heat of fusion = 80 cal/g or 3.34×10^2 J/g; heat of vaporization = 540 cal/g or 2.26×10^3 J/g)

(A) water at 90°C
(B) water at 100°C
(C) steam at 100°C
(D) Part of the water would be vaporized to steam.
(E) All of the water would be vaporized to steam.

61. How many moles of electrons are needed to electroplate a deposit of 0.5 mole of silver from a silver nitrate solution?

(A) 0.5
(B) 1
(C) 27
(D) 54
(E) 108

62. All of the following statements about carbon dioxide are true EXCEPT:

(A) It can be prepared by the action of acid on $CaCO_3$.
(B) It is used in fire extinguishers.
(C) It dissolves slightly in water at room temperature.
(D) It sublimes rather than melts at 20°C and 1 atm pressure.
(E) It is a product of photosynthesis in plants.

63. Three moles of H_2 and 3 moles of I_2 are introduced into a liter box at a temperature of 490°C. What will the K_{eq} expression be for this reaction? ($K_{eq} = 45.9$)

(A) $K_{eq} = \dfrac{[H_2][I_2]}{[HI]}$

(B) $K_{eq} = \dfrac{[HI]}{[H_2][I_2]}$

(C) $K_{eq} = \dfrac{2x}{(x)(x)}$

(D) $K_{eq} = \dfrac{(2x)^2}{(3-x)^2}$

(E) $K_{eq} = \dfrac{(3-x)^2}{(2x)^2}$

64. If the following reaction has achieved equilibrium in a closed system:

$$N_2O_4(g) \rightleftharpoons 2\ NO_2(g)$$

which of the following is (are) increased by decreasing the size of the container?

I. The value of K_{eq}
II. The concentration of $N_2O_4(g)$
III. The rate of the reverse reaction

(A) I only
(B) III only
(C) I and II only
(D) II and III only
(E) I, II, and III

65. Which of the following correctly completes this nuclear reaction: ${}^{14}_{7}N + {}^{4}_{2}He \rightarrow \cdots + {}^{1}_{1}H$?

(A) ${}^{17}_{8}O$
(B) ${}^{16}_{9}O$
(C) ${}^{17}_{8}N$
(D) ${}^{17}_{7}N$
(E) ${}^{16}_{8}O$

66. How many grams of NaCl will be needed to make 100 milliliters of 2 M solution?

(A) 5.85
(B) 11.7
(C) 29.2
(D) 58.5
(E) 117

67. How many grams of H_2SO_4 are in 1,000 grams of a 10% solution? (1 mol of H_2SO_4 = 98 g)

(A) 1.0
(B) 9.8
(C) 10
(D) 98
(E) 100

68. If 1 mole of ethyl alcohol in 1,000 grams of water depresses the freezing point by 1.86° Celsius, what will be the freezing point of a solution of 1 mole of ethyl alcohol in 500 grams of water?

(A) −0.93°C
(B) −1.86°C
(C) −2.79°C
(D) −3.72°C
(E) −5.58°C

69. Which nuclear reaction shows the release of a beta particle?

(A) ${}^{235}_{92}U + {}^{1}_{0}n \rightarrow {}^{93}_{36}Kr + {}^{140}_{56}Ba + 3\,{}^{1}_{0}n$
(B) ${}^{210}_{84}Po \rightarrow {}^{206}_{82}Pb + {}^{4}_{2}He$
(C) ${}^{14}_{6}C \rightarrow {}^{14}_{7}N + {}^{0}_{-1}e$
(D) ${}^{106}_{47}Ag + {}^{0}_{-1}e \rightarrow {}^{106}_{46}Pd$
(E) ${}^{38}_{19}K \rightarrow {}^{38}_{18}Ar + {}^{0}_{+1}e$

STOP

If you finish before one hour is up, you may go back to check your work or complete unanswered questions.

Answers and Explanations for Test 4

1. **(A)** The volume of a gas increases as temperature increases provided that pressure remains constant. This is a direct proportion. Heating a balloon is a good example.

2. **(C)** The volume of a gas decreases as the pressure is increased provided that the temperature is held constant. This is shown by the inversely proportional curve in (C). Pressure increase on a closed cylinder is a good example.

3. **(E)** The graph shows that there is a starting quantity in solution, and a slight positive slope to the right indicates a directly proportional change in the solubility as temperature rises.

4. **(A)** This is the definition of any molecule.

5. **(B)** A commercial cake mix is a mixture of ingredients.

6. **(C)** This is the definition of an isotope.

7. **(E)** An acid salt contains one or more H atoms in the salt formula separating a positive ion and the hydrogen-bearing negative ion. For example, Na_2SO_4 is a *normal* salt and $NaHSO_4$ is an *acid* salt because of the presence of H in the hydrogen sulfate ion.

8. **(C)** An atom with atomic number 9 would have a 2,7 electron configuration, which matches the outer energy level of iodine.

9. **(B)** Three $(OH)^-$ ions each have 2 atoms = 6 atoms plus one A1 = 7.

10. **(A)** pH from 0 to 6 is acid, 7 neutral, 8 to 14 basic. Most acid is 1.

11. **(E)** Molarity is defined as moles of solute/liter of solution.

12. **(A)** Gas densities can be expressed in grams/liter.

13. **(C)** Molar mass is not affected by pressure and temperature.

14. **(A)** If the density of a gas is known, the mass of 1 L can be multiplied by 22.4 to find the molecular mass because 1 mol occupies 22.4 L at STP.

15. **(A)** Buffers resist changes in pH.

16. **(B)** Color change is the function of indicators.

17. **(C)** On the pH scale, from 0 to 6 is acid and 7 is neutral.

18. **(D)** On the pH scale, from 8 to 14 is basic.

19. **(A)** $Zn(s) + 2HCl(aq) \rightarrow ZnCl_2(aq) + H_2(g)$ is the reaction that occurs.

20. **(C)** Only CO_2, with a molecular mass of 44, is heavier than air, for which the assigned molecular mass is 29.

21. **(E)** $2KClO_3 \rightarrow 2KCl + 3O_2(g)$ is the reaction that occurs.

22. **(C)** CO_2 is slightly soluble in water, forming carbonic acid, H_2CO_3, which is a weak acid.

23. **(B)** NH_3 is very soluble in water, and forms a solution of the weak ammonium hydroxide base.

101. **(T, F)** The assertion is true, but the degree of motion of gas molecules is directly related to the temperature.

102. **(T, T, CE)** Assertion and reason are true; an electron can be treated as either an electromagnetic wave or a bundle of negative charge.

103. **(T, T, CE)** A homologous series increases each member by a constant number of carbons and hydrogens. Examples are the alkane, alkene, and alkyne series, which each increase the chain by a CH_2 group. The reason is true and does explain the assertion.

104. **(T, F)** The nuclear charge of an active metallic ion is greater than that of the electron cloud. The reason is false.

105. **(T, T, CE)** A negative heat of formation indicates that the reaction is exothermic and the enthalpy change is negative.

106. **(T, F)** Water is a polar molecule because there is unequal, not equal, sharing of bonding electrons.

107. **(T, T, CE)** This is a function of a catalyst—to speed up a reaction without permanently changing itself. Assertion and reason are true.

108. **(F, T)** The Cu is losing electrons and thus being oxidized; the assertion is false. It is furnishing electrons and thus is a reducing agent; the reason is true.

109. **(F, T)** $H_2 = 2$, He = 4 (molecular mass); then inversely $\sqrt{4} : \sqrt{2} = 2 : \sqrt{2}$ is the rate of diffusion of hydrogen to helium. The assertion is false; the reason, true.

110. **(T, T, CE)** Since the gas is being heated at constant pressure, it expands. The temperatures are converted to kelvins (K) by adding 273° to the Celsius readings. The fraction must be $\frac{373}{273}$ and this will increase the volume.

111. **(F, T)** Gibbs free energy is useful in indicating the conditions under which a chemical reaction will occur. It is not related to solubility. It is true

that, generally speaking, solubility of a solute increases with an increase in the temperature of the solvent.

112. **(F, F)** Water is $\frac{1}{9}$ hydrogen and $\frac{8}{9}$ oxygen by weight. Both assertion and reason are false.

113. **(T, T, CE)** Four grams of ice would require 4×80 cal/g = 320 cal or $4 \times 3.34 \times 10^2$ J/g = 1.34×10^3 J to melt the ice.

114. **(F, T)** The reaction is: $2H_2 + O_2 \rightarrow 2H_2O$. The coefficients of this gaseous reaction show that 2 liters of hydrogen react with 1 liter of oxygen. This would leave 1 liter of unreacted oxygen. The limiting factor is the hydrogen.

115. **(T, F)** The reason why water is a good solvent is false.

116. **(T, T, CE)** The density of a gas at STP is found by dividing the molecular mass by 22.4 L. NH_3 has a gram-molecular mass of N = 14 + 3H = 3 or a total of 17 g. The gram-molecular mass of Ar is 40 g. The density of each can be found by dividing by 22.4 L, but obviously the density of the ammonia will be smaller.

24. **(C)** The energy necessary to get the reaction started, which is the activation energy, is shown at *C*.

25. **(E)** $2KClO_3 \rightarrow 2KCl + 3O_2(g)$ shows that 2 mol of $KClO_3$ yield 3 mol of O_2.

$$3 \text{ mol} \times \frac{22.4 \text{ L}}{1 \text{ mol}} = 67.2 \text{ L}$$

26. **(E)** The atomic number gives the number of protons in the nucleus and the total number of electrons. The mass number indicates the total number of protons and neutrons in the nucleus—for Na, 23 (11 protons + 12 neutrons).

27. **(B)** A catalyst can speed up a reaction by lowering the activation energy needed to start the reaction and then keep it going.

28. **(B)** Density = $\frac{\text{Mass}}{\text{Volume}}$. For gases this is expressed as grams per liter. Since 1 gram-molecular mass of a gas occupies 22.4 L, 17 g/22.4 L = 0.76 g/L.

29. **(B)** Choice III is made up of elements from extreme sides of the Periodic Table and will therefore form ionic bonds.

30. **(B)** Only III is a ring hydrocarbon of the aromatic series.

31. **(C)** Since Li is higher in Group 1 than Na, and K is higher than Cs, they have smaller radii and hence higher ionization energies. Al is to the right of Na and therefore has a higher ionization energy.

32. **(D)** Only II and III are correctly balanced. To be correct, I should have only $3e^-$. Equations must be balanced for both numbers of atoms and charge.

33. **(E)** These answers are based on the fact that
34. **(C)** the total of the assigned oxidation numbers times their occurrence for all the atoms in a
35. **(D)** compound is zero.

36. **(B)** This orbital configuration shows 6 electrons in the third energy level. The atom would like to gain $2e^-$ to fill the $3p$ and thereby gain a −2 oxidation number.

37. **(C)** With this structure, the atom would tend to lose these electrons and get a +2 charge.

38. **(C)** Assembled, the apparatus would look like this:

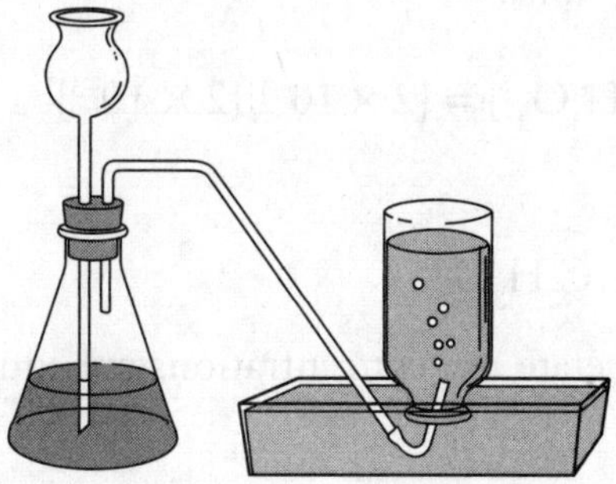

and could be used to prepare a gas by reacting a solid with a liquid.

39. **(D)** The setup depends on the property of insolubility of the gas collected over water.

40. **(B)** An acid plus zinc would also be needed to prepare hydrogen.

41. **(E)** The pressure in the bottle would be less than atmospheric pressure by the Hg equivalent height of the 30 mm of water above the level in the collecting pan. This is calculated as 40.8 mm water/(13.6 mm water/1 mm Hg) and must be subtracted from atmospheric pressure. The other adjustment is to subtract the vapor pressure of water that is in the hydrogen gas since it was collected over water. This pressure is given as 30.0 mm Hg. Subtracting each of these from 730 mm Hg, the given atmospheric pressure, you have 730 mm − 40.8 mm/13.6 − 30.0 mm.

42. **(B)** The equilibrium shifts in the direction that tends to relieve the stress and thus regain equilibrium.

43. **(C)** The thermal reaction shows 2 mol of hydrogen reacting, or 4 g. Therefore, 8 g would release twice the amount of energy:

$$2 \times 483.6 \text{ kJ} = 967.2 \text{ kJ}.$$

44. **(B)** The reaction potential calculation would be:

$$3(Zn^{2+} + 2e^- \rightarrow Zn^0) \qquad E^0 = -0.76\ V$$

$$2(Au^0 \rightarrow Au^{3+} + 3e^-) \qquad E^0 = -1.42\ V$$

$$3\ Zn^{2+} + 2Au^0 \rightarrow 3Zn^0 + 2Au^{3+} \qquad E^0 = -2.18\ V$$

Note: The standard potentials are *not* multiplied by the coefficients in calculating the E^0 for the reaction.

45. **(D)** Since $Ag^+ + 1e^- \rightarrow Ag^0$, 1 mol of electrons yields 1 mol of silver; 1 mol silver $= 6.02 \times 10^{23}$ atoms 4×108 g/mol = 432 g

46. **(E)** $5\ M = \frac{5\ mol}{L}$, and since HCl ionizes completely there would be 5 mol of H^+ and 5 mol of Cl^- ions.

47. **(D)** $K_{sp} = [Ag^+][C_2H_3O_2^-] = [2 \times 10^{-3}][2 \times 10^{-3}]$

(Since

$AgC_2H_3O_2 \rightleftharpoons Ag^+ + C_2H_3O_2^-$

the silver ion and acetate ion concentrations are equal.)

$K_{sp} = 4 \times 10^{-6}$

48. **(B)** It is the explanation for the observed high boiling point and high freezing point of water compared with hydrogen sulfide.

49. **(D)** $pH = -\log [H^+]$

$= -\log [10^{-5}] = -[-5] = 5$

50. **(B)** 1 mol H_2SO_4 contains 1 molar mass of sulfur, that is, 32 g.

51. **(B)** $N_2O = 44$ g/mol

$(2 \times 14 + 16 = 44)$

1 mol of a gas occupies 22.4 L, so 44 g/22.4 L = 1.99 g/L.

52. **(D)** $CH_2 = 14$

$(12 + 2 = 14$ molecular mass)

$56 \div 14 = 4$

Then $4 \times CH_2 = C_4H_8$

53. **(C)** Only insoluble gases can be collected in this way.

54. **(C)** HCl is very soluble in water and denser than air, so it is suited to the No. 2 collection method.

55. **(D)** Ca = 40
C = 12
3O = 48

100 g/mol

56. **(D)** Gram-molecular mass of H_2 is 2 g. 4 g is 2 mol, and each mole occupies 22.4 L. 2 × 22.4 = 44.8 L.

57. **(D)**
$$\overset{x \text{ mol}}{2KOH} + \overset{196\text{ g}}{H_2SO_4} \rightarrow K_2SO_4 + 2H_2O$$
$$\text{2 mol} \quad \text{98 g}$$

$$\frac{x \text{ mol}}{2 \text{ mol}} = \frac{196 \text{ g}}{98 \text{ g}}$$

Then x = 4.0 mol.

58. **(C)**
$$\underset{1\text{ L}}{\overset{22.4\text{ L}}{N_2(g)}} + 3H_2(g) \rightarrow \underset{2\text{ L}}{\overset{x\text{ L}}{2NH_3(g)}}$$

$$\frac{22.4 \text{ L}}{1 \text{ L}} = \frac{x \text{ L}}{2 \text{ L}}.$$

Then x = 44.8 L.

59. **(C)**
$$\underset{32\text{ g}}{\overset{64\text{ g}}{S}} + O_2 \rightarrow \underset{22.4\text{ L}}{\overset{x\text{ L}}{SO_2}}$$

$$\frac{64 \text{ g}}{32 \text{ g}} = \frac{x \text{ L}}{22.4 \text{ L}}$$ or

$$64 \text{ g} \times \frac{1 \text{ mol S}}{32 \text{ g}} \times \frac{1 \text{ mol } SO_2}{1 \text{ mol S}} = 2 \text{ mol } SO_2$$

$$2 \text{ mol } SO_2 \times \frac{22.4 \text{ L } SO_2}{1 \text{ mol } SO_2} = 44.8 \text{ L } SO_2$$

Then x = 44.8 L.

60. **(D)** 5 g ice to water = 5 × 80 cal = 400 cal. 1 g at 4° can go to 100°C as water and absorb 1 cal/°C. Then 400 cal – (100° – 4°) = 400 – 96 = 304 cal. 304 cal can change $\frac{304\text{ cal}}{540\text{ cal/g}}$ or 0.56 g of water to steam. There obviously is not enough heat to vaporize all the water. If done in joules, 5 g of ice to water = $5 \times 3.34 \times 10^2$ g/J = 1.67×10^3 J.

1 g at 4°C to 100°C = Δtemp =96°

$96° \times 4.18$ J/g/°C = 17.3 J/g = 1.73×10^1J = 0.0173×10^3 J

1.67×10^3J – 0.0173×10^3 J = 1.65×10^3 J left for vaporization. Since 1 g requires 2.26×10^3 J for vaporization, (D) is the answer.

61. **(A)** Since Ag^+ gains $1e^-$ to become Ag^0, 0.5 mol requires 0.5 mol of electrons.

62. **(E)** CO_2 is a reactant in photosynthesis, not a product. The reaction is

$$6CO_2 + 6H_2O \rightarrow C_6H_{12}O_6 + 6O_2(g)$$
simple sugar

or

$$6CO_2 + 5H_2O \rightarrow C_6H_{12}O_5 + 6O_2(g)$$
cellulose

63. **(D)** $K_{eq} = \frac{[HI]^2}{[H_2][I_2]}$

Let x = moles of H_2 and also of I_2 that combine to form HI.

Then at equilibrium $[H_2] = 3 - x$, $[I_2] = 3 - x$, $[HI] = 2x$.

$$\text{Then } K_{eq} = \frac{(2x)^2}{(3-x)(3-x)} \text{ or } = \frac{(2x)^2}{(3-x)^2}$$

64. **(D)** In a closed system, decreasing the size of the container will cause the pressure to increase. When pressure is applied to an equilibrium involving gases, the reaction that lowers the pressure by decreasing the number of molecules will increase in rate. In this reaction, the rate of the reverse reaction, in which 2 molecules are decreased to 1, increases, thus reducing pressure while also increasing the concentration of N_2O_4. Thus, II and III are true.

65. **(A)** This is Rutherford's famous artificial transmutation experiment, done in 1919.

66. **(B)** $2\text{ M} = \dfrac{2\text{ mol}}{1{,}000\text{ mL}}$

$$2\text{ mol of NaCl} = 2 \times 58.5\text{ g} = 117.0\text{ g}$$

$$2\text{ M} = 117\text{ g}/1{,}000\text{ mL},$$

$$\text{so } \frac{117\text{ g}}{1{,}000\text{ mL}} = \frac{x\text{ g}}{100\text{ mL}},$$

$$x = 11.7\text{ g}$$

67. **(E)** Percent is by mass, so 10% is $0.1 \times 1{,}000$ g or 100 g.

68. **(D)** First find the molality. 1 mol in 500 g = 2 mol in 1,000 g. Then $2 \times 1.86°\text{C} = 3.72°\text{C}$ drop from 0°C or −3.72°C, the new freezing point.

69. **(C)** The nuclear reactions shown release: (A) a neutron, (B) an alpha particle, (C) a beta particle, (D) no particles, (E) a positron.

CALCULATING YOUR SCORE

Your score on Practice Test 4 can now be computed manually. The actual test will be scored by machine, but the same method is used to arrive at the raw score. You get one point for each correct answer. For each wrong answer, you lose one-fourth of a point. Questions that you omit or that have more than one answer are not counted. On your answer sheet mark all correct answers with a "C" and all incorrect answers with an "X".

Determining Your Raw Test Score

Total the number of correct answers you have recorded on your answer sheet. It should be the same as the total of all the numbers you place in the block in the lower left corner of each area of the Subject Area summary in the next section.

A. Enter the total number of correct answers here: ________
   Now count the number of wrong answers you recorded on your answer sheet.
B. Enter the total number of wrong answers here: ________
   Multiply the number of wrong answers in B by 0.25.
C. Enter that product here: ________
   Subtract the result in C from the total number of right answers in A.
D. Enter the result of your subtraction here: ________
E. Round the result in D to the nearest whole number: ________.
   This is your raw test score.

Conversion of Raw Scores to Scaled Scores

Your raw score is converted by the College Board into a scaled score. The College Board scores range from 200 to 800. This conversion is done to ensure that a score earned on any edition of a particular SAT Subject Test in Chemistry is comparable to the same scaled score earned on any other edition of the same test. Because some editions of the tests may be slightly easier or more difficult than others, scaled scores are adjusted so that they indicate the same level of performance regardless of the edition of the test taken and the ability of the group that takes it. Consequently, a specific raw score on one edition of a particular test will not necessarily translate to the same scaled score on another edition of the same test.

Because the practice tests in this book have no large population of scores with which they can be scaled, scaled scores cannot be determined.

Results from previous SAT Chemistry tests appear to indicate that the conversion of raw scores to scaled scores GENERALLY follows this pattern:

Raw Score	Scaled Score	Raw Score	Scaled Score
85–75	800–800	30–25	560–540
75–70	800–780	25–20	540–520
70–65	780–750	20–15	520–490
65–60	750–720	15–10	490–460
60–55	720–700	10–5	460–430
55–50	700–680	5–0	430–410
50–45	680–650	0 to –5	410–380
45–40	650–620	–5 to –10	380–360
40–35	620–590	–10 to –15	360–330
35–30	590–560		

Note that this scale provides only a *general idea* of what a raw score may translate into on a scaled score range of 800–200. Scaling on every test is usually slightly different. Some students who had taken the SAT Subject Test in Chemistry after using this book had reported that they have scored slightly higher on the SAT test than on the practice tests in this book. They *all* reported that preparing well for the test paid off in a better score!

DIAGNOSING YOUR NEEDS

After taking Practice Test 4, check your answers against the correct ones. Then fill in the chart below.

In the space under each question number, place a check if you answered that question correctly.

EXAMPLE:

If your answer to question 5 was correct, place a check in the appropriate box.

Next, total the check marks for each section and insert the number in the designated block. Now do the arithmetic indicated, and insert your percent for each area.

Subject Area*	(✓) Questions Answered Correctly
I. Atomic Theory and Structure, including periodic relationships ☐ No. of checks ÷ 8 × 100 = _____%	6, 8, 102, 107, 26, 31, 36, 37
II. Nucleonics ☐ No. of checks ÷ 2 × 100 = _____%	65, 69
III. Chemical Bonding and Molecular Structure ☐ No. of checks ÷ 9 × 100 = _____%	4, 9, 106, 29, 33, 34, 35, 46, 48
IV. States of Matter and Kinetic Molecular Theory of Gases ☐ No. of checks ÷ 7 × 100 = _____%	1, 2, 14, 101, 109, 110, 28
V. Solutions, including concentration units, solubility, and colligative properties ☐ No. of checks ÷ 5 × 100 = _____%	3, 11, 66, 67, 68
VI. Acids and Bases ☐ No. of checks ÷ 6 × 100 = _____%	7, 10, 15, 16, 17, 18
VII. Oxidation-Reduction and Electrochemistry ☐ No. of checks ÷ 6 × 100 = _____%	108, 112, 32, 44, 45, 61
VIII. Stoichiometry ☐ No. of checks ÷ 12 × 100 = _____%	113, 114, 116, 25, 50, 51, 52, 55, 56, 57, 58, 59
IX. Reaction Rates ☐ No. of checks ÷ 2 × 100 = _____%	107, 27

**The subject areas have been expanded to identify specific areas in the text.*

Subject Area*	(✓) Questions Answered Correctly					
X. Equilibrium		42	47	49	63	64
☐ No. of checks ÷ 5 × 100 = _____%						
XI. Thermodynamics: energy changes in chemical reactions, randomness, and criteria for spontaneity		105	111	24	43	60
☐ No. of checks ÷ 5 × 100 = _____%						
XII. Descriptive Chemistry: physical and chemical properties of elements and their familiar compounds; organic chemistry; periodic properties	5	12	13	19	20	21
	22	23	103	115	30	62
☐ No. of checks ÷ 12 × 100 = _____%						
XIII. Laboratory: equipment, procedures, observations, safety, calculations, and interpretation of results	38	39	40	41	53	54
☐ No. of checks ÷ 6 × 100 = _____%						

**The subject areas have been expanded to identify specific areas in the text.*

To develop a study plan, refer to pages 33–35.

APPENDIXES

Modern Periodic

KEY

1 (+1, −1) **H** Hydrogen 1.00794 $1s^{1}$

- Common oxidation states
- Atomic number
- Element symbol
- Element name
- Atomic mass (or mass number of longest-lived isotope)
- Electron configuration

Note: Atomic masses are based on carbon-12 = 12.000...u

s-block — GROUP 1, 2

d-block — Transition Elements — GROUP 3–9

Period	1	2	3	4	5	6	7	8	9
1	1 (+1, −1) **H** Hydrogen 1.00794 $1s^{1}$								
2	3 (+1) **Li** Lithium 6.941 [He]$2s^{1}$	4 (+2) **Be** Beryllium 9.01218 [He]$2s^{2}$							
3	11 (+1) **Na** Sodium 22.98977 [Ne]$3s^{1}$	12 (+2) **Mg** Magnesium 24.305 [Ne]$3s^{2}$							
4	19 (+1) **K** Potassium 39.0983 [Ar]$4s^{1}$	20 (+2) **Ca** Calcium 40.078 [Ar]$4s^{2}$	21 (+3) **Sc** Scandium 44.9559 [Ar]$3d^{1}4s^{2}$	22 (+2, +3, +4) **Ti** Titanium 47.88 [Ar]$3d^{2}4s^{2}$	23 (+2, +3, +4, +5) **V** Vanadium 50.9415 [Ar]$3d^{3}4s^{2}$	24 (+2, +3, +6) **Cr** Chromium 51.996 [Ar]$3d^{5}4s^{1}$	25 (+2, +3, +4, +7) **Mn** Manganese 54.9380 [Ar]$3d^{5}4s^{2}$	26 (+2, +3) **Fe** Iron 55.847 [Ar]$3d^{6}4s^{2}$	27 (+2, +3) **Co** Cobalt 58.9332 [Ar]$3d^{7}4s^{2}$
5	37 (+1) **Rb** Rubidium 85.4678 [Kr]$5s^{1}$	38 (+2) **Sr** Strontium 87.62 [Kr]$5s^{2}$	39 (+3) **Y** Yttrium 88.9059 [Kr]$4d^{1}5s^{2}$	40 (+4) **Zr** Zirconium 91.224 [Kr]$4d^{2}5s^{2}$	41 (+3, +5) **Nb** Niobium 92.9064 [Kr]$4d^{4}5s^{1}$	42 (+3, +6) **Mo** Molybdenum 95.94 [Kr]$4d^{5}5s^{1}$	43 (+4, +6, +7) **Tc** Technetium (98) [Kr]$4d^{5}5s^{2}$	44 (+3) **Ru** Ruthenium 101.07 [Kr]$4d^{7}5s^{1}$	45 (+3) **Rh** Rhodium 102.9055 [Kr]$4d^{8}5s^{1}$
6	55 (+1) **Cs** Cesium 132.9054 [Xe]$6s^{1}$	56 (+2) **Ba** Barium 137.33 [Xe]$6s^{2}$	57 (+3) **La** Lanthanum 138.9055 [Xe]$5d^{1}6s^{2}$	72 (+4) **Hf** Hafnium 178.49 [Xe]$4f^{14}5d^{2}6s^{2}$	73 (+5) **Ta** Tantalum 180.9479 [Xe]$4f^{14}5d^{3}6s^{2}$	74 (+6) **W** Tungsten 183.85 [Xe]$4f^{14}5d^{4}6s^{2}$	75 (+4, +6, +7) **Re** Rhenium 186.207 [Xe]$4f^{14}5d^{5}6s^{2}$	76 (+3, +4) **Os** Osmium 190.2 [Xe]$4f^{14}5d^{6}6s^{2}$	77 (+3, +4) **Ir** Iridium 192.22 [Xe]$4f^{14}5d^{7}6s^{2}$
7	87 (+1) **Fr** Francium (223) [Rn]$7s^{1}$	88 (+2) **Ra** Radium 226.0254 [Rn]$7s^{2}$	89 (+3) **Ac** Actinium 227.0278 [Rn]$6d^{1}7s^{2}$	104 **Rf** Rutherfordium (261) [Rn]$5f^{14}6d^{2}7s^{2}$	105 **Db** Dubnium (262) [Rn]$5f^{14}6d^{3}7s^{2}$	106 **Sg** Seaborgium (263) [Rn]$5f^{14}6d^{4}7s^{2}$	107 **Bh** Bohrium (262) [Rn]$5f^{14}6d^{5}7s^{2}$	108 **Hs** Hassium (265) [Rn]$5f^{14}6d^{6}7s^{2}$	109 **Mt** Meitnerium (266) [Rn]$5f^{14}6d^{7}7s^{2}$

Lanthanide Series	58 (+3, +4) **Ce** Cerium 140.12 [Xe]$4f^{1}5d^{1}6s^{2}$	59 (+3) **Pr** Praseodymium 140.9077 [Xe]$4f^{3}6s^{2}$	60 (+3) **Nd** Neodymium 144.24 [Xe]$4f^{4}6s^{2}$	61 (+3) **Pm** Promethium (145) [Xe]$4f^{5}6s^{2}$	62 (+2, +3) **Sm** Samarium 150.36 [Xe]$4f^{6}6s^{2}$
Actinide Series	90 (+4) **Th** Thorium 232.0381 [Rn]$6d^{2}7s^{2}$	91 (+4, +5) **Pa** Protactinium 231.0359 [Rn]$5f^{2}6d^{1}7s^{2}$	92 (+3, +4, +5, +6) **U** Uranium 238.0289 [Rn]$5f^{3}6d^{1}7s^{2}$	93 (+3, +4, +5, +6) **Np** Neptunium 237.048 [Rn]$5f^{4}6d^{1}7s^{2}$	94 (+3, +4, +5, +6) **Pu** Plutonium (244) [Rn]$5f^{6}7s^{2}$

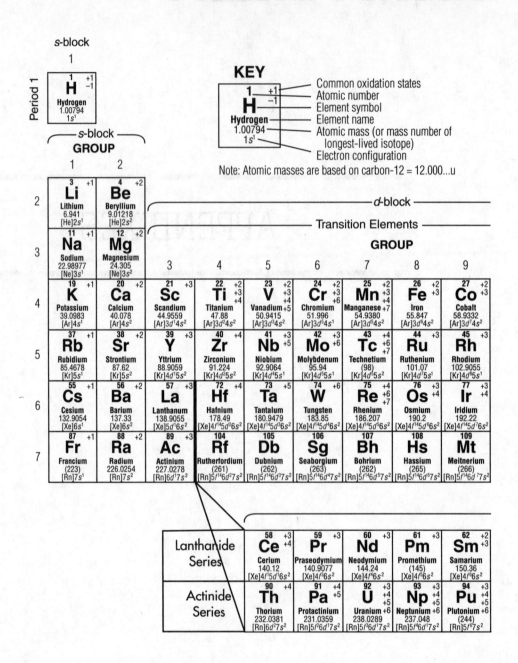

Table

p-block

GROUP

10	11	12	13	14	15	16	17	18
								2 He Helium 4.00260 $1s^2$ 0
			5 B Boron 10.81 [He]$2s^22p^1$ +3	6 C Carbon 12.011 [He]$2s^22p^2$ −4 +2 +4	7 N Nitrogen 14.0067 [He]$2s^22p^3$ −3 −2 −1 +1 +2 +3 +4 +5	8 O Oxygen 15.9994 [He]$2s^22p^4$ −2	9 F Fluorine 18.998403 [He]$2s^22p^5$ −1	10 Ne Neon 20.1797 [He]$2s^22p^6$ 0
			13 Al Aluminum 26.98154 [Ne]$3s^23p^1$ +3	14 Si Silicon 28.0855 [Ne]$3s^23p^2$ −4 +2 +4	15 P Phosphorus 30.97376 [Ne]$3s^23p^3$ −3 +3 +5	16 S Sulfur 32.066 [Ne]$3s^23p^4$ −2 +4 +6	17 Cl Chlorine 35.453 [Ne]$3s^23p^5$ −1 +1 +3 +5 +7	18 Ar Argon 39.948 [Ne]$3s^23p^6$ 0
28 Ni Nickel 58.69 [Ar]$3d^84s^2$ +2 +3	29 Cu Copper 63.546 [Ar]$3d^{10}4s^1$ +1 +2	30 Zn Zinc 65.39 [Ar]$3d^{10}4s^2$ +2	31 Ga Gallium 69.72 [Ar]$3d^{10}4s^24p^1$ +3	32 Ge Germanium 72.61 [Ar]$3d^{10}4s^24p^2$ −4 +2 +4	33 As Arsenic 74.9216 [Ar]$3d^{10}4s^24p^3$ −3 +3 +5	34 Se Selenium 78.96 [Ar]$3d^{10}4s^24p^4$ −2 +4 +6	35 Br Bromine 79.904 [Ar]$3d^{10}4s^24p^5$ −1 +1 +5	36 Kr Krypton 83.80 [Ar]$3d^{10}4s^24p^6$ 0 +2
46 Pd Palladium 106.42 [Kr]$4d^{10}$ +2 +4	47 Ag Silver 107.8682 [Kr]$4d^{10}5s^1$ +1	48 Cd Cadmium 112.41 [Kr]$4d^{10}5s^2$ +2	49 In Indium 114.82 [Kr]$4d^{10}5s^25p^1$ +3	50 Sn Tin 118.710 [Kr]$4d^{10}5s^25p^2$ +2 +4	51 Sb Antimony 121.757 [Kr]$4d^{10}5s^25p^3$ −3 +3 +5	52 Te Tellurium 127.60 [Kr]$4d^{10}5s^25p^4$ −2 +4 +6	53 I Iodine 126.9045 [Kr]$4d^{10}5s^25p^5$ −1 +1 +5 +7	54 Xe Xenon 131.29 [Kr]$4d^{10}5s^25p^6$ 0 +2 +4 +6
78 Pt Platinum 195.08 [Xe]$4f^{14}5d^96s^1$ +2 +4	79 Au Gold 196.9665 [Xe]$4f^{14}5d^{10}6s^1$ +1 +3	80 Hg Mercury 200.59 [Xe]$4f^{14}5d^{10}6s^2$ +1 +2	81 Tl Thallium 204.383 [Xe]$4f^{14}5d^{10}6s^26p^1$ +1 +3	82 Pb Lead 207.2 [Xe]$4f^{14}5d^{10}6s^26p^2$ +2 +4	83 Bi Bismuth 208.9804 [Xe]$4f^{14}5d^{10}6s^26p^3$ +3 +5	84 Po Polonium (209) [Xe]$4f^{14}5d^{10}6s^26p^4$ +2 +4	85 At Astatine (210) [Xe]$4f^{14}5d^{10}6s^26p^5$	86 Rn Radon (222) [Xe]$4f^{14}5d^{10}6s^26p^6$ 0
110 Uun Ununnilium (269) [Rn]$5f^{14}6d^87s^2$	111 Uuu Unununium (272) [Rn]$5f^{14}6d^97s^2$	112 Uub Ununbium (277) [Rn]$5f^{14}6d^{10}7s^2$						

f-block

63 Eu Europium 151.96 [Xe]$4f^76s^2$ +2 +3	64 Gd Gadolinium 157.25 [Xe]$4f^75d^16s^2$ +3	65 Tb Terbium 158.9254 [Xe]$4f^96s^2$ +3	66 Dy Dysprosium 162.50 [Xe]$4f^{10}6s^2$ +3	67 Ho Holmium 164.9304 [Xe]$4f^{11}6s^2$ +3	68 Er Erbium 167.26 [Xe]$4f^{12}6s^2$ +3	69 Tm Thulium 168.9342 [Xe]$4f^{13}6s^2$ +3	70 Yb Ytterbium 173.04 [Xe]$4f^{14}6s^2$ +2 +3	71 Lu Lutetium 174.967 [Xe]$4f^{14}5d^16s^2$ +3
95 Am Americium (243) [Rn]$5f^77s^2$ +3 +4 +5 +6	96 Cm Curium (247) [Rn]$5f^76d^17s^2$ +3	97 Bk Berkelium (247) [Rn]$5f^97s^2$ +3 +4	98 Cf Californium (251) [Rn]$5f^{10}7s^2$ +3	99 Es Einsteinium (252) [Rn]$5f^{11}7s^2$	100 Fm Fermium (257) [Rn]$5f^{12}7s^2$	101 Md Mendelevium (258) [Rn]$5f^{13}7s^2$	102 No Nobelium (259) [Rn]$5f^{14}7s^2$	103 Lr Lawrencium (262) [Rn]$5f^{14}6d^17s^2$

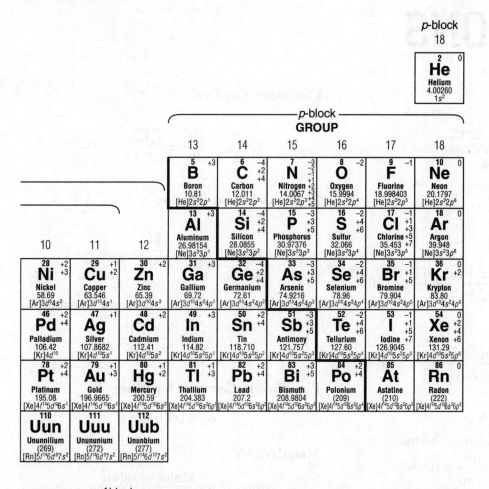

SOME IMPORTANT EQUATIONS

Density (*d*)

$$\text{Density} = \frac{\text{Mass}}{\text{Volume}}$$

$$d = \frac{m}{v}$$

Percent Error

$$\text{Percent error} = \frac{\text{Measured value} - \text{Accepted value}}{\text{Accepted value}} \times 100\%$$

Percent Yield

$$\text{Percent yield} = \frac{\text{Actual yield}}{\text{Expected yield}} \times 100\%$$

Percentage Composition

$$\text{Percentage composition by mass} = \frac{\text{Mass of element}}{\text{Mass of compound}} \times 100\%$$

Boyle's Law

$$P_1V_1 = P_2V_2$$

$$PV = k$$

Charles's Law

$$V_1T_2 = V_2T_1$$

Dalton's Law of Partial Pressures

$$P_T = p_a + p_b + p_c + \ldots$$

Ideal Gas Law

$$PV = nRT$$

Combined Gas Law

$$\frac{P_1V_1}{T_1} = \frac{P_2V_2}{T_2}$$

Graham's Law of Effusion

$$\frac{r_1}{r_2} = \sqrt{\frac{M_2}{M_1}}$$

General Equation for Equilibrium Constant

$$K_{eq} = \frac{[\mathrm{C}]^c[\mathrm{D}]^d}{[\mathrm{A}]^a[\mathrm{B}]^b}$$

Equilibrium Constant for Water

$$K_w = [H^+][OH^-]$$

Molarity (*M*)

$$\text{Molarity} = \frac{\text{Moles of solute}}{\text{Liters of solution}}$$

Titration

$$M_aV_a = M_bV_b$$

Molality (*m*)

$$\text{Molality} = \frac{\text{Moles of solute}}{\text{Kilograms of solvent}}$$

Dilution

$$M_1V_1 = M_2V_2$$

Boiling Point Elevation

$$\Delta T_b = K_b m$$

where K_b is the molal boiling-point-elevation constant

Freezing Point Depression

$$\Delta T_f = K_f m$$

where K_f is the molal boiling-point-elevation constant

Rate of Reaction

$$\text{Rate} = k[\text{A}]^x[\text{B}]^y$$

where [A] and [B] are molar concentrations of reactants, and k is a rate constant

Hess's Law

$$\Delta H_{\text{net}} = \Delta H_1 + \Delta H_2$$

Entropy Change

$$\Delta S = S_{\text{products}} - S_{\text{reactants}}$$

Gibbs Free Energy

$$\Delta G = \Delta H - T\Delta S$$

Some Useful Tables

A

Standard Energies of Formation of Compounds at 1 atm and 298 K

Compound	Heat (Enthalpy) of Formation* ($\Delta H_f°$)		Free Energy of Formation ($\Delta G_f°$)	
	kJ/mol	kcal/mol	kJ/mol	kcal/mol
Aluminum oxide $Al_2O_3(s)$	–1676.5	–400.5	–1583.1	–378.2
Ammonia $NH_3(g)$	–46.0	–11.0	–16.3	–3.9
Barium sulfate $BaSO_4(s)$	–1473.9	–352.1	–1363.0	–325.6
Calcium hydroxide $Ca(OH)_2(s)$	–986.6	–235.7	–899.2	–214.8
Carbon dioxide $CO_2(g)$	–393.9	–94.1	–394.7	–94.3
Carbon monoxide $CO(g)$	–110.5	–26.4	–137.3	–32.8
Copper(II) sulfate $CuSO_4(s)$	–771.9	–184.4	–662.2	–158.2
Ethane $C_2H_6(g)$	–84.6	–20.2	–33.1	–7.9
Ethene (ethylene) $C_2H_4(g)$	52.3	12.5	68.2	16.3
Ethyne (acetylene) $C_2H_2(g)$	226.9	54.2	209.3	50.0
Hydrogen fluoride $HF(g)$	–271.3	–64.8	–273.3	–65.3
Hydrogen iodide $HI(g)$	26.4	6.3	1.7	0.4
Iodine chloride $ICl(g)$	18.0	4.3	–5.4	–1.3
Lead(II) oxide $PbO(s)$	–215.6	–51.5	–188.4	–45.0
Methane $CH_4(g)$	–74.9	–17.9	–50.7	–12.1
Magnesium oxide $MgO(s)$	–601.9	–143.8	–569.7	–136.1
Nitrogen(II) oxide $NO(g)$	90.4	21.6	86.7	20.7
Nitrogen(IV) oxide $NO_2(g)$	33.1	7.9	51.5	12.3
Potassium chloride $KCl(s)$	–437.0	–104.4	–409.4	–97.8
Sodium chloride $NaCl(s)$	–411.5	–98.3	–384.3	–91.8
Sulfur dioxide $SO_2(g)$	–296.8	–70.9	–300.1	–71.7
Water $H_2O(g)$	–242.0	–57.8	–228.6	–54.6
Water $H_2O(\ell)$	–285.9	–68.3	–237.3	–56.7

*Minus sign indicates an exothermic reaction.

Sample equations:

$$2Al(s) + \frac{3}{2}O_2(g) \rightarrow Al_2O_3(s) + 400.5\text{ kcal}$$

$$2Al(s) + \frac{3}{2}O_2(g) \rightarrow Al_2O_3(s) \quad \Delta H = -400.5\text{ kcal/mol}$$

B

Heats of Reaction at 1 atm and 298 K

Reaction	ΔH kJ	ΔH kcal
$CH_4(g) + 2O_2(g) \rightarrow CO_2(g) + 2H_2O(\ell)$	–890.8	–212.8
$C_3H8(g) + 5O_2(g) \rightarrow 3CO_2(g) + 4H_2O(\ell)$	–2221.1	–530.6
$CH_3OH(\ell) + \frac{3}{2}O_2(g) \rightarrow CO_2(g) + 2H_2O(\ell)$	–726.7	–173.6
$C_6H_{12}O_6(s) + 6O_2(g) \rightarrow 6CO_2(g) + 6H_2O(\ell)$	–2804.2	–669.9
$CO(g) + \frac{1}{2}O_2(g) \rightarrow CO_2(g)$	–283.4	–67.7
$C_8H_{18}(\ell) + \frac{25}{2}O_2(g) \rightarrow 8CO_2(g) + 9H_2O(\ell)$	–5453.1	–1302.7
$KNO_3(s) \xrightarrow{H_2O} K^+(aq) + NO_3^-(aq)$	34.7	8.3
$NaOH(s) \xrightarrow{H_2O} Na^+(aq) + OH^-(aq)$	–44.4	–10.6
$NH_4Cl(s) \xrightarrow{H_2O} NH_4^+(aq) + Cl^-(aq)$	14.7	3.5
$NH_4NO_3(s) \xrightarrow{H_2O} NH_4^+(aq) + NO_3^-(aq)$	25.5	6.1
$NaCl(s) \xrightarrow{H_2O} Na^+(aq) + Cl^-(aq)$	3.8	0.9
$KClO_3(s) \xrightarrow{H_2O} K^+(aq) + ClO_3^-(aq)$	41.4	9.9
$LiBr(s) \xrightarrow{H_2O} Li^+(aq) + Br^-(aq)$	–49.0	–11.7
$H^+(aq) + OH^-(aq) \rightarrow H_2O(\ell)$	–57.8	–13.8

C

Symbols Used In Nuclear Chemistry

alpha particle	${}^{4}_{2}He$	α
beta particle (electron)	${}^{0}_{-1}e$	β^-
gamma radiation		γ
neutron	${}^{1}_{0}n$	n
proton	${}^{1}_{1}H$	p
deuteron	${}^{2}_{1}H$	
triton	${}^{3}_{1}H$	
positron	${}^{0}_{+1}e$	β^+

D

Physical Constants and Conversion Factors

Name	Symbol	Value(s)	Units
Avogadro number	N_A	6.02×10^{23} mol	
Charge of electron	e	1.60×10^{-19} C	coulomb
Electron-volt	eV	1.60×10^{-19} J	joule
Speed of light	c	3.00×10^{8} m/s	meters/second
Planck's constant	h	6.63×10^{-34} J · s	joule-second
		1.58×10^{-37} kcal · s	kilocalorie-second
Universal gas constant	R	0.0821 L · atm/mol · K	liter-atmosphere/mole-kelvin
Atomic mass unit	μ	1.66×10^{-24} g	gram
Volume standard, liter	L	1×10^{3} cm³ = 1 dm³	cubic centimeters, cubic decimeter
Standard pressure, atmosphere	atm	101.3 kPa	kilopascals
		760 mm Hg	millimeters of mercury
		760 torr	torr
Heat equivalent, kilocalorie	kcal	4.18×10^{3} J	joules

Physical Constants for H_2O

Specific heat . . . . 4.18 J/g K
Molal freezing point depression . . . . 1.86°C/m
Molal boiling point elevation . . . . 0.52°C/m
Heat of fusion . . . . 79.72 cal/g, 333.6 J/g
Heat of vaporization . . . . 539.4 cal/g, 2,259 J/g

E

Standard (SI) Units

Symbol	Name	Quantity
m	meter	length
kg	kilogram	mass
Pa	pascal	pressure
K	kelvin	thermodynamic temperature
mol	mole	amount of substance
J	joule	energy, work, quantity of heat
s	second	time
C	coulomb	quantity of electricity
V	volt	electric potential, potential difference
L	liter	volume

F

Prefixes Used with SI Units

Prefix	Symbol	Meaning	Scientific Notation
exa-	E	1,000,000,000,000,000,000	10^{18}
peta-	P	1,000,000,000,000,000	10^{15}
tera-	T	1,000,000,000,000	10^{12}
giga-	G	1,000,000,000	10^{9}
mega-	M	1,000,000	10^{6}
kilo-	k	1,000	10^{3}
hecto-	h	100	10^{2}
deka-	da	10	10^{1}
—	—	1	10^{0}
deci-	d	0.1	10^{-1}
centi-	c	0.01	10^{-2}
milli-	m	0.001	10^{-3}
micro-	µ	0.000 001	10^{-6}
nano-	n	0.000 000 001	10^{-9}
pico-	p	0.000 000 000 001	10^{-12}
femto-	f	0.000 000 000 000 001	10^{-15}
atto-	a	0.000 000 000 000 000 001	10^{-18}

The Chemical Elements

(Atomic masses in this table are based on the atomic mass of carbon-12 being exactly 12.)

Name	Symbol	Atomic Number	Atomic Mass	Name	Symbol	Atomic Number	Atomic Mass
Actinium	Ac	89	(227)	Mendelevium	Md	101	(258)
Aluminum	Al	13	26.98	Mercury	Hg	80	200.59
Americium	Am	95	(243)	Molybdenum	Mo	42	95.94
Antimony	Sb	51	121.75	Neilsbohrium	Ns	107	(262)
Argon	Ar	18	39.95	Neodymium	Nd	60	144.24
Arsenic	As	33	74.92	Neon	Ne	10	20.18
Astatine	At	85	(210)	Neptunium	Np	93	237.05
Barium	Ba	56	137.34	Nickel	Ni	28	58.71
Berkelium	Bk	97	(247)	Niobium	Nb	41	92.90
Beryllium	Be	4	9.01	Nitrogen	N	7	14.01
Bismuth	Bi	83	208.98	Nobelium	No	102	(259)
Bohrium	Bh	107	(262)	Osmium	Os	76	190.2
Boron	B	5	10.81	Oxygen	O	8	16.00
Bromine	Br	35	79.90	Palladium	Pd	46	106.4
Cadmium	Cd	48	112.40	Phosphorus	P	15	30.97
Caesium	Cs	55	132.91	Platinum	Pt	78	195.09
Calcium	Ca	20	40.08	Plutonium	Pu	94	(244)
Californium	Cf	98	(251)	Polonium	Po	84	(210)
Carbon	C	6	12.01	Potassium	K	19	39.10
Cerium	Ce	58	140.12	Praseodymium	Pr	59	140.90
Chlorine	Cl	17	35.45	Promethium	Pm	61	(145)
Chromium	Cr	24	52.00	Protactinium	Pa	91	231.04
Cobalt	Co	27	58.93	Radium	Ra	88	(226)
Copper	Cu	29	63.55	Radon	Rn	86	(222)
Curium	Cm	96	(247)	Rhenium	Re	75	186.2
Dabnium		105	(262)	Rhodium	Rh	45	102.91
Dysprosium	Dy	66	162.50	Rubidium	Rb	37	85.47
Einsteinium	Es	99	(254)	Ruthenium	Ru	44	101.07
Erbium	Er	68	167.26	Rutherfordium	Rf	104	(261)
Europium	Eu	63	151.96	Samarium	Sm	62	150.35
Fermium	Fm	100	(257)	Scandium	Sc	21	44.95
Fluorine	F	9	19.00	Seaborgium	Sg	106	(266)
Francium	Fr	87	(223)	Selenium	Se	34	78.96
Gadolinium	Gd	64	157.25	Silicon	Si	14	28.09
Gallium	Ga	31	69.72	Silver	Ag	47	107.89
Germanium	Ge	32	72.59	Sodium	Na	11	22.99
Gold	Au	79	196.97	Strontium	Sr	38	87.62
Hafnium	Hf	72	178.49	Sulfur	S	16	32.06
Hassium	Hs	108	(269)	Tantalum	Ta	73	180.95
Helium	He	2	4.00	Technetium	Tc	43	(99)
Holmium	Ho	67	164.93	Tellurium	Te	52	127.60
Hydrogen	H	1	1.008	Terbium	Tb	65	158.92
Indium	In	49	114.82	Thallium	Tl	81	204.37
Iodine	I	53	126.90	Thorium	Th	90	232.03
Iridium	Ir	77	192.2	Thulium	Tm	69	168.93
Iron	Fe	26	55.85	Tin	Sn	50	118.69
Krypton	Kr	36	83.80	Titanium	Ti	22	47.90
Lanthanum	La	57	138.91	Tungsten	W	74	183.85
Lawrencium	Lr	103	(262)	Uranium	U	92	238.03
Lead	Pb	82	207.19	Vanadium	V	23	50.94
Lithium	Li	3	6.94	Xenon	Xe	54	131.30
Lutetium	Lu	71	174.97	Ytterbium	Yb	70	173.04
Magnesium	Mg	12	24.31	Yttrium	Y	39	88.91
Manganese	Mn	25	54.94	Zinc	Zn	30	65.37
Meitnerium	Mt	109	(268)	Zirconium	Zr	40	91.22

**A number in parentheses is the mass number of the most stable isotope.*

Glossary

(See Index for additional references)

absolute temperature Temperature measured on the absolute scale, which has its origin at absolute zero. *See also* **Kelvin scale.**

absorption The process of taking up by capillary, osmotic, chemical, or solvent action, as a sponge absorbs water.

acid A water solution that has an excess of hydrogen ions; an acid turns litmus paper pink or red, has a sour taste, and neutralizes bases to form salts.

acid anhydride A nonmetallic oxide that, when placed in water, reacts to form an acid solution.

acid salt A salt formed by replacing part of the hydrogen ions of a dibasic or tribasic acid with metallic ions. Examples: $NaHSO_4$, NaH_2PO_4.

actinide series The series of radioactive elements starting with actinium, No. 89, and ending with lawrencium, No. 103.

activated charcoal A specially treated and finely divided form of carbon, which possesses a high degree of adsorption.

activation energy The minimum energy necessary to start a reaction.

adsorption The adhesion (in an extremely thin layer) of the molecules of gases, of dissolved substances, or of liquids to the surfaces of solid or liquid bodies with which they come into contact.

alcohol An organic hydroxyl compound formed by replacing one or more hydrogen atoms of a hydrocarbon with an equal number of hydroxyl (OH) groups.

aldehyde An organic compound formed by dehydrating oxidized alcohol; contains the characteristic —CHO group.

alkali Usually, a strong base, such as sodium hydroxide or potassium hydroxide.

alkaline Referring to any substance that has basic properties.

alkyl A substituent obtained from a saturated hydrocarbon by removing one hydrogen atom. Examples: methyl (CH_3^-), ethyl ($C_2H_5^-$).

allotropic forms Forms of the same element that differ in their crystalline structures.

alloy A substance composed of two or more metals, which are intimately mixed; usually made by melting the metals together.

alpha particles Positively charged helium nuclei.

alum A double sulfate of a monovalent metal and a trivalent metal, such as K_2SO_4 $Al_2(SO_4)_3$ $24H_2O$; also, a common name for commercial aluminum sulfate.

amalgam An alloy of mercury and another metal.

amine A compound such as CH_3NH_2, derived from ammonia by substituting one or more hydrocarbon radicals for hydrogen atoms.

amino acid One of the "building blocks" of proteins; contains one or more NH_2^- groups that have replaced the same number of hydrogen atoms in an organic acid.

amorphous Having no definite crystalline structure.

amphoteric Referring to a hydroxide that may have either acidic or basic properties, depending on the substance with which it reacts.

analysis The breaking down of a compound into two or more simpler substances.

anhydride A compound derived from another compound by the removal of water; it will combine with water to form an acid (acid anhydride) or a base (basic anhydride).

anhydrous Containing no water.

anion An ion or particle that has a negative charge and thus is attracted to a positively charged anode.

anode The electrode in an electrolytic cell that has a positive charge and attracts negative ions.

antichlor A substance used to remove the excess of chlorine in the bleaching process.

aromatic compound A compound whose basic structure contains the benzene ring; it usually has an odor.

atmosphere The layer of gases surrounding the earth; also, a unit of pressure (1 atm = approx. 760 mm of Hg or torr).

atom The smallest particle of an element that retains the properties of that element and can enter into a chemical reaction.

atomic energy *See* **nuclear energy**, a more accurate term.

atomic mass (relative atomic mass or atomic weight) The average mean value of the isotopic masses of the atoms of an element. It indicates the relative mass of the element as compared with that of carbon-12, which is assigned a mass of exactly 12 atomic mass units.

atomic mass unit One twelfth of the mass of a carbon-12 atom; equivalent to 1.660531×10^{-27} kilogram (abbreviation: amu or μ).

atomic number The number that indicates the order of an element in the periodic system; numerically equal to the number of protons in the nucleus of the atom, or the number of negative electrons located outside the nucleus of the atom.

atomic radius One-half the distance between adjacent nuclei in the crystalline or solid phase of an element; the distance from the atomic nucleus to the valence electrons.

atomic weight *See* **atomic mass.**

Aufbau Principle The principle that states that an electron occupies the lowest energy orbital that can receive it.

Avogadro's hypothesis *See under* **laws.**

Avogadro's number The number of molecules in 1 gram-molecular volume of a substance, or the number of atoms in 1 gram-atomic mass of an element; equal to 6.022169×10^{23}. *See also* **mole.**

barometer An instrument, invented by Torricelli in 1643, used for measuring atmospheric pressure.

base A water solution that contains an excess of hydroxide ions; a proton acceptor; a base turns litmus paper blue and neutralizes acids to form salts.

basic anhydride A metallic oxide that forms a base when placed in water.

beta particles High-speed, negatively charged electrons $_{-1}^{0}e$ or $_{-1}^{0}\beta$ emitted in radiation.

binary Referring to a compound composed of two elements, such as H_2O.

boiling point The temperature at which the vapor pressure of a liquid equals the atmospheric pressure.

bond energy The energy needed to break a chemical bond and form a neutral atom.

bonding The union of atoms to form compounds or molecules by filling their outer shells of electrons. This can be done through giving and taking electrons (ionic) or by sharing electrons (covalent).

Boyle's Law *See under* **laws.**

brass An alloy of copper and zinc.

breeder reactor A nuclear reactor in which more fissionable material is produced than is used up during operation.

Brownian movement Continuous zigzagging movement of colloidal particles in a dispersing medium, as viewed through an ultramicroscope.

buffer A substance that, when added to a solution, makes changing the pH of the solution more difficult.

calorie A unit of heat; the amount of heat needed to raise the temperature of 1 gram of water 1 degree on the Celsius scale.

calorimeter An instrument used to measure the amount of heat liberated or absorbed during a change.

carbonated water Water containing dissolved carbon dioxide.

carbon dating The use of radioactive carbon-14 to estimate the ages of ancient materials, such as archeological or paleontological specimens.

catalyst A substance that speeds up or slows down a reaction without being permanently changed itself.

cathode The electrode in an electrolytic cell that is negatively charged and attracts positive ions.

cathode rays Streams of electrons given off by the cathode of a vacuum tube.

cation An ion that has a positive charge.

Celsius scale A temperature scale divided into 100 equal divisions and based on water freezing at 0° and boiling at 100°. Synonymous with centigrade.

chain reaction A reaction produced during nuclear fission when at least one neutron from each fission produces another fission, so that the process becomes self-sustaining without additional external energy.

Charles's Law *See under* **laws.**

chemical change A change that alters the atomic structures of the substances involved and results in different properties.

chemical property A property that determines how a substance will behave in a chemical reaction.

chemistry The science concerned with the compositions of substances and the changes that they undergo.

colligative property A property of a solution that depends primarily on the concentration, not the type, of particles present.

colloids Particles larger than those found in a solution but smaller than those in a suspension.

Combining Volumes *See Gay-Lussac's, under* **laws.**

combustion A chemical action in which both heat and light are given off.

compound A substance composed of elements chemically united in definite proportions by weight.

condensation (a) A change from gaseous to liquid state; (b) the union of like or unlike molecules with the elimination of water, hydrogen chloride, or alcohol.

Conservation of Energy *See under* **laws.**

Conservation of Matter *See under* **laws.**

control rod In a nuclear reactor, a rod of a certain metal such as cadmium, which controls the speed of the chain reaction by absorbing neutrons.

coordinate covalence Covalence in which both electrons in a pair come from the same atom.

covalent bonding Bonding accomplished through the sharing of electrons so that atoms can fill their outer shells.

critical mass The smallest amount of fissionable material that will sustain a chain reaction.

critical temperature The temperature above which no gas can be liquefied, regardless of the pressure applied.

crystalline Having a definite molecular or ionic structure.

crystallization The process of forming definitely shaped crystals when water is evaporated from a solution of the substance.

cyclotron A device used to accelerate charged particles to high energies for bombarding the nuclei of atoms.

Dalton's Law of Partial Pressures *See under* **laws.**

decomposition The breaking down of a compound into simpler substances or into its constituent elements.

Definite Composition *See under* **laws.**

dehydrate To take water from a substance.

dehydrating agent A substance able to withdraw water from another substance, thereby drying it.

deliquescence The absorption by a substance of water from the air, so that the substance becomes wet.

denatured alcohol Ethyl alcohol that has been "poisoned" in order to produce (by avoiding federal tax) a cheaper alcohol for industrial purposes.

density The mass per unit volume of a substance; the mathematical formula is $D = m/V$, where D = density, m = mass, and V = volume.

destructive distillation The process of heating an organic substance, such as coal, in the absence of air to break it down into solid and volatile products.

deuterium An isotope of hydrogen, sometimes called heavy hydrogen, with an atomic weight of 2.

dew point The highest temperature at which water vapor condenses out of the air.

dialysis The process of separation of a solution by diffusion through a semipermeable membrane.

diffusion The process whereby gases or liquids intermingle freely of their own accord.

dipole-dipole attraction A relatively weak force of attraction between polar molecules; a component of van der Waals forces.

displacement A change by which an element takes the place of another element in a compound.

dissociation (ionic) The separation of the ions of an ionic compound due to the action of a solvent.

distillation The process of first vaporizing a liquid and then condensing the vapor back into a liquid, leaving behind the nonvolatile impurities.

double bond A bond between atoms involving two electron pairs. In organic chemistry: unsaturated.

double displacement A reaction in which two chemical substances exchange ions with the formation of two new compounds.

dry ice Solid carbon dioxide.

ductile Capable of being drawn into thin wire.

effervescence The rapid escape of excess gas that has been dissolved in a liquid.

efflorescence The loss by a substance of its water of hydration on exposure to air at ordinary temperatures.

effusion The flow of a gas through a small aperture.

Einstein equation $E = mc^2$, which relates mass to energy; E = energy in ergs, m = mass in grams, and c = velocity of light, 3×10^{10} centimeters/second.

electrode A terminal of an electrolytic cell.

electrode potential The difference in potential between an electrode and the solution in which it is immersed.

electrolysis The process of separating the ions in a compound by means of electrically charged poles.

electrolytic cell A cell in which electrolysis is carried out.

electrolyte A liquid that will conduct an electric current.

electron A negatively charged particle found outside the nucleus of the atom; it has a mass of 9.109×10^{-28} gram.

electron dot symbol *See* **Lewis dot symbol.**

electronegativity The numerical expression of the relative strength with which the atoms of an element attract valence electrons to themselves; the higher the number, the greater the attraction.

electron volt A unit for expressing the kinetic energy of subatomic particles; the energy acquired by an electron when it is accelerated by a potential difference of 1 volt; equals 1.6×10^{-12} erg or 23.1 kilocalories/mole (abbreviation: eV).

electroplating Depositing a thin layer of (usually) a metallic element on the surface of another metal by electrolysis.

element One of the more than 100 "building blocks" of which all matter is composed. An element consists of atoms of only one kind and cannot be decomposed further by ordinary chemical means.

empirical formula A formula that shows only the simplest ratio of the numbers and kinds of atoms, such as CH_4.

emulsifying agent A colloidal substance that forms a film about the particles of two immiscible liquids, so that one liquid remains suspended in the other.

emulsion A suspension of fine particles or droplets of one liquid in another, the two liquids being immiscible in each other; the droplets are surrounded by a colloidal (emulsifying) agent.

endothermic Referring to a chemical reaction that results in an overall absorption of heat from its surroundings.

energy The capacity to do work. In every chemical change energy is either given off or taken in. Forms of energy are heat, light, motion, sound, and electrical, chemical, and nuclear energy.

enthalpy The heat content of a chemical system.

entropy The measure of the randomness or disorder that exists in a system.

equation A shorthand method of showing the changes that take place in a chemical reaction.

equilibrium The point in a reversible reaction at which the forward reaction is occurring at the same rate as the opposing reaction.

erg A unit of energy or work done by a force of 1 dyne (1/980 g of force) acting through a distance of 1 centimeter; equals 2.4×10^{-11} kilocalorie.

ester An organic salt formed by the reaction of an alcohol with an organic (or inorganic) acid.

esterification A chemical reaction between an alcohol and an acid, in which an ester is formed.

ether An organic compound containing the –O– group.

eudiometer A graduated glass tube into which gases are placed and subjected to an electric spark; used to measure the individual volumes of combining gases.

evaporation The process in which molecules of a liquid (or a solid) leave the surface in the form of vapor.

exothermic Referring to a chemical reaction that results in the giving off of heat to its surroundings.

Fahrenheit scale The temperature scale that has 32° as the freezing point of water and 212° as the boiling point.

fallout The residual radioactivity from an atmospheric nuclear test, which eventually settles on the surface of the earth.

Faraday's Law *See under* **laws.**

filtration The process by which suspended matter is removed from a liquid by passing the liquid through a porous material.

First Law of Thermodynamics *See under* **laws.**

fission A nuclear reaction that releases energy because of the splitting of large nuclei into smaller ones.

fixation of nitrogen Any process for converting atmospheric nitrogen into compounds, such as ammonia and nitric acid.

flame The glowing mass of gas and luminous particles produced by the burning of a gaseous substance.

flammable Capable of being easily set on fire; combustible (same as *inflammable*).

fluorescence Emission by a substance of electromagnetic radiation, usually visible, as the immediate result of (and only during) absorption of energy from another source.

fluoridation Addition of small amounts of fluoride (usually NaF) to drinking water to help prevent tooth decay.

flux In metallurgy: a substance that helps to melt and remove the solid impurities as slag. In soldering: a substance that cleans the surface of the metal to be soldered. In nucleonics: the concentration of nuclear particles or rays.

formula An expression that uses the symbols for elements and subscripts to show the basic makeup of a substance.

formula mass The sum of the atomic mass units of all the atoms (or ions) contained in a formula.

fractional crystallization The separation of the components in a mixture of dissolved solids by evaporation according to individual solubilities.

fractional distillation The separation of the components in a mixture of liquids having different boiling points by vaporization.

free energy *See* **Gibbs free energy**.

freezing point The specific temperature at which a given liquid and its solid form are in equilibrium.

fuel Any substance used to furnish heat by combustion. *See also* **nuclear fuel.**

fuel cell A device for converting an ordinary fuel such as hydrogen or methane directly into electricity.

functional group A group of atoms that characterizes certain types of organic compounds, such as —OH for alcohols, and that reacts more or less independently.

fusion A nuclear reaction that releases energy because of the union of smaller nuclei to form larger ones.

fusion melting Changing a solid to the liquid state by heating.

galvanizing Applying a coating of zinc to iron or steel to protect the latter from rusting.

gamma rays A type of radiation consisting of high-energy waves that can pass through most materials. Symbol: γ

gas A phase of matter that has neither definite shape nor definite volume.

Gay-Lussac's Law *See under* **laws.**

Gibbs free energy Changes in Gibbs free energy, ΔG, are useful in indicating the conditions under which a chemical reaction will occur. The equation is $\Delta G = \Delta H - T\Delta S$, where ΔH = change in enthalpy and ΔS = change in entropy. If ΔG is negative, the reaction will proceed spontaneously to equilibrium.

glass An amorphous, usually translucent substance consisting of a mixture of silicates. Ordinary glass is made by fusing together silica and sodium carbonate and lime; the various forms of glass contain many other silicates.

Graham's Law *See under* **laws.**

gram A unit of weight in the metric system; the weight of 1 milliliter of water at 4°C (abbreviation: g).

gram-atomic mass The atomic mass, in grams, of an element.

gram-formula weight The formula weight, in grams, of a substance.

group A vertical column of elements in the periodic table that generally have similar properties.

half-life The time required for half of the mass of a radioactive substance to disintegrate.

half-reaction One of the two parts, either the reduction part or the oxidation part, of a redox reaction.

halogen Any of the five nonmetallic elements (fluorine, chlorine, bromine, iodine, astatine) that form part of Group 17 of the Periodic Table.

heat A form of molecular energy; it passes from a warmer body to a cooler one.

heat capacity (specific heat) The quantity of heat, in calories, needed to raise the temperature of 1 gram of a substance 1 degree on the Celsius scale.

heat of formation The quantity of heat either given off or absorbed in the formation of 1 mole of a substance from its elements.

heat of fusion The amount of heat, in calories, required to melt 1 gram of a solid; for water, 80 calories.

heat of vaporization The quantity of heat needed to vaporize 1 gram of a liquid at constant temperature and pressure; for water at 100°C, 540 calories.

heavy water (deuterium oxide, D_2O) Water in which the hydrogen atoms are replaced by atoms of the isotope of hydrogen, deuterium.

Henry's Law *See under* **laws.**

Hess's Law *See under* **laws.**

homogeneous Uniform; having every portion exactly like every other portion.

homologous Alike in structure; referring to series of organic compounds, such as hydrocarbons, in which each member differs from the next by the addition of the same group.

humidity The amount of moisture in the air.

hybridization The combination of two or more orbitals to form new orbitals.

hydrate A compound that has water molecules included in its crystalline makeup.

hydride Any binary compound containing hydrogen, such as HCl.

hydrogenation A process in which hydrogen is made to combine with another substance, usually organic, in the presence of a catalyst.

hydrogen bond A weak chemical linkage between the hydrogen of one polar molecule and the oppositely charged portion of a closely adjacent molecule.

hydrolysis Of carbohydrates: the action of water in the presence of a catalyst upon one carbohydrate to form simpler carbohydrates. Of salts: a reaction involving the splitting of water into its ions by the formation of a weak acid, a weak base, or both.

hydronium ion A hydrated ion, $H_2O \cdot H^+$ or H_3O^+.

hydroponics Growing plants without the use of soil, as in nutrient solution or in sand irrigated with nutrient solution.

hydroxyl Referring to the —OH radical.

hygroscopic Referring to the ability of a substance to draw water vapor from the atmosphere to itself and become wet.

hypothesis A possible explanation of the nature of an action or phenomenon; a hypothesis is not as completely developed as a theory.

Ideal Gas Law *See under* **laws**.

immiscible Referring to the inability of two liquids to mix.

indicator A dye that shows one color in the presence of the hydrogen ion (acid) and a different color in the presence of the hydroxyl ion (base).

inertia The property of matter whereby it remains at rest or, if in motion, remains in motion in a straight line unless acted upon by an outside force.

ion An atom or a group of combined atoms that carries one or more electric charges.

Examples: NH_4^+, OH^-.

ionic bonding The bonding of ions due to their opposite charges.

ionic equation An equation showing a reaction among ions.

ionization The process in which ions are formed from neutral atoms.

ionization equation An equation showing the ions set free from an electrolyte.

isomerization The rearrangement of atoms in a molecule to form isomers.

isomers Two or more compounds having the same percentage composition but different arrangements of atoms in their molecules and hence different properties.

isotopes Two or more forms of an element that differ only in the number of neutrons in the nucleus and hence in their mass numbers.

IUPAC International Union of Pure and Applied Chemistry, an organization that establishes standard rules for naming compounds.

joule The SI unit of work or of energy equal to work done; 1 joule = 0.2388 calorie; 1 calorie = 4.18 joule.

Kelvin scale A temperature scale based on water freezing at 273 and boiling at 373 Kelvin units; its origin is absolute zero. Synonymous with *absolute scale.*

kernel (atomic) The nucleus and all the electron shells of an atom except the outer one; usually designated by the symbol for the atom.

ketone An organic compound containing the —CO— group.

kilocalorie A unit of heat; the amount of heat needed to raise the temperature of 1 kilogram of water 1 degree on the Celsius scale.

kindling temperature The temperature to which a given substance must be raised before it ignites.

Kinetic-Molecular Theory The theory that all molecules are in motion; this motion is most rapid in gases, less rapid in liquids, and very slow in solids.

lanthanide series The "rare earth" series of elements starting with lanthanum, No. 57, and ending with lutetium, No. 71.

law (in science) A generalized statement about the uniform behavior in natural processes.

laws

Avogadro's Equal volumes of gases under identical conditions of temperature and pressure contain equal numbers of particles (atoms, molecules, ions, or electrons).

Boyle's The volume of a confined gas is inversely proportional to the pressure to which it is subjected, provided that the temperature remains the same.

Charles's The volume of a confined gas is directly proportional to the absolute temperature, provided that the pressure remains the same.

Combining Volumes See *Gay-Lussac's* under **laws**.

Conservation of Energy Energy can be neither created nor destroyed, so that the energy of the universe is constant.

Conservation of Matter Matter can be neither created nor destroyed (or weight remains constant in an ordinary chemical change).

Dalton's When a gas is made up of a mixture of different gases, the pressure of the mixture is equal to the sum of the partial pressures of the components.

Definite Composition A compound is composed of two or more elements chemically combined in a definite ratio by weight.

Faraday's During electrolysis, the weight of any element liberated is proportional (1) to the quantity of electricity passing through the cell, and (2) to the equivalent weight of the element.

First Law of Thermodynamics The total energy of the universe is constant and cannot be created or destroyed.

Gay-Lussac's The ratio between the combining volumes of gases and the product, if gaseous, can be expressed in small whole numbers.

Graham's The rate of diffusion (or effusion) of a gas is inversely proportional to the square root of its molecular mass.

Henry's The solubility of a gas (unless the gas is very soluble) is directly proportional to the pressure applied to the gas.

Hess's If a series of reactions are added together, the enthalpy change for the total reaction is the sum of the enthalpy changes for the individual steps.

Ideal Gas Any gas that obeys the gas laws perfectly. No such gas actually exists.

Multiple Proportions When any two elements, A and B, combine to form more than one compound, the different masses of B that unite with a fixed mass of A bear a small whole-number ratio to each other.

Periodic The chemical properties of elements vary periodically with their atomic numbers.

Second Law of Thermodynamics Heat cannot, of itself, pass from a cold body to a hot body.

Le Châtelier's Principle If a stress is placed on a system in equilibrium, the system will react in the direction that relieves the stress.

lepton An elementary particle; the electron and neutrino are believed to consist of leptons.

Lewis dot symbol The chemical symbol (kernel) for an atom, surrounded by dots to represent its outer level electrons. Examples: K ·, Sr :.

liquid A phase of matter that has a definite volume but takes the shape of the container.

liquid air Air that has been cooled and compressed until it liquefies.

litmus An organic substance, obtained from the lichen plant and used as an indicator; it turns red in acidic solution and blue in basic solution.

London force The weakest of the van der Waals forces between molecules. These weak, attractive forces become apparent only when the molecules approach one another closely (usually at low temperatures and high pressure). They are due to the way the positive charges of one molecule attract the negative charges of another molecule because of the charge distribution at any one instant.

luminous Emitting a steady, suffused light.

malleable Capable of being hammered or pounded into thin sheets.

manometer A U-tube (containing mercury or some other liquid) used to measure the pressure of a confined gas.

mass The quantity of matter that a substance possesses; it can be measured by its resistance to a change in position or motion, and is not related to the force of gravity.

mass number The nearest whole number to the combined atomic mass of the individual atoms of an isotope when that mass is expressed in atomic mass units.

mass spectograph A device for determining the masses of electrically charged particles by separating them into distinct streams by means of magnetic deflection.

matter A substance that occupies space, has mass, and cannot be created or destroyed easily.

melting The change in phase of a substance from solid to liquid.

melting point The specific temperature at which a given solid changes to a liquid.

meson Any unstable, elementary nuclear particle having a mass between that of an electron and that of a proton.

metal (a) An element whose oxide combines with water to form a base; (b) an element that readily loses electrons and acquires a positive valence.

metallurgy The process involved in obtaining a metal from its ores.

meter The basic unit of length in the metric system; defined as 1,650,763.73 times the wavelength of krypton-86 when excited to give off an orange-red spectral line.

MeV A unit for expressing the kinetic energy of subatomic particles; equals 10^6 electron volts.

micron One thousandth of a millimeter (abbreviation: μ).

mineral An inorganic substance of definite composition found in nature.

miscible Referring to the ability of two liquids to mix with one another.

mixture A substance composed of two or more components, each of which retains its own properties.

moderator A substance such as graphite, paraffin, or heavy water used in a nuclear reactor to slow down neutrons.

molal solution A solution containing 1 mole of solute in 1,000 grams of solvent (indicated by m).

molar mass The mass arrived at by the addition of the atomic masses of the units that make up a molecule of an element or compound. Expressed in grams/mole, the molar mass of a gaseous substance at STP occupies a molar volume equal to 22.4 liters.

molar solution A solution containing 1 mole of solute in 1,000 milliliters of solution (indicated by M).

mole A unit of quantity that consists of 6.02×10^{23} particles.

molecular mass The sum of the masses of all the atoms in a molecule of a substance.

molecular theory *See* **Kinetic-Molecular Theory.**

molecule The smallest particle of a substance that retains the physical and chemical properties of that substance.

Example: He, Br_2, H_2O

monobasic acid An acid having only one hydrogen atom that can be replaced by a metal or a positive radical.

mordant A chemical, such as aluminum sulfate, used for fixing colors on textiles.

multiple proportions *See under* **laws.**

nascent (atomic) Referring to an element in the atomic form as it has just been liberated in a chemical reaction.

neutralization The union of the hydrogen ion of an acid and the hydroxyl ion of a base to form water.

neutron A subatomic particle found in the nucleus of the atom; it has no charge and has the same mass as the proton.

neutron capture A nuclear reaction in which a neutron attaches itself to a nucleus; a gamma ray is usually emitted simultaneously.

nitriding A process in which ammonia or a cyanide is used to produce case-hardened steel; a nitride is formed instead of a carbide.

nitrogen fixation Any process by which atmospheric nitrogen is converted into a compound such as ammonia or nitric acid.

noble gas A gaseous element that has a complete outer level of electrons; any of a group of rare gases (helium, neon, argon, krypton, xenon, and radon) that exhibit great stability and very low reaction rates.

noble gas structure The outer energy level electron configuration characteristic of the inert gases—two electrons for helium; eight electrons for all others.

nonelectrolyte A substance whose solution does not conduct a current of electricity.

nonmetal (a) An element whose oxide reacts with water to form an acid; (b) an element that takes on electrons and acquires a negative valence.

nonpolar compound A compound in whose molecules the atoms are arranged symmetrically so that the electric charges are uniformly distributed.

normal salt A salt in which all the hydrogen of the acid has been displaced by a metal.

normal solution A solution that contains 1 gram of H^+ (or its equivalent: 17 g of OH^-, 23 g of Na^+, 20 g of Ca^{2+}, etc.) in 1 liter of solution (indicated by N).

nuclear energy The energy released by spontaneously or artificially produced fission, fusion, or disintegration of the nuclei of atoms.

nuclear fuel A substance that is consumed during nuclear fission or fusion.

nuclear reaction Any reaction involving a change in nuclear structure.

nuclear reactor A device in which a controlled chain reaction of fissionable material can be produced.

nucleonics The science that deals with the constituents and all the changes in the atomic nucleus.

nucleus The center of the atom, which contains protons and neutrons.

nuclide A species of atom characterized by the constitution of its nucleus.

orbital A subdivision of a nuclear shell; it may contain none, one, or two electrons.

ore A natural mineral substance from which an element, usually a metal, may be obtained with profit.

organic acid An organic compound that contains the –COOH group.

organic chemistry The branch of chemistry dealing with carbon compounds, usually those found in nature.

oxidation The chemical process by which oxygen is attached to a substance; the process of losing electrons.

oxidation number (state) A positive or negative number representing the charge that an ion has or an atom appears to have when its electrons are counted according to arbitrarily accepted rules: (1) electrons shared by two unlike atoms are counted with the more electronegative atom; (2) electrons shared by two like atoms are divided equally between the atoms.

oxidation potential An electrode potential associated with the oxidation half-reaction.

oxidizing agent A substance that (a) gives up its oxygen readily, (b) removes hydrogen from a compound, (c) takes electrons from an element.

ozone An allotropic and very active form of oxygen, having the formula O_3.

paraffin series The methane series of hydrocarbons.

pascal The SI unit of pressure, equal to 1 newton per square meter.

pasteurization Partial sterilization of a substance, such as milk, by heating to approximately 65°C for $^1/_2$ hour.

Pauli Exclusion Principle Each electron orbital of an atom can be filled by only two electrons, which have opposite spins.

period A horizontal row of elements in the Periodic Table.

Periodic Law *See under* **laws.**

petroleum (meaning "oil from stone") A complex mixture of gaseous, liquid, and solid hydrocarbons obtained from the earth.

pH A numerical expression of the hydrogen or hydronium ion concentration in a solution; defined as $-\log [H^+]$, where $[H^+]$ is the concentration of hydrogen ions, in moles per liter.

phenolphthalein An organic indicator; it is colorless in acid solution and red in the presence of OH^- ions.

photosynthesis The reaction taking place in all green plants that produces glucose from carbon dioxide and water under the catalytic action of chlorophyll in the presence of light.

physical change A change that does not involve any alteration in chemical composition.

physical property A property of a substance arrived at through observation of its smell, taste, color, density, and so on, which does not relate to chemical activity.

pi bond A bond between *p* orbitals.

pile A general term for a nuclear reactor; specifically, a graphite-moderated reactor in which uranium fuel is distributed throughout a "pile" of graphite blocks.

pitchblende A massive variety of uraninite that contains a small amount of radium.

plasma Very hot ionized gases.

polar covalent bond A bond in which electrons are closer to one atom than to another. *See also* **polar molecule.**

polar dot structure Representation of the arrangements of electrons around the atoms of a molecule in which the polar characteristics are shown by placing the electrons closer to the more electronegative atom.

polar molecule A molecule that has differently charged areas because of unequal sharing of electrons.

polyatomic ion A group of chemically united atoms that react as a unit and have an electric charge.

polymerization The process of combining several molecules to form one large molecule (polymer). (a) Additional polymerization: The addition of unsaturated molecules to each other. (b) Condensation polymerization: The reaction of two molecules by loss of a molecule of water.

positron A positively charged particle of electricity with about the same weight as the electron.

potential energy Energy due to the position of a body or to the configuration of its particles.

precipitate An insoluble compound formed in the chemical reaction between two or more substances in solution.

proteins Large, complex organic molecules, with nitrogen an essential part, found in plants and animals.

proton A subatomic particle found in the nucleus that has a positive charge.

qualitative analysis A term applied to the methods and procedures used to determine any or all of the constituent parts of a substance.

quantitative analysis A term applied to the methods and procedures used to determine the definite quantity or percentage of any or all of the constituent parts of a substance.

quenching Cooling a hot piece of metal rapidly, as in water or oil.

radiation The emission of particles and rays from a radioactive source; usually alpha and beta particles and gamma rays.

radioactive Referring to substances that have the ability to emit radiations (alpha or beta particles or gamma rays).

radioisotope An isotopes that is radioactive, such as uranium-235.

reactant A substance involved in a reaction.

reaction A chemical transformation or change. The four basic types are combination (synthesis), decomposition (analysis), single replacement or single displacement, and double replacement or double displacement.

reaction potential The sum of the oxidation potential and reduction potential for a particular reaction.

reagent Any chemical taking part in a reaction.

recrystallization A series of crystallizations, repeated for the purpose of greater purification.

redox A shortened name for a reaction that involves reduction and oxidation.

reducing agent From an electron standpoint, a substance that loses its valence electrons to another element; a substance that is readily oxidized.

reduction A chemical reaction that removes oxygen from a substance; a gain of electrons.

reduction potential An electrode potential associated with a reduction half-reaction.

refraction (of light) The bending of light rays as they pass from one material into another.

relative humidity The ratio, expressed in percent, between the amount of water vapor in a given volume of air and the amount the same volume can hold when saturated at the same temperature.

resonance The phenomenon in a molecular structure that exhibits properties between those of a single bond and those of a double bond and thus possesses two or more alternative structures.

reversible reaction Any reaction that reaches an equilibrium, or that can be made to proceed from right to left as well as from left to right.

roasting Heating an ore (usually a sulfide) in an excess of air to convert the ore to an oxide, which can then be reduced.

salt A compound, such as NaCl, made up of a positive metallic ion and a negative nonmetallic ion or radical.

saturated solution A solution that contains the maximum amount of solute under the existing temperature and pressure.

Second Law of Thermodynamics *See under* **laws**.

shell A level outside the nucleus of an atom that electrons can inhabit without expending energy.

sigma bond A bond between *s* orbitals or between an *s* orbital and another kind of orbital.

significant figures All the certain digits, that is, those recorded in a measurement, plus one uncertain digit.

slag The product formed when the flux reacts with the impurities of an ore in a metallurgical process.

solid A phase of matter that has a definite size and shape.

solubility A measure of the amount of solute that will dissolve in a given quantity of solvent at a given temperature.

solute The material that is dissolved to make a solution.

solution A uniform mixture of a solute in a solvent.

solvent The dispersing substance that allows the solute to go into solution.

specific gravity (mass) The ratio between the mass of a certain volume of a substance and the mass of an equal volume of water (or, in the case of gases, an equal volume of air); expressed as a single number.

specific heat The ratio between the number of calories needed to raise the temperature of a certain mass of a substance 1 degree on the Celsius scale and the number of calories needed to raise the temperature of the same mass of water 1 degree on the Celsius scale.

spectroscope An instrument used to analyze light by separating it into its component wavelengths.

spectrum The image formed when radiant energy is dispersed by a prism or grating into its various wavelengths.

spinthariscope A device for viewing through a microscope the flashes of light made by particles from radioactive materials against a sensitized screen.

spontaneous combustion (ignition) The process in which slow oxidation produces enough heat to raise the temperature of a substance to its kindling temperature.

stable Referring to a substance not easily decomposed or dissociated.

standard conditions An atmospheric pressure of 760 millimeters or torr or 1 atmosphere (mercury pressure) and a temperature of 0°C (273 K)(abbreviation: STP).

stratosphere The upper portion of the atmosphere, in which the temperature changes but little with altitude, and clouds of water never form.

strong acid (or base) An acid (or a base) capable of a high degree of ionization in water solution. Example: sulfuric acid (sodium hydroxide).

structural (graphic) formula A pictorial representation of the atomic arrangement of a molecule.

sublime To vaporize directly from the solid to the gaseous state, and then condense back to the solid.

substance A single kind of matter, element, or compound.

substitution product A product formed by the substitution of other elements or radicals for hydrogen atoms in hydrocarbons.

sulfation An accumulation of lead sulfate on the plates and at the bottom of a (lead) storage cell.

supersaturated solution A solution that contains a greater quantity of solute than is normally possible at a given temperature.

suspension A mixture of finely divided solid material in a liquid, from which the solid settles on standing.

symbol A letter or letters representing an element of the periodic table. Examples: O, Mn.

synthesis The chemical process of forming a substance from its individual parts.

Système International d'Unités The modernized metric system of measurements universally used by scientists. There are seven base units: kilogram, meter, second, ampere, kelvin, mole, and candela.

temperature The intensity or the degree of heat of a body, measured by a thermometer.

tempering The heating and then rapid cooling of a metal to increase its hardness.

ternary Referring to a compound composed of three different elements, such as H_2SO_4.

theory An explanation used to interpret the "mechanics" of nature's actions; a theory is more fully developed than a hypothesis.

thermochemical equation An equation that includes values for the calories absorbed or evolved.

thermoplastic Capable of being softened by heat; may be remolded.

thermosetting Capable of being permanently hardened by heat and pressure; resistant to the further effects of heat.

tincture An alcoholic solution of a substance, such as a tincture of iodine.

torr A unit of pressure defined as 1 millimeter of mercury; 1 torr equals 133.32 pascals.

tracer A minute quantity of radioactive isotope used in medicine and biology to study chemical changes within living tissues.

transmutation Conversion of one element into another, either by bombardment or by radioactive disintegration.

tribasic acid An acid that contains three replaceable hydrogen atoms in its molecule, such as H_3PO_4.

tritium A very rare, unstable, "triple-weight" hydrogen isotope (H^3) that can be made synthetically.

Tyndall effect The scattering of a beam of light as it passes through a colloidal material.

ultraviolet light The portion of the spectrum that lies just beyond the violet; therefore of short wavelength.

U.S.P. (United States Pharmacopeia) chemicals Chemicals certified as having a standard of purity that demonstrates their fitness for use in medicine.

valence The combining power of an element; the number of electrons gained, lost, or borrowed in a chemical reaction.

valence electrons The electrons in the outermost level or levels of an atom that determine its chemical properties.

van der Waals forces Weak attractive forces existing between molecules.

vapor The gaseous phase of a substance that normally exists as a solid or liquid at ordinary temperatures.

vapor pressure The pressure exerted by a vapor given off by a confined liquid or solid when the vapor is in equilibrium with its liquid or solid form.

volatile Easily changed to a gas or a vapor at relatively low pressure.

volt A unit of electrical potential or voltage, equal to the difference of potential between two points in a conducting wire carrying a constant current of 1 ampere when the power dissipated between these two points is equal to 1 watt (abbreviation: V).

volume The amount of three-dimensional space occupied by a substance.

VSEPR The valence shell electron pair repulsion model. It expresses the non-90° variations in bond angles for *p* orbitals in the outer energy levels of atoms in molecules because of electron repulsions.

water of hydration Water that is held in chemical combination in a hydrate and can be removed without essentially altering the composition of the substance. *See also* **Hydrate**.

weak acid (or base) An acid (or base) capable of being only slightly ionized in an aqueous solution. Example: acetic acid (ammonium hydroxide).

weak electrolyte A substance that, when dissolved in water, ionizes only slightly and hence is a poor conductor of electricity.

weight The measure of the force with which a body is attracted toward Earth by gravity.

work The product of the force exerted on a body and the distance through which the force acts; expressed mathematically by the equation $W = Fs$, where W = work, F = force, and s = distance.

X-rays Penetrating radiations, of extremely short wavelength, emitted when a stream of electrons strikes a solid target in a vacuum tube.

zeolite A natural or synthesized silicate used to soften water.

Index

THE FOLLOWING DOCUMENTATION APPLIES IF YOU PURCHASED

SAT Subject Test Chemistry, 9th Edition.

Please disregard this information if your edition does not contain the CD-ROM.

How to Use the CD-ROM

The software is not installed on your computer; it runs directly from the CD-ROM. Barron's CD-ROM includes an "autorun" feature that automatically launches the application when the CD is inserted into the CD-ROM drive. In the unlikely event that the autorun feature is disabled, follow the manual launching instructions below.

Windows®

Insert the CD-ROM and the program should launch automatically. If the software does not launch automatically, follow the steps below.

1. Click on the Start button and choose "My Computer."
2. Double-click on the CD-ROM drive, which will be named **SAT_CHEM**.
3. Double-click **SAT_CHEM.exe** application to launch the program.

Macintosh®

1. Insert the CD-ROM.
2. Double-click the CD-ROM icon.
3. Double-click the **SAT_CHEM** icon to start the program.

SYSTEM REQUIREMENTS

The program will run on a PC with:
Windows® Intel® Pentium II 450 MHz
or faster, 128MB of RAM
1024 X 768 display resolution
Windows 2000, XP, Vista
CD-ROM Player

The program will run on a Macintosh® with:
PowerPC® G3 500 MHz
or faster, 128MB of RAM
1024 X 768 display resolution
Mac OS X v.10.1 through 10.4
CD-ROM Player

NOTE: It is important that you read the first section of this book about the SAT Subject Area Test in Chemistry. It gives specific information about the types of questions asked and the directions for each type along with other very important aspects of the test. Take the tests in the book first. Then, for extra practice, take the tests on the CD-ROM. Note, however, that the time limits given for the tests on the CD-ROM are not correct. The actual test is 60 minutes long and is not timed by section. To get the most out of your studying, use the **Timed** mode to learn how to pace yourself. Allow 15 minutes for Part A, 13 minutes for Part B, and 32 minutes for Part C. Or, set an alarm clock for exactly one hour.